Values of fundamental constants

Quantity	Symbol	Value
Speed of light in vacuum	c_0	3.00×10^8 m/s
Gravitational constant	G	6.6738×10^{-11} N$\cdot$m^2/kg^2
Avogadro's number	N_A	6.0221413×10^{23}
Boltzmann's constant	k_B	1.381×10^{-23} J/K
Charge on electron	e	1.60×10^{-19} C
Electric constant	ϵ_0	$8.85418782 \times 10^{-12}$ C^2/(N$\cdot$m^2)
Magnetic constant	μ_0	$4\pi \times 10^{-7}$ T$\cdot$m/A
Planck's constant	h	6.626×10^{-34} J$\cdot$s
Electron mass	m_e	9.11×10^{-31} kg
Proton mass	m_p	1.6726×10^{-27} kg
Neutron mass	m_n	1.6749×10^{-27} kg
Atomic mass unit	amu	1.6605×10^{-27} kg

Other useful numbers

Number or quantity	Value
π	3.1415927
e	2.7182818
1 radian	57.2957795°
Absolute zero ($T = 0$)	-273.15 °C
Average acceleration g due to gravity near Earth's surface	9.8 m/s^2
Speed of sound in air at 20 °C	343 m/s
Density of dry air at atmospheric pressure and 20 °C	1.29 kg/m^3
Earth's mass	5.97×10^{24} kg
Earth's radius (mean)	6.38×10^6 m
Earth–Moon distance (mean)	3.84×10^8 m

Eric Mazur • Daryl Pedigo
Additional Contributions from
Peter A. Dourmashkin • Ronald J. Bieniek

Practice of Physics

Part 1
Custom Edition for the University of Minnesota

Taken from:
Principles & Practice of Physics
by Eric Mazur and Daryl Pedigo

Pearson Learning Solutions, 501 Boylston Street, Suite 900, Boston, MA 02116
A Pearson Education Company
www.pearsoned.com

Printed in the United States of America

1 2 3 4 5 6 7 8 9 10 V064 18 17 16 15 14

000200010271907258

SL

ISBN 10: 1-269-94358-8
ISBN 13: 978-1-269-94358-1

Brief Contents

Chapter 1   Foundations   1

Chapter 2   Motion in One Dimension   16

Chapter 3   Acceleration   34

Chapter 4   Momentum   51

Chapter 5   Energy   69

Chapter 6   Principle of Relativity   86

Chapter 7   Interactions   105

Chapter 8   Force   124

Chapter 9   Work   145

Chapter 10   Motion in a Plane   165

Chapter 11   Motion in a Circle   189

Chapter 12   Torque   209

Chapter 13   Gravity   232

Chapter 14   Special Relativity   248

Chapter 15   Periodic Motion   267

Appendix A: Notation   A-1

Appendix B: Mathematics Review   A-11

Appendix C: Useful Data   A-17

Answers to Selected Odd-Numbered Questions and Problems   A-21

Credits   C-1

Getting the Most Out of *Practice*

The ideas of physics explain phenomena all around you—from static electricity in your laundry to how your smartphone knows your location. Some of these principles are straightforward, while others are not immediately obvious. Understanding physics requires not only understanding the concepts but also being able to *apply* these concepts to new situations. Transferring your knowledge to new situations requires advanced reasoning skills, such as judging which concepts are relevant to the problem at hand and then devising a plan to solve the problem. The *Practice* text puts the ideas you learned in the *Principles* text into practice. It teaches you how to make quantitative assumptions and use reason and strategy to solve problems. After you have read your assigned chapter in the *Principles* text, then turn to the corresponding chapter in your *Practice* text and work your way through the chapter.

Here's how the various parts of each chapter will help you become a competent problem solver:

Begin with the **Chapter Summary,** which highlights the major relationships covered in the chapter. Use it to refresh your understanding before you do your homework or as a quick study tool before class.

The **Review Questions** are designed to make sure you understand the primary points of each section. The answers to these questions are on the last page of the chapter. If you have difficulty answering the Review Questions, go back and re-read the appropriate *Principles* section.

You cannot determine whether the answer you obtain for a problem is reasonable if you have no idea what constitutes a reasonable magnitude for the answer. The **Developing a Feel** exercises help you develop both order-of-magnitude estimation skills as well as a "gut feeling" for the range of values appropriate for the physical quantities introduced in each chapter.

A series of **Worked Problems** provide detailed examples of how to approach, solve, and evaluate problems. Each Worked Problem is followed by a related **Guided Problem,** which guides you to solve a similar problem by yourself. The answers to the Guided Problems are on the last page of the chapter.

The **Questions and Problems** are designed as homework or review for examinations. They are provided in a range of difficulty levels, some with more conceptual emphasis and some with more quantitative emphasis.

Together, the *Principles & Practice of Physics* combine to create a learning tool that was developed with you, the student, at the center. We have worked hard to create accurate, physically plausible scenarios that will help you develop the type of problem-solving skills that will serve you not only in your physics course but also in your future career.

Eric Mazur
Harvard University

Daryl Pedigo
University of Washington

PRACTICE
Foundations

Chapter Summary 3
Review Questions 5
Developing a Feel 6
Worked and Guided Problems 7
Questions and Problems 10
Answers to Review Questions 15
Answers to Guided Problems 15

Organization of this book

The material for this course is presented in two volumes. The first volume (*Principles*) is aimed at guiding you in developing a solid understanding of the principles of physics. The second volume (the one you are reading now, *Practice*) provides a variety of questions and problems that allow you to apply and sharpen your understanding of physics.

Each chapter in the *Practice* volume contains specific aids and exercises targeted to the physics discussed in the corresponding *Principles* chapter. These categories are ordered so that the earlier materials support the later ones:

1. **Chapter summary.** The summary is just what the name implies, a condensed record of the key elements from the corresponding *Principles* chapter.
2. **Review questions.** A set of simple questions probes your understanding of the basic material. You should be able to answer these questions without trouble after having read the corresponding *Principles* chapter.
3. **Developing a feel.** These estimation problems are designed to exercise your ability to grasp the scope of the world around you, following the approach outlined in the section of the same title in the *Principles* volume.
4. **Worked and guided problems.** This section presents a series of paired example problems: a worked example followed by a similar problem presented with only

a few guiding hints and questions. The idea is that if you understand the methods of the worked example, then you should be able to adapt those methods (and perhaps add a twist) to solve the guided problem. At the beginning of this section you will find a copy of any **Procedure boxes** from the *Principles* volume.

5. **Questions and problems.** After working your way to this point, you should be ready to solve some problems on your own. These problems are both conceptual and quantitative, organized by chapter section and labeled with a rough "degree of difficulty" scale, indicated by one, two, or three blue dots. One-dot problems are fairly straightforward and usually involve only one major concept. Two-dot problems typically require you to put together two or more ideas from the chapter, or even to combine some element of the current chapter with material from other chapters. Three-dot problems are more challenging, or even a bit tricky. Some of these are designated as "CR" (context-rich), a category that is described later in this chapter of the *Practice* volume. Context-rich problems are in the **Additional Problems** at the end of this section, together with other general problems.
6. **Answers and solutions** to Review Questions and Guided Problems are at the end of the chapter.

Chapter Summary

The scientific method (Section 1.1)

Concepts The **scientific method** is an iterative process for going from observations to a hypothesis to an experimentally validated theory. If the predictions made by a hypothesis prove accurate after repeated experimental tests, the hypothesis is called a **theory** or a **law,** but it always remains subject to additional experimental testing.

Quantitative tools

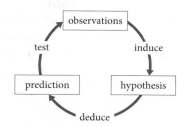

Symmetry (Section 1.2)

Concepts An object exhibits **symmetry** when certain operations can be performed on it without changing its appearance. Important examples are *translational symmetry* (movement from one location to another), *rotational symmetry* (rotation about a fixed axis), and *reflection symmetry* (reflection in a mirror). The concept of symmetry applies both to objects and to physical laws.

Some basic physical quantities and their units (Sections 1.3, 1.4, 1.6)

Concepts **Length** is a distance or extent in space. The SI (International System) base unit of length is the **meter** (m).

Time is a property that allows us to determine the sequence in which related events occur. The SI base unit of time is the **second** (s). The **principle of causality** says that whenever event A causes an event B, all observers see event A happen before event B.

Density is a measure of how much of some substance there is in a given volume.

Quantitative Tools If there are N objects in a volume V, then the *number density n* of these objects is

$$n \equiv \frac{N}{V}. \tag{1.3}$$

If an object of mass m occupies a volume V, then the *mass density ρ* of this object is

$$\rho \equiv \frac{m}{V}. \tag{1.4}$$

To convert one unit to an equivalent unit, multiply the quantity whose unit you want to convert by one or more appropriate *conversion factors*. Each conversion factor must equal one, and any combination of conversion factors used must cancel the original unit and replace it with the desired unit. For example, converting 2.0 hours to seconds, we have

$$2.0 \; \cancel{h} \times \frac{60 \; \cancel{min}}{1 \; \cancel{h}} \times \frac{60 \; s}{1 \; \cancel{min}} = 7.2 \times 10^3 \; s.$$

Representations (Section 1.5)

Concepts Physicists use many types of representations in making models and solving problems. Rough sketches and detailed diagrams are generally useful, and often crucial, to this process. Graphs are useful for visualizing relationships between physical quantities. Mathematical expressions represent models and problems concisely and permit the use of mathematical techniques.

Quantitative Tools As you construct a model, begin with a simple visual representation (example: represent a cow with a dot) and add details as needed to represent additional features that prove important.

Significant digits (Section 1.7)

Concepts The **significant digits** in a number are the digits that are reliably known.

Quantitative Tools If a number contains no zeros, then all the digits shown are significant: 345 has three significant digits; 6783 has four significant digits.

For numbers that contain zeros:

- Zeros between two nonzero digits are significant: 4.03 has three significant digits.
- Trailing digits to the right of the decimal point are significant: 4.9000 has five significant digits.
- Leading zeros before the first nonzero digit are not significant: 0.000 175 has three significant digits.
- In this book, trailing zeros in numbers without a decimal point are significant: 8500 has four significant digits.

The number of significant digits in a product or quotient is the same as the number of significant digits in the input quantity that has the *fewest significant digits:* $0.10 \times 3.215 = 0.32$.

The number of decimal places in a sum or difference is the same as the number of decimal places in the input quantity that has the *fewest decimal places:* $3.1 + 0.32 = 3.4$.

Solving problems and developing a feel (Sections 1.8, 1.9)

Concepts **Strategy for solving problems:**

1. **Getting started.** Analyze and organize the information and determine what is being asked of you. A sketch or table is often helpful. Decide which physics concepts apply.
2. **Devise plan.** Determine the physical relationships and equations necessary to solve the problem. Then outline the steps you think will lead to the solution.
3. **Execute plan.** Carry out the calculations, and then check your work using the following points:
    Vectors/scalars used correctly?
    Every question answered?
    No unknown quantities in answers?
    Units correct?
    Significant digits justified?
4. **Evaluate result.** Determine whether or not the answer is reasonable.

To develop a feel for the approximate size of a calculated quantity, make an **order-of-magnitude** estimate, which means a calculation rounded to the nearest power of ten.

Quantitative Tools Determining order of magnitude

Example 1:

4200 is 4.200×10^3.

Round the coefficient 4.200 to 10 (because it is greater than 3), so that 4.200×10^3 becomes $10 \times 10^3 = 1 \times 10^4$.

The order of magnitude is 10^4.

Example 2:

0.027 is 2.7×10^{-2}.

Round the coefficient 2.7 to 1 (because it is less than 3), so that 2.7×10^{-2} becomes 1×10^{-2}.

The order of magnitude is 10^{-2}.

Strategy to compute order-of-magnitude estimates:

- Simplify the problem.
- Break it into smaller parts that are easier to estimate.
- Build your estimate from quantities that you know or can easily obtain.

Review Questions

Answers to these questions can be found at the end of this chapter.

1.1 The scientific method

1. What is a common definition of *physics*? That is, what is physics about?
2. Briefly describe the scientific method and what it involves.
3. Name some skills that are useful in doing science.
4. Describe the difference between the two types of reasoning involved in doing science.

1.2 Symmetry

5. What does symmetry mean in physics?
6. What are two types of symmetry that are demonstrated in the reproducibility of experimental results?

1.3 Matter and the universe

7. In physics, what is the definition of *universe*?
8. What does expressing a value to an order of magnitude mean? Why would we express values in this way?
9. To what order of magnitude should you round the numbers 2900 and 3100? Explain why your answers are different for the two numbers.
10. Physicists study phenomena that extend over what range of sizes? Over what range of time intervals?

1.4 Time and change

11. What does the phrase *arrow of time* mean?
12. What principle that relates events depends on the arrow of time? State this principle, and briefly explain what it means.

1.5 Representations

13. When solving physics problems, what are the advantages of making simplified visual representations of the situations?

14. What is the purpose of the *Concepts* part of each chapter in the *Principles* volume of this book? What is the purpose of the *Quantitative Tools* part of each *Principles* chapter?

1.6 Physical quantities and units

15. What two pieces of information are necessary to express any physical quantity?
16. What are the seven SI base units and the physical quantities they represent?
17. What concept does density represent?
18. What is the simplest way to convert a quantity given in one unit to the same quantity given in a different unit?

1.7 Significant digits

19. Explain the difference between number of digits, number of decimal places, and number of significant digits in a numerical value. Illustrate your explanation using the number 0.037 20.
20. What is the difference between leading zeros and trailing zeros? Which ones are considered significant digits?
21. How many significant digits are appropriate in expressing the result of a multiplication or division?
22. How many significant digits are appropriate in expressing the result of an addition or subtraction?

1.8 Solving problems

23. Summarize the four-stage problem-solving procedure used in this book.
24. When checking calculations, what do the letters in the acronym VENUS stand for?

1.9 Developing a feel

25. What are the benefits of making order-of-magnitude estimates?

Developing a Feel

Make an order-of-magnitude estimate of each of the following quantities. Letters in parentheses refer to hints below. Use them as needed to guide your thinking.

(You determine the order of magnitude of any quantity by writing it in scientific notation and rounding the coefficient in front of the power of ten to 1 if it is equal to or less than 3 or to 10 if it is greater than 3. Then write the answer as a power of ten without a coefficient in front. Also remember to express your answer in SI units.)

1. The width of your index finger (C, H)
2. The length of a commercial airliner (E, A)
3. The height of a stack containing 1,000,000 one-dollar bills (G, M)
4. The area of your bedroom floor (J, X)
5. The number of jelly beans needed to fill a 1-gallon jar (F, L)
6. The mass of the air in a typical one-family house (J, D, N)
7. The mass of water needed to fill the passenger compartment of a midsize car (P, I)
8. The mass of a mountain (K, V, S, P)
9. The mass of water in a city swimming pool (T, P)
10. The number of cups of coffee consumed yearly in the United States (R, W)
11. The number of people in the world who are eating at the instant you are reading this question (Q, Z)
12. The number of automobile mechanics in California (B, U, O, Y)

Hints

A. What is the distance between rows?
B. What is the population of California?
C. What is the width of a finger in inches?
D. How many bedroom-size rooms are there in a typical one-family house?
E. How many rows of seats are there in an airliner?
F. What are the dimensions of a jelly bean?
G. How thick is a ream of paper?
H. How many inches make 1 meter?
I. What are the length, width, and height of a car's passenger compartment in meters?
J. What are the length, width, and height of your bedroom in feet?
K. What size and shape could model a mountain?
L. What is the volume of a gallon jar in SI units?
M. How many sheets of paper are there in a ream?
N. What is the mass density of air?
O. How many hours of maintenance does a car need each year?
P. What is the mass density of water?
Q. What is the population of the world?
R. What is the adult population of the United States?
S. How does the mass density of rock compare with that of water?

T. What are the dimensions of a city pool?
U. How many cars are there in California?
V. What volume results from your model?
W. How many cups of coffee does the average American drink daily?
X. How many square feet are there in 1 square meter?
Y. How many hours does an automobile mechanic work each year?
Z. What fraction of your day do you spend eating?

Key (all values approximate)

A. 1 m; B. 4×10^7 people; C. 0.5 in.; D. 8 rooms;
E. 4×10^1 rows; F. $(1 \times 10^{-2} \text{ m}) \times (1 \times 10^{-2} \text{ m}) \times (2 \times 10^{-2} \text{ m})$;
G. 5×10^{-2} m; H. 4×10^1 in./m; I. $2 \text{ m} \times 2 \text{ m} \times 1 \text{ m}$;
J. $(1 \times 10^1 \text{ ft}) \times (2 \times 10^1 \text{ ft}) \times (1 \times 10^1 \text{ ft})$; K. a cone 1 mile high with a 1-mile base radius; L. 4 quarts $\approx 4 \text{ L} = 4 \times 10^{-3} \text{ m}^3$;
M. 5×10^2 sheets; N. 1 kg/m^3; O. 6 h; P. 1×10^3 kg/m^3;
Q. 7×10^9 people; R. 2×10^8 people; S. $5 \times$ larger;
T. $(7 \text{ m}) \times (2 \times 10^1 \text{ m}) \times (2 \text{ m})$; U. 3×10^7 cars; V. $4 \times 10^9 \text{ m}^3$;
W. 1 cup; X. $1 \times 10^1 \text{ ft}^2/\text{m}^2$; Y. 2×10^3 h; Z. 0.1

Worked and Guided Problems

Procedure: Solving Problems

Although there is no set approach when solving problems, it helps to break things down into several steps whenever you are working a physics problem. Throughout this book, we use the four-step procedure summarized here to solve problems. For a more detailed description of each step, see *Principles* Section 1.8.

1. **Getting started.** Begin by carefully analyzing the information given and determining in your own words what question or task is being asked of you. Organize the information by making a sketch of the situation or putting data in tabular form. Determine which physics concepts apply, and note any assumptions you are making.
2. **Devise plan.** Decide what you must do to solve the problem. First determine which physical relationships or equations you need, and then determine the order in which you will use them. Make sure you have a sufficient number of equations to solve for all unknowns.
3. **Execute plan.** Execute your plan, and then check your work for the following five important points:

   Vectors/scalars used correctly?
   Every question asked in problem statement answered?

No unknown quantities in answers?
Units correct?
Significant digits justified?

As a reminder to yourself, put a checkmark beside each answer to indicate that you checked these five points.

4. **Evaluate result.** There are several ways to check whether an answer is reasonable. One way is to make sure your answer conforms to what you expect based on your sketch and the information given. If your answer is an algebraic expression, check to be sure the expression gives the correct trend or answer for special (limiting) cases for which you already know the answer. Sometimes there may be an alternative approach to solving the problem; if so, use it to see whether or not you get the same answer. If any of these tests yields an unexpected result, go back and check your math and any assumptions you made. If none of these checks can be applied to your problem, check the algebraic signs and order of magnitude.

These examples involve material from this chapter but are not associated with any particular section. Some examples are worked out in detail; others you should work out by following the guidelines provided.

Worked Problem 1.1 Solar hydrogen

The mass of the Sun is 1.99×10^{30} kg, its radius is 6.96×10^8 m, and its composition by mass is 71.0% hydrogen (H). The mass of a hydrogen atom is 1.67×10^{-27} kg. Calculate (*a*) the average mass density and (*b*) the average number density of the hydrogen atoms in the Sun.

❶ GETTING STARTED To calculate mass and number densities we need to determine the volume of the Sun. We assume the Sun is perfectly spherical so that we have a formula for its volume. We are given the mass of the Sun and the percentage of hydrogen, so we can determine the mass of hydrogen in the Sun and the number of atoms required to provide this mass.

❷ DEVISE PLAN Mass density is mass per unit volume, $\rho = m/V$, and number density is number per unit volume, $n = N/V$. The (assumed spherical) solar volume is $V = \frac{4}{3}\pi R^3$. The solar radius is given, so we need either the number of hydrogen atoms or the mass of hydrogen in the Sun in order to proceed. The mass of hydrogen is 71.0% of the Sun's mass, so we use that value to compute the mass density first. Then the number N of hydrogen atoms is the mass of all the hydrogen atoms divided by the mass of a single atom. We can use this number to calculate the number density n, or we can simply note that n is equal to the mass density of hydrogen divided by the mass of a single hydrogen atom.

❸ EXECUTE PLAN
(*a*) For the mass density, we have

$$\rho_H = \frac{m_H}{\frac{4}{3}\pi R_{Sun}^3} = \frac{0.710 m_{Sun}}{\frac{4}{3}\pi R_{Sun}^3} = \frac{(0.710)(1.99 \times 10^{30}\ \text{kg})}{\frac{4}{3}\pi(6.96 \times 10^8\ \text{m})^3}$$

$$= 1.00 \times 10^3\ \text{kg/m}^3.$$

(*b*) The number density of the hydrogen atoms is

$$n_H = \frac{\rho_H}{m_{H\,atom}} = \frac{1.00 \times 10^3\ \text{kg/m}^3}{1.67 \times 10^{-27}\ \text{kg}} = 5.99 \times 10^{29}\ \text{atoms/m}^3.$$

Using VENUS to check, we have
Vectors/scalars: all quantities are scalars ✔
Every question answered: mass density, ✔ number density ✔
No unknown quantities in answers: none ✔
Units correct: kg/m³ for mass density, ✔ atoms/m³ for number density ✔
Significant digits: three in each answer because all given quantities have three ✔

PRACTICE

❹ **EVALUATE RESULT** We calculated a hydrogen mass density equal to the mass density of water. Because hydrogen is a gas, you may think that this mass density is unreasonably large and that the answer should be about equal to the value found for helium gas in *Principles* Exercise 1.6, about $0.2 \text{ kg}/\text{m}^3$. However, because the gas in the Sun is highly compressed, a mass density several orders of magnitude larger is not unreasonable. But this value is the mass density of water! Does that make any sense? Well, water vapor certainly has a much smaller mass and number density than liquid water. If the hydrogen atoms in the Sun were squeezed together as closely as the molecules in liquid water, we might expect their mass density to be of the same order of magnitude. We might also compare our answer to the average mass density of the Sun obtained by using the data provided in the problem statement. Assuming a spherical Sun, we obtain

$$\rho_{\text{Sun}} = \frac{m_{\text{Sun}}}{\frac{4}{3}\pi R_{\text{Sun}}^3} = 1.4 \times 10^3 \text{ kg}/\text{m}^3.$$

So it seems reasonable that our hydrogen mass density is of the same order of magnitude as the average mass density.

Our result for the number density is also several orders of magnitude larger than the number density of helium found in *Principles* Exercise 1.6, which is what we might expect for a very dense object like the Sun.

Note: When we are dealing with quantities completely outside our everyday experience, a quick check in a reference book or a couple of independent online sources might be needed to remove any lingering doubts. Doing so, we obtain an average solar mass density consistent with the calculation above.

Guided Problem 1.2 Solar oxygen

Oxygen atoms make up 0.970% of the Sun's mass, and each one has a mass of 2.66×10^{-26} kg. Calculate the average mass density of oxygen in the Sun and the average number density of the oxygen atoms. Use information given in Worked Problem 1.1 as needed.

❶ **GETTING STARTED**
1. How much of the plan of Worked Problem 1.1 can be used?

❷ **DEVISE PLAN**
2. What is the definition of mass density? Number density?
3. Do you have enough information to compute these values?

❸ **EXECUTE PLAN**
4. What is the mass of oxygen in the Sun?
5. How is the number density of oxygen related to its mass density and to the mass of an oxygen atom?

❹ **EVALUATE RESULT**
6. Are your answers consistent with the ones in Worked Problem 1.1?

Worked Problem 1.3 Rod volume

A cylindrical rod is 2.58 m long and has a diameter of 3.24 in. Calculate its volume in cubic meters.

❶ **GETTING STARTED** We know that the volume of a cylinder is length times cross-sectional area. We are given the length, and we can use the diameter value given to calculate area. Because the diameter is in inches, however, we must convert to the SI equivalent.

❷ **DEVISE PLAN** First we convert inches to meters via the conversion factors 25.4 mm = 1 in. (Equation 1.5) and 1 m = 1000 mm. Then we use the SI values for length (ℓ) and area (A) in the formula for volume: $V = A\ell = \pi R^2 \ell$.

❸ **EXECUTE PLAN** In meters, the radius, (3.24 in.)/2, is

$$1.62 \text{ in.} \times \frac{25.4 \text{ mm}}{1 \text{ in.}} \times \frac{1 \text{ m}}{1000 \text{ mm}} = 4.115 \times 10^{-2} \text{ m}.$$

Note: We use four significant digits here. Because the values given all have three significant digits, the final answer must have three also.

In intermediate steps, though, we carry an extra digit to avoid accumulating rounding errors.

The volume of the rod is thus

$$V = \pi R^2 \ell = \pi (4.115 \times 10^{-2} \text{ m})^2 (2.58 \text{ m}) = 1.37 \times 10^{-2} \text{ m}^3.$$

Using VENUS to check, we have
Vectors/scalars: all quantities are scalars ✔
Every question answered: volume ✔
No unknown quantities in answer: none ✔
Units correct: cubic meters ✔
Significant digits: three in answer because all given quantities have three ✔

❹ **EVALUATE RESULT** Although this rod is about 2.5 m long, its radius is only 1.62 in., which is about 40 mm, or 0.040 m. We can make an order-of-magnitude estimate for comparison by treating the rod as a rectangular block of square cross section 8 cm on a side, with length 3 m. The result is $(8 \times 10^{-2} \text{ m})(8 \times 10^{-2} \text{ m})(3 \text{ m}) \approx 10^{-2} \text{ m}^3$. Therefore a volume of about 10^{-2} m^3 is reasonable.

Guided Problem 1.4 Box volume

A box measures 1420 mm by 2.75 ft by 87.8 cm. Express its volume in cubic meters.

❶ **GETTING STARTED**
1. Is the approach of Worked Problem 1.3 useful?

❷ **DEVISE PLAN**
2. What is the relationship between the given dimensions and the volume of the box?

3. Which of the given quantities do you need to convert to SI units?
4. Can you use any of the values given in SI units in the form given?

❸ **EXECUTE PLAN**
5. What are your conversion factors?

❹ **EVALUATE RESULT**

Worked Problem 1.5 Working with digits

Express the result of each calculation both to the proper number of significant digits and as an order of magnitude:

(a) $(42.003)(1.3 \times 10^4)(0.007\,000)$

(b) $(42.003)(13,000)(0.007\,000)$

(c) $\dfrac{170.08\pi}{32.6}$

(d) $113.7540 - 0.08$

❶ GETTING STARTED We have two products, a quotient, and a difference involving quantities that have different numbers of significant digits. The number of significant digits in each answer depends on the number of significant digits in the given values.

❷ DEVISE PLAN To express each answer to the proper number of significant digits, we use the rules given in the *Principles* volume. The number of significant digits in a product or quotient is the same as the number of significant digits in the input quantity that has the *fewest* significant digits. The number of decimal places in a sum or difference is the same as the number of decimal places in the input quantity that has the *fewest* decimal places.

To express our answers as orders of magnitude, we write each in scientific notation, round the coefficient either down to 1 (coefficients ≤ 3) or up to 10 (coefficients > 3), and then write the answer as a power of ten without the coefficient.

❸ EXECUTE PLAN
(a) In $(42.003)(1.3 \times 10^4)(0.007000)$, the first factor has five significant digits, the second has two significant digits, and the third has four significant digits. Therefore the product can have only two significant digits: 3.8×10^3. The coefficient 3.8 is greater than 3, and so we round it to 10, making the order-of-magnitude answer $10 \times 10^3 = 10^4$.

(b) This time the middle factor has five significant digits, so we adjust our calculation to allow four (due to the third factor) significant digits: 3.822×10^3. Of course, the order of magnitude is unchanged.

(c) In $170.08\pi/32.6$, the denominator has three significant digits, the first factor in the numerator has five significant digits, and $\pi \, (=3.14159\ldots)$ has as many significant digits as our calculator shows. So the result can have only three significant digits: 16.4. To express this value as an order of magnitude, we must write it in scientific notation, 1.64×10^1. Because 1.64 is less than 3, we round it to 1, making the order-of-magnitude answer $1 \times 10^1 = 10^1$.

(d) The value that has the fewest decimal places is 0.08, meaning the difference must be reported to two decimal places: 113.67. To express this result as an order of magnitude, we write 1.1367×10^2 and round the 1.1367 down to 1, making the order-of-magnitude answer $1 \times 10^2 = 10^2$.

Checking by VENUS gives
Vectors/scalars: all quantities are scalars ✔
Every question answered: all results reported both to the correct number of significant digits and as an order of magnitude ✔
No unknowns in answers: none ✔
Units correct: no units given ✔
Significant digits: all significant digits are correct ✔

❹ EVALUATE RESULT We can check that each answer has the correct order of magnitude.

(a) $(42.003)(1.3 \times 10^4)(0.007\,000)$ is about $40 \times 13,000 \times 0.01 = 5200$, order of magnitude $5.2 \times 10^3 \approx 10 \times 10^3 = 10^4$, consistent with our answer.

(b) The same result holds.

(c) $(200\pi)/30 = 600/30 = 20 = 2.0 \times 10^1 \approx 1 \times 10^1 = 10^1$

(d) $100 - 0 = 100 = 10^2$

Guided Problem 1.6 Digits on your own

Express the result of each calculation both to the proper number of significant digits and as an order of magnitude:

(a) $(205)(0.0041)(489.623)$

(b) $\dfrac{(190.8)(0.407\,500)}{\pi}$

(c) $6980.035 + 0.2$

❶ GETTING STARTED
1. Are the techniques of Worked Problem 1.5 useful? Sufficient?

❷ DEVISE PLAN
2. For parts *a* and *b*, how many significant digits does each number contain?
3. For part *c*, which value limits the number of decimal places allowed in the answer?

❸ EXECUTE PLAN
4. How do you convert each answer to an order-of-magnitude number?

❹ EVALUATE RESULT

Worked Problem 1.7 Oceans

Make an order-of-magnitude estimate of the percent of Earth's mass that is contained in the oceans.

❶ GETTING STARTED About 70% of Earth's surface is covered by oceans, but the entire volume of Earth contributes to its mass. To obtain a percentage we need to know the mass of the oceans and the mass of Earth. The latter we might look up or recall, but the

former requires a computation based on mass density and volume. We must devise a simple model for the volume of the oceans and perhaps also for Earth.

❷ DEVISE PLAN A spherical Earth coated with a thin shell of water covering 70% of the surface seems a reasonable first try. The volume of the oceans is then 70% of the surface area of this sphere multiplied by the ocean depth. The mass of the oceans then involves the mass density of ocean water, the radius of Earth squared, and the average depth of the oceans. We need to estimate the value of each of these and then divide the mass of the oceans by the mass of Earth in order to obtain the desired percentage. That means we must estimate the mass of Earth, too, and this is also related to the radius of Earth. Some factors may cancel out if we express both masses in terms of mass densities.

❸ **EXECUTE PLAN** The mass of Earth, including oceans, is $m_E = \rho_E V_E = \rho_E(\frac{4}{3}\pi R_E^3)$. The area of the oceans is $0.70A_E$. Assuming the oceans have an average depth d, we can approximate their volume as surface area A_o times depth d: $V_o = A_o d = 0.70A_E d$. Their mass is therefore $m_o = \rho_o V_o = \rho_o(0.70A_E d) = \rho_o(0.70)(4\pi R_E^2)d$. The fraction f of Earth's mass contained in the oceans is

$$f = \frac{m_o}{m_E} = \frac{\rho_o(0.70)(4\pi R_E^2)d}{\rho_E(\frac{4}{3}\pi R_E^3)} = 2.1\frac{d}{R_E}\frac{\rho_o}{\rho_E}.$$

There are still four quantities to estimate, but at least all the squaring and cubing of values are eliminated! The ocean's average depth d is about a mile, or 1.6 km. The radius of Earth is about 4000 mi, or 6400 km. The mass density of the ocean is about the same as that of fresh water. The mass density of Earth's solid surface materials (things like rocks) must be a few times greater than

the mass density of water because rocks sink readily. Earth's interior must have a considerably larger mass density than this because gravity compresses matter near the center. So we can estimate that the average mass density of Earth is about five times that of water, and therefore the ratio ρ_o/ρ_E is about $1/5$. This makes the fraction f

$$f = (2.1)\left(\frac{1.6\text{ km}}{6400\text{ km}}\right)\frac{1}{5} = 1.1 \times 10^{-4} \approx 10^{-4}.$$

Therefore about $1/100$ of 1% of Earth's mass is contained in the oceans.

Check by VENUS:
  Vectors/scalars: all quantities are scalars ✔
  Every question answered: the percent is calculated ✔
  No unknowns in answers: all quantities known or estimated ✔
  Units correct: the answer has no units, it is a percentage ✔
  Significant digits: order-of-magnitude estimate ✔

❹ **EVALUATE RESULT** The depth of the oceans is much less than the radius of Earth, and water is considerably less dense than the solid material of Earth. We therefore should expect the oceans to contain a very small percent of Earth's mass. If you look up the actual value, you find that our estimate is within a factor of 2 or so.

Guided Problem 1.8 Roof area

One suggestion for reducing the use of fossil fuels is to cover the roofs of all buildings in the United States with solar collectors. Make an order-of-magnitude estimate of the combined surface area, in square kilometers, of all these solar collectors.

❶ **GETTING STARTED**
  1. What simple shape can you use to approximate the United States?
  2. Is it reasonable to assume that buildings are primarily in cities?

❷ **DEVISE PLAN**
  3. What is the approximate area of the United States?
  4. What percent of that area is occupied by cities?

❸ **EXECUTE PLAN**

❹ **EVALUATE RESULT**

Questions and Problems

For instructor-assigned homework, go to MasteringPhysics®

Solving context-rich problems

The problems labeled CR are *context-rich* problems—problems that are more like those in the everyday world. These questions are typically embedded in a short narrative and often do not specify what variables you must calculate to answer the question. So, rather than asking What is the mass of …? (telling you that you must calculate the mass of some object), a context-rich problem may ask Do you accept the bet? (leaving it to you to determine what to calculate). The problem typically contains extraneous information, and you may need to supply some missing information, either by estimation or by looking up values.

Like all problems, context-rich problems should be solved using the four-step procedure outlined in the

Procedure box in this chapter. Because context-rich problem statements are never broken down into parts, the first two steps ("Getting started" and "Devise plan") are particularly important.

Context-rich problems will sharpen the skills you need to solve everyday problems, where you need to move toward a goal, but the path you must take is not immediately clear (if it were clear, there wouldn't be a problem). The information available to you may be a bit sketchy or contradictory, there may be more than one way to approach a situation, and some ways may be more fruitful than others, but that may not be clear at the outset. So consider context-rich problems as an opportunity to stretch your problem-solving abilities.

PRACTICE

Dots indicate difficulty level of problems: • = *easy,* •• = *intermediate,* ••• = *hard;* CR = *context-rich problem.*

1.1 The scientific method

1. In a discussion of what holds airplanes aloft, a classmate offers this hypothesis: "Airplanes are held up by an undetectable force field produced by magnets." Which word in this statement is most likely to keep the statement from satisfying the criteria for a scientific hypothesis? •

2. An advertisement for a food product states that it contains 50% less fat per serving than a competing product. What assumptions are you making if you accept the claim as valid? •

3. You are asked to predict the next item in the following sequence of integers: 1, 2, 3. If your prediction is 4, what assumptions did you make? ••

4. Checkpoint 1.1 of the *Principles* volume states that two coins together have a value of 30 cents but one of them is not a nickel. If the checkpoint said instead that "neither coin is worth 5 cents," what hidden assumptions might prevent you from obtaining a solution? ••

5. In a 4-by-4 Sudoku puzzle with one 2-by-2 subsquare filled in as shown in Figure P1.5, how many ways are there to complete the puzzle? (In Sudoku, the digits 1, 2, 3, 4 must appear only once in each row and once in each column of the 4-by-4 square and only once in each subsquare.) •••

Figure P1.5

4	3		
1	2		

1.2 Symmetry

6. How many axes of reflection symmetry does the triangle in Figure P1.6 have? Restrict your answer to axes that lie in the plane of the triangle. •

Figure P1.6

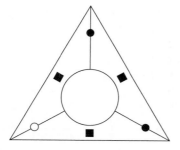

7. How many axes of rotational symmetry does a cone have? •

8. Three identical coins are positioned on a grid as shown in Figure P1.8. Where would you place a fourth coin to form a coin arrangement that has both reflection symmetry and a 90° rotational symmetry? •

Figure P1.8

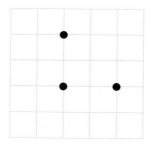

9. Describe the reflection symmetry of each of the letters in the word below: ••

T A B L E S

10. Which types of geometrical symmetry mentioned in the *Principles* volume does a sphere have? ••

11. How many axes of reflection symmetry does a cube have? How many axes of rotational symmetry? ••

12. Suppose you have a square of paper with a single vertical line down the middle and a single horizontal line across the middle. The upper left square is blank. The lower left and upper right squares are blue. The lower right square has a red dot in its center. How many axes of reflection symmetry does the square have? Restrict your answer to axes that lie in the plane of the square. ••

13. A cube is placed on a table with a green side facing up. You know that the side facing down is also green but has a red dot in the middle of the green face. From where you are sitting, you can see that two adjacent vertical sides are one red and the other blue. What is the maximum number of axes of reflection symmetry possible? •••

1.3 Matter and the universe

14. Traveling at 299,792,458 m/s, how far does light travel in 78 years? •

15. The Sun is approximately 93 million miles from Earth. (*a*) What is this distance in millimeters? (*b*) How many Earths could fit side by side in this distance? (See Worked Problem 1.7 for the radius of Earth.) •

16. *Principles* Figure 1.9 tells us that a human contains 10^{29} atoms and a blue whale contains 10^{32} atoms. Use these values to determine the ratio of the length of a blue whale to the height of a human. ••

17. A gastrotrich is a small aquatic organism with a life span of about 3 days, and a giant tortoise has an average life span of 100 years. Make an order-of-magnitude estimate of the number of gastrotrich lifetimes that are equal to one tortoise lifetime. In your calculation, use an order-of-magnitude value for the number of days in 1 year. ••

18. You measure the diameter of a drop of water to be 3 mm. Make an order-of-magnitude estimate of how many such drops are contained in all the water in your body. ••

19. If you could stack copies of your physics textbook until they reached the Moon, what is the order of magnitude of the number of books you would need? ••

20. How many water molecules are there in a swimming pool that is 15 m long, 8.5 m wide, and 1.5 m deep? ••

21. Cube 1 has side ℓ_1 and volume V_1. Cube 2 has side $\ell_2 = 2\ell_1$. (a) When making order-of-magnitude estimates, by how many orders of magnitude is the volume of the second cube greater than the volume of the first cube? (b) Does the answer depend on the numerical value of ℓ_1? ••

22. Earth requires approximately 365 days to orbit the Sun along a path that can be approximated by a circle with a radius of 1.50×10^8 km. Make an order-of-magnitude estimate of the length of time it would take light to move once around this orbit. ••

23. While admiring a tree, you notice that the leaves form a more or less continuous spherical shell, with relatively few leaves growing in the shell interior. You estimate that the diameter of this shell is about 30 m. Each leaf is about 5 in. long and 3 in. wide. Make an order-of-magnitude estimate of the number of leaves on this tree. •••

1.4 Time and change

24. A human generation is about 30 years, and the age of the universe is 10^{17} s. How many human generations have there been since the beginning of the universe? (Ignore the fact that humans did not exist during most of this time.) •

25. You hear a peal of thunder and go to the window. You then see a flash of lightning. Is it reasonable to assume that the thunder caused the lightning? •

26. A second is defined as the duration of 9.19×10^9 periods of the radiation emitted by a cesium atom. What is the duration in seconds of a single period for this atom? •

27. While watching a railroad crossing, you observe that a crossing barrier is lowered about 30 s before a train passes. This happens every time for a long while, until suddenly after many times the barrier is not lowered but a train passes anyway! Discuss the causal relationship between the passing of a train and the lowering of the barrier. Does the single unusual occurrence make a difference to your answer? ••

28. Assume the radiation from the cesium atom mentioned in Problem 26 moves at the speed of light. How many meters does the radiation travel in the length of time corresponding to the period calculated in Problem 26? ••

1.5 Representations

29. Translate this statement into a mathematical expression, using any symbols you wish, but include a "key" to the meaning of each symbol: A certain type of energy of an object is equal to the object's mass times the speed of light squared. •

30. A planar shape is formed by joining the ends of four straight-line segments, two with length ℓ and two with length 2ℓ. No loose ends are allowed, but crossings are allowed, and only two segment ends may join at any single point. The ends of identical-length segments cannot join each other. What shapes are possible? ••

31. A planar shape is formed from four straight-line segments of length ℓ that do not cross. One end of segment 1 is connected to one end of segment 2 such that they make a 30° angle. The other end of segment 2 is connected to one end of segment 3

such that they also make a 30° angle. Finally, the other end of segment 3 is connected to one end of segment 4, forming another 30° angle. What is the distance between the unconnected ends of segments 1 and 4? ••

32. You are arranging family members for a photograph, and you want them to stand in order of increasing height. Your uncle is half a foot shorter than your aunt, who is taller than your cousin. Your grandmother is 2 in. shorter than your grandfather. Your brother is 1 in. taller than your aunt and 3 in. taller than your cousin. Your grandfather is 1 cm taller than your aunt. List the relatives in order of increasing height. ••

33. The Jupiter–Sun distance is 778 million kilometers, and the Earth–Sun distance is 150 million kilometers. Suppose an imaginary line from Jupiter to the Sun forms a right angle with an imaginary line from Earth to the Sun. How long would it take light to travel from the Sun to a spaceship halfway along the straight-line path between Jupiter and Earth? ••

34. During a physics lab, you and your partner keep track of the position of a small cart while it is moving along a slope. The lab instructions ask you to prepare (a) a pictorial representation of the experiment, (b) a table recording the position of the cart measured every 2.00 seconds from its initial position, and (c) a graph that shows the position of the cart on the vertical axis and time on the horizontal axis. You produce the sketch shown in Figure P1.34. What should your table and your graph look like? ••

Figure P1.34

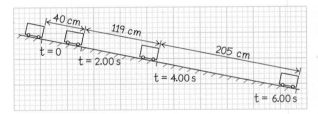

35. The graph in Figure P1.35 shows a relationship between two quantities: position x, measured in meters, and time t, measured in seconds. Describe the relationship using (a) a verbal expression and (b) a mathematical expression. •••

Figure P1.35

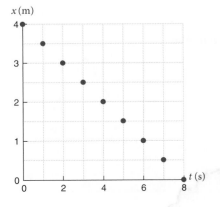

36. You have four equilateral triangles, two red and two blue. You want to arrange them so that each blue triangle shares a side with the other blue triangle and with both red triangles, and each red triangle shares a side with the other red triangle and with both blue triangles. How can this be done? •••

1.6 Physical quantities and units

37. The world record in the long jump was recently set at 8.95 m. What is this distance in inches? •

38. Planes fly at an altitude of 35,000 ft above sea level. What is this distance in miles? In meters? •

39. A metal block of uniform mass density is cut into two pieces. How does the density of each piece compare with the density of the original block (*a*) if each piece has half the volume of the original block and (*b*) if one piece has one-third of the original volume? •

40. The speed of light is 2.9979×10^8 m/s. Convert this speed to miles per second, inches per nanosecond, and kilometers per hour. ••

41. Suppose you measure the mass and volume of two solid stones. Stone 1 has a mass of 2.90×10^{-2} kg and a volume of 10.0 cm^3. Stone 2 has a mass of 2.50×10^{-2} kg and a volume of 7.50 cm^3. Is it likely that the stones are made from the same material? ••

42. The Kentucky Derby horserace is 1.000 mile and 440 yards long. Express this distance in feet. ••

43. You are doing a calculation in which you must add two quantities *A* and *B*. You know that *A* can be expressed as $A = \frac{1}{2}at^2$, where *a* has units of meters per second squared and *t* has units of seconds. No matter what the expression for *B* is, what must the units of *B* be? ••

44. Both *mass* and *volume* can be used to describe the amount of matter present in an object. What physical quantity relates these two concepts? ••

45. The average radius of Earth is 6371 km. Give order-of-magnitude estimates of what the mass of Earth would be if the planet has the mass density of (*a*) air (≈ 1.2 kg/m^3), (*b*) 5515 kg/m^3, and (*c*) an atomic nucleus ($\approx 10^{18}$ kg/m^3). ••

46. Suppose $x = ay^{3/2}$, where $a = 7.81$ μg/Tm. Determine the value of *y* when $x = 61.7$ (Eg·fm^2)/(ms^3). Express the result in scientific notation and simplify the units. (Hint: Refer to *Principles* Table 1.3, and note that SI prefixes are never used to multiply powers of units. For example, the abbreviation cm^2 means $(10^{-2}$ m$)^2$, not 10^{-2} m^2, and ns^{-1} is 1/ns or 10^9 s^{-1}, not 10^{-9} s^{-1}. Also note that m can stand for meter or for milli, depending on the context.) •••

1.7 Significant digits

47. (*a*) What is the speed of light in vacuum expressed to three significant digits? (*b*) What is the speed of light in vacuum squared expressed to three significant digits? (*c*) Is the numerical value in part *b* the square of the numerical value in part *a*? Why or why not? •

48. Express the distance of the Kentucky Derby in kilometers to the same number of significant digits needed to answer Problem 42. •

49. You have 245.6 g of sugar and wish to divide it evenly among six people. If you calculate how much sugar each person receives, how many significant digits does your answer have? •

50. You are recording your car's gas mileage. Your odometer measures to the tenth of a mile. The gas station pump displays gallons of gasoline dispensed to the thousandth of a gallon. Given these levels of precision, is there any difference in the precision of your calculation when you drive 40.0 miles or 400.0 miles? ••

51. The odometer on your car says 35,987.1 km. To the correct number of significant digits, what should the odometer read after you drive 47.00 m? ••

52. A certain brand of soft drink contains 34 mg of caffeine per 355-mL serving. If one mole of caffeine has a mass of 194.19 g and you drink an average of two servings of this drink a day, how many caffeine molecules do you ingest in one year? ••

53. You work in a hospital and are preparing a saline solution for a patient. This solution must have a concentration of 0.15 mol of NaCl (mass of one mole, 58.44 gram) per liter of solution. You have lots of volume-measuring devices, but the only mass-measuring tool you have is a balance that measures mass to the nearest 0.1 g. What is the minimum amount of solution you should mix? ••

54. You find a container of an unknown clear liquid. You are interested in its contents, but you have no chemical analysis devices at hand. You do have a good electronic scale, so you measure out 25.403 g of the liquid and then pour this liquid into a graduated cylinder, with the level indicated in Figure P1.54. What is the mass density of this liquid? ••

Figure P1.54

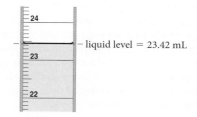

liquid level = 23.42 mL

55. Suppose that, in a laboratory exercise, you add 0.335 g of liquid to a 145.67-g ceramic bowl. After you heat the liquid for 25.01 s, you measure the mass of the bowl and liquid together to be 145.82 g. What was the average mass of liquid that evaporated each second? •••

1.8 Solving problems

56. A test question gives the mass density of seawater and asks what volume is occupied by 1.0×10^3 kg of seawater. Someone answers 0.9843 m. Without knowing the mass density value given, explain what is wrong with this answer. •

57. You measure a watch's hour and minute hands to be 8.0 mm and 11.3 mm long, respectively. In one day, by how much does the distance traveled by the tip of the minute hand exceed the distance traveled by the tip of the hour hand? ••

58. An Olympic running track consists of two straight sections, each 84.39 m long, and two semicircular ends, each with a radius of 36.80 m as measured from the inside lane (lane 1). There are eight lanes, each 1.22 m wide. If runners did not start staggered (so that each runner covers the same distance), by how much would the distance traveled by a runner in lane 8 exceed the distance traveled by a runner in lane 1 over one lap? ••

59. You have three coins that are identical in appearance. One is counterfeit, and its mass is different from the mass of the two legitimate coins. You also have an equal-arm balance (a device that compares the masses of two objects and "balances" if they are equal). Describe how you can determine which is the counterfeit coin and whether it is heavier or lighter than the legitimate coins by making only two measurements with the balance. ●●

60. A certain type of rice has an average grain length and diameter of 6 mm and 2 mm, respectively. One cup of this rice, after being cooked, contains about 785 food calories. (a) How many grains are there in 1 cup of rice? (b) How many calories are there in one grain? (c) How many cups of uncooked rice are needed to provide 2000 food calories to each of four adults? ●●●

1.9 Developing a feel

61. What is the order of magnitude of the number of moles of matter in the observable universe? ●

62. You add about 10^2 raisins to a loaf of bread you are baking. When fully baked, the loaf is hemispherical with a radius of about 8 in. What is the order of magnitude of the average distance between neighboring raisins? ●

63. A tree cut down for lumber has a diameter of 0.80 m over its length of 32 m. About how many 6-ft-long, 2-in. by 2-in. boards can be made from this tree? ●

64. Estimate the order of magnitude of the number of times the letter d occurs in this text. ●●

65. Estimate the combined length of all the hairs on the head of a person who has a a full head of shoulder-length hair. ●●

66. An urban legend making the rounds in quantitative reasoning classes is that a student paid for four years at a private university in the United States exclusively with redeemed deposits on soda cans collected from people's trash. Estimate whether this is possible. ●●

67. Computer hard drives have increased dramatically in capacity since their introduction in the 1950s. The first commercial hard drive had a capacity of about 3.8 MB stored on 50 circular platters, each 610 mm in diameter. In the early part of this century, hard drives approached 1.0 TB of storage using just five platters, each with a diameter of about 90 mm. By what order-of-magnitude factor does the storage capacity per unit area of a 21st-century drive exceed that of an original drive? ●●

Additional Problems

68. Express the speed $1.082\,43 \times 10^{19}$ nm/y in meters per second. ●

69. A 200-sheet stack of loose-leaf paper is 2.75 in. thick. What is the thickness in millimeters of one sheet? ●

70. A water tank is filled to its 50×10^3-L capacity. (a) What is the tank's volume in cubic millimeters? (b) What is the mass of the water in milligrams? (c) If you drank eight average-sized glasses of water a day, how long would the water in the tank last? ●●

71. You are working on an exhibit of the atom at a science museum. You want the exhibit to have a model of the atom in which the nucleus is big enough for visitors to see, about 500 mm in diameter. The model needs to fit inside a square room with walls about 25 m long. Will it fit if you make the nucleus this size? ●●

72. Recent experimental values for the universal gravitational constant, the mass and mean radius of Earth, and the standard gravitational acceleration are

$$G = 6.6738 \times 10^{-11}\ \mathrm{m^3 \cdot kg^{-1} \cdot s^{-2}}$$

$$M_\mathrm{E} = 5.9736 \times 10^{24}\ \mathrm{kg}$$

$$R_\mathrm{E} = 6.378\,140 \times 10^6\ \mathrm{m}$$

$$g = 9.806\,65\ \mathrm{m \cdot s^{-2}}$$

Assuming a spherical Earth, these quantities are related by the expression $gR_\mathrm{E}^2 = GM_\mathrm{E}$. (a) To how many significant digits is this relation satisfied using the recent data? (b) Can you round off all the data to the same number of significant digits before multiplying and still satisfy this relation to that number of significant digits? If so, what numbers of significant digits work? ●●

73. You are in charge of safety at a local park. A tire swing hangs from a tree branch by a 5-m rope, and the bottom of the tire is 1 m from the ground directly beneath it. The land is sloped at a 12° angle toward a pond. From experience, you know that even an energetic user will be unlikely to cause the swing to make a maximum angle exceeding 30° with respect to the vertical. You are certain that parents will want reassurance about the safety of the tire swing. ●●● CR

74. While working for a space agency, you are tasked with ensuring that the oxygen needs of human beings on long voyages in outer space are met. Your boss hands you a data table and suggests that you consider the amount of oxygen a person might need in 1 year as well as the storage requirements for this oxygen at room temperature and atmospheric pressure. ●●● CR

Volume of one breath	4.5 L
Oxygen content of air, by volume	20.95%
Percent oxygen absorbed per breath	25%
Average breathing rate	15 breaths/min
Mass density of air at room temperature and atmospheric pressure	1.0 kg/m^3

75. Neutron stars can have a mass on the order of 10^{30} kg, but they have relatively small radii, on the order of tens of kilometers. (a) What order of magnitude is the mass density of a neutron star? (b) How many orders of magnitude larger is this than the mass density of Earth? Of water? (c) If water had the mass density of a neutron star, what order of magnitude of mass would be contained in the soda in a full 2-L bottle? ●●●

Answers to Review Questions

1. Physics is the study of matter and motion in the universe. It is about understanding the unifying patterns that underlie all phenomena in nature.

2. The scientific method is an iterative process that develops validated theories to explain our observations of nature. It involves observing some phenomenon, formulating a hypothesis from the observations, making predictions based on the hypothesis, and validating the predictions by running experiments to test them.

3. Some useful skills are interpreting observations, recognizing patterns, making and recognizing assumptions, reasoning logically, developing models, and using the models to make predictions.

4. Inductive reasoning is arguing from the specific to the general; deductive reasoning is arguing from the general to the specific.

5. Symmetry means that the appearance of an object, process, or law is not changed by a specific operation, such as rotation or reflection.

6. Translational symmetry in space, in which different observers at different locations get the same value for a given measurement, and translational symmetry in time, in which an observer gets the same value for a given measurement taken at different instants, are two types.

7. The *universe* is the totality of matter and energy plus the space and time in which all events occur.

8. An order of magnitude is a value rounded to the nearest power of ten. Using orders of magnitude gives you a feel for a quantity and is a key skill in any quantitative field.

9. The order of magnitude of 2900 is 10^3, and the order of magnitude appropriate for 3100 is 10^4. This is because the first digit 3 is used as the demarcation between rounding up and rounding down. On a logarithmic scale, the base-10 logarithm of 3 is about halfway between the logarithm of 1 and the logarithm of 10.

10. The size scale ranges from the subatomic (10^{-16} m or smaller) to the size of the universe (10^{26} m or larger). The time scale ranges from a hundredth of an attosecond (10^{-20} s) or smaller to the age of the universe (10^{17} s).

11. Time flows in a single, irreversible direction, from past to present to future.

12. The principle of causality states that if an event A causes an event B, all observers see A happening before B. This means that if event A is observed to occur after event B, then A cannot be the cause of B.

13. Making simplified visual representations such as sketches, graphs, or tables helps you to establish a clear mental image of the situation, relate it to past experience, interpret its meaning and consequences, focus on essential features, and organize more relevant information than you can keep track of in your head.

14. The *Concepts* part develops the conceptual framework of the topics covered in the chapter. The *Quantitative Tools* part develops the mathematical framework for these topics.

15. A numerical value and an appropriate unit of measurement are needed.

16. The SI base units are meter for length, second for time, kilogram for mass, ampere for electric current, kelvin for temperature, mole for quantity of substance, and candela for luminous intensity.

17. Density is the concept of how much there is of some substance in a given volume.

18. Multiply the quantity by a conversion factor, a fraction in which the numerator is a number and the desired unit and the denominator is the equivalent value expressed in the given unit.

19. The number of digits is all the digits written to express a numerical value. The number of decimal places is the number of digits to the right of the decimal point. The number of significant digits is the number of digits that are reliably known. The number 0.037 20 has 6 digits, the last 4 of them significant, and 5 decimal places.

20. Leading zeros are any that come before the first nonzero digit in a number. Trailing zeros are any that come after the last nonzero digit. No leading zero is significant. All trailing zeros to the right of the decimal point are significant. Trailing zeros to the left of the decimal point may or may not be significant.

21. The number of significant digits in the result should be the same as the number of significant digits in the input quantity that has the fewest significant digits.

22. The number of decimal places in the result should be the same as the number of decimal places in the input quantity that has the fewest decimal places.

23. Getting started: Identify problem, visualize situation, organize relevant information, clarify goal.
    Devise plan: Figure out what to do by developing a strategy and identifying physical relationships or equations you can use.
    Execute plan: Proceed stepwise through the plan, performing all the mathematical, algebraic, and computational operations necessary to reach the goal; check calculations.
    Evaluate result: Consider whether the answer is reasonable, makes sense, reproduces known results in limiting or special cases, or can be confirmed by an alternative approach.

24. Vectors and scalars used correctly? Every question answered? No unknown quantities in answers? Units correct? Significant digits justified?

25. They allow you to develop a feel for a problem without getting too involved in the details. They help you explore relationships between physical quantities, consider alternative approaches, make simplifying assumptions, and evaluate answers obtained by more rigorous methods.

Answers to Guided Problems

Guided Problem 1.2

$$\rho_O = \frac{0.0097 m_{Sun}}{\frac{4}{3}\pi R_{Sun}^3} = 13.7 \text{ kg/m}^3;$$

$$n_O = \frac{\rho_O}{m_{O\,atom}} = 5.14 \times 10^{26} \text{ atoms/m}^3$$

Guided Problem 1.4 $V = Wh\ell = 1.05 \text{ m}^3$

Guided Problem 1.6 (*a*) 4.1×10^2, 10^3; (*b*) 24.75, 10^1; (*c*) 6980.2, 10^4

Guided Problem 1.8 About 10^5 km², assuming that buildings cover about 1%–2% of the land and that the United States is a rectangle with dimensions 1000 mi × 3000 mi (1600 km × 4800 km)

2

PRACTICE
Motion in One Dimension

Chapter Summary   17
Review Questions   18
Developing a Feel   19
Worked and Guided Problems   20
Questions and Problems   25
Answers to Review Questions   33
Answers to Guided Problems   33

PRACTICE

Chapter Summary

Average speed and average velocity (Sections 2.2–2.4, 2.6, 2.7)

Concepts The **distance traveled** is the accumulated distance covered by an object along the path of its motion, without regard to direction.

The x component of an object's **displacement** is the change in its x coordinate.

The **average speed** of an object is the distance it travels divided by the time interval it takes to travel that distance.

The x component of an object's **average velocity** is the x component of its displacement divided by the time interval taken to travel that displacement. The magnitude of the average velocity is not necessarily equal to the average speed.

Quantitative Tools The **distance** d between two points x_1 and x_2 is

$$d = |x_1 - x_2|. \tag{2.5}$$

The x component of the **displacement** Δx of an object that moves from a point x_i to a point x_f is

$$\Delta x = x_f - x_i. \tag{2.4}$$

The x component of the object's **average velocity** is

$$v_{x,\mathrm{av}} = \frac{\Delta x}{\Delta t} = \frac{x_f - x_i}{t_f - t_i}. \tag{2.14}$$

Vectors and scalars (Sections 2.5, 2.6)

Concepts A **scalar** is a physical quantity that is completely specified by a number and a unit of measure. A **vector** is a physical quantity that is completely specified by a number, a unit of measure, and a direction. The number and unit of measure together are called the **magnitude** of the vector.

A **unit vector** has magnitude one and no units.

To **add** two vectors, place the tail of the second vector in the sum at the tip of the first vector; the vector representing the sum runs from the tail of the first vector to the tip of the second. To **subtract** one vector from another, reverse the direction of the vector being subtracted and add the reversed vector to the vector from which you are subtracting.

Quantitative Tools The unit vector $\hat{\imath}$ has magnitude one and points along the x axis in the direction of increasing x.

A vector $\vec{b}$ that points in the x direction and has an x component b_x can be written in **unit vector notation** as

$$\vec{b} = b_x\,\hat{\imath}. \tag{2.2}$$

Position, displacement, and velocity as vectors (Sections 2.6, 2.7)

Concepts The **position vector**, or **position**, of a point is drawn from the origin of a coordinate system to the point.

The **displacement vector** of an object is the change in its position vector. It is the vector that points from the tip of the original position vector to the tip of the final position vector.

Quantitative Tools The position $\vec{r}$ of a point that has x coordinate x is

$$\vec{r} = x\,\hat{\imath}. \tag{2.9}$$

The displacement vector $\Delta\vec{r}$ of an object moving along the x axis is

$$\Delta\vec{r} = \vec{r}_f - \vec{r}_i = (x_f - x_i)\hat{\imath}. \tag{2.7, 2.8, 2.10}$$

The average velocity of an object moving along the x axis is

$$\vec{v}_{\mathrm{av}} = \frac{\Delta\vec{r}}{\Delta t} = \frac{x_f - x_i}{t_f - t_i}\hat{\imath}. \tag{2.14, 2.15}$$

Additional properties of velocity (Sections 2.8, 2.9)

When the velocity of an object is constant, its **position-versus-time graph** is a straight line having nonzero slope and its **velocity-versus-time graph** is a horizontal line.

The **instantaneous velocity** of an object is its velocity at any particular instant.

The slope of an object's $x(t)$ curve at a given instant is numerically equal to the x component of the object's velocity at that instant. The area under the object's $v_x(t)$ curve during a time interval is the x component of the object's displacement during that interval.

The x component of an object's instantaneous velocity is the derivative of the object's x coordinate with respect to time:

$$v_x = \frac{dx}{dt}. \tag{2.22}$$

Review Questions

Answers to these questions can be found at the end of this chapter.

2.1 From reality to model

1. A friend constructs a graph of position as a function of frame number for a film clip of a moving object. She then challenges you to construct a graph identical to hers. She tells you that she measured distances in millimeters, but what two additional pieces of information must she give you?
2. How do you determine whether an object is moving or at rest?

2.2 Position and displacement

3. Suppose you have a film clip showing the motion of several objects. If you want to calibrate distances for a graph, what information do you need to know about at least one of the objects?
4. Explain how the information you named in Review Question 3 can be used to calibrate a graph.
5. What is the purpose of using the phrase *x component of* when describing some physical quantities?
6. How do you represent displacement in a drawing or diagram?

2.3 Representing motion

7. What does *interpolation* mean with regard to plotting data points?
8. Describe how to determine the x component of the position of an object at a specific instant, given (*a*) a graph of position x as a function of time t and (*b*) an equation for $x(t)$.

2.4 Average speed and average velocity

9. In an $x(t)$ curve, what is the significance of a steep slope as opposed to a gentle slope? What is the significance of a curve that slopes downward as you move from left to right along the time axis as opposed to a curve that slopes upward as you move from left to right along that axis?
10. What does it signify physically if the x component of an object's average velocity is negative over some time interval?

2.5 Scalars and vectors

11. What are the defining characteristics of a scalar? Of a vector?
12. What is the relationship between the magnitude of a vector and the vector's x component?
13. What is the purpose of the unit vector $\hat{\imath}$?

2.6 Position and displacement vectors

14. What is the mathematical meaning of the symbol Δ, the Greek capital letter delta?

15. Is displacement a scalar or a vector? Is distance a scalar or a vector?
16. Can distance be negative? Can distance traveled be negative?
17. Under what conditions is the x component of the displacement negative?
18. How is distance traveled calculated when motion occurs in three segments: first in one direction along the x axis, then in the opposite direction, and finally in the initial direction?
19. Describe the graphical procedure for adding two vectors and for subtracting one vector from another.
20. If you multiply a vector $\vec{A}$ by a scalar c, is the result a scalar or a vector? If the result is a scalar, what is its magnitude? If the result is a vector, what are its magnitude and direction? What if the scalar $c = 0$?

2.7 Velocity as a vector

21. Is average speed a scalar or a vector? Is average velocity a scalar or a vector?
22. What properties does velocity have that make it a vector?
23. How is an object's average velocity related to its displacement during a given time interval?

2.8 Motion at constant velocity

24. What is the shape of the $x(t)$ curve for an object moving at constant velocity? What is the shape of the $v_x(t)$ curve for this object?
25. An object travels at a constant velocity of 10 m/s north in the time interval from $t = 0$ to $t = 8$ s. What additional information must you know in order to determine the position of the object at $t = 5$ s?
26. In a velocity-versus-time graph, what is the significance of the area under the curve for any time interval $t_f - t_i$?

2.9 Instantaneous velocity

27. What feature of the $x(t)$ curve for a moving object gives the object's x component of velocity at a given instant?
28. Under what conditions is the average velocity over a time interval equal to the instantaneous velocity at every instant in the interval?
29. What mathematical relationship allows you to compute the x component of an object's velocity at some instant, given the object's x component of position as a function of time?

Developing a Feel

Make an order-of-magnitude estimate of each of the following quantities. Letters in parentheses refer to hints below. Use them as needed to guide your thinking.

1. The height of a 20-story apartment building (D)
2. The distance light travels during a human life span (B, N)
3. The displacement (from your mouth) of an (indigestible) popcorn kernel as it passes through your body, and the distance traveled by the same kernel (F, O)
4. The time interval within which a batter must react to a fast pitch before it reaches home plate in professional baseball (C, H)
5. The time interval needed to drive nonstop from San Francisco to New York City by the most direct route (G, K)
6. The distance traveled when you nod off for 2 s while driving on the freeway (K)
7. The average speed of an airliner on a flight from San Francisco to New York City (G, Q)
8. The average speed of a typical car in the United States in one year (not just while it's running) (E)
9. The time interval for a nonstop flight halfway around the world from Paris, France, to Auckland, New Zealand (J and item 7 above)
10. The number of revolutions made by a typical car's tires in one year (L, E)
11. The maximum speed of your right foot while walking (A, M, P)
12. The thickness of rubber lost during one revolution of a typical car tire (I, R, L, S)

Hints

A. What is your average walking speed?
B. What is the speed of light?
C. What is the speed of a fastball thrown by a professional pitcher?
D. What is the height of each story in an apartment building?
E. What distance does a typical car travel during one year?
F. When you are sitting upright, how far above the chair seat is your mouth?
G. What is the distance between San Francisco and New York City?
H. What is the distance from the pitcher's mound to home plate?
I. What thickness of rubber is lost during the lifetime of a car tire?
J. What is the circumference of Earth?
K. What is a typical freeway speed?
L. What is the circumference of a car tire?
M. For what time interval is your right foot at rest if you walk for 2 min?
N. What is a typical human life span?

O. What is the length of the digestive tract in an adult person?
P. If you walk 10 m in a straight line, what is the displacement of your right foot?
Q. How much elapsed time does a flight from San Francisco to New York City require?
R. How many miles of service does a car tire provide?
S. How many revolutions does a car tire make in traveling 1 m?

Key (all values approximate)

A. 2 m/s; B. 3×10^8 m/s; C. 4×10^1 m/s; D. 4 m;
E. 2×10^7 m; F. 1 m; G. 5×10^6 m; H. 2×10^1 m;
I. 1×10^{-2} m; J. 4×10^7 m; K. 3×10^1 m/s; L. 2 m; M. 1 min;
N. 2×10^9 s; O. 7 m; P. 1×10^1 m; Q. 2×10^4 s; R. 8×10^7 m;
S. 0.5 rev/m

Worked and Guided Problems

These examples involve material from this chapter but are not associated with any particular section. Some examples are worked out in detail; others you should work out by following the guidelines provided.

Worked Problem 2.1 Shopping hunt

A frantic shopper performs the following sequence of movements along a supermarket aisle in search of a special item:

    (1) He moves eastward at a constant speed of 3.0 m/s for 10 s,
    (2) stops for 5.0 s,
    (3) walks slowly in the same direction at a constant speed of 0.50 m/s for 20 s, and
    (4) immediately turns around and rushes westward at a constant speed of 4.0 m/s for 9.0 s.

(*a*) Plot the shopper's position as a function of time during the 44-s interval. (*b*) What is the *x* component of his average velocity from $t = 0$ to $t = 35$ s? From $t = 0$ to $t = 44$ s? (*c*) What distance does he travel during the 44-s interval? (*d*) What is his average speed during this interval?

1 GETTING STARTED We know that the shopper first moves in a straight line in one direction, then stops for a while, then moves in the same direction again at a slower speed, and then switches direction. To visualize the motion, we make a sketch that represents the velocity in each of the four segments (Figure WG2.1).

Figure WG2.1

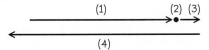

We know that the speed and direction of travel affect the shopper's position at any given instant. Let us say that all the motion is along an *x* axis. Our first task is to plot the $x(t)$ curve from the starting position, based on the information about his speed and direction of motion from that instant. In addition, we need to determine his average velocity during two time intervals, the distance traveled in 44 s, and the average speed in the 44-s interval.

We summarize the given information to see which quantities we know. We need unambiguous symbols for the quantities in the various segments. We shall use the segment number as a subscript to identify the final instant and final position of each segment; for example, t_3 is the final instant of the third segment, and at this instant the shopper is at position x_3. Choosing the positive *x* direction to be eastward, we can list this information from the problem statement:

$v_{x,1} = +3.0$ m/s during interval from $t_0 = 0$ to $t_1 = 10$ s

$v_{x,2} = 0$ during interval from $t_1 = 10$ s to $t_2 = 15$ s

$v_{x,3} = +0.50$ m/s during interval from $t_2 = 15$ s to $t_3 = 35$ s

$v_{x,4} = -4.0$ m/s during interval from $t_3 = 35$ s to $t_4 = 44$ s.

Although it would be tempting to label the initial instant t_i rather than t_0, we recognize that there are several segments to this motion and that the initial instant of motion in one segment is the final

instant of motion in the previous segment. Thus t_i in our general kinematic formulas has different values in the different segments and is not always equal to zero. To avoid confusion, therefore, we designate our starting instant by something other than t_i; we choose t_0. Likewise we choose t_4 rather than t_f as the ending instant of the fourth segment to avoid confusion with the symbol t_f used in our general equations as the final instant of each segment.

Because one task is to plot position as a function of time from the starting position, we choose the starting coordinate to be $x_0 = 0$.

2 DEVISE PLAN In order to make our position-versus-time graph for part *a*, we need to determine the shopper's final position for each segment of the motion. The initial position for each successive segment is equal to the final position for the previous segment. We can connect the start and end positions in any segment with a straight line (constant slope) because the velocity is constant in each segment. For motion at constant velocity, we can use Eq. 2.19 to calculate the position x_f at the end of any time interval:

$$x_f = x_i + v_x(t_f - t_i). \tag{1}$$

For part *b*, we can divide the displacement between the two specified instants by the corresponding time interval to get the *x* component of the average velocity (Eq. 2.14):

$$v_{x,\text{av}} = \frac{x_f - x_i}{t_f - t_i}. \tag{2}$$

For part *c*, we get the distance traveled in 44 s by adding the distances traveled in the four segments, noting that distance traveled is a magnitude. For part *d*, the average speed is the distance traveled in 44 s divided by that time interval. Of our four tasks, the only "hard" part is getting the displacement in each time interval for our position-versus-time graph.

3 EXECUTE PLAN (*a*) To plot position versus time, we need to calculate the position at the end of each segment of the motion. To use Eq. 1, we note that in the first segment the initial position is $x_i = x_0 = 0$, the velocity in the *x* direction is $v_x = v_{x,1} = +3.0$ m/s, the initial instant is $t_i = t_0 = 0$, and the final instant is $t_f = t_1 = 10$ s. Substituting these values into Eq. 1, we compute the shopper's final position $x_f = x_1$ after 10 s (segment 1)

$$x_1 = x_0 + v_{x,1}(t_1 - t_0) = 0 + (+3.0 \text{ m/s})(10 \text{ s} - 0) = +30 \text{ m}.$$

We repeat this procedure for the other three segments, using the final position of the preceding segment as the initial position of the current segment:

$$x_2 = x_1 + v_{x,2}(t_2 - t_1) = +30 \text{ m} + (0)(15 \text{ s} - 10 \text{ s}) = +30 \text{ m}$$

$$x_3 = +30 \text{ m} + (+0.50 \text{ m/s})(35\text{s} - 15 \text{ s}) = +40 \text{ m}$$

$$x_4 = +40 \text{ m} + (-4.0 \text{ m/s})(44 \text{ s} - 35 \text{ s}) = +4.0 \text{ m}.$$

Now we plot these positions in Figure WG2.2 and connect them with straight lines. ✔

Figure WG2.2

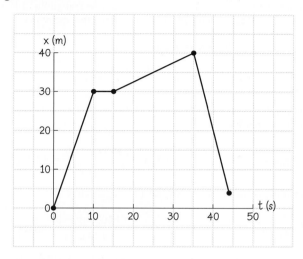

(b) From Eq. 2, the x component of the average velocity for the first 35 s, which runs to the end of segment 3, is the x component of the displacement during that time interval divided by the time interval:

$$v_{x,03} = \frac{x_3 - x_0}{t_3 - t_0} = \frac{+40 \text{ m} - 0}{35 \text{ s} - 0} = +1.1 \text{ m/s.} ✔$$

This result is positive because the overall travel is eastward. The x component of the shopper's average velocity over the whole trip (to the end of segment 4) is

$$v_{x,04} = \frac{x_4 - x_0}{t_4 - t_0} = \frac{+4.0 \text{ m} - 0}{44 \text{ s} - 0} = +0.091 \text{ m/s.} ✔ \qquad (3)$$

(c) The distance traveled in 44 s is the sum of the distances traveled in the various segments. In this case, because each segment involves motion in a single direction, the distance traveled is the same as the distance—that is, the magnitude of the displacement. We can use the symbol d with appropriate subscripts to represent the distances traveled in the segments:

$$\begin{aligned}
d_{04} &= d_{01} + d_{12} + d_{23} + d_{34} \\
&= |x_1 - x_0| + |x_2 - x_1| + |x_3 - x_2| + |x_4 - x_3| \\
&= |+30 \text{ m} - 0| + |+30 \text{ m} - (+30 \text{ m})| + \\
&\quad |+40 \text{ m} - (+30 \text{ m})| + |+4.0 \text{ m} - (+40 \text{ m})| \\
&= 30 \text{ m} + 0 + 10 \text{ m} + 36 \text{ m} = 76 \text{ m.} ✔
\end{aligned}$$

(d) The average speed is the distance traveled in 44 s divided by that time interval:

$$v_{av} = v_{04} = \frac{d_{04}}{\Delta t_{04}} = \frac{d_{04}}{t_4 - t_0} = \frac{76 \text{ m}}{44 \text{ s}}$$

$$= 1.7 \text{ m/s (between } t_0 \text{ and } t_4). ✔ \qquad (4)$$

Remember that the checkmarks (✔) in our calculations show that VENUS, the mnemonic for checking our work, has been applied.

❹ EVALUATE RESULT It is not unreasonable for a frantic shopper to move 76 m in 44 s. This is covering a distance that is a little shorter than the length of a football field in a time interval somewhat less than a minute, which is within the realm of possibility for a person pressed for time.

The signs of our position values agree with the positions we drew in Figure WG2.1: All four x values are on the positive side of the x axis (because the shopper always stays east of the origin). Notice that the average speed of 1.7 m/s (Eq. 4) is very different from the magnitude of the average velocity, 0.091 m/s (Eq. 3). This is to be expected because average speed is based on distance traveled, which is the sum of the magnitudes of the individual displacements, while average velocity is based on displacement, whose magnitude is much less than the distance traveled because of the shopper's backtracking.

Guided Problem 2.2 City driving

You need to drive to a grocery store that is 1.0 mi west of your house on the same street on which you live. There are five traffic lights between your house and the store, and on your trip you reach all five of them just as they change to red. While you are moving, your average speed is 20 mi/h, but you have to wait 1 min at each light. (a) How long does it take you to reach the store? (b) What is your average velocity for the trip? (c) What is your average speed?

❶ GETTING STARTED

1. Draw a diagram that helps you visualize all the driving and stopped segments and your speed in each segment. Does it matter where the traffic lights are located?

2. You start and stop, speed up and slow down. What does the average speed signify in this case, and how is it related to your displacement in each segment?

❷ DEVISE PLAN

3. During how long a time interval are you moving? During how long a time interval are you stopped at the lights?

4. What is your displacement for the trip?

5. How can you apply the answers to questions 3 and 4 to obtain your average velocity?

6. What is the distance traveled, and how is it related to your average speed?

7. How are average velocity and average speed related in this case?

❸ EXECUTE PLAN

❹ EVALUATE RESULT

8. Are your answers plausible, and is your result within the range of your expectations?

Worked Problem 2.3 Head start

Two runners in a 100-m race start from the same place. Runner A starts as soon as the starting gun is fired and runs at a constant speed of 8.00 m/s. Runner B starts 2.00 s later and runs at a constant speed of 9.30 m/s.

(*a*) Who wins the race? (*b*) At the instant she crosses the finish line, how far is the winner ahead of the other runner? (In your calculations, take the race length, 100 m, to be an exact value.)

❶ **GETTING STARTED** Although runner A has a 2.00-s head start, runner B runs faster than runner A. We want to know who first reaches the finish line, which is at $x_f = +100$ m. We begin by making a position-versus-time graph for both runners. Because both runners have constant speeds, their $x(t)$ curves are straight lines, with both starting at the same position ($x_i = 0$) but at the different initial instants $t_{A,i}$ and $t_{B,i}$ (Figure WG2.3). Note that the faster runner's curve has the steeper slope.

Figure WG2.3

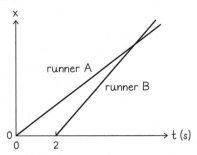

We see from our drawing that at some instant the two $x(t)$ curves intersect, meaning that the runners are at the same position at that instant, but we do not know whether this instant is before or after the slower runner A crosses the 100-m mark on the track.

❷ **DEVISE PLAN** Because we want to know who crosses the finish line first, we can determine the instant t_f at which each runner reaches the position $x_f = +100$ m using Eq. 2.18: $x_f - x_i = v_x(t_f - t_i)$.

The runner who has the lower value for t_f is the winner. Once we know who wins and at what instant she reaches the finish line, we can find out where the other runner is at that instant and then determine the distance between them. Although both runners start at $x_i = 0$, their start times are not the same, and we must account for this. When runner A leaves the starting position ($x_{A,i} = 0$), the timer's stopwatch reads $t_{A,i} = 0$. When runner B leaves the same starting position ($x_{B,i} = 0$), the stopwatch reads $t_{B,i} = 2.00$ s.

❸ **EXECUTE PLAN** (*a*) We need to solve for the instant t_f at which each runner reaches the finish line at $x_f = 100$ m. Each runner moves

in the direction of the positive x axis, so the x component of the velocity for each of them is positive: $v_x = +v$, where v is the particular runner's speed. Using $\Delta t = t_f - t_i$, we can rearrange Eq. 2.18 to express t_f, the desired quantity, in terms of the other quantities:

$$x_f - x_i = v_x\Delta t = (+v)(t_f - t_i)$$

$$t_f = t_i + \frac{x_f - x_i}{v}.$$

The reading on the timer's stopwatch for runner A as she reaches the 100-m mark is

$$t_{A,f} = t_{A,i} + \frac{x_{A,f} - x_{A,i}}{v_{Ax}} = 0 + \frac{(+100 \text{ m}) - 0}{+8.00 \text{ m/s}} = 12.5 \text{ s}.$$

As runner B crosses the finish line, the stopwatch reads

$$t_{B,f} = t_{B,i} + \frac{x_{B,f} - x_{B,i}}{v_{Bx}} = 2.00 \text{ s} + \frac{(+100 \text{ m}) - 0}{+9.30 \text{ m/s}} = 12.8 \text{ s}.$$

Runner A wins because she crosses the finish line first. ✔

(*b*) To determine their distance apart at the instant runner A crosses the finish line, when the stopwatch reads $t_{A,f} = 12.5$ s, we have to compute $x_{B,f}$, the position of runner B at that instant. From Eq. 2.19, we have

$$x_{B,f} = x_{B,i} + v_{Bx}(t_{A,f} - t_{B,i})$$

$$= 0 + (+9.30 \text{ m/s})(12.5 \text{ s} - 2.00 \text{ s}) = +97.6 \text{ m}.$$

Their distance apart when runner A crosses the finish line is

$$d = |x_{B,f} - x_{A,f}| = |+97.6 \text{ m} - (+100 \text{ m})| = 2.4 \text{ m}. ✔$$

❹ **EVALUATE RESULT** The values we obtained for $t_{A,f}$ and $t_{B,f}$ are both positive, as they must be because negative values would represent instants before the race started. The time intervals of approximately 12 s to run 100 m are reasonable.

We can also solve the second part of this problem in a different way. Runner B is catching up with runner A at the rate of 9.30 m/s − 8.00 m/s = 1.30 m/s. Runner A has a 2.00-s head start, and so she is (8.00 m/s)(2.00 s) = 16.0 m ahead when runner B starts. Runner A wins the race with a time of 12.5 s; runner B has then been running for 12.5 s − 2.0 s = 10.5 s and has therefore closed the 16.0-m gap by (1.30 m/s)(10.5 s) = 13.6 m; that leaves her 16.0 m − 13.6 m = 2.4 m behind runner A. That result is the same as the one we obtained above.

Guided Problem 2.4 Race rematch

If the two runners in Worked Problem 2.3 want to cross the finish line together in a rematch, and runner A again starts at the instant the starting gun is fired, how many seconds should runner B delay in order to catch runner A right at the finish line?

❶ **GETTING STARTED**

1. Make a graph that shows the new situation, in which both runners reach 100 m at the same instant t_f.

❷ **DEVISE PLAN**

2. What type of motion occurs in this race? What equations can you use to describe this motion?

3. What physical quantity must you determine? What algebraic symbol is associated with this variable?

❸ **EXECUTE PLAN**

❹ **EVALUATE RESULT**

4. After solving for the desired starting time for runner B, how can you use the result of Worked Problem 2.3 to evaluate your answer?

Worked Problem 2.5 Airplane velocity

An airplane travels in a straight line from one airport to another. It flies at a constant speed of 600 km/h for half the distance and then at a constant speed of 800 km/h for the remainder of the distance in order to reach the destination at the scheduled time. What is its average velocity for the entire trip?

❶ GETTING STARTED To picture the situation, we draw a motion diagram. We are not given the distance between the airports, and so we have to use the variable d to represent this distance, as shown in Figure WG2.4.

Figure WG2.4

$$v = 600 \text{ km/h} \qquad v = 800 \text{ km/h}$$

$$\longmapsto \text{———} d/2 \text{ ———} \longmapsto \longmapsto \text{———} d/2 \text{ ———} \longmapsto$$

To get the average velocity, we must divide the displacement by the time interval required for the trip, but neither of these values is given! So we have to devise a way to relate these two unknown quantities to the average velocity. The distance traveled in each segment of the trip is the same, but the segment with the slower speed requires a longer time interval. With this insight, we can draw a position-versus-time graph for the plane's motion (Figure WG2.5) that shows the two segments, where we choose the positive x axis to be in the direction of the plane's motion. We include a dashed line to represent a plane that leaves the first airport at the same instant our plane leaves, travels at the average speed of our plane for its whole trip, and arrives at the second airport at the same instant our plane arrives. The slope of this dashed line represents the average velocity that we seek.

Figure WG2.5

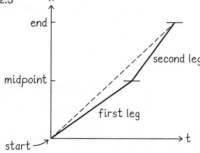

❷ DEVISE PLAN If we know the x component of displacement, we can get the x component of the average velocity from Eq. 2.14:

$$v_{x,\text{av}} \equiv \frac{\Delta x}{\Delta t} = \frac{x_f - x_i}{t_f - t_i}.$$

The x component of the displacement is the sum of the x components of the two (equal) displacements during the two segments of the flight. We can use the symbol x_m to represent the x coordinate of the midpoint position:

$$\Delta x = x_f - x_i = (x_f - x_m) + (x_m - x_i) = \Delta x_2 + \Delta x_1 = \Delta x_1 + \Delta x_2.$$

Furthermore, we know that $\Delta x_1 = \Delta x_2 = +\frac{1}{2}d$. The overall time interval is the sum of the two (unequal) time intervals: $\Delta t = \Delta t_1 + \Delta t_2$. The x component of the velocity during each segment is equal to the change in position divided by the time interval for that segment. Knowing this allows us to express the unknown time intervals in terms of the equal (but unknown) distances.

❸ EXECUTE PLAN For the entire trip:

$$v_{x,\text{av}} = \frac{\Delta x}{\Delta t} = \frac{\Delta x_1 + \Delta x_2}{\Delta t_1 + \Delta t_2}.$$

Although the time intervals are unknown, we can express them in terms of the two equal x components of the displacements:

$$\Delta t_1 = \frac{\Delta x_1}{v_{x,1}} \quad \text{and} \quad \Delta t_2 = \frac{\Delta x_2}{v_{x,2}}.$$

Substituting these expressions for Δt into the preceding equation gives

$$v_{x,\text{av}} = \frac{\Delta x_1 + \Delta x_2}{\left(\dfrac{\Delta x_1}{v_{x,1}}\right) + \left(\dfrac{\Delta x_2}{v_{x,2}}\right)} = \frac{\left(+\frac{1}{2}d\right) + \left(+\frac{1}{2}d\right)}{\left(\dfrac{+\frac{1}{2}d}{v_{x,1}}\right) + \left(\dfrac{+\frac{1}{2}d}{v_{x,2}}\right)}$$

$$= \frac{1}{\left(\dfrac{\frac{1}{2}}{v_{x,1}}\right) + \left(\dfrac{\frac{1}{2}}{v_{x,2}}\right)} = \frac{2}{\left(\dfrac{1}{v_{x,1}}\right) + \left(\dfrac{1}{v_{x,2}}\right)}.$$

The x components of the velocities in both segments are positive—that is, equal to the speeds: $v_{x,1} = +v_1$ and $v_{x,2} = +v_2$. Substituting in the numerical values gives us

$$v_{x,\text{av}} = \frac{2}{\left(\dfrac{1}{+600 \text{ km/h}}\right) + \left(\dfrac{1}{+800 \text{ km/h}}\right)} = +686 \text{ km/h}.$$

Because the problem asks for the average velocity, which is a vector, we multiply the x component of the velocity by the unit vector to get the desired result:

$$\vec{v}_{\text{av}} = v_{x,\text{av}}\hat{\imath} = (+686 \text{ km/h})\hat{\imath}. \checkmark$$

❹ EVALUATE RESULT We expect the x component of velocity to be positive because the velocities in both segments are in the positive x direction. The value of the answer is plausible because it lies between the two velocities at which the airplane actually flies.

Using the two constant speeds 600 km/h and 800 km/h means we ignored the short time interval needed to take off and reach cruising speed and the short time interval needed to slow down for landing. These two time intervals are much shorter than the time interval needed for the flight, and so our simplification is a reasonable one.

PRACTICE

Guided Problem 2.6 Wrong way

A helicopter pilot at an airport is told to fly a distance d at speed v directly east to pick up a stranded hiker. However, he sees no hiker when he arrives. Radioing the control tower, he learns that the hiker is actually a distance d west of the airport. The pilot turns around and heads straight for the hiker at a speed that is 50% faster than his speed during the initial segment of this wrong-way trip. (a) What is the x component of the helicopter's average velocity for the trip from takeoff to rendezvous with the hiker? Assume the x axis points east and has its origin at the airport. (b) What is the helicopter's average velocity for the trip?

❶ GETTING STARTED

1. Draw a diagram representing the motion during the trip, indicating the position of the airport, the turning point, and the rendezvous location.
2. Label the eastward and westward segments with symbols you can use to set up appropriate equations. Then indicate how your chosen symbols relate to the given information, d and v.
3. Add labeled vector arrows to your diagram to indicate the velocity in each segment.
4. Would a graph of $x(t)$ be useful?

❷ DEVISE PLAN

5. How does this problem differ from Worked Problem 2.5? How are the two problems similar?

6. Because no numerical values are given, what physical quantities do you expect to be represented by algebraic symbols in your answer?
7. Be careful about the signs of the x component of the displacement and the x component of the velocity in each segment of the trip. Is Δx in the eastward segment positive or negative? Is v_x in this segment positive or negative? What about Δx and v_x in the westward segment? How can these quantities be expressed in terms of v and d?
8. What time interval is needed to fly the eastward segment? What time interval for the westward segment? If you don't get positive values for both intervals, check your signs for Δx and v_x.
9. How can you combine this information to determine the x component of the average velocity and then the average velocity in vector form?

❸ EXECUTE PLAN

❹ EVALUATE RESULT

10. Is the sign on $v_{x,\text{av}}$ what you expect based on the displacement of the helicopter?

Worked Problem 2.7 Uphill putt

The position of a golf ball moving uphill along a gently sloped green is given as a function of time t by $x(t) = p + qt + rt^2$, with the positive direction up the green and with $p = +2.0$ m, $q = +8.0$ m/s, and $r = -3.0$ m/s^2.
   (a) Calculate the x component of the velocity and the speed of the ball at $t = 1.0$ s and at $t = 2.0$ s.
   (b) What is the ball's average velocity during the time interval from 1.0 s to 2.0 s?

❶ GETTING STARTED
We are asked to calculate velocity, a vector. Equation 2.25 defines velocity as the time derivative of position: $\vec{v} = d\vec{r}/dt$. It is easiest to work with components of vectors and then express the final result as a vector. In this case, the components are x and v_x, which we can use to construct the vectors $\vec{x} = x\,\hat{\imath}$ and $\vec{v} = v_x\,\hat{\imath}$. According to Eq. 2.22, the x component of the velocity equals the time derivative of the ball's x coordinate:

$$v_x(t) = \frac{dx}{dt}. \tag{1}$$

The problem statement gives this x coordinate as a function of time, and so we have the needed information.

❷ DEVISE PLAN
For part a, we need to get v_x through our knowledge of the mathematical expression for the golf ball's position $x(t)$. We can then evaluate v_x at the specified instants to get numerical values. For part b, the x component of the average velocity can be determined using Eq. 2.14:

$$v_{x,\text{av}} = \frac{\Delta x}{\Delta t} = \frac{x_f - x_i}{t_f - t_i}. ✔ \tag{2}$$

All we need to compute in this equation is Δx because we know that the time interval Δt is 1.0 s.

❸ EXECUTE PLAN
(a) To determine the x component of the velocity as a function of time, we use Eq. 1:

$$v_x(t) = \frac{d}{dt}[p + qt + rt^2] = q + 2rt$$
$$= +8.0 \text{ m/s} + 2(-3.0 \text{ m/s}^2)t$$
$$= +8.0 \text{ m/s} - (6.0 \text{ m/s}^2)t. ✔$$

Evaluating this at the specified instants yields

$$v_x(1.0 \text{ s}) = +8.0 \text{ m/s} - (6.0 \text{ m/s}^2)(1.0 \text{ s}) = +2.0 \text{ m/s} ✔$$
$$v_x(2.0 \text{ s}) = +8.0 \text{ m/s} - (6.0 \text{ m/s}^2)(2.0 \text{ s}) = -4.0 \text{ m/s}. ✔$$

The desired speeds are the magnitudes of these velocity components, 2.0 m/s and 4.0 m/s. ✔
(b) For the average velocity, we need the initial and final coordinates of the motion in the specified time interval for use in Eq. 2:

$$x_i = x(1.0 \text{ s}) = 2.0 \text{ m} + (8.0 \text{ m/s})(1.0 \text{ s}) - (3.0 \text{ m/s}^2)(1.0 \text{ s})^2$$
$$= +7.0 \text{ m}$$
$$x_f = x(2.0 \text{ s}) = 2.0 \text{ m} + (8.0 \text{ m/s})(2.0 \text{ s}) - (3.0 \text{ m/s}^2)(2.0 \text{ s})^2$$
$$= +6.0 \text{ m}.$$

The average velocity is then

$$\vec{v}_{\text{av}} = v_{x,\text{av}}\hat{\imath} = \left(\frac{x_f - x_i}{t_f - t_i}\right)\hat{\imath} = \frac{(+6.0 \text{ m}) - (+7.0 \text{ m})}{(2.0 \text{ s}) - (1.0 \text{ s})}\hat{\imath}$$
$$= (-1.0 \text{ m/s})\hat{\imath} \quad \text{(between 1.0 s and 2.0 s). ✔}$$

④ **EVALUATE RESULT** The negative signs for the x component of the velocity at 2.0 s and the x component of the average velocity mean that the ball, whose initial velocity was positive, changed direction during the motion because of the upward slope of the green. This is consistent with the fact that the position x at 2.0 s is a smaller positive number than the position x at 1.0 s. It is also reassuring that the x component of the average velocity computed from the displacement for the interval from 1.0 s to 2.0 s is between the values we calculated for the instantaneous velocity at 1.0 s and 2.0 s.

Guided Problem 2.8 You're it!

Six children—call them A, B, C, D, E, and F—are running up and down the street playing tag. Their positions as a function of time are plotted in Figure WG2.6. The street runs east-west, and the positive x direction is eastward. Use Figure WG2.6 to answer these questions: (*a*) In which direction is each child moving? (*b*) Which, if any, of the children are running at constant velocity? Of those who are, is the x component of their velocity positive or negative? (*c*) Which, if any, of the children are not moving at constant velocity? Are they speeding up or slowing down? (*d*) Which child has the highest average speed? The lowest average speed? (*e*) Does any child pass another child during the time interval shown in the graph?

Figure WG2.6

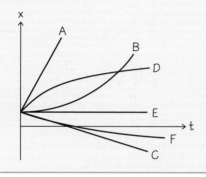

① **GETTING STARTED**

1. Use the curves in Figure WG2.6 to describe the motion of each child in words.

② **DEVISE PLAN**

2. How can you determine, from an $x(t)$ curve, the direction of a child's motion at any instant, and how is that information related to velocity and to speed?
3. What does motion at constant velocity look like on a position-versus-time graph? What does a changing-velocity situation look like on the graph?
4. How can you tell which child has the highest speed and which the lowest speed?
5. What feature in the graph would indicate that one child has passed another? In other words, what is the relationship of their positions at the instant one passes another?

③ **EXECUTE PLAN**

④ **EVALUATE RESULT**

Questions and Problems

For instructor-assigned homework, go to MasteringPhysics® (MP)

Dots indicate difficulty level of problems: • = *easy,* •• = *intermediate,* ••• = *hard;* CR = *context-rich problem.*

2.1 From reality to model

1. What minimum information must be extracted from a film clip of a moving object in order to quantify the object's motion? •

2. The sequence in Figure P2.2 represents a ball rolling into a wall and bouncing off of it. The ball is 10 mm in diameter. Make a graph showing the distance from the leading edge of the ball to the closest part of the wall (using the wall as the origin) as it changes from frame to frame. ••

Figure P2.2

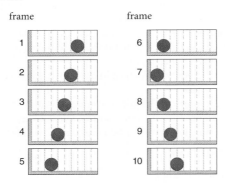

3. The sequence in Figure P2.3 represents a ball that is initially held above the ground. In the first frame the ball is released. In subsequent frames the ball falls, bounces on the ground, rises, and bounces again. The ball is 10 mm in diameter. Make a plot showing the position of this ball relative to the ground. ••

Figure P2.3

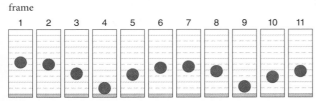

4. Figure P2.4 shows a graph of position versus frame number from a film clip of a moving object. Describe this motion from beginning to end, and state any assumptions you make. ••

Figure P2.4

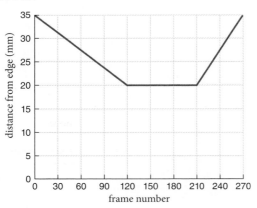

5. Your class observed several different objects in motion along different lines. Figure P2.5 shows some of the graphs other students made of the events. They have labeled the horizontal axis "time" and the vertical axis "position," but they have not marked any points along those axes and have not specified an origin. Which of the graphs could be describing the same motion? ••

Figure P2.5

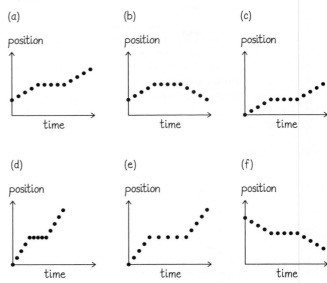

2.2 Position and displacement

6. If an object's initial position is $x_i = +6.5$ m and its final position is $x_f = +0.23$ m, what is the x component of its displacement? •

7. You walk 3.2 km to the supermarket and then back home. What is your distance traveled? What is your displacement? •

8. A 2000-m race is held on a 400-m oval track. From start to finish, what is the displacement of the winner? •

9. You are shown a film clip of a dog running in front of a blank wall. The observations are to be plotted using the horizontal and vertical axes labeled with seconds and meters, respectively. How many different graphs of this motion with respect to time could be produced if: (*a*) No additional information is given? (*b*) An origin is given? (*c*) Both the origin and the length of the dog are given? (*d*) An origin, the length of the dog, and the time interval between frames of the film clip are all given? ••

10. Assume you have a film clip of someone walking from left to right. You draw a position-versus-time graph of the motion and choose your origin to be the left edge of the frame. A friend chooses to take the right edge of the frame as the origin. Both of you choose the positive x direction to be from left to right. (*a*) How would your graphs differ? (*b*) Would the two be equally good representations of reality? ••

PRACTICE

11. Suppose the vertical axis in Figure P2.11 was calibrated in inches rather than in meters and the horizontal axis in minutes rather than seconds. How would the shape of the curve change? ••

Figure P2.11

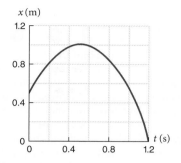

12. In the graph in Figure P2.12, determine (*a*) the displacement of the object and (*b*) the distance the object traveled. ••

Figure P2.12

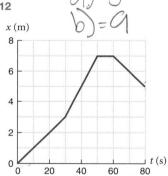

13. You walk four blocks east along 12th Street, then two blocks west, then one block east, then five blocks east, then seven blocks west. Let the *x* axis point east and have its origin at your starting point. If all blocks are equal in size, what is the *x* component of your displacement, in blocks? ••

14. In the foreground of a side-view picture of a table, the legs touch the floor at a point that is 12 mm from the bottom of the picture, and the tabletop is 65 mm from the bottom of the picture. One end of the tabletop is 14 mm from the left edge of the picture, and the other end is 99 mm from the left edge. Estimate the actual length of the tabletop. •••

2.3 Representing motion

15. Figure P2.15 shows the position of a swimmer in a race as a function of time. Describe this motion. •

Figure P2.15

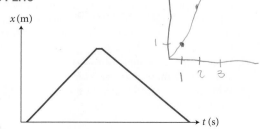

16. In the morning, a hiker at the bottom of a mountain heads up the trail toward the top. At the same instant, another hiker at the top of the mountain heads down the same trail toward the bottom. Each hiker arrives at her destination by the end of the day. Explain why, no matter what happens to either hiker along the way (lunch breaks, turnarounds to fetch dropped pocket knives, or changes of pace), they meet on the trail sometime during the day. •

17. Will interpolation between known data points always give an accurate continuous path? If so, explain why. If not, give a counterexample. •

18. Figure P2.18 shows the motion of an object as a function of time. How long did it take the object to get from the position $x = 2.0$ m to the position $x = 3.0$ m? Is there only one correct answer? ••

Figure P2.18

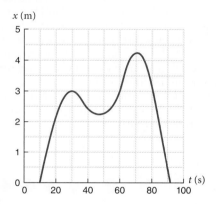

19. Figure P2.19 is based on a multiple-flash photographic sequence of a ball rolling from left to right on a soft surface. Sketch a graph of the ball's position as a function of time. ••

Figure P2.19

20. The position of an object is given by $x(t) = p + qt + rt^2$, with $p = +0.20$ m, $q = -2.0$ m/s, and $r = +2.0$ m/s². (*a*) Draw a graph of this motion from $t = 0$ to $t = 1.2$ s. (*b*) What is the displacement of the object in the interval from $t = 0$ to $t = 0.50$ s? (*c*) What maximum distance from the origin does the object reach during the interval in part *a*? ••

21. After a rocket is launched at $t = 0$, its position is given by $x(t) = qt^3$, where q is some positive constant. (*a*) Sketch a graph of the rocket's position as a function of time. (*b*) What is the rocket's displacement during the interval from $t = T$ to $t = 3T$? ••

22. Consider the position function $x(t) = p + qt + rt^2$ for a moving object, with $p = +3.0$ m, $q = +2.0$ m/s, and $r = -5.0$ m/s². (*a*) What is the value of $x(t)$ at $t = 0$? (*b*) At what value of t does $x(t)$ have its maximum value? (*c*) What is the value of $x(t)$ at this instant? (*d*) Draw a position-versus-time graph. (*e*) Describe the behavior of an object that is represented by this function. (*f*) How far has the object traveled in these intervals: From $t = 0$ to $t = 0.50$ s? From $t = 0$ to $t = 1.0$ s? From $t = 0.50$ s to $t = 1.0$ s? ••

PRACTICE

23. The motion of some object is described by the equation $x(t) = at - b\sin(ct)$, where $a = 1.0$ m/s, $b = 2.0$ m, and $c = 4\pi$ s^{-1}. This motion is being observed by four students, and measurements of the position of this object begin at time $t = 0$. Students A, B, C, and D measure every 1.0 s, 0.5 s, 0.25 s, and 0.1 s, respectively. Draw four graphs of the measurements made by the four different students. Include at least ten data points in each graph. Do the four students agree on the type of motion that is taking place? ●●●

2.4 Average speed and average velocity

24. In the Midwest, you sometimes see large marks painted on the highway shoulder. How can police patrols flying overhead use these marks to check for speeders? ●

25. Calculate the average speed for the runners in the following races: (a) 100 m in 9.84 s, (b) 200 m in 19.32 s, (c) 400 m in 43.29 s, (d) 1500 m in 3 min, 27.37 s, (e) 10 km in 26 min, 38.08 s, (f) marathon (26 mi, 385 yd) in 2 h, 6 min, 50 s. ●

26. (a) Can two cars traveling in opposite directions on a highway have the same speed? (b) Can they have the same velocity? ●

27. Figure P2.27 is based on a multiple-flash photographic sequence, taken at equal time intervals, of a ball rolling on a smooth surface from right to left. (a) Argue that the ball moves with at least two different speeds during the motion represented in the sequence. (b) During which part of the motion does the ball have the larger speed? ●

Figure P2.27

28. Can the average speed of an object moving in one direction ever be larger than the object's maximum speed? ●●

29. Figure P2.29 shows position as a function of time for two cars traveling along the same highway. (a) At what instant(s) are the cars next to each other? (b) At what instant(s) are they traveling at the same speed? ●●

Figure P2.29

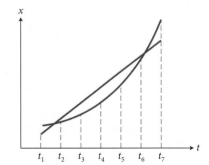

30. Figure P2.30 is based on two multiple-flash photographic sequences of a hockey puck sliding on ice. Sequence a was shot at 30 flashes per second, and sequence b was shot at 20 flashes per second. Compare the speeds of the puck in the two cases. ●●

Figure P2.30

(a)

(b)

31. A cyclist takes 10 min to ride from point A to point B and then another 10 min to continue on from point B to point C, all along a straight line. If you know that the average speed on the ride from A to B was faster than the average speed on the ride from B to C, what, if anything, can you conclude about the position of point B relative to points A and C? ●●

32. In a road rally race, you are told to drive half the trip at 25 m/s and half the trip at 35 m/s. It's not clear from the directions whether this means to drive half the *time* at each speed or drive half the *distance* at each speed. Which would yield the shorter travel time for the entire trip? ●●

33. A bicycle racer rides from a starting marker to a turnaround marker at 10 m/s. She then rides back along the same route from the turnaround marker to the starting marker at 16 m/s. (a) What is her average speed for the whole race? (b) A friend of yours gets an answer of 13 m/s. What is most likely the source of his error? ●●

34. You walk 1.25 km from home to a restaurant in 20 min, stay there for 1.0 h, and then take another 20 min to walk back home. (a) What is your average speed for the trip? (b) What is your average velocity? ●●

35. You are going to visit your grandparents, who live 500 km away. As you drive on the freeway, your speed is a constant 100 km/h. Half an hour after you leave home, your brother discovers that you forgot your wallet. He jumps into his car and speeds after you. The two of you arrive at your grandparents' house at the same instant. What is your brother's average speed for the trip? ●●

36. You and your brother both leave your house at the same instant and drive in separate cars along a straight highway to a nearby lake. After 10 min, you are both 3.0 km from your house. You are now driving at 100 km/h, and you continue at this constant speed; your brother is going faster than this. After another 20 min, your brother arrives at the lake, but you are still 5.0 km away. (a) How far is it from your house to the lake? (b) What is your brother's average speed for the trip? (c) Your brother beats you to the lake by what time interval? ●●●

2.5 Scalars and vectors

37. You are standing on a sidewalk that runs east-west. Consider these instructions I might give you: (1) Walk 15 steps along the sidewalk and stop. (2) Walk 15 steps westward on the sidewalk and stop. Do my instructions unambiguously determine your final location in each case? ●

38. What is the x component of (a) $(+3 \text{ m})\hat{\imath}$, (b) $(+3 \text{ m/s})\hat{\imath}$, and (c) $(-3 \text{ m/s})\hat{\imath}$? ●

39. What is the magnitude of (a) $(+3 \text{ m})\hat{\imath}$, (b) $(+3 \text{ m/s})\hat{\imath}$, and (c) $(-3 \text{ m/s})\hat{\imath}$? ●

40. Vectors $\vec{A}$ and $\vec{B}$ each have a magnitude of 5 m and point to the left. Vector $\vec{A}$ begins at the origin, while vector $\vec{B}$ begins at a location 8 m to the right of the origin. (*a*) If the positive *x* axis is pointed to the right, what is the *x* component of each vector? (*b*) If the positive *x* axis is pointed to the left, what is the *x* component of each vector? ••

41. Vector $\vec{A}$ points to the right, as does the positive *x* axis. (*a*) Express this vector in unit vector notation. (*b*) Now flip $\vec{A}$ to the opposite direction. Express it in unit vector notation. (*c*) Keep the vector in its new direction and flip the *x* axis so that the positive direction is to the left. Express the vector in unit vector notation. ••

2.6 Position and displacement vectors

42. Consider two vectors along the *x* axis, one with *x* component $A_x = +3$ m and the other with *x* component $B_x = -5$ m. What are (*a*) $\vec{A} + \vec{B}$ and (*b*) $\vec{A} - \vec{B}$? •

43. You stop to rest while climbing a vertical 10-m pole. With the origin at the level of your head and with the positive *x* direction upward, as shown in Figure P2.43, what are (*a*) the *x* coordinate of the pole's tip and (*b*) the tip's position vector? •

Figure P2.43

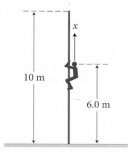

44. The height *x* above the ground of a vertically launched projectile is given by $x(t) = pt - qt^2$, with $p = 42$ m/s and $q = 4.9$ m/s^2. (*a*) At what instant is the projectile at a height of 20 m? (*b*) What is the meaning of the two solutions obtained in part *a*? (*c*) Sketch a graph of the *x* component of the projectile's velocity as a function of time. ••

45. Figure P2.45 shows the *x* coordinate as a function of time for a moving object. What is the object's *x* coordinate (*a*) at $t = 0$, (*b*) $t = 0.20$ s, and (*c*) $t = 1.2$ s? What is the object's displacement (*d*) between $t = 0$ and $t = 0.20$ s, (*e*) between $t = 0.20$ s and $t = 1.2$ s, and (*f*) between $t = 0$ and $t = 1.2$ s? What is the distance traveled by the object (*g*) between $t = 0$ and $t = 0.80$ s, (*h*) between $t = 0.80$ s and $t = 1.2$ s, and (*i*) between $t = 0$ and $t = 1.2$ s? ••

Figure P2.45

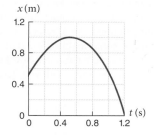

46. Arrange three displacement vectors, of magnitudes 2 m, 5 m, and 7 m, so that their sum is (*a*) $(+10 \text{ m})\hat{\imath}$, (*b*) $(-4 \text{ m})\hat{\imath}$, and (*c*) 0. ••

47. The direction of vector $\vec{A}$ is opposite the direction of the unit vector $\hat{\imath}$. Vector $\vec{B}$ has half the magnitude of $\vec{A}$, and $\vec{A} - \vec{B}$ is a vector of magnitude $\frac{3}{2} A$. Express $\vec{B}$ in terms of $\vec{A}$. ••

48. You have to deliver some 5.0-kg packages from your home to two locations. You drive for 2.0 h at 25 mi/h due east (call this segment 1 of your trip), then turn around and drive due west for 30 min at 20 mi/h (segment 2). Use a coordinate system with the positive *x* axis aimed toward the east and the origin at your home. (*a*) What is your position vector at the instant you reach the end of segment 1? (*b*) What is your position vector at the instant you reach the end of segment 2? (*c*) Calculate your displacement during segment 2. (*d*) Calculate your displacement for the whole trip. (*e*) What is the distance traveled? (*f*) Draw to scale the position vectors you found in parts *a* and *b*. (*g*) Use vector addition to determine the displacement vectors asked for in parts *c* and *d*. Do your results in part *g* agree with the values you calculated in parts *c* and *d*? ••

49. (*a*) In Figure P2.49, what vector must you add to $\vec{A}$ to get $\vec{C}$? (*b*) What vector must you subtract from $\vec{A}$ to get $\vec{C}$? Sketch your answers on a copy of the figure to confirm your results. ••

Figure P2.49

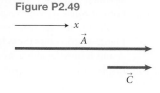

2.7 Velocity as a vector

50. You drive due east at 40 km/h for 2.0 h and then stop. (*a*) What is your speed during the trip? (*b*) Is speed a scalar or a vector? (*c*) How far have you gone? Is distance a scalar or a vector? (*d*) Write a vector expression for your position after you stop (in the form "*N* units of distance in the *Q* direction"). (*e*) Write a vector expression for your velocity during the trip, assuming east is the $+x$ direction. •

51. For the motion represented in Figure P2.45, calculate (*a*) the object's average velocity between $t = 0$ and $t = 1.2$ s, and (*b*) its average speed during this same time interval. (*c*) Why is the answer to part *a* different from the answer to part *b*? ••

52. Figure P2.52 is the position-versus-time graph for a moving object. What is the object's average velocity (*a*) between $t = 0$ and $t = 1.0$ s, (*b*) between $t = 0$ and $t = 4.0$ s, and (*c*) between $t = 3.0$ s and $t = 6.0$ s? (*d*) What is its average speed between $t = 3.0$ s and $t = 6.0$ s? (*e*) Draw to scale the velocity vector for the intervals $t = 3.0$ s to $t = 3.5$ s, $t = 3.5$ s to $t = 4.0$ s, $t = 4.0$ s to $t = 5.0$ s, and $t = 5.0$ s to $t = 6.0$ s. (*f*) What is the sum of all the vectors you drew in part *e*? ••

Figure P2.52

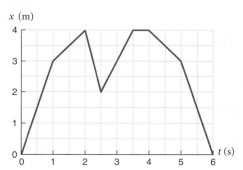

53. You normally drive a 12-h trip at an average speed of 100 km/h. Today you are in a hurry. During the first two-thirds of the distance, you drive at 108 km/h. If the trip still takes 12 h, what is your average speed in the last third of the distance? •••

2.8 Motion at constant velocity

54. A cart starts at position $x = -2.073$ m and travels along the x axis with a constant x component of velocity of -4.02 m/s. What is the position of the cart after 0.103 s? •

55. A bug on a windowsill walks at 10 mm/s from left to right for 120 mm, slows to 6.0 mm/s and continues rightward for another 3.0 s, stops for 4.0 s, and then walks back to its starting position at 8.0 mm/s. Draw a graph (*a*) of the x coordinate of the bug's displacement as a function of time and (*b*) of the x component of its velocity as a function of time. •

56. You and a friend work in buildings four equal-length blocks apart, and you plan to meet for lunch. Your friend strolls leisurely at 1.2 m/s, while you like a brisker pace of 1.6 m/s. Knowing this, you pick a restaurant between the two buildings at which you and your friend will arrive at the same instant if both of you leave your respective buildings at the same instant. In blocks, how far from your building is the restaurant? ••

57. Figure P2.57 shows the velocity-versus-time graphs for objects A and B moving along an x axis. Which object has the greater displacement over the time interval shown in the graph? ••

Figure P2.57

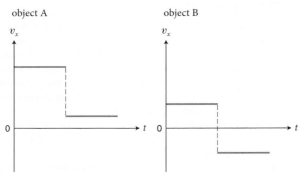

58. Figure P2.58 shows the x component of the velocity as a function of time for objects A and B. Which object has the greater displacement over the time interval shown in the graph? ••

Figure P2.58

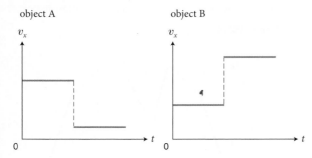

59. An object moving along an x axis starts out at $x = -10$ m. Using its velocity-versus-time graph in Figure P2.59, draw a graph of the object's x coordinate as a function of time. ••

Figure P2.59

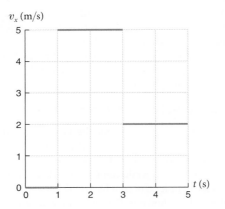

60. You and a friend ride bicycles to school. Both of you start at the same instant from your house, you riding at 10 m/s and your friend riding at 15 m/s. During the trip your friend has a flat tire that takes him 12 min to fix. He then continues the trip at the same speed of 15 m/s. If the distance to school is 15 km, which of you gets to school first? ••

61. You are going on a bicycle ride with a friend. You start 3.0 min ahead of her from her carport, pedal at 5.0 m/s for 10 min, then stop and chat with a neighbor for 5.0 min. As you chat, your friend pedals by, oblivious to you. You notice that she's forgotten to bring the lunch basket you plan to share, and you immediately start to pedal back to her house to get it, now moving at 10 m/s. When you are halfway back, your friend catches up with you, having herself remembered about the basket. (*a*) What was the magnitude of her average velocity for the trip up to that instant? (*b*) What was the magnitude of her average velocity between the instant she passed you and the neighbor and the instant she caught up with you? •••

62. You and your roommate are moving to a city 320 mi away. Your roommate drives a rental truck at a constant 60 mi/h, and you drive your car at 70 mi/h. The two of you begin the trip at the same instant. An hour after leaving, you decide to take a short break at a rest stop. If you are planning to arrive at your destination a half hour before your roommate gets there, how long can you stay at the rest stop before resuming your drive? •••

63. You are jogging eastward at an average speed of 2.0 m/s. Once you are 2.0 km from your home, you turn around and begin jogging westward, back to your house. At one point on your return trip, you look to the north and see a friend. You run northward at a greater speed of 3.0 m/s to catch your friend. After jogging next to each other for a time (at 3.0 m/s), you turn around (southward) and return at 3.0 m/s to your original path, and resume jogging at 2.0 m/s westward. If you return home 40 min after you left, how far north did you run while catching and running with your friend? •••

2.9 Instantaneous velocity

64. Which of these quantities depend on the choice of origin in a coordinate system: position, displacement, speed, average velocity, instantaneous velocity? •

65. A dragster's position as a function of time is given by $x(t) = bt^{3/2}$, where $b = 30.2 \text{ m/s}^{3/2}$. Calculate the x component of its velocity at 1.0 s and at 4.0 s. •

66. A mouse runs along a baseboard in your house. The mouse's position as a function of time is given by $x(t) = pt^2 + qt$, with $p = 0.40 \text{ m/s}^2$ and $q = -1.20 \text{ m/s}$. Determine the mouse's average velocity and average speed (a) between $t = 0$ and $t = 1.0$ s and (b) between $t = 1.0$ s and $t = 4.0$ s. ••

67. Car A is spotted passing car B just east of Westerville at exactly 2:00 p.m. The same two cars are then spotted next to each other just west of Easterville at exactly 3:00 p.m. If car B had a constant velocity of 30 m/s eastward for the whole trip, defend the proposition that car A had to have a velocity of 30 m/s eastward at some instant during that hour. ••

68. The motion of an electron is given by $x(t) = pt^3 + qt^2 + r$, with $p = -2.0 \text{ m/s}^3$, $q = +1.0 \text{ m/s}^2$, and $r = +9.0$ m. Determine its velocity at (a) $t = 0$, (b) $t = 1.0$ s, (c) $t = 2.0$ s, and (d) $t = 3.0$ s. ••

69. The car in Figure P2.69 passes a bright streetlight at constant speed, casting a shadow on a wall on the other side of the street. For simplicity, assume that car and light are at the same height. (a) Which is greater: the average speed of the car or the average speed of the shadow's leading edge? (b) Is there any instant at which the speed of the shadow's leading edge is the same as the speed of the car? If so, what is the car's position relative to the streetlight at that instant? If not, why not? ••

Figure P2.69

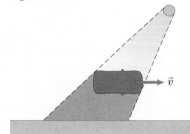

70. The position of a 6.0-kg shopping cart rolling down a ramp is given by $x(t) = p + qt^2$, with $p = +1.50$ m and $q = +2.00 \text{ m/s}^2$. What is the x component of the cart's average velocity (a) between $t = 2.00$ s and $t = 3.00$ s, (b) between $t = 2.00$ s and $t = 2.10$ s, and (c) between $t = 2.00$ s and $t = 2.01$ s? (d) Compute the limit of the average velocity between $t_i = 2.00$ s and $t_f = 2.00$ s $+ \Delta t$ as Δt approaches zero. (e) Show that your result agrees with what is expected by taking the time derivative of position. •••

Additional Problems

71. You leave Fort Worth, Texas, at 2:38 p.m. and arrive in Dallas at 3:23 p.m., covering a distance of 58 km. What is your average speed (a) in meters per second and (b) in miles per hour? •

72. In a footrace between two runners, is it possible for the second-place finisher to have a greater speed at the finish line than the winner? •

73. You wish to describe the position of the base of the pole in Figure P2.43 using the indicated coordinate system. What are (a) the position coordinate of the base, (b) the position vector of the base, and (c) the magnitude of this position vector? •

74. Runners P, Q, and R run a 5-km race in 15, 20, and 25 min, respectively, each at a constant speed. When runner Q crosses the 1-km mark, what is the distance, to the nearest meter, between runners P and R? Assume that times and distances are correct to at least four significant digits. ••

75. Runners A, B, and C run a 100-m race, each at a constant speed. Runner A takes first place, beating runner B by 10 m. Runner B takes second place, beating runner C by 10 m. By what time interval does runner A beat runner C? ••

76. At $t = 0$, car A passes a milepost at constant speed v_A. Car B passes the same milepost at constant speed $v_B > v_A$ after a time interval Δt has elapsed. (a) In terms of v_A, v_B, and Δt, at what instant does car B catch up to car A? (b) How far past the milepost does this happen? ••

77. You and your friend are running at a long racetrack. You pass the starting line while running at a constant 4.0 m/s. Fifteen seconds later, your friend passes the starting line while running at 6.0 m/s in the same direction, and at the same instant you increase your speed to 5.0 m/s. (a) How long does it take for your friend to catch up with you? (b) How far are the two of you from the starting line when your friend catches up with you? ••

78. You drive an old car on a straight, level highway at 45 mi/h for 10 mi, and then the car stalls. You leave the car and, continuing in the direction in which you were driving, walk to a friend's house 2.0 mi away, arriving 40 min after you begin walking. What is your average speed during the whole trip? ••

79. The position of a moving car is given by $x(t) = c\sqrt{t}$. Over the interval $t = 10$ s to $t = 20$ s, is the average speed greater than, equal to, or less than the instantaneous speed (a) at $t = 15$ s and (b) at $t = 20$ s? ••

80. The following equations give the x component of the position for four objects as functions of time:

    object 1: $x(t) = a$, where $a = 5$ m

    object 2: $x(t) = bt + c$, where $b = +4$ m/s and $c = -1$ m

    object 3: $x(t) = et^2 + ft$, where $e = +5 \text{ m/s}^2$ and $f = -9$ m/s

    object 4: $x(t) = gt^2 + h$, where $g = -3 \text{ m/s}^2$ and $h = +12$ m.

    (a) Which objects have a velocity that changes with time? (b) Which object is at the origin at the earliest instant, and what is that instant? (c) What is that object's velocity 1 s after the instant you calculated in part b? ••

81. A furniture mover is lifting a small safe by pulling on a rope threaded through the pulley system shown in Figure P2.81. (a) What is the ratio of the vertical distance the safe moves to the length of the rope pulled by the mover? (b) What is the ratio of the speed of the safe to the speed of the segment of rope pulled by the mover? ••

Figure P2.81

82. Two steamrollers begin 100 m apart and head toward each other, each at a constant speed of 1.00 m/s. At the same instant, a fly that travels at a constant speed of 2.20 m/s starts from the front roller of the southbound steamroller and flies to the front roller of the northbound one, then turns around and flies to the front roller of the southbound once again, and continues in this way until it is crushed between the steamrollers in a collision. What distance does the fly travel? ●●

83. Consider a 2.0-kg object that moves along the x axis according to the expression $x(t) = ct^3$, where $c = +0.120$ m/s^3. (a) Determine the x component of the object's average velocity during the interval from $t_i = 0.500$ s to $t_f = 1.50$ s. (b) Repeat for the interval from $t_i = 0.950$ s to $t_f = 1.05$ s. (c) Show that your results approach the x component of the velocity at $t = 1.00$ s if you continue to reduce the interval by factors of ten. Use all significant digits provided by your calculator at each step. ●●

84. The position of a yo-yo as a function of time is given by $x(t) = A \cos(pt + q)$, where $A = 0.60$ m, $p = \frac{1}{2}\pi$ s^{-1}, and $q = \frac{1}{2}\pi$. (a) Plot this function at 17 equally spaced instants from $t = 0$ to $t = 8.0$ s. (b) At what instants is the velocity zero? (c) Plot the x component of the velocity as a function of time over the time interval from $t = 0$ to $t = 8$ s. ●●●

85. Zeno, a Greek philosopher and mathematician, was famous for his paradoxes, one of which can be paraphrased as follows: A runner has a race of length d to run. At the instant that is a time interval Δt after the start, he is a distance $\frac{1}{2}d$ from the finish line. After an additional time interval of $\frac{1}{2}\Delta t$ (elapsed time is $\Delta t + \frac{1}{2}\Delta t$), he has traveled a distance of $\frac{3}{4}d$. He then is a distance $\frac{1}{4}d$ from the finish line. At elapsed time $\Delta t + \frac{1}{2}\Delta t + \frac{1}{4}\Delta t$, he is a distance $\frac{1}{8}d$ from the finish line, and so on. This description of the motion is an infinite series of smaller and smaller displacements. The paradox is that this analysis suggests that the runner will never make it to the finish line, even though we know he does. (a) What is the runner's speed? (b) What time interval is needed for the trip? (c) Describe the resolution to this seeming paradox. ●●●

86. Four traffic lights on a stretch of road are spaced 300 m apart. There is a 10-s lag time between successive green lights: The second light turns green 10 s after the first light turns green, the third light turns green 10 s after the second light does (and thus 20 s after the first light turns green), and so on. Each light stays green for 15 s. Having watched this sequence for some time, you understand that there should be one best constant speed to drive to make all four green lights. You do not want to be delayed, but you are a cautious driver and want to drive as slowly as you can but still make it through all the lights on a single cycle. ●●● CR

87. Hare and Tortoise of Aesop's fable fame are having a rematch, a mile-long race. Hare has planned more carefully this time. Five minutes into the race, he figures he can take a 40-min nap and still win easily because it takes him only 10 min to cover 1 mi

and because he thinks that Tortoise's top speed is 1.0 mi/h. However, Tortoise has been working out, and as he passes napping Hare, he suddenly increases his speed to $\frac{5}{3}$ mi/h. When Hare wakes up, he sees Tortoise way ahead and runs at full speed to try to catch up. ●●● CR

88. Two runners are in a 100-m race. Runner A can run this distance in 12.0 s, but runner B takes 13.5 s on a good day. To make the race interesting, runner A starts behind the starting line. If she wants the race to end in a tie, how far behind the starting line should she position herself? ●●●

89. You are on planet Dither, whose inhabitants often change their minds on how to choose a reference axis. At time $t = 0$, while standing 2.0 m to the left of the origin of a reference axis for which the positive direction points to the left, you launch a toy car that moves with increasing speed to the right. The car's x coordinate (x component of the position) is given by $x(t) = p + qt + rt^2$, where $p = +2.0$ m, $q = -3.0$ m/s, and $r = -4.0$ m/s^2. (a) What is the equation of the car's x coordinate if the reference axis is flipped so that the positive direction points to the right? (b) What is the equation of the car's x coordinate if the axis is kept in its original orientation (leftward is the positive direction) but the origin is moved to where you are standing? (c) Draw a position-versus-time graph for each of the three combinations of axis direction and origin location. (d) What is the x component of the car's displacement at $t = 4.0$ s in each of these three axis/origin combinations? (e) What is the x component of the car's velocity at this instant in each combination? (f) Explain why there is no physical difference in the answers to parts d and e even though some of the mathematical expressions you get look different from one another. ●●●

90. Your dream job as a 12-year-old was to sit at a computer at NASA mission control and guide the motion of the Mars rover across the Martian surface 2.0×10^8 km away. Communication signals to steer the rover travel at the speed of light between Earth and Mars, and the rover's top speed is 2.0 m/min. In your dream, your job was to make sure, based on television images beamed back from the craft, that the rover does not run off a Martian cliff. The shift supervisor recommended that your first task be to determine how far ahead in the rover's field of view you have to watch out for obstacles in order to avoid a disaster. ●●● CR

91. You are a driver for Ace Mining Company. The boss insists that, every hour on the hour, a loaded truck leaves the mine at 90.0 km/h, carrying ore to a mill 630 km away. She also insists that an empty truck traveling at 105 km/h leaves the mill every hour on the half hour to return to the mine. One day you begin to count the empty trucks that you encounter, per hour, on your trip to the mill. The next day you decide that loaded trucks are more important, counted on your return trip. ●●● CR

Answers to Review Questions

1. If your graph is to be identical, she must tell you what she chose as her reference axis and her origin.
2. An object is moving if its position changes with time. It is at rest if its position remains the same as time passes.
3. You need to know the actual size of the object.
4. Use one dimension of the object's actual size to determine what real-world length 1 mm on a frame of the film clip corresponds to. If the object is, for instance, a car 3 m long and the car length on the frame is 2 mm, you know that every 1-mm distance moved by the objects in the film clip represents a real-world distance of 1.5 m. Then use the conversion factor $(1.5 \text{ m})/(1 \text{ mm})$ to convert any distance each object moves on the film clip to a real-world distance in meters, and plot the distances (or quantities derived from them) on your graph.
5. The phrase reminds you that those quantities are vectors and are measured with respect to some specific x axis.
6. Displacement is a vector, so it is represented by an arrow drawn from the initial position to the final position.
7. *Interpolation* means drawing a smooth curve through a set of data points plotted on a graph. The region of the curve between any two adjacent points tells us, for the infinite number of points between the two plotted points, the numerical value of the quantity being plotted.
8. (*a*) Locate the instant on the time axis and draw a vertical line that passes through that value and intersects the curve. Then draw a horizontal line that passes through the intersection point and extends to the position axis. (*b*) Solve the equation for x, substitute the specific time value wherever the symbol t occurs in the equation, and calculate the numerical value of x.
9. A steep slope indicates a larger speed than that associated with a gentle slope. A downward slope indicates velocity in the negative x direction; an upward slope indicates velocity in the positive x direction.
10. The object moved in the negative x direction during that time interval.
11. A scalar is a mathematical quantity that is completely specified by a number and a unit of measure. A vector is a quantity that must be specified by a direction in addition to a number and a unit of measure.
12. The magnitude of a vector is a number and a unit of measure equal to the absolute value of the x component of the vector.
13. The purpose of $\hat{\imath}$ is to define the direction of the positive x axis.
14. The symbol Δ means "the change in" the variable that immediately follows the symbol Δ. It is the difference between the final value of that variable and its initial value.

15. Displacement is a vector; distance is a scalar.
16. Distance can never be negative because it is defined as the absolute value of the difference between two positions. Distance traveled can never be negative because it is a sum of distances (all positive) between successive positions in the segments of a trip.
17. The x component of the displacement is negative when the direction of the displacement is opposite the direction of the positive x axis.
18. Distance traveled is found by adding the three distances (which are all positive by definition).
19. You add two vectors by placing the tail of the second vector at the tip of the first vector. The sum is the vector that runs from the tail of the first vector to the tip of the second. You subtract one vector from another by reversing the direction of the vector being subtracted and then adding this reversed vector to the other vector.
20. The result is a vector of magnitude $|c\vec{A}| = |c||\vec{A}| = |c|A$. Its direction is the same as the direction of $\vec{A}$ if c is positive and opposite the direction of $\vec{A}$ if c is negative. If c is zero, the result is the *zero vector*, which has no direction.
21. Average speed is a scalar; average velocity is a vector.
22. Velocity has a magnitude (the moving object's speed) and a direction (which way the object is going).
23. The average velocity during some time interval is the displacement divided by the time interval.
24. The $x(t)$ curve is a straight line with a constant slope that is not zero, and the $v_x(t)$ curve is a horizontal line.
25. You need to know its position at some instant during this interval, usually the initial position.
26. The area under the curve gives the x component of the displacement during the time interval.
27. The x component of velocity at a given instant is the slope of the tangent to the $x(t)$ curve at that instant.
28. They are equal when the velocity is constant.
29. The derivative of the position with respect to time $v_x = dx/dt$.

Answers to Guided Problems

Guided Problem 2.2 (*a*) 8 min; (*b*) 7.5 mi/h, west; (*c*) 7.5 mi/h
Guided Problem 2.4 1.8 s
Guided Problem 2.6 (*a*) $-0.43v$; (*b*) $-0.43v\hat{\imath}$, with $\hat{\imath}$ pointing west
Guided Problem 2.8 (*a*) A, B, D east; C, F west; E standing still; (*b*) A, C, E; x component of velocity is positive for A, negative for C, zero for E; (*c*) B speeding up; D, F slowing down; (*d*) A highest average speed, E lowest average speed; (*e*) B passes D

3

PRACTICE
Acceleration

Chapter Summary   35

Review Questions   36

Developing a Feel   37

Worked and Guided Problems   38

Questions and Problems   42

Answers to Review Questions   50

Answers to Guided Problems   50

PRACTICE

Chapter Summary

Accelerated motion (Sections 3.1, 3.4, 3.5, 3.8)

Concepts If the velocity of an object is changing, the object is **accelerating.** The x component of an object's **average acceleration** is the change in the x component of its velocity divided by the time interval during which this change takes place.

The x component of the object's **instantaneous acceleration** is the x component of its acceleration at any given instant.

A **motion diagram,** which is constructed as described in the Procedure box on page 38, shows the positions of a moving object at equally spaced time intervals.

Quantitative tools The x component of the **average acceleration** is

$$a_{x,\text{av}} \equiv \frac{\Delta v_x}{\Delta t} = \frac{v_{x,\text{f}} - v_{x,\text{i}}}{t_\text{f} - t_\text{i}}. \tag{3.1}$$

The x component of the **instantaneous acceleration** is

$$a_x = \frac{dv_x}{dt} = \frac{d^2x}{dt^2}. \tag{3.23}$$

The x component of the change in velocity over a time interval is given by

$$\Delta v_x = \int_{t_\text{i}}^{t_\text{f}} a_x(t)dt. \tag{3.27}$$

The x component of the displacement over a time interval is given by

$$\Delta x = \int_{t_\text{i}}^{t_\text{f}} v_x(t)dt. \tag{3.28}$$

Motion with constant acceleration (Section 3.5)

If an object has constant acceleration, the $v_x(t)$ curve is a straight line that has a nonzero slope and the $a_x(t)$ curve is a horizontal line.

If an object moves in the x direction with constant acceleration a_x starting at $t = 0$, with initial velocity $v_{x,\text{i}}$ at initial position x_i, its x coordinate at any instant t is given by

$$x(t) = x_\text{i} + v_{x,\text{i}}t + \tfrac{1}{2}a_x t^2. \tag{3.11}$$

The x component of its instantaneous velocity is given by

$$v_x(t) = v_{x,\text{i}} + a_x t \tag{3.12}$$

and the x component of its final velocity is given by

$$v_{x,\text{f}}^2 = v_{x,\text{i}}^2 + 2a_x \Delta x. \tag{3.13}$$

Free fall and projectile motion (Sections 3.2, 3.3, 3.6)

An object subject only to gravity is in **free fall.** All objects in free fall near the surface of Earth have the same acceleration, which is directed downward. We call this acceleration the **acceleration due to gravity** and denote its magnitude by the letter g.

An object that is launched but not self-propelled is in **projectile motion.** Once it is launched, it is in free fall. The path it follows is called its **trajectory.**

The magnitude g of the downward acceleration due to gravity is

$$g = |\vec{a}_{\text{free fall}}| = 9.8 \text{ m/s}^2 \quad \text{(near Earth's surface).} \tag{3.14}$$

Motion on an inclined plane (Section 3.7)

An object moving up or down an inclined plane on which friction is negligible has a constant acceleration that is directed parallel to the surface of the plane and points downward along the surface.

When friction is negligible, the x component of acceleration a_x for an object moving on an inclined plane that rises at an angle θ above the horizontal is

$$a_x = +g \sin \theta \tag{3.20}$$

when the x axis is directed downward along the plane.

Review Questions

Answers to these questions can be found at the end of this chapter.

3.1 Changes in velocity

1. What is the difference between velocity and acceleration?
2. Does *nonzero acceleration* mean the same thing as *speeding up*?
3. Does the acceleration vector always point in the direction in which an object is moving? If so, explain why. If not, describe a situation in which the direction of the acceleration is not the same as the direction of motion.
4. How is the curvature of an $x(t)$ curve related to the sign of the x component of acceleration?

3.2 Acceleration due to gravity

5. A cantaloupe and a plum fall from kitchen-counter height at the same instant. Which hits the floor first?
6. Is it correct to say that a stone dropped from a bridge into the water speeds up as it falls because the acceleration due to gravity increases as the stone gets closer to Earth?

3.3 Projectile motion

7. You toss a rock straight up. Compare the acceleration of the rock at the instant just after it leaves your hand with its acceleration at the instant just before it lands back in your hand, which has remained at the point of release.
8. You throw a ball straight up. What is the ball's acceleration at the top of its trajectory?

3.4 Motion diagrams

9. List the information that should be included in a motion diagram.
10. What is the purpose of a motion diagram?

3.5 Motion with constant acceleration

11. What can you say about a train's acceleration if its $v(t)$ curve is (*a*) a straight line that is not parallel to the *t* axis and (*b*) a horizontal line that is parallel to the *t* axis?

12. For an object experiencing constant acceleration, the expression for position as a function of time is $x(t) = x_i + v_{x,i}t + \frac{1}{2}a_x t^2$. Explain, in terms of the area under the $v(t)$ curve for the object, why the acceleration term includes the factor $\frac{1}{2}$.
13. For constant acceleration, describe the relationship among displacement, initial and final velocities, and acceleration when the time variable is algebraically eliminated.

3.6 Free-fall equations

14. You throw a ball straight up and then hold your hand at the release position. Compare the time interval between the release of the ball and its arrival at its highest position to the time interval between leaving its highest position and returning to your hand.
15. (*a*) For an object released from rest, describe how the distance the object falls varies with time. (*b*) Describe how the object's speed varies with time.

3.7 Inclined planes

16. How is distance traveled related to the amount of time needed to travel that distance for a ball rolling down an inclined plane?
17. In what way does the motion of an object rolling down an inclined plane resemble that of an object in free fall?
18. On which of the following, if any, does the magnitude of the acceleration of a ball rolling down an inclined plane depend: angle of incline, speed of ball, direction of motion?

3.8 Instantaneous acceleration

19. For what type of motion is it important to distinguish between instantaneous and average acceleration?
20. What does each of the following represent: (*a*) slope of an $x(t)$ curve at a given point on the curve, (*b*) curvature of an $x(t)$ curve at a given point on the curve, (*c*) slope of a $v(t)$ curve at a given point on the curve, (*d*) area under an $a(t)$ curve, (*e*) area under a $v(t)$ curve?

Developing a Feel

Make an order-of-magnitude estimate of each of the following quantities. Letters in parentheses refer to hints below. Use them as needed to guide your thinking.

1. The time interval needed for a hailstone to fall to the ground from the altitude of a cruising airliner (H, F, O, E, G)
2. The speed of a ball released from the top of a 100-story skyscraper just before the ball hits the ground (H, F, R, O, V)
3. The speed at which a hailstone hits the ground after falling from the altitude of a cruising airliner (H, F, R, O, G)
4. The magnitude of the average acceleration of a fast sports car as it accelerates from rest to freeway cruising speed (D, P, O, C)
5. The magnitude of the average acceleration of a passenger car cruising on the freeway when it has to make an emergency stop (D, A, P, K)
6. The magnitude of a car's average acceleration moving at a typical city speed when it crashes into a stalled dump truck (R, M, K, S)
7. The magnitude of your average acceleration while getting up to speed in a sprint run (D, O, Q, T)
8. The magnitude of your average acceleration while getting up to speed on a bicycle (D, I, O, U)
9. The magnitude of your acceleration while landing feet first on the ground after jumping off a wall and falling for 1 s (R, N, J)
10. The magnitude of the average acceleration of an airliner speeding up for takeoff (D, L, B)

Hints

A. How many seconds does it take a car to stop from cruising speed?
B. How many seconds does it take an airliner to lift off after starting from rest?
C. What minimum time interval is needed to reach freeway cruising speed?
D. What expression relates change in velocity, average acceleration, and the time interval over which the velocity change occurs?
E. How are distance traveled, acceleration, and time interval related for this type of motion?
F. What type of motion is this?
G. What is the cruising altitude of an airliner?
H. What must you assume about the effect of air resistance?
I. What is your maximum speed on a bicycle?
J. What is your speed just before your feet hit the ground?
K. What is the final speed?
L. What is an airliner's takeoff speed?
M. What is the typical speed of a car traveling on an uncongested city street?
N. When your feet hit the ground, what maximum vertical displacement could the center of your body travel before stopping?
O. What is the initial speed?

P. What is the cruising speed of a car on a freeway?
Q. What is your sprinting speed?
R. How are initial speed, final speed, acceleration, and displacement related for this type of motion?
S. What distance do you travel while a car accelerates during a crash?
T. How many seconds does it take you to reach sprinting speed?
U. How many seconds does it take you to reach top speed on a bicycle?
V. What is the height of a 100-story skyscraper?

Key (all values approximate)

A. 4 s; B. 3×10^1 s; C. 4 s; D. Eq. 3.1: $a_{x,av} = \dfrac{\Delta v_x}{\Delta t}$; E. for an object starting from rest, (distance traveled) = |displacement| = $|a_x(\Delta t)^2/2|$; F. free-fall motion; G. 1×10^4 m; H. assume that ignoring air resistance does not affect your calculation; I. 7 m/s; J. 1×10^1 m/s; K. 0; L. 6×10^1 m/s; M. 1×10^1 m/s; N. 1 m; O. assume $v_i = 0$; P. 3×10^1 m/s; Q. 5 m/s; R. Eq. 3.13: $v_{x,f}^2 = v_{x,i}^2 + 2a_x\Delta x$; S. 1 m; T. 2 s; U. 5 s; V. 4×10^2 m

PRACTICE

Worked and Guided Problems

Procedure: Analyzing motion using motion diagrams

When you solve motion problems, it is important to begin by making a diagram that summarizes what you know about the motion.

1. Use dots to represent the moving object at equally spaced time intervals. If the object moves at constant speed, the dots are evenly spaced; if the object speeds up, the spacing between the dots increases; if the object slows down, the spacing decreases.
2. Choose an *x* (position) axis that is convenient for the problem. Most often this is an axis that (*a*) has its origin at the initial or final position of the object and (*b*) is oriented in the direction of motion or acceleration.
3. Specify the position and velocity at all relevant instants. In particular, specify the *initial conditions*—position

and velocity at the beginning of the time interval of interest—and the *final conditions*—position and velocity at the end of that time interval. Also specify all positions where the velocity reverses direction or the acceleration changes. Label any unknown parameters with a question mark.

4. Indicate the acceleration of the object between all the instants specified in step 3.
5. To consider the motion of more than one object, draw separate diagrams side by side, one for each object, using one common *x* axis.
6. If the object reverses direction, separate the motion diagram into two parts, one for each direction of travel.

These examples involve material from this chapter but are not associated with any particular section.
Some examples are worked out in detail; others you should work out by following the guidelines provided.

Worked Problem 3.1 Speeding up

A woman driving at the speed limit in a 25-mi/h zone enters a zone where the speed limit is 45 mi/h. She accelerates at a constant rate and reaches the new speed limit in 6.00 s. What distance does the car travel during that acceleration?

❶ GETTING STARTED This is a constant-acceleration problem. Let the car's direction of motion be the positive *x* direction. We sketch a motion diagram (Figure WG3.1) and on it indicate the initial and final velocities v_i and v_f, the time interval $\Delta t = t_f - t_i$, and the unknown distance *d* the car travels in the time interval Δt.

Figure WG3.1

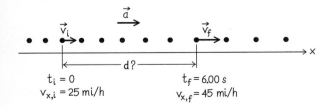

$t_i = 0$
$v_{x,i} = 25$ mi/h

$t_f = 6.00$ s
$v_{x,f} = 45$ mi/h

❷ DEVISE PLAN Because the car moves in one direction only, the distance *d* it travels is equal to the absolute value of its displacement: $d = |\Delta x|$. We must therefore determine Δx. Knowing that displacement is equal to the area under a constant-acceleration $v(t)$ curve, we sketch one that matches *Principles* Figure 3.17*b* (Figure WG3.2):

Figure WG3.2

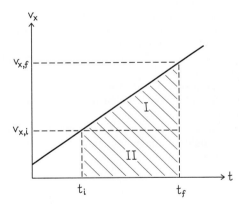

Here the shaded area representing Δx as the car's speed changes from $v_{x,i}$ to $v_{x,f}$ in the time interval Δt.

❸ EXECUTE PLAN The shaded area is most easily analyzed as a triangle (I) plus a rectangle (II). The areas are

$$I = \tfrac{1}{2}(v_{x,f} - v_{x,i})(t_f - t_i); \qquad II = v_{x,i}(t_f - t_i).$$

Combining areas I and II, we obtain the car's displacement:

$$\Delta x = \tfrac{1}{2}(v_{x,i} + v_{x,f})\Delta t.$$

After we convert units, this expression allows us to calculate

$$d = |\Delta x|$$

to three significant digits:

$$\frac{25.0 \text{ mi}}{\text{h}} \times \frac{1609 \text{ m}}{1 \text{ mi}} \times \frac{1 \text{ h}}{3600 \text{ s}} = 11.2 \text{ m/s};$$

$$\frac{45.0 \text{ mi}}{\text{h}} \times \frac{1609 \text{ m}}{1 \text{ mi}} \times \frac{1 \text{ h}}{3600 \text{ s}} = 20.1 \text{ m/s}$$

$$d = |\Delta x| = \tfrac{1}{2}(11.2 \text{ m/s} + 20.1 \text{ m/s})(6.00 \text{ s}) = 93.9 \text{ m}. ✔$$

❹ **EVALUATE RESULT** Accelerating from 25 mi/h to 45 mi/h in just less than 94 m (a bit more than 300 ft) is about what you might expect based on your experience driving cars.

Guided Problem 3.2 Slowing down

You are driving at 45 mi/h in a 30-mi/h zone. Spotting a police officer ahead, you brake at a constant rate to 27 mi/h, traveling 275 ft in so doing. What time interval do you need to reach your final speed?

❶ **GETTING STARTED**

1. What quantities do you know? What quantity must you compute?
2. Draw a motion diagram. What kind of motion is involved?

❷ **DEVISE PLAN**

3. How is this problem similar to Worked Problem 3.1? How is it different?
4. What are the numerical values of the known quantities in SI units?

❸ **EXECUTE PLAN**

❹ **EVALUATE RESULT**

Worked Problem 3.3 Average and instantaneous

You love projectile motion, and you throw baseballs, rocks, and other small objects almost every day. It occurs to you that, for constant-acceleration motion, there might be a special relationship between average velocity and instantaneous velocity. Standing on a platform with your shoulders at height h_{launch} above the ground, you launch a ball upward at speed v_i and observe that the ball rises to a height h_{max} above the ground and then falls to the ground.

❶ **GETTING STARTED** This is a context-rich problem (see *Practice* Chapter 1) because there is no explicit question. And there are no numbers either! We must therefore generate as much information as we can about average and instantaneous velocities, and hope that a relationship between average velocity and instantaneous velocity reveals itself. We begin with a motion diagram and then write the motion equations.

Figure WG3.3 is the motion diagram for the ball. We separate the motion into upward and downward portions for clarity in the

diagram; this separation is not required in the algebraic analysis, however, because the acceleration is the same throughout the motion.

❷ **DEVISE PLAN** The instantaneous velocity values other than $v_{x,i}$ can be obtained from an equation such as Eq. 3.12 with $-g$ in place of a_x:

$$v_x(t) = v_{x,i} - gt. \tag{1}$$

However, we might also use Eq. 3.13, which involves instantaneous velocity and displacement but not a time interval, again with $-g$ in place of a_x:

$$v_{x,f}^2 = v_{x,i}^2 - 2g\Delta x. \tag{2}$$

There is also Eq. 3.11 for position:

$$x(t) = x_i + v_{x,i}t + \tfrac{1}{2}a_x t^2,$$

but this expression does not contain any velocity information other than the initial velocity. We tentatively reject it.

The time interval over which the motion takes place, with t_i set at 0 and the subscript on t_f dropped, is

$$\Delta t = t_f - t_i = t, \tag{3}$$

and t can be obtained from Eq. 1.

Our task is to obtain a relationship between instantaneous velocity v and average velocity v_{av}. Average velocity is defined as the displacement Δx divided by the time interval Δt over which the displacement occurs (see Chapter 2). We should be able to combine the three equations above with our knowledge of the ball's initial and final positions to get an expression for average velocity in terms of instantaneous velocities.

❸ **EXECUTE PLAN** The displacement is the difference in the ball's final and initial positions, both of which are known. We choose the origin at ground level and up as the positive x direction: $x_i = h_{\text{launch}}$, $x_f = 0$:

$$\Delta x = x_f - x_i = 0 - h_{\text{launch}} = -h_{\text{launch}}.$$

Figure WG3.3

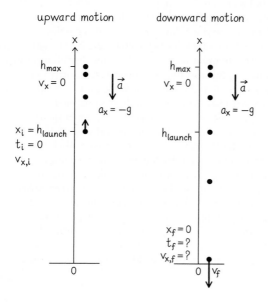

We use Eq. 2 to get the ball's final speed at the instant just before it hits the ground:

$$v_{x,f}^2 = v_{x,i}^2 - 2g\Delta x = v_{x,i}^2 + 2gh_{launch}$$

$$|v_{x,f}| = \sqrt{v_{x,i}^2 + 2gh_{launch}}. \tag{4}$$

We do not yet know the initial speed needed for Eq. 4, but it must be related to the ball's maximum height h_{max} because a faster throw should make the ball go higher. Let's use Eq. 2 again to obtain this relationship, evaluating between the initial height h_{launch} and the maximum height h_{max} (because the ball's speed is zero at h_{max}):

$$0 = v_{x,i}^2 - 2g(h_{max} - h_{launch}),$$

so that

$$|v_{x,i}| = v_i = \sqrt{2g(h_{max} - h_{launch})}.$$

Substituting this expression for $v_{x,i}$ into Eq. 4 gives for the final speed of the ball at the instant before it hits the ground:

$$|v_{x,f}| = \sqrt{v_{x,i}^2 + 2gh_{launch}} = \sqrt{2g(h_{max} - h_{launch}) + 2gh_{launch}}$$

$$= \sqrt{2gh_{max} - 2gh_{launch} + 2gh_{launch}}$$

$$|v_{x,f}| = \sqrt{2gh_{max}}.$$

Note that this is a speed, which means we should take the positive square root. We know, though, from our choice of positive x axis upward that the final velocity is directed downward and hence is negative.

We can get the time interval Δt by combining Eqs. 1 and Eq. 3:

$$\Delta t = t = \frac{v_{x,f} - v_{x,i}}{-g}.$$

The average velocity in terms of our expressions for Δx and Δt is

$$v_{av} = \frac{\Delta x}{\Delta t} = \frac{-h_{launch}}{\left(\dfrac{v_{x,f} - v_{x,i}}{-g}\right)} = \frac{gh_{launch}}{v_{x,f} - v_{x,i}}. \tag{5}$$

This is nice, but is there a way to get the right side entirely in terms of instantaneous velocities? Yes, there is. If we substitute $-h_{launch}$ for Δx in Eq. 2 and rearrange, we have

$$gh_{launch} = \tfrac{1}{2}(v_{x,f}^2 - v_{x,i}^2).$$

Substituting the term on the right into Eq. 5 gives

$$v_{x,av} = \frac{\tfrac{1}{2}(v_{x,f}^2 - v_{x,i}^2)}{(v_{x,f} - v_{x,i})} = \tfrac{1}{2}(v_{x,f} + v_{x,i}). ✔ \tag{6}$$

This expression tells us that the average velocity during a chosen time interval is equal to the numerical average of the initial and final instantaneous velocities for that interval. In this example, because $v_{x,f}$ is negative and of larger magnitude than v_i, which we know from Eq. 4, the average velocity is negative.

This relationship between average and instantaneous velocities seems to be quite general because it does not rely on any values

specific to this problem. Of course, it applies only to cases of constant acceleration because Eqs. 1 and 2 are only valid for constant acceleration.

❹ **EVALUATE RESULT** There are no numbers to check, so we seek another way to solve the problem. Perhaps we can visualize this general result graphically. What sort of graph should we draw? A graph of instantaneous velocity versus time comes to mind for several reasons. First, our answer involves instantaneous velocity. Second, from this graph we can obtain displacement (area under the curve) and hence determine average velocity. Third, because velocity is a linear function of time in constant-acceleration problems (Eq. 3.12), this graph is much easier to sketch than the quadratic graph of position versus time (Eq. 3.11). (The acceleration-versus-time graph is even easier to sketch because it is a straight line parallel to the time axis, but it provides less insight.)

Figure WG3.4 shows the velocity-versus-time graph for the ball. Because we chose the positive direction to be upward in our motion diagram, the $v(t)$ curve is a straight line with negative slope (equal to $-g$). The curve extends along the time axis until the ball strikes the ground.

Figure WG3.4

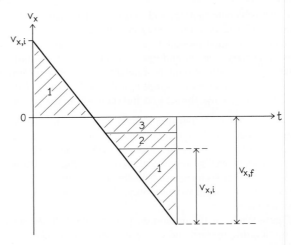

Consider the labeled areas between the $v(t)$ curve and the t axis in Figure WG3.4. Suppose we ask, What (constant) average velocity would produce the same displacement in the same time interval as the varying velocity we had in this problem? As we saw in Chapter 2, at constant velocity the $v(t)$ curve is a straight line parallel to the t axis. In order to produce the same displacement in a given time interval, the area under the constant-velocity curve between any two values t_f and t_i must match the area under the actual $v(t)$ curve for our tossed ball between those same values of t_f and t_i.

Note that the two areas labeled 1 are equal in size and opposite in sign. This means that they cancel when added, and so the net area under the curve is area 2 + area 3. Both these areas are negative in our example, leading to a combined negative area, as we expect for an object with negative average velocity.

Suppose we choose the dividing line between areas 2 and 3 such that these areas have the same height. By construction, area 2 + area 3 has height $-(|v_{x,f}| - |v_{x,i}|)$. Note that $|v_{x,i}| = +v_{x,i}$, while $|v_{x,f}| = -v_{x,f}$, so we can write the height as $-(-v_{x,f} - v_{x,i}) = (v_{x,f} + v_{x,i})$. Note that this is a negative height, as expected. Splitting the height equally between areas 2 and 3 gives each height $\frac{1}{2}(v_{x,f} + v_{x,i})$. Now imagine moving area 2 to the location shown in Figure WG3.5. The result is a rectangle 2 + 3 of height $\frac{1}{2}(v_{x,f} + v_{x,i})$ and length Δt. Thus the area under the $v(t)$ curve, which is the ball's displacement, is equal to the area of rectangle 2 + 3:

$$\Delta x = \frac{1}{2}(v_{x,f} + v_{x,i})\Delta t,$$

which means

$$v_{x,av} = \frac{\Delta x}{\Delta t} = \frac{1}{2}(v_{x,f} + v_{x,i}), \checkmark \qquad (6)$$

which is the same as Eq. 6. This same geometrical construction works for initial and final velocities that are either negative or positive (try it!), as long as the curve is a line of constant slope (constant acceleration).

Because our graphical and algebraic methods lead to the same results, we have confidence in the answer.

Figure WG3.5

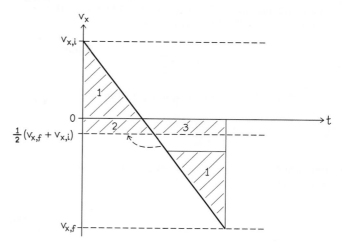

Guided Problem 3.4 Double launch

You throw a ball upward from height h_{launch} as in Worked Problem 3.3 and it rises to height $h_{max} = 2h_{launch}$ before falling to the ground. At the instant you release the ball, a friend standing on a platform at height h_{max} simultaneously throws an identical ball with half the upward velocity you used. Determine (a) which ball hits the ground first, (b) the final speed of each ball, and (c) the time interval between the instant one ball lands and the instant the other ball lands.

❶ GETTING STARTED

1. Draw a motion diagram and a velocity-versus-time graph for each ball.
2. Which technique(s) of Worked Problem 3.3 can you use?

❷ DEVISE PLAN

3. Where is your friend's ball at the instant your ball reaches its maximum height h_{max}?

❸ EXECUTE PLAN

4. Express your answers in terms of the given information.

❹ EVALUATE RESULT

Worked Problem 3.5 Inclined track

Your physics instructor prepares a laboratory exercise in which you will use a modern version of Galileo's inclined plane to determine acceleration due to gravity. In the experiment, an electronic timer records the time interval required for a cart initially at rest to descend 1.20 m along a low-friction track inclined at some angle θ with respect to the horizontal.

(a) In preparation for the experiment, you must obtain an equation from which you can calculate g on the basis of these measurements. What is that equation?

(b) To make it possible to check the students' measurements quickly, the instructor breaks the class into five groups and assigns one value of θ to each group. If no mistakes are made, these five θ values yield time intervals of 0.700, 0.800, 0.900, 1.00, and 1.20 s. What are the five θ values?

❶ GETTING STARTED The cart undergoes constant acceleration, from rest, on an inclined plane. We know how to analyze this type of motion, and we know how the acceleration at any given incline angle is related to the acceleration g due to gravity. We sketch a motion diagram (Figure WG3.6), representing a cart moving down an inclined plane, and choose the positive x direction as pointing down the track.

Figure WG3.6

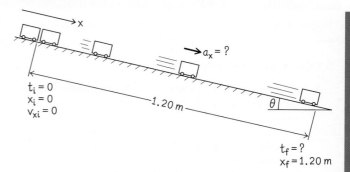

❷ DEVISE PLAN The equation asked for in part a might be based on Eq. 3.20, $a_x = +g \sin \theta$, but you will not be measuring a_x values directly in this experiment. You will measure displacements Δx and time intervals Δt, which means we need an expression that gives acceleration in terms of these two variables. Equation 3.11 comes to mind, which we can manipulate so that a_x is expressed in terms of Δx and Δt. In part b, we can use this result to obtain the five θ values.

❸ **EXECUTE PLAN** (*a*) Equation 3.11 yields $a_x = 2(x_f - x_i)/t^2 = 2\Delta x/t^2$. Because in the derivation for Eq. 3.11 t_i was taken to be zero, the t^2 is actually $(\Delta t)^2$, so that we have

$$a_x = \frac{2\Delta x}{(\Delta t)^2}.$$

Substituting this expression for a_x in Eq. 3.20 yields

$$\frac{2\Delta x}{(\Delta t)^2} = g \sin \theta,$$

from which we obtain the expression for g to be used in the experiment:

$$g = \frac{2\Delta x}{(\Delta t)^2 \sin \theta}. ✔ \tag{1}$$

(*b*) Manipulation of Eq. 1 gives the instructor

$$\sin \theta = \frac{2\Delta x}{g(\Delta t)^2}$$

$$\theta = \sin^{-1}\left(\frac{2\Delta x}{g(\Delta t)^2}\right). \tag{2}$$

Before evaluating this expression five times, the instructor calculates the constant quantity $2\Delta x/g = 0.2449$ and stores it in her calculator. Substitution of $\Delta t = 0.700, 0.800, 0.900, 1.00,$ and 1.20 s into Eq. 2 yields the angles of incline she assigned to the five groups: 30.0°, 22.5°, 17.6°, 14.2°, and 9.79°. ✔

❹ **EVALUATE RESULT** The numerical values for the angles are reasonable: Larger angles are associated with smaller time intervals. Even the shortest interval is considerably longer than the time interval needed for an object to fall freely from a height of 1.2 m, as expected.

Guided Problem 3.6 Another inclined track

At another university, in a laboratory exercise similar to that described in Worked Problem 3.5, students measure the angle of incline θ of a low-friction track, the (nonzero) initial and final speeds of a cart as it descends between two positions on the track, and the distance between those two positions. For an angle of incline of 10.0°, one group obtains the values $v_i = 0.820$ m/s and $v_f = 1.65$ m/s for a distance of 0.608 m. On the basis of these data, what value do these students obtain for g?

❶ **GETTING STARTED**
1. Draw a motion diagram. What kind of motion are we dealing with?
2. Choose an appropriate x direction and origin.

❷ **DEVISE PLAN**
3. How is this problem similar to Worked Problem 3.5? How is it different?

❸ **EXECUTE PLAN**

❹ **EVALUATE RESULT**
4. Consider whether small or large differences in measured data would be needed to produce a result of 9.80 m/s².

Questions and Problems

For instructor-assigned homework, go to MasteringPhysics® (MP)

Dots indicate difficulty level of problems: • = *easy,* •• = *intermediate,* ••• = *hard;* CR = *context-rich problem.*

3.1 Changes in velocity

1. Figure P3.1 is based on a multiple-flash photographic sequence of an object moving from left to right on a track, as seen from above. The time intervals between successive flashes are all the same. During which portion(s) of the motion is the object (*a*) speeding up and (*b*) slowing down? Explain how you know in each case. (*c*) How would your answers change if the object were moving from right to left? •

Figure P3.1

•• •   •   •   •   • ••

2. A car is traveling north. What are the direction of its acceleration and the direction of its velocity (*a*) if it is speeding up and (*b*) if it is slowing down? •

3. Figure P3.3 shows a series of photographs of a racehorse taken in 1877 by Eadweard Muybridge (1830–1904). Muybridge used multiple, equally spaced cameras triggered sequentially at equal time intervals. Is the horse accelerating? How can you tell? •

Figure P3.3

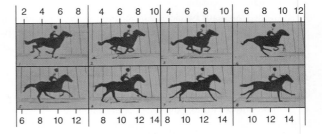

4. Figure P3.4 shows a graph of position as a function of time for an object moving along a horizontal surface. At which of the labeled points is the object speeding up? •

Figure P3.4

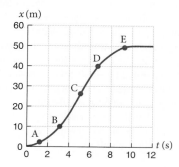

5. While you are driving on a highway, a friend in another car passes you. You accelerate at a constant rate to catch up. When you do catch up, are your two cars going the same speed? Use a graph to support your answer. ••

6. The position of a person pacing in a hall is given by the $x(t)$ curve in Figure P3.6. During which time interval(s) is the acceleration (a) positive and (b) negative? (c) Is the acceleration ever zero during the 4-s interval shown? ••

Figure P3.6

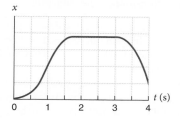

7. You enter an elevator on the ground (first) floor and ride it to the 19th floor. Describe your acceleration at the different stages of this trip. ••

8. Figure P3.8 shows the position curves for two carts, A and B, moving on parallel tracks along a horizontal surface. The instant when the two carts are at the same distance from their starting point is indicated by point P. Which cart has the greater acceleration at that instant? ••

Figure P3.8

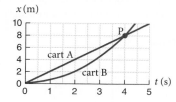

9. Two cars are headed south on the highway at different speeds. Starting when the faster car pulls alongside the slower car, one of the cars accelerates northward for 5.0 s and the other accelerates southward for 5.0 s. At the end of that 5.0-s interval, the two cars have the same speed. Which car is now ahead? •••

3.2 Acceleration due to gravity

10. Is it true that a pebble released from rest off a bridge falls 9.8 m in the first second of its fall? •

11. (a) What is the average speed, over the first 1.0 s of its motion, of a pebble released from rest off a bridge? (b) What is the pebble's average speed over the second 1.0 s of its motion? (c) What is its average speed during these first 2.0 s of its motion? •

12. You toss a (wrapped) sandwich to a friend leaning out of a window 10 m above you, throwing just hard enough for it to reach her. At the same instant, she drops a silver dollar to you. Do the dollar and the sandwich pass each other at a position 5 m above you, more than 5 m above you, or less than 5 m above you? ••

13. Suppose that the acceleration due to gravity near Earth's surface was cut in half (to about 5 m/s^2). How would this affect the graphs shown in *Principles* Figure 3.6, assuming that everything else about the experiment remains the same? Sketch the new curves. ••

3.3 Projectile motion

14. A photographer shows you a multiple-flash photographic sequence of a ball traveling vertically (Figure P3.14). (a) If the ball is traveling downward, what is the correct orientation of the picture? (b) If the ball is traveling upward, what is the correct orientation of the picture? (c) According to the evidence provided by the picture, what is the direction of the ball's acceleration in parts a and b? •

Figure P3.14

15. A cannonball is shot straight up at an initial speed of 98 m/s. What are its velocity and its speed after (a) 5.0 s, (b) 10 s, (c) 15 s, and (d) 20 s? •

16. A coin flipped in the air from elbow height lands on the ground 1.8 s later. Did the coin reach its highest position 0.9 s after it started moving up, earlier than that, or later than that? •

17. Is it possible for an object to have (a) zero velocity and nonzero acceleration or (b) nonzero velocity and zero acceleration? If you answer yes in either part, give examples to support your answer. ••

18. How would graphs c and d in *Principles* Figure 3.8 be different if the same experiment was conducted on the surface of the Moon, where the acceleration due to gravity is six times less than it is on Earth? ••

19. You throw snowballs down to the sidewalk from the roof of a building. Which technique, if either, makes the snowballs land with more speed: throwing them straight down as hard as you can or throwing them straight up just as hard? (Ignore air resistance.) ••

20. You are standing by a window and see a ball, thrown from below, moving up past the window. The ball is visible for a time interval Δt_{up}. On its way back down the ball passes the window again, remaining visible for a time interval Δt_{down}. Neglecting the effects of air resistance, is $\Delta t_{up} > \Delta t_{down}$, $\Delta t_{up} = \Delta t_{down}$, $\Delta t_{up} < \Delta t_{down}$, or is it impossible to tell without more information? Briefly explain your reasoning, using graphs. ••

21. For an object in free fall, the curve of its velocity as a function of time is a straight line. How would this curve be different when air resistance is not negligible? •••

22. When air resistance is ignored, it is straightforward to calculate the travel time interval for a Ping-Pong ball tossed up into the air and caught on its way down because the acceleration can be taken to be a constant 9.8 m/s² downward. However, a more accurate result is obtained when air resistance is taken into account. When this is done, is the magnitude of the ball's acceleration greater than, less than, or the same as when air resistance is ignored (*a*) as the ball rises and (*b*) as it falls? (*c*) Do you expect the travel time interval obtained in the more accurate analysis to be greater than, less than, or the same as in the simplified case? •••

3.4 Motion diagrams

23. Draw a motion diagram for a car that starts from rest and accelerates at 4.0 m/s² for 10 s. •

24. A car accelerating from rest at constant acceleration reaches a speed of 30 km/h in 5.0 s. Draw a motion diagram for the car. •

25. The motion of a cart moving along a horizontal surface is described by the motion diagram shown in Figure P3.25a. The position of the cart is measured every 0.5 s. Asked to suggest a qualitative graph of velocity versus time that would correspond to this motion, three of your classmates draw the graphs shown in Figures P3.25b–d. Which graph is correct? •

Figure P3.25

(*a*)

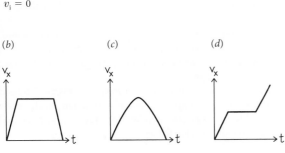

$t_i = 0$
$x_i = 0$
$v_i = 0$

(*b*)   (*c*)   (*d*)

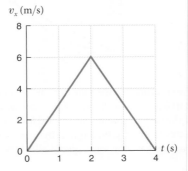

26. Draw a motion diagram for a car that starts from rest and accelerates at 5.0 m/s² for 6.0 s, then travels with constant speed for 10 s, and then slows to a stop with constant acceleration in 4.0 s. ••

27. A ball you throw straight up has a speed of 30 m/s when it leaves your hand. Draw a motion diagram for the ball up to the instant at which it has a speed of 30 m/s again. Indicate the instant at which the ball reaches its highest position. ••

28. Figure P3.28 shows motion diagrams for two cars, A and B, beginning a race. The diagrams show the position of each car at instants separated by equal time intervals. Both cars drive up to the starting line and then begin to accelerate. (*a*) Which car has the greater velocity at instant 6? How can you tell? (*b*) Which car has the greater velocity at instant 11? How can you tell? (*c*) Which car has the greater acceleration after instant 6? How can you tell? ••

Figure P3.28

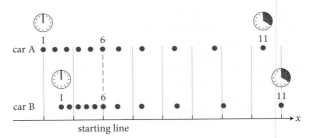

3.5 Motion with constant acceleration

29. You start your car from rest and accelerate at a constant rate along a straight path. Your speed is 20 m/s after 1.0 min. (*a*) What is your acceleration? (*b*) How far do you travel during that 1.0 min? •

30. You and your little brother are rolling toy cars back and forth to each other across the floor. He is sitting at $x = 0$, and you are at $x = 4.0$ m. You roll a car toward him, giving it an initial speed of 2.5 m/s. It stops just as it reaches him in 3.0 s. (*a*) In what direction is the car's acceleration? (*b*) What is the car's average acceleration? •

31. An electron is accelerated from rest to 3.0×10^6 m/s in 5.0×10^{-8} s. (*a*) What distance does the electron travel in this time interval? (*b*) What is its average acceleration? •

32. Figure P3.32 shows a graph of velocity as a function of time for a cart moving along a horizontal surface. When asked to describe the motion that resulted in this graph, a student states, "The cart first moved forward with a constant speed, reached a maximum distance at time 2 seconds, then turned around and came back to its initial position at time 4 seconds." What is your assessment of this comment? •

Figure P3.32 v_x (m/s)

33. (*a*) How do you determine an object's displacement from a velocity-versus-time graph? (*b*) What distance does a car travel as its speed changes from 0 to 20 m/s in 10 s at constant acceleration? (*c*) The *x* component of the average velocity of a particle moving along the *x* axis at constant acceleration is $v_{x,\text{av}} = \frac{1}{2}(v_{x,\text{i}} + v_{x,\text{f}})$ (see Worked Problem 3.3). Is the *x* component of the average velocity also given by this expression when the acceleration is not constant? Use a velocity-versus-time graph to support your answer. ••

34. On a freeway entrance ramp, you accelerate your 1200-kg car from 5 m/s to 20 m/s over 500 m. Having an open lane, you continue to accelerate at the same (constant) rate for another 500 m. Is your final speed 35 m/s, less than 35 m/s, or greater than 35 m/s? ••

35. In an introductory physics laboratory, a student drops a steel ball of radius 15 mm, and a device records its position as a function of time. The clock on the device is set so that $t = 0$ at the instant the ball is dropped. What is the ball's displacement (*a*) between 0.15 s and 0.25 s and (*b*) between 0.175 s and 0.275 s? ••

36. Which car has the greater acceleration magnitude: one that accelerates from 0 to 10 m/s in 50 m or one that accelerates from 10 m/s to 20 m/s in 50 m? ••

37. An electron in an old-fashioned television picture tube is accelerated, at a constant rate, from 2.0×10^5 m/s to 1.0×10^7 m/s in a 12-mm-long "electron gun." (*a*) What is the acceleration of the electron? (*b*) What time interval is needed for the electron to travel the length of the gun? ••

38. In a car moving at constant acceleration, you travel 250 m between the instants at which the speedometer reads 40 km/h and 60 km/h. (*a*) How many seconds does it take you to travel the 250 m? (*b*) What is your acceleration? ••

39. Based on the $v(t)$ curve in Figure P3.39, explain whether each of these statements is necessarily true for the time interval shown: (*a*) The acceleration is constant. (*b*) The object passes through the position $x = 0$. (*c*) The object has zero velocity at some instant. (*d*) The object is always moving in the same direction. ••

Figure P3.39

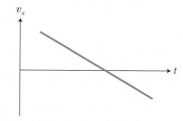

40. In October 1997, Andy Green broke the sound barrier on land in a jet-powered car traveling at 763 mi/h over a 1.0-mi course. As it arrived at the beginning of this measured mile, Green's car had accelerated from rest to 763 mi/h over a 5.0-mi stretch. (*a*) What was Green's average acceleration in that 5.0 mi? (*b*) Calculate the time interval needed for Green to reach his maximum speed. ••

41. You jump on your bicycle, ride at a constant acceleration of 0.60 m/s² for 20 s, and then continue riding at a constant velocity for 200 m. You then slow to a stop with a constant acceleration over 10 m. (*a*) What distance do you travel? (*b*) Calculate the time interval needed for your trip. (*c*) What is your average velocity? ••

42. In rush-hour traffic, the car in front of you suddenly puts on the brakes. You apply your brakes 0.50 s later. The accelerations of the two cars are the same. Does the distance between the two cars remain constant, decrease, or increase? ••

43. Two cars are moving at 97 km/h, one behind the other, on a rural road. A deer jumps in front of the lead car, and its driver slams on the brakes and stops. What minimum initial distance between the rear of the lead car and the front of the second car is required if the second car is to stop before hitting the lead car? Assume that the acceleration is the same for both cars and that the driver of the second car begins braking 0.50 s after the lead car begins braking. •••

44. As a physics instructor hurries to the bus stop, her bus passes her, stops ahead, and begins loading passengers. She runs at 6.0 m/s to catch the bus, but the door closes when she's still 8.0 m behind the door, and the bus leaves the stop at a constant acceleration of 2.0 m/s². She has missed her bus, but as a physics exercise she keeps running at 6.0 m/s until she draws even with the bus door. (*a*) Calculate Δt from the instant the bus leaves the stop to the instant the instructor draws even with the door. (*b*) At the instant she draws even, how fast is she moving? (*c*) At the instant she draws even, how fast is the bus moving? (*d*) Plot $v(t)$ for the bus and the instructor. (*e*) Plot $x(t)$ for the bus and the instructor. •••

45. The day after the incident described in Problem 44, the instructor finds herself in the same situation. This time, she tries a harder physics exercise. She keeps running at a constant 6.0 m/s after drawing even with the bus door and pulls ahead for a while, but the accelerating bus soon overtakes her. By what maximum distance does she get ahead of the door? •••

3.6 Free-fall equations

46. (*a*) How many seconds does it take a pebble released from rest off a bridge to fall 9.8 m? (*b*) What is the pebble's speed when it has fallen 9.8 m? •

47. Calculated using the value $g = 9.8$ m/s², the time interval required for an object released from rest at some arbitrary point A above the ground to reach the ground is Δt. If g were 4.9 m/s², by what factor would Δt differ from the value calculated using $g = 9.8$ m/s²? •

48. A mortar-style fireworks shell is launched upward at 35 m/s. Draw a motion diagram showing the position of the shell and its velocity vector at $t = 0$, 1 s, 2 s, 3 s, and 4 s, and draw a graph of $x(t)$. •

49. The distance a ball thrown upward and released at $t = 0$ travels between $t = 3.0$ s and $t = 4.0$ s is the same as the distance the ball travels between $t = 2.0$ s and $t = 3.0$ s. How is this possible? ••

50. You toss a ball straight up, and it reaches a maximum height h above the launch position. If you want to double the length of the time interval during which the ball stays in the air above the launch position, to what maximum height do you have to throw it? ••

51. (*a*) With what minimum speed must a ball be thrown straight up in order to reach a height of 25 m above the launch position? (*b*) How many seconds does the ball take to reach this height? ••

52. A snowshoer falls off a ridge into a snow bank 3.6 m below and penetrates 0.80 m into the snow before stopping. What is her average acceleration in the snow bank? ••

53. You are on the second floor of a building under construction, laying bricks of dimensions 63 mm × 89 mm × 170 mm. At your command, a coworker below tosses bricks up to you, and you catch them when they have risen 5.0 m above the launch point. He tosses the bricks consistently so that, if untouched by you, they would rise to a maximum height of 6.0 m above the launch point. (*a*) Soon you both get good at this, and you know exactly how many seconds after you yell "3, 2, 1, throw!" you should reach out and catch the brick on its way up. What is the length of this time interval if your coworker throws without delay at your command? (*b*) If you miss a brick on the way up, during what remaining time interval can you reach out and catch the brick on its way down? ••

54. A rocket ignited on the ground travels vertically upward with an acceleration of magnitude 4*g*. A spent rocket stage detaches from the payload after the rocket has accelerated for 5.0 s. With what speed does the spent stage hit the ground? ••

55. Show that the distances traveled by a falling rock in successive 1-s intervals after release are in the ratio 1:3:5:7.... ••

56. A hot-air balloon takes off from the ground traveling vertically with a constant upward acceleration of magnitude *g*/4. After time interval Δ*t*, a crew member releases a ballast sandbag from the basket attached to the balloon. How many seconds does it take the sandbag to reach the ground? •••

57. A rock dropped from the top of a building takes 0.50 s to fall the last 50% of the distance from the top to the ground. How tall is the building? •••

58. A hot-air balloon of diameter 10 m rises vertically at a constant speed of 12 m/s. A passenger accidentally drops his camera from the railing of the basket when it is 18 m above the ground. If the balloon continues to rise at the same speed, how high is the railing when the camera hits the ground? •••

59. A ball tossed vertically upward from the ground next to a building passes the bottom of a window 1.8 s after being tossed and passes the top of the window 0.20 s later. The window is 2.0 m high from top to bottom. (*a*) What was the ball's initial velocity? (*b*) How far is the bottom of the window from the launch position? (*c*) How high does the ball rise above the launch position? •••

3.7 Inclined planes

60. What is the magnitude of the acceleration of a 65-kg skier moving down a hill that makes a 45° angle with the horizontal? Ignore friction. •

61. Starting from rest, a cart takes 1.25 s to slide 1.80 m down an inclined low-friction track. What is the angle of incline of the track with respect to the horizontal? •

62. Imagine that in his experiment with balls rolling down an inclined plane, Galileo had considered balls that were given an initial speed. Would he have reached different conclusions about the ratio of distance and square of time? •

63. A worker releases boxes at the top of a ramp. From the bottom of the ramp the boxes slide 10 m across the floor to a barrier wall. If the ramp is at an elevation angle of 20°, how long must the ramp be if the boxes are to reach the wall 2.0 s after leaving the ramp? Ignore friction. ••

64. A man steps outside one winter day to go to work. His icy driveway is 8.0 m long from top to mailbox, and it slopes downward at 20° from the horizontal. He sets his briefcase on the ice at the top while opening the garage, and the briefcase slides down the driveway. Ignore friction. (*a*) What is its acceleration? (*b*) How many seconds does it take to get halfway to the mailbox? (*c*) How many seconds does it take to reach the mailbox? (*d*) What is its speed at the instant it reaches the mailbox? ••

65. You and a friend ride what are billed as the "world's longest slides" at a county fair. Your slide is 100 m long, and your trip takes 10 s, including any effect of friction. Your friend chooses a taller, 150-m-long slide made from the same material as yours and with the same angle of incline. (*a*) What is the magnitude of your acceleration down the slide? (*b*) What is the magnitude of your friend's acceleration down the slide? (*c*) How many seconds does it take your friend to get to the bottom? (*d*) What is your speed when you hit the bottom? (*e*) What is your friend's speed when she reaches the bottom? ••

66. Two children at a playground slide from rest down slides that are of equal height but are inclined at different angles with respect to the horizontal (Figure P3.66). Ignoring friction, at height *h* above the ground, which child has (*a*) the greater acceleration and (*b*) the higher speed? ••

Figure P3.66

67. You are playing air hockey with a friend. The puck is sitting at rest in his goal when he suddenly lifts his end of the table by 0.50 m. The puck slides down the tilted surface into your goal, 2.4 m away. Ignore friction. (*a*) How many seconds does it take the puck to reach your goal? (*b*) With what speed does the puck hit your goal? ••

68. A block has an initial speed of 6.0 m/s up an inclined plane that makes an angle of 37° with the horizontal. Ignoring friction, what is the block's speed after it has traveled 2.0 m? ••

69. A box is at the lower end of a very slippery ramp of length ℓ that makes a nonzero angle θ with the horizontal. A worker wants to give the box a quick shove so that it reaches the top of the ramp. (*a*) How fast must the box be going after the shove for it to reach its goal? (Ignore the distance traveled while the shove is executed.) (*b*) What is its speed halfway up the ramp? ••

70. A skier is at the top of a run that consists of two slopes having different inclines (Figure P3.70). The skier lets go, beginning the run with essentially zero speed. Ignoring friction, what are (*a*) his speed at the end of the lower slope and (*b*) his average acceleration over the entire run? ••

Figure P3.70

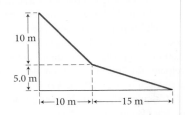

71. A ball is projected vertically upward from an initial position 5.0 m above the ground (Figure P3.71). At the same instant, a cube is released from rest down an ice-covered incline, from a height not necessarily equal to 5.0 m. The two objects reach the ground at the same instant, and both have a final speed of 15 m/s. What is the angle of the incline? ●●●

Figure P3.71

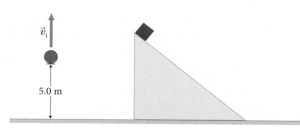

72. You hold a puck at the top of an ice-covered ramp inclined at 60° with respect to the vertical. Your friend stands nearby on level ground and holds a ball at the same height h above ground as the puck. If the puck and the ball are released from rest at the same instant, what is the ratio of the time interval it takes the ball to reach the ground to the time interval it takes the puck to reach the end of the ramp? ●●●

73. A child on a sled slides down an icy slope, starting at a speed of 2.5 m/s. The slope makes a 15° angle with the horizontal. After sliding 10 m down the slope, the child enters a flat, slushy region, in which she slides for 2.0 s with a constant negative acceleration of -1.5 m/s^2 with respect to her direction of motion. She then slides up another icy slope that makes a 20° angle with the horizontal. How far up the second slope does she slide? ●●●

3.8 Instantaneous acceleration

74. You throw a ball straight up with an initial speed of 10 m/s. (a) What is the ball's instantaneous acceleration at instant t_1, just after it leaves your hand; at instant t_2, the top of its trajectory; and at instant t_3, just before it hits the ground? (b) What is its average acceleration for the upward portion of its journey (t_1 to t_2) and for the entire trip (t_1 to t_3)? ●

75. A particle is accelerated such that its position as a function of time is given by $\vec{x} = bt^3\hat{\imath}$, with $b = 1.0$ m/s^3. What is the particle's acceleration as a function of time? ●

76. Figure P3.76 shows graphs of the x component of acceleration as a function of time for two different carts rolling along a flat horizontal table. In which case is the change in the x component of velocity greater over the time interval shown? ●

Figure P3.76

(a)

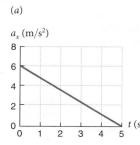

(b)

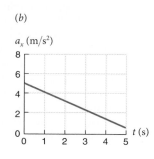

77. The position of a cart on a low-friction track can be represented by the equation $x(t) = b + ct + et^2$, where $b = 4.00$ m, $c = 6.00$ m/s, and $e = 0.200$ m/s^2. (a) Is the cart accelerating? If so, is the acceleration constant? (b) What is the average velocity between $t_i = 0.200$ s and $t_f = 0.400$ s? (c) What is the velocity at $t = 0.200$ s and at $t = 0.400$ s? (d) What is the average acceleration between $t_i = 0.200$ s and $t_f = 0.400$ s? (e) What is the acceleration at $t = 0.200$ s and at $t = 0.400$ s? ●●

78. A particle moves in the x direction according to the equation $x(t) = bt^3 + ct^2 + d$, where $b = 4.0$ m/s^3, $c = -10$ m/s^2, and $d = 20$ m. (a) What are its instantaneous velocity and instantaneous acceleration at $t = 2.0$ s? (b) What are its average velocity and average acceleration in the interval $t = 2.0$ s to $t = 5.0$ s? ●●

79. A rocket's x component of acceleration is given by $a_x = bt$, where $b = 1.00$ m/s^3. The rocket begins accelerating from rest at $t = 0$. (a) What is its x component of acceleration at $t = 10.0$ s? (b) What is its x component of velocity at $t = 10.0$ s? (c) What is its x component of average acceleration for the 10.0-s interval? (d) How far does it travel in the 10.0-s interval? ●●

80. The acceleration of a particular car during braking has magnitude bt, where t is the time in seconds from the instant the car begins braking, and $b = 2.0$ m/s^3. If the car has an initial speed of 50 m/s, how far does it travel before it stops? ●●

81. A car is 12 m from the bottom of a ramp that is 8.0 m long at its base and 6.0 m high (Figure P3.81). The car moves from rest toward the ramp with an acceleration of magnitude 2.5 m/s^2. At some instant after the car begins moving, a crate is released from rest from some position along the ramp. The crate and car reach the bottom of the ramp at the same instant and at the same speed. (a) At what distance d up the ramp was the crate released? (b) How many seconds after the car started was the crate released? ●●●

Figure P3.81

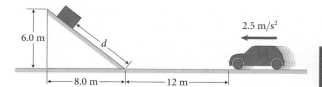

82. In a laboratory experiment, a sphere of diameter 8.0 mm is released from rest at $t = 0$ at the surface of honey in a jar, and the sphere's downward speed v when it travels in the honey is found to be given by $v = v_{max}(1 - e^{-t/\tau})$, where $v_{max} = 0.040$ m/s and $\tau = 0.50$ s. (a) Obtain an expression for $a(t)$. (b) Draw graphs for $v(t)$ and $a(t)$ for the time interval 0 to 2.0 s. (c) Obtain an expression for $x(t)$, choosing the positive x axis as downward, and draw the graph for this function. (d) Use your $x(t)$ graph to determine the time interval needed for the sphere to reach the bottom of the container if the surface of the honey is 0.10 m above the bottom of the jar. ●●●

83. A video recording of a moving object shows the object moving up and down with a velocity given by $v_x(t) = v_{max}\cos(\omega t)$, where $v_{max} = 1.20$ m/s and $\omega = 0.15$ s^{-1} and where the positive x direction is upward. (a) Derive an expression for $a(t)$, the object's acceleration as a function of time. (b) If the object's position at $t = 0$ is $x(0) = 0$, derive an expression for $x(t)$, its position as a function of time. ●●●

Additional Problems

84. The motion of a particle in the x direction can be described by the equation $x(t) = bt^2 + ct + d$, where $b = 0.35$ m/s^2, $c = 6.0$ m/s, and $d = 30$ m. (a) What is the particle's acceleration at $t = 10$ s? (b) What is its velocity at that instant? •

85. It takes you 7.0 m to brake to a panic stop from a speed of 9.0 m/s. Using the same acceleration, how far do you go as you brake to a panic stop from a speed of 27 m/s? •

86. A person standing on a building ledge throws a ball vertically from a launch position 45 m above the ground. If it takes 2.0 s for the ball to hit the ground, (a) with what initial speed was the ball thrown and (b) in which direction was it thrown? ••

87. A bullet fired straight through a board 0.10 m thick strikes the board with a speed of 480 m/s and emerges with a speed of 320 m/s. (a) What is the x component of the average acceleration of the bullet while in the board? (b) What is the time interval between the instant the bullet strikes the front face of the board and the instant it exits the back face? (c) Assuming the average acceleration you calculated in part a, what minimum thickness must the board have in order to stop the bullet? ••

88. During a blackout, you are trapped in a tall building. You want to call rescuers on your cell phone, but you can't remember which floor you're on. You pry open the doors to the elevator shaft, drop your keys down the shaft, and hear them hit bottom at ground level 3.27 s later. (a) Making a simple calculation and remembering that the ground floor is numbered floor 1, you determine which floor you're on. What floor is that? (b) Before you make the phone call, however, you realize you've forgotten that the sound of the impact at the bottom travels up the shaft at 340 m/s, and so you redo your calculation. What floor number do you tell the rescuers? ••

89. Starting from rest, a motorcycle moves with constant acceleration along a highway entrance ramp 0.22 km long. The motorcycle enters the traffic stream on the highway 14.0 s later at 31 m/s. (a) What is the magnitude of the acceleration along the ramp? (b) How fast is the motorcycle moving 9.5 s after it begins to accelerate? (c) How far along the ramp is the motorcycle 9.5 s after it begins to accelerate? ••

90. Hold a dollar bill with its short sides parallel to the floor, and have a friend hold the thumb and forefinger of one hand on either side of the bottom edge, with the thumb and finger about 20 mm apart. Offer your friend the following deal: He can keep the dollar if he can catch it between his thumb and finger after you release it. Most people fail to catch the bill. Use this information (and your measurement of the length of the long side of a dollar bill) to calculate a typical minimum reaction time. ••

91. A mortar-style fireworks shell must be detonated at a height of 100 m. (a) If the shell is to detonate at the top of its trajectory, what must its launch speed be? (b) For how many seconds must the detonation fuse burn if it is lit at the instant of launch? ••

92. A rock dropped from the top of a building travels 30 m in the last second before it hits the ground. How high is the building? ••

93. A ball launched directly upward from the ground at an initial speed of 24.5 m/s hits the ground 5.0 s later. What are (a) the ball's position 1.0, 2.0, 3.0, and 4.0 s after launch; (b) its velocity 1.0, 2.0, 3.0 and 4.0 s after launch; (c) its average velocity while it is in the air; and (d) its average speed for the entire trip? ••

94. After landing on Mars, you drop a marker from the door of your landing module and observe that it takes 2.1 s to fall to the ground. When you dropped the marker from the module door on Earth, it took 1.3 s to hit the ground. What is the magnitude of the acceleration due to gravity near the surface of Mars? ••

95. A very elastic rubber ball dropped from a height of 2.1 m rebounds to 88% of its original height. If the ball is in contact with the floor for 0.013 s, what is its average acceleration during that interval? ••

96. You are standing on a bridge holding a rock 15.0 m above the water. (a) How many seconds after you let go of the rock do you hear it splash into the water? Take the speed of sound to be 340 m/s. (b) What is the rock's speed just before it hits the water? (c) What is its velocity at that instant? ••

97. An elevator cable breaks, and the elevator car falls four floors (from rest) before the emergency brakes kick in. If the car falls one more floor before it stops, what is its average acceleration during braking? ••

98. A car initially traveling with velocity $\vec{v}_i$ slows down at constant acceleration and stops. (a) Sketch a graph of $v_x(x)$, the car's x component of velocity as a function of position. (b) If the car travels a distance ℓ during this braking, after what displacement does it have half its original speed? ••

99. A theorist has predicted that certain forms of matter exhibit a gravitational repulsion. On a planet made of this matter, the "antigravitational acceleration" points upward, away from the planet's surface, and any object released near the ground moves not in free-fall motion but rather in "free-rise motion." Suppose that on this planet, a projectile launched downward from a cliff at an initial speed of 20 m/s is observed 10 s later moving upward at 40 m/s. (a) What is the free-rise acceleration on this planet? (b) At what instant after launch is the projectile's speed zero? ••

100. A spaceship lifts off vertically with constant acceleration and reaches a speed of 300 km/h in 1.0 min. What are the ship's (a) average acceleration, (b) velocity at $t = 30$ s, and (c) altitude at $t = 60$ s? ••

101. From a promontory overhanging a lake, you throw a stone that enters the water vertically at 15.0 m/s. You hear the splash 2.68 s after you release the stone. Given that the speed of sound in air is 340 m/s: (a) What was the stone's initial speed? (b) Did you throw the stone down or up? (c) From what height above the lake surface did you release the stone? ••

102. Your state's *Driver Handbook* says, "When the vehicle ahead of you passes a certain point such as a sign, count 'one-thousand-one, one-thousand-two, one-thousand-three.' This takes about three seconds. If you pass the same point before you finish counting, you are following too closely." In your car, you are following a truck on the highway, both of you moving at 100 km/h. You are keeping a 3.0-s gap between your front bumper and the rear of the truck. Suddenly a large box appears on the road beneath the rear of the truck. Apparently the truck's axles were tall enough to allow it to pass right over the object, but yours are not! It takes you 0.75 s to react and hit your brakes. (a) What is the minimum constant acceleration required if you are not to hit the box? (b) What would your answer be if you were keeping a 2.0-s gap between your car and the truck, as is sometimes recommended as safe, under optimal driving conditions, for alert, experienced drivers with good reflexes? Compare and comment on your answers in parts a and b. ••

103. A falling rock hits the ground at a speed of 16 m/s. How many seconds were needed for the rock to fall the last 12 m? ••

104. You want to determine whether a tall tree will fit onto your 75-m-long logging trailer after you cut it down. You throw a rock straight up with a speed great enough that the rock just reaches the top of the tree, releasing the rock 2.0 m above the ground. You observe that the time interval from the instant you release the rock to the moment it hits the ground is 5.0 s. ••• CR

105. Your cousin is always bragging about how high his model rockets go, so you go shopping. Unfortunately none of the brands specify altitude, but you do find three models in your price range with comparable technical information. "Cloudscraper" promises a 3.3 s fuel burn time and 2.9*g* vertical acceleration. "Stratosphere" offers 3.0 s at 3.2*g*, and "Astronaut" claims 2.7 s at 3.6*g*. ••• CR

106. You are enjoying a coffee break with your cousin in her highway patrol car when a sports car speeds by at 135 km/h. Your cousin tells you to hold her coffee as she starts the car, and she offers to bet you ten dollars that she can catch the speeder before he reaches the state line. You know that the state line is 2.0 km away along the straight and level highway, and you have read that police cruisers can reach a top speed of 210 km/h within 15 s from a resting start. As the car begins to move, you realize that 5.0 s has already elapsed and that you must decide soon. ••• CR

107. At the drag races, three physics students note that the winning dragster in one category has a speed of 215 mi/h at the end of its quarter-mile run. The run takes 6.30 s, and the car,
initially at rest, starts exactly when the timing clock starts. Using these data, and with *Principles* Chapter 3 as their guide, the students calculate the magnitude of the dragster's acceleration. Although their physics instructor verifies that they have made no numerical errors, the students get three different values for the acceleration. (*a*) What are those values? (*b*) One of the students notices an interesting connection among the three values they obtained. What is that connection? (*c*) Is there a reasonable explanation for the discrepancies found in part *a*? •••

108. You are touring a distant planet on which the magnitude of the free-fall acceleration is 65.0% of what it is on Earth. For a little excitement, you jump off a precipice 500 m above the planet's surface. After 5.00 s of free fall, you ignite the jet-pack on your back, changing your acceleration to some new, constant value for the remainder of the fall. You reach the ground 26.0 s after igniting the jet-pack. At what speed do you hit the ground? •••

109. A classmate leaves a message on your voice mail betting that you cannot throw a stone high enough so it lands on the roof of a 20-m-high building. As you stare out of your window pondering whether to accept the challenge, the well in the courtyard suddenly gives you an idea. You drop a stone into the well and note that you hear a splash 4.0 s later. You repeat the experiment with another stone, but this time you throw the stone down as fast as you can. This time the splash comes 3.0 s after the stone leaves your hand. You carry out a quick calculation and then you call back your friend. ••• CR

Answers to Review Questions

1. Velocity measures displacement (change in position) per unit of time; acceleration measures change in velocity per unit of time.
2. No. *Accelerating* means moving with changing velocity. This includes speeding up as well as any other nonconstant velocity, such as slowing down.
3. No. An object that is slowing down along a straight path has an acceleration vector that points in the direction opposite the direction of motion.
4. The x component of acceleration is positive if the curvature is upward and negative if the curvature is downward. No curvature means no acceleration.
5. They hit at the same instant. If we ignore air resistance, all freely falling objects fall with the same constant acceleration.
6. No. While the stone is falling through the air, its acceleration is a constant 9.8 m/s^2 straight down. Its velocity (and, in this case, its speed) increases because of this constant acceleration. (Once in the water, the stone experiences a different acceleration because the water resistance is much greater than the air resistance, which is negligibly small during the stone's fall.)
7. Because air resistance is negligible for a rock tossed in the air, the acceleration is 9.8 m/s^2 downward at both instants. The acceleration due to gravity is constant near Earth's surface, and the rock experiences that acceleration at all instants during its round trip.
8. The acceleration at the top is 9.8 m/s^2 downward, just as it is during the rest of the flight. The fact that $v = 0$ at the top of the trajectory doesn't mean $a = 0$. The ball is headed upward the instant before reaching the top and headed downward an instant later. This change in velocity direction means that the ball must be accelerating at the top, and the only acceleration operating here is that due to gravity.
9. A motion diagram should use dots to represent the position of the moving object at equally spaced time intervals. The spacing of the dots is crucial: Small spacing means low speed, large spacing means high speed. The diagram should include numerical labels (including units) for known quantities as well as information about the direction of travel (represented by an arrow), if known. It should also include a coordinate axis showing a defined positive direction and, for constant-acceleration situations, an arrow showing the acceleration direction.
10. A motion diagram summarizes the information known about a motion. It provides a visual representation that helps with the mathematical solution to a kinematics problem. The diagram organizes your information and helps you break down the problem into parts.
11. (*a*) Its acceleration is constant. (*b*) Its acceleration is zero.
12. The $\frac{1}{2}$ is included because the area of a triangle is one-half base times height, and with a linear $v(t)$ curve the area under any segment involves a triangle and (in most cases) a rectangle. These areas could be combined into a trapezoid, but doing so would obscure the fact that the triangular portion represents the slope of the $v(t)$ curve (and the slope gives us the object's acceleration).
13. For constant acceleration, the difference in the squares of the final and initial x components of velocity is equal to twice the product of the displacement and the acceleration (Eq. 3.13).
14. Both time intervals are the same. The $x(t)$ curve representing the trajectory is symmetrical about the instant at which the ball reaches its highest position.
15. (*a*) Quadratically (Eq. 3.11); (*b*) linearly (Eq. 3.12).
16. The ratio of the distance traveled to the square of the elapsed time is constant for a ball released from rest rolling down an inclined plane.
17. In both cases, the acceleration is constant and the motion follows a straight path. The motion diagrams are similar, although the numerical values of the acceleration differ.
18. The acceleration magnitude depends on only the angle of incline (Eq. 3.20).
19. Motion with nonconstant acceleration.
20. (*a*) Velocity at that value of t; (*b*) the second time derivative of position, which is the acceleration; (*c*) acceleration at that value of t; (*d*) the integral of a with respect to t, which is the change in velocity during the time interval (Eq. 3.27); (*e*) the integral of v with respect to t, which is the displacement during the time interval (Eq. 3.28).

Answers to Guided Problems

Guided Problem 3.2 5.2 s

Guided Problem 3.4 (*a*) Friend's ball first; (*b*) yours: $|v_\text{f}| = \sqrt{4gh_\text{launch}}$,

friend's: $|v_\text{f}| = \sqrt{4.5gh_\text{launch}}$; (*c*) $\Delta t = (2 - \sqrt{2})\sqrt{\dfrac{h_\text{launch}}{g}}$

Guided Problem 3.6 9.71 m/s^2

4
Practice
Momentum

Chapter Summary 52

Review Questions 53

Developing a Feel 54

Worked and Guided Problems 55

Questions and Problems 61

Answers to Review Questions 68

Answers to Guided Problems 68

PRACTICE

Chapter Summary

Inertia (Sections 4.1 – 4.3, 4.5)

Concepts *Friction* is the resistance to motion that one surface encounters when moving over another surface. In the absence of friction, objects moving along a horizontal track keep moving without slowing down.

Inertia is a measure of an object's tendency to resist a change in its velocity. Inertia is determined entirely by the type of material of which the object is made and by the amount of that material contained in the object. Inertia is related to *mass*, and for this reason we use the symbol m to represent it. The SI unit of inertia is the **kilogram** (kg).

Quantitative Tools If an object of unknown inertia m_u collides with an inertial standard of inertia m_s, the ratio of the inertias is related to the changes in the velocities by

$$\frac{m_u}{m_s} \equiv -\frac{\Delta v_{sx}}{\Delta v_{ux}}. \tag{4.1}$$

Systems and momentum (Sections 4.4, 4.6, 4.7, 4.8)

Concepts A **system** is any object or group of objects that can be separated, in our minds, from the surrounding environment. The **environment** is everything that is not part of the system. You can choose the system however you want, but once you decide to include a certain object in the system, that object must remain a part of the system throughout your analysis.

A system for which there are no external interactions is called an **isolated system.**

An **extensive quantity** is one whose value is proportional to the size or "extent" of the system. An **intensive quantity** is one that does not depend on the extent of the system.

A **system diagram** shows the initial and final conditions of a system.

Quantitative Tools The **momentum** $\vec{p}$ of an object is the product of its inertia and velocity:

$$\vec{p} \equiv m\vec{v}. \tag{4.6}$$

The momentum of a system of objects is the sum of the momenta of its constituents:

$$\vec{p} \equiv \vec{p}_1 + \vec{p}_2 + \cdots. \tag{4.11, 4.23}$$

Conservation of momentum (Sections 4.4, 4.8)

Concepts Any extensive quantity that cannot be created or destroyed is said to be **conserved,** and the amount of any *conserved* quantity in an isolated system is *constant*. Momentum is a conserved quantity, and therefore the momentum of an isolated system is constant. The momentum can be transferred from one object to another in the system, but the momentum of the system cannot change.

Quantitative Tools The momentum of an isolated system is constant:

$$\Delta \vec{p} = \vec{0}. \tag{4.17}$$

Another way to say this is that for an isolated system, the initial momentum is equal to the final momentum:

$$\vec{p}_i = \vec{p}_f. \tag{4.22}$$

The **impulse** $\vec{J}$ delivered to a system is equal to the change in momentum of the system:

$$\vec{J} = \Delta \vec{p}. \tag{4.18}$$

For an isolated system, $\vec{J} = \vec{0}$.

Review Questions

Answers to these questions can be found at the end of this chapter.

4.1 Friction

1. Describe several ways of minimizing friction between a surface and an object moving on that surface. Is it possible for a surface to be completely frictionless?

4.2 Inertia

2. Two standard carts collide on a horizontal, low-friction track. How does the change in the velocity of one cart compare with that of the other?

3. Carts A and B collide on a horizontal, low-friction track. Cart A has twice the inertia of cart B, and cart B is initially motionless. How does the change in the velocity of A compare with that of B?

4.3 What determines inertia?

4. An iron cube and an iron sphere contain the same volume of material. How do their inertias compare?

5. Two objects of identical volume and shape are made of different materials: iron and wood. Are their inertias identical?

4.4 Systems

6. What is a system?

7. What is the key difference between an extensive quantity and an intensive quantity?

8. What four processes can change the value of an extensive quantity in a system?

9. What does it mean if an extensive quantity is a conserved quantity?

4.5 Inertial standard

10. Using a pair of standard carts and a baseball, how would you determine the inertia of the baseball?

11. Can the inertia of any object be negative?

12. Which has greater inertia: 1 kg of feathers or 1 kg of lead?

4.6 Momentum

13. Which has greater momentum: a flying bumblebee or a stationary train? Which has greater inertia?

14. A 3-g bullet can knock a wooden block off a fence as easily as a 140-g baseball can. How is this possible?

15. Can the momentum of any object be negative?

4.7 Isolated systems

16. What is the meaning of the word *interaction* in physics?

17. For a given system, what is the distinction between external and internal interactions, and why is the distinction important?

18. What is an isolated system, and of what use is an isolated system?

4.8 Conservation of momentum

19. Are these two statements equivalent? (*a*) The momentum of an isolated system remains constant in time. (*b*) Momentum is conserved.

20. Suppose the 1-kg international inertial standard cylinder were lost or destroyed. How would this affect conservation of momentum?

21. What is impulse, and how is it related to momentum?

22. Compare and contrast the *conservation of momentum* and the *momentum law*.

Developing a Feel

Make an order-of-magnitude estimate of each of the following quantities. Letters in parentheses refer to hints below. Use them as needed to guide your thinking.

1. The inertia of a suitcase filled with physics textbooks (H, L, Q)
2. The ratio of the inertias of an object and a bowling ball if the object, initially at rest on a bowling alley, doesn't move noticeably after being hit by the bowling ball (C, M, B)
3. The magnitude of the momentum of a tennis ball as it crosses the net (F, R)
4. The magnitude of the momentum of a bowling ball moving toward the pins (A, M)
5. The magnitude of the momentum of a baseball pitched by a major-league pitcher (I, S)
6. The magnitude of the momentum of a typical marathon runner (D, O, V)
7. The magnitude of the change in momentum of a baseball bat that reverses the velocity of the baseball of question 5 (G, N)
8. The magnitude of the change in momentum of a stationary bowling ball hit by the tennis ball of question 3 (B, G, N)
9. The magnitude of the change in momentum of a car that leaves the freeway to stop for gas (E, J, T)
10. The magnitude of the change in velocity of a car that hits a stationary deer (E, P, K, U, N)

Hints

A. What is the inertia of a bowling ball?
B. How does the object of less inertia behave after a collision with an object of greater inertia?
C. What is the largest nonzero speed that is too small to be noticed during a few seconds of observation?
D. What is the inertia of a marathon runner?
E. What is the inertia of a typical car?
F. What is the inertia of a tennis ball?
G. What is the magnitude of the change in momentum of the object with less inertia?
H. What is the inertia of a single physics textbook?
I. What is the inertia of a baseball?
J. What is the speed of the car while on the freeway?
K. What is the inertia of a typical deer?
L. What is the volume of a typical suitcase?
M. What is the typical speed of a bowling ball moving toward the pins?
N. What do you know about the momentum of this system of two objects?
O. What time interval is required to run a marathon?
P. What is the likely speed of a car on a road with deer crossings?

Q. What is the volume of a single physics book?
R. What is the speed of a tennis serve?
S. What is the speed of a pitched baseball?
T. What is the speed of a car when fueling?
U. With what speed would the deer rebound? (First consider a moving deer bouncing off a stationary car by analogy to question 8.)
V. What is the length of a marathon?

Key (all values approximate)

A. 7 kg; B. it bounces back, approximately reversing its velocity; C. 1×10^{-3} m/s; D. 6×10^1 kg; E. 2×10^3 kg; F. 6×10^{-2} kg; G. because the motion reverses, about twice the magnitude of the object's initial momentum; H. 3 kg; I. 0.2 kg; J. 3×10^1 m/s; K. 5×10^1 kg; L. 0.1 m³; M. 7 m/s; N. it remains approximately constant; O. 3 h or more; P. less than half freeway speed, perhaps 1×10^1 m/s; Q. 3×10^{-3} m³; R. 5×10^1 m/s; S. 4×10^1 m/s; T. 0; U. at a speed somewhat slower than twice the initial speed of the car; V. 4×10^1 km

Worked and Guided Problems

Procedure: Choosing an isolated system

When you analyze momentum changes in a problem, it is convenient to choose a system for which no momentum is transferred into or out of the system (an isolated system). To do so, follow these steps:

1. Separate all objects named in the problem from one another.
2. Identify all possible interactions among these objects and between these objects and their environment (the air, Earth, etc.).
3. Consider each interaction individually and determine whether it causes the interacting objects to accelerate. Eliminate any interaction that does not affect (or has

only a negligible effect on) the objects' accelerations during the time interval of interest.

4. Choose a system that includes the object or objects that are the subject of the problem (for example, a cart whose momentum you are interested in) in such a way that none of the remaining interactions cross the system boundary. Draw a dashed line around the objects in your choice of system to represent the system boundary. None of the remaining interactions should cross this line.
5. Make a system diagram showing the initial and final states of the system and its environment.

These examples involve material from this chapter but are not associated with any particular section. Some examples are worked out in detail; others you should work out by following the guidelines provided.

Worked Problem 4.1 Jump ship

In a game of canoe-ball on a calm lake, an athlete stands at the front end of a canoe at rest, facing the shore. A player on shore throws a 1.8-kg ball that arrives at the canoe with a speed of 2.5 m/s. The rules of the game demand that the athlete hold on to the ball and immediately dive into the water, which he does, horizontally, off the front of the canoe. His inertia is 60 kg, and the inertia of the canoe is 80 kg. Friends on shore determine that his speed in the dive is 1.2 m/s. How fast is the canoe moving after he jumps off?

❶ **GETTING STARTED** Because we are given the canoe's inertia, we can determine its speed once we know the magnitude of its momentum ($v = p/m$). The crucial point is that the momentum exchange is both between the canoe and the athlete and between the ball and the athlete. However, some interaction between the water and the canoe is also possible. Because the event we are analyzing happens during a short time interval and because we know that water is "slippery," we choose to ignore any effect that resistance from the water might have on the canoe's motion. With this simplification, the system comprising the ball, canoe, and athlete is isolated.

We draw a system diagram (Figure WG4.1) including the relevant velocity information and using subscripts to keep track of our objects: b for ball, a for athlete, c for canoe. We also use the subscripts i for initial quantities and f for final quantities to keep track of the temporal order of events.

Figure WG4.1

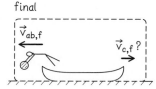

❷ **DEVISE PLAN** Because momentum is a conserved quantity and because the system is isolated, the system's initial momentum

must be equal to the system's final momentum. We notice that the momentum is not zero before the catch because the ball is in motion. We also note that, after the dive begins, the athlete and ball can be treated as a single object because they move with a common velocity. We label this composite object with the subscript ab. We have to account for all the contributions to the system's momentum in the initial state (just before the dive) and in the final state (just after the dive), and set the two equal. It is best to draw initial and final momentum representations to keep track of all the objects in the system and to show the direction we choose for the positive x axis (Figure WG4.2). We draw the final momentum vector for the canoe a bit larger than the vector for the athlete plus ball because the system momentum must be in the same direction and of the same magnitude in the initial and final pictures. Note that we choose the direction of the incoming ball as the positive x direction.

Figure WG4.2

initial   final

$\vec{p}_{b,i}$ $\vec{p}_{a,i} = \vec{0}$ $\vec{p}_{ab,f}$ $\vec{p}_{c,f}$

$\vec{p}_{c,i} = \vec{0}$

$\longrightarrow x$

With this figure as a guide, we write an equation setting the initial and final momenta equal to each other:

$$\vec{p}_{a,i} + \vec{p}_{b,i} + \vec{p}_{c,i} = \vec{p}_{ab,f} + \vec{p}_{c,f}$$

$$\vec{0} + m_b\vec{v}_{b,i} + \vec{0} = (m_a + m_b)\vec{v}_{ab,f} + m_c\vec{v}_{c,f}.$$

Because we know all the inertias, the three initial velocities, and the final velocity of the athlete-ball combination, we have all we need to get the final speed of the canoe.

❸ EXECUTE PLAN Recalling that vector components are signed quantities whose signs depend on our choice for the positive x direction, which is in the direction of the incoming ball, we write the equation in component form:

$$m_b v_{bx,i} = (m_a + m_b)v_{abx,f} + m_c v_{cx,f}.$$

In terms of speeds, this becomes

$$m_b v_{b,i} = (m_a + m_b)(-v_{ab,f}) + m_c v_{c,f}.$$

We isolate our unknown, $v_{c,f}$, and then substitute values for the quantities we know:

$$m_c v_{c,f} = m_b v_{b,i} - (m_a + m_b)(-v_{ab,f})$$

$$v_{c,f} = \frac{m_b v_{b,i} + (m_a + m_b)v_{ab,f}}{m_c} \qquad (1)$$

$$v_{c,f} = \frac{(1.8 \text{ kg})(2.5 \text{ m/s}) + (60 \text{ kg} + 1.8 \text{ kg})(1.2 \text{ m/s})}{80 \text{ kg}}$$

$$= \frac{78.7 \text{ kg} \cdot \text{m/s}}{80 \text{ kg}} = 0.983 \text{ m/s}.$$

The final speed of the canoe, to two significant digits, is

$$v_{c,f} = 0.98 \text{ m/s.} ✔$$

❹ EVALUATE RESULT The numerical value of our result is not unreasonable; that is, it is of the order of magnitude we expect for the speed of a human-powered boat. If there were significant water resistance, it would reduce the canoe's speed, but this reduction would take significantly longer than the jump, as you may know from pushing a boat off. For this reason the assumption that water resistance can be neglected during the time interval of interest is justified.

We should also check to see whether our algebraic answer behaves in the way we expect. Equation 1 shows that the canoe's final speed is a sum of positive terms. If Eq. 1 allowed a negative answer for $v_{c,f}$, we would have to rethink our choice of signs for velocity components. The canoe's final x component of momentum $p_{cx,f}$ must be positive for this reason: Because the initial momentum of the system is positive, the final momentum must also be positive (but the final momentum of the athlete-ball combination is negative). This is consistent with our intuition that the boat moves to the right in Figure WG4.1 as the athlete dives to the left.

Furthermore, we see that Eq. 1 implies that the canoe's final speed would increase if the ball had greater inertia, if it came in at a higher speed, if the athlete had greater inertia, or if he dove at a higher speed. This matches our qualitative understanding that each of these changes would increase either the initial (positive) momentum of the system or the final (negative) momentum of the athlete-ball combination.

Guided Problem 4.2 A pie in the face

At a carnival, a super-sized, 1.0-kg cream pie is thrown with a speed of 5.0 m/s into the face of a 60-kg clown at rest on roller skates. After the pie strikes his face and sticks there, how fast does the clown roll backward?

❶ GETTING STARTED
1. Select an isolated system and sketch a system diagram for the pie-clown collision.
2. What object(s) is (are) in motion before the collision? After the collision?

❷ DEVISE PLAN
3. What do you know about the momentum of your system?
4. Express your answer to question 3 quantitatively and count the unknowns.

❸ EXECUTE PLAN

❹ EVALUATE RESULT
5. Is your answer unreasonably large or small?

Worked Problem 4.3 Exploding bicycle pump

You are riding your bike at a steady 4.0 m/s when suddenly you swerve to avoid a pothole. Your bicycle pump falls from its bracket, strikes a rock at the road's edge, and explodes into three pieces. You watch the pump body sail ahead of you, then hit the ground and stop 0.50 s after the explosion. The pump was horizontal when it exploded, and its length was aligned parallel to the road. Retracing your path, you measure the distance from the pump body to the rock as 3.2 m. Just beside the rock is the pump handle. These two pieces were easy to find, but the third piece—a thin metal spring—is going to be hard to spot. If the inertias are 0.40 kg for the pump body, 0.25 kg for the handle, and 0.20 kg for the spring, where should you look for the spring?

❶ GETTING STARTED We begin with a sketch of the situation (Figure WG4.3). We are given position information for two of the

three pieces, but that alone will not locate the third piece. We also have the inertia of each piece, and we know that the velocity of the pump before the explosion was the same as that of the bike.

Figure WG4.3

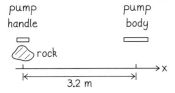

❷ DEVISE PLAN We consider the three pieces into which the pump disintegrates as our system. For the time interval between the instant after the pump hits the rock and the instant before the pieces land, the system is isolated and so the momentum of the three-piece system does not change. The best we can do, given the information at hand, is to assume that all three pieces hit the ground 0.50 s after the explosion. This allows us to use Eq. 4.19:

$$\vec{p}_f = \vec{p}_i \quad \text{(isolated system)}.$$

There are three objects that move as one (pump) in the initial state but separately (body, handle, spring) in the final state. During the time interval of interest, it is reasonable to assume that all objects move parallel to the road, and so we choose the positive x direction to be the direction of initial travel, from the rock toward the pump body in Figure WG4.3. We arbitrarily locate the origin of the x axis at the rock. The resulting component equation is

$$p_{px,i} = p_{bx,f} + p_{hx,f} + p_{sx,f}$$

$$(m_b + m_h + m_s)v_{px,i} = m_b v_{bx,f} + m_h v_{hx,f} + m_s v_{sx,f}. \quad (1)$$

We know the x component of the initial velocity, but we do not know the x components of the final velocities, and so we need more information. Because the pump handle lies beside the rock, we conclude that the x component of the velocity of the pump handle was zero immediately after the explosion. The pump body lies 3.2 m from the rock, and we know that it traveled this far in 0.50 s. We can determine the x component of its final velocity immediately after the explosion from this information. That leaves only the x component of the final velocity of the spring unknown, which we should be able to compute from the momentum equation. With that information we can compute how far the spring traveled in 0.50 s, and hence where to look for it.

③ EXECUTE PLAN The x component of the velocity of the pump body after the explosion is

$$v_{bx,f} = \frac{+3.2 \text{ m}}{0.50 \text{ s}} = +6.4 \text{ m/s}.$$

Next we solve Eq. 1 for $v_{sx,f}$ and substitute our known values:

$$(m_b + m_h + m_s)v_{px,i} = m_b v_{bx,f} + m_h v_{hx,f} + m_s v_{sx,f}$$

$$v_{sx,f} = \frac{(m_b + m_h + m_s)v_{px,i} - m_b v_{bx,f} - m_h v_{hx,f}}{m_s}$$

$$= \frac{(0.40 \text{ kg} + 0.25 \text{ kg} + 0.20 \text{ kg})(+4.0 \text{ m/s})}{0.20 \text{ kg}}$$

$$- \frac{(0.40 \text{ kg})(+6.4 \text{ m/s}) + (0.25 \text{ kg})(0)}{0.20 \text{ kg}}$$

$$= +4.2 \text{ m/s}.$$

The spring traveled for 0.50 s at 4.2 m/s, and so we should find it by looking at about

$$x_{s,f} = 0 + (+4.2 \text{ m/s})(0.50 \text{ s}) = +2.1 \text{ m}$$

from the rock in the initial direction of travel. ✔

④ EVALUATE RESULT The pump was moving in the x direction prior to the explosion, and so the momentum of the pieces should carry at least some of them down the road in the direction of the initial motion. The spring is located at a position between the body and the handle, which is not unreasonable. The value is of the correct order of magnitude.

Guided Problem 4.4 Space maneuvers

The commander of the starship *Enterprise* is stranded in his orbital shuttle craft, which rests a few kilometers from its docking station. Fortunately, he has two cargo pods that can be ejected (in a direction away from the docking station) with an explosive charge, and he hopes that carrying out this maneuver will move the shuttle craft back to the docking station. Pod 1 has inertia m_1, pod 2 has inertia $m_2 < m_1$, and both pods eject with speed v. Which will get him to the docking station more quickly: (*a*) ejecting first pod 1 and then pod 2, (*b*) ejecting first pod 2 and then pod 1, or (*c*) ejecting both at the same instant?

❶ GETTING STARTED

1. Choose an isolated system for option *a*, and sketch a system diagram showing the initial and final conditions of the system, building your drawing up as you read each sentence of the problem statement. Do you need a separate diagram for each option?

2. Identify the unknown quantity you are after. What physical principle can be applied to determine this quantity?

❷ DEVISE PLAN

3. Do you expect that the order in which the pods are jettisoned makes a difference?

4. How many moving objects are there in choice *a*, choice *b*, and choice *c*?

5. Add a representation of the momentum vectors before and after the explosive charges are fired in each of your system diagrams.

6. Write the appropriate equation(s) for each case and count the unknowns.

❸ EXECUTE PLAN

❹ EVALUATE RESULT

7. Rank your three answers in order of increasing speed and determine whether or not the order is reasonable based on your conceptual understanding of momentum.

Worked Problem 4.5 Forensic physics

Your friend from law enforcement claims that, in certain cases, a piece of an object hit by gunfire can move toward the shooter rather than away from the shooter. You decide to investigate whether this is possible by firing a target rifle at some melons. Suppose that in one of your tests an 8.0-g bullet is fired at 400 m/s toward a 1.20-kg melon several meters away, splitting the melon into two pieces of unequal size. The bullet lodges in the smaller piece and propels it forward (that is, in the direction the bullet originally traveled) at 9.2 m/s. If the combined inertia of this piece and the lodged bullet is 0.45 kg, determine the final velocity of the larger piece.

❶ GETTING STARTED As usual, we begin by identifying an isolated system. What about one made up of the bullet and melon? There may be some interaction with the ground or the support upon which the melon sits, but if we choose our time interval to start immediately before the bullet strikes the melon and end immediately after the smaller piece breaks loose, we can ignore this interaction. During this very short time interval, we can treat the system as isolated.

Next we sketch a system diagram showing the initial and final conditions of the system (Figure WG4.4). We arbitrarily choose the direction in which the smaller piece moves (to the right in the figure) as our positive x direction. We don't know which way the larger piece moves, and so we draw it as going forward, too. Our calculation will let us know whether this guess is right or wrong. The momentum of the system remains constant during this collision because the system is isolated.

Figure WG4.4

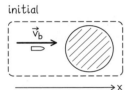 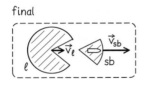

initial   final

❷ DEVISE PLAN We can translate the information given into an equation that requires no change in the momentum of the system.

There are two objects in the system in the final condition, but they are different from the ones in the initial condition. We designate the inertia of the bullet as m_b and the x component of its velocity as v_{bx}, and we note that the intact melon had an initial speed $v_{m,i} = 0$ and inertia m_m. We have

$$\Delta \vec{p} = 0 \Rightarrow \vec{p}_i = \vec{p}_f$$

$$m_b v_{bx,i} + m_m v_{mx,i} = (m_s + m_b)v_{sbx,f} + m_\ell v_{\ell x,f}$$

where the subscript s denotes the smaller piece, the subscript ℓ denotes the larger piece, and the subscript sb denotes the combination of the bullet and the smaller piece. We are looking for $v_{\ell x,f}$, and because all our other variables have known values, the planning is finished.

❸ EXECUTE PLAN We isolate the desired unknown and then insert numerical values. Recall that all of the velocity components in our diagram are in the positive direction.

$$m_\ell v_{\ell x,f} = m_b v_{bx,i} + m_m v_{mx,i} - (m_s + m_b)v_{sbx,f}$$

$$v_{\ell x,f} = \frac{m_b(+|\vec{v}_{b,i}|) - (m_s + m_b)(+|\vec{v}_{sb,f}|)}{m_\ell}$$

$$= \frac{(0.0080 \text{ kg})(400 \text{ m/s}) - (0.45 \text{ kg})(9.2 \text{ m/s})}{1.20 \text{ kg} - (0.45 \text{ kg} - 0.0080 \text{ kg})}$$

$$= -1.240 \text{ m/s} = -1.2 \text{ m/s.} ✔$$

❹ EVALUATE RESULT The negative sign means that the larger piece moves in the negative x direction, which means that our initial guess for the direction of this motion was wrong. Note, however, that the algebra nicely informed us of the incorrect assumption without pain or confusion. The larger piece really does move back toward the rifle. The magnitude of this velocity is fairly small, as we would expect, but it certainly would be noticeable.

Guided Problem 4.6 Bullet impact

A rifle is fired twice in succession at a target that is free to slide on a frozen pond, and both bullets embed in the target. In terms of the relevant inertias and the rifle's muzzle velocity, compute the velocity of the target after the first impact and after the second impact.

❶ GETTING STARTED
1. Should you choose one system for the entire process or a separate system for each impact?
2. Draw the system diagram(s) you need.
3. What physical principle governs the target's motion?

❷ DEVISE PLAN
4. Carefully consider the time sequence of events when deciding how to apply the labels "initial" and "final."

5. Remember to account for the inertia of the embedded bullets.
6. What equation(s) can you write to describe the physics in each segment of the motion?

❸ EXECUTE PLAN
7. Must you solve each equation separately, or can you combine them into one equation for the target's final speed?

❹ EVALUATE RESULT
8. Are the sign and magnitude of your answer what you expect?

Worked Problem 4.7 Grain dump slows freight

The problem on your new job at the grain elevator is that, as a series of coupled, empty freight cars coast past the grain-loading chute, the grain added to the cars slows the train down. The result is that there is more grain in the rear cars (which pass under the chute when the train is moving slowly) than in the front cars (which pass under the chute when the train is moving quickly). Your boss does not want to incur the expense of a locomotive to move the train at a constant speed. She wants to know how much more grain ends up in the last car than in the first car, and which variables affect the amount of grain in different cars. She hands you a chart of typical coasting speeds, loading rates, empty-car inertias, numbers of cars to be loaded, and so on. You start by assuming that the cars, when empty, are all identical and that all motion occurs along a horizontal straight line (the train tracks).

❶ GETTING STARTED We begin as usual with a sketch (Figure WG4.5). Note that the sketch shows the situation at an instant when a car in the middle of the train is being loaded rather than some instant when the first car or last car is taking on grain. Choosing this intermediate instant should make the sketch relevant to our analysis throughout the loading process.

Figure WG4.5

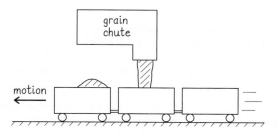

There is a lot going on: Grain is falling into the cars, increasing their inertia; the motion is slowing, allowing more grain into each car as the train moves forward, and the additional grain in the cars changes the train's motion more noticeably with time. We have no numerical values as clues, and so we will have to derive some general expressions. The amount of grain in each car will have some relationship to the speed of that car as it passes under the chute, and so we must obtain a relationship involving speed or velocity. We will also need to know how the time interval a given car spends being loaded depends on how much time has passed since the first car began loading. The problem statement implies that the rate of grain release from the chute is constant. This implication is consistent with the fact that the rear cars end up with more grain, and so we assume a constant loading rate. The final goal is to obtain an expression for how long a time interval each car spends under the chute. Given that and a constant loading rate, we can determine the amount of grain in each car and so compare the amounts in the first and last cars.

❷ DEVISE PLAN The system of cars plus grain is isolated if the chute and the track produce no horizontal impulse on the system. We therefore assume that the grain falls vertically into each car and that friction between the track rails and the wheels of the cars is negligible. Under these conditions, the momentum of the system must remain constant, which means that the reduction in velocity exactly balances the increase in the train's inertia. Our analysis begins with N initially empty cars (combined inertia m_i) at the instant $t = 0$ coasting with velocity $\vec{v}_i$.

Equation 4.19 is a good starting point:

$$\vec{p}_i = \vec{p}_f$$
$$m_i\vec{v}_i = m_f\vec{v}_f$$

We seek a general relationship that is valid throughout the loading process, so we represent the final instant by an arbitrary instant t:

$$m_i\vec{v}_i = m(t)\vec{v}(t). \qquad (1)$$

Both $m(t)$ and $\vec{v}(t)$ are unknown, which means we need another equation involving either inertia or velocity.

The grain is added in small uniform increments, and the variables that contribute to the momentum of each piece of the system (inertias, velocities) continually change with time. Perhaps we can divide the loading process into very short time intervals, then analyze the inertia change during each interval and integrate over time to obtain an expression for the inertia as a function of time. Substituting the expression we obtain into Eq. 1, we can isolate $v(t)$. Then we must extract information about the time interval each car spends under the chute. This is equivalent to asking, during what time interval does a car move a distance equal to its length ℓ? That suggests we integrate $v(t)$ with respect to time to obtain displacement. We can then define an x axis running along the track and isolate t for the cases $\Delta x = \ell, 2\ell, \ldots, N\ell$. Given these loading time intervals, we can determine the amount of grain in each car, and we will see which variables have an effect on the loading process.

❸ EXECUTE PLAN We denote the constant loading rate by $dm/dt = \lambda$ (kg/s). Because we know the initial inertia of the train m_i, we can integrate to obtain an expression for its final inertia at an arbitrary instant t:

$$\int_{m_i}^{m(t)} dm = \int_0^t \lambda dt \Rightarrow m(t) - m_i = \lambda t$$

$$m(t) = m_i + \lambda t.$$

Solving Eq. 1 for $\vec{v}(t)$ and then substituting $m_i + \lambda t$ for $m(t)$, we obtain a general expression for $\vec{v}(t)$:

$$\vec{v}(t) = \frac{m_i\vec{v}_i}{m_i + \lambda t}.$$

Defining the positive x direction to be along the track in the direction of the initial motion allows us to express this as a component equation:

$$v_x(t) = \frac{m_i v_{x,i}}{m_i + \lambda t}.$$

What remains is to extract from this expression the time interval each car spends under the chute. Each car has length ℓ, and so we need an expression for the displacement that can be set equal to an integer number of car lengths. Noting that displacement is the integral of velocity with respect to time, we obtain

$$\Delta x = \int v_x(t)dt = m_i v_{x,i} \int \frac{dt}{(m_i + \lambda t)}.$$

We can either look this integral up or calculate it with a substitution of variables: $u = (m_i + \lambda t)$; $du = \lambda dt$. The result is a natural logarithm:

$$\Delta x = \frac{m_i v_{x,i}}{\lambda} \ln(m_i + \lambda t) + C.$$

To eliminate the constant of integration, we pick specific limits of integration. We start the integral at $t = 0$, as the first car just reaches the loading area, and end it just as each car leaves the loading area. That means we need several upper limits, one for each car.

For now we use t to represent an arbitrary final instant:

$$\Delta x = \frac{m_i v_{x,i}}{\lambda} \ln(m_i + \lambda t) \Big|_0^t = \frac{m_i v_{x,i}}{\lambda} \ln\left(\frac{m_i + \lambda t}{m_i}\right).$$

We can extract the time variable by isolating the logarithm and then taking an exponential on both sides:

$$\frac{\lambda \Delta x}{m_i v_{x,i}} = \ln\left(\frac{m_i + \lambda t}{m_i}\right)$$

$$e^{\lambda \Delta x / m_i v_{x,i}} = \left(\frac{m_i + \lambda t}{m_i}\right) = 1 + \frac{\lambda t}{m_i}$$

$$\Delta t = t - 0 = t = \frac{m_i}{\lambda}(e^{\lambda \Delta x / m_i v_{x,i}} - 1). \tag{2}$$

Now we can obtain the time interval Δt required to fill one car, two cars, or all the cars in the train by substituting the appropriate length ($\ell, 2\ell, \ldots, N\ell$) for the displacement Δx. Suppose each of the N cars in the train has an inertia m_{car} when empty. Then $Nm_{car} = m_i$. The first car is under the chute for a time interval Δt_1, found with $\Delta x = \ell$:

$$\Delta t_1 = \frac{Nm_{car}}{\lambda}(e^{\lambda \ell / Nm_{car} v_{x,i}} - 1).$$

The amount of grain loaded during an interval Δt is given by

$$\Delta m = \int dm = \int_0^t \lambda \, dt = \lambda \Delta t.$$

The amount loaded into the first car is thus

$$\Delta m_1 = Nm_{car}(e^{\lambda \ell / Nm_{car} v_{x,i}} - 1).$$

For each subsequent car, we obtain the time interval from $t = 0$ until the car leaves the loading area, then subtract the time interval used to fill the previous cars. Thus the second car is under the chute for a time interval $\Delta t_2 - \Delta t_1$, where Δt_2 is the time interval required for the train to travel a distance $\Delta x = 2\ell$:

$$\Delta t_2 = \frac{Nm_{car}}{\lambda}(e^{\lambda 2\ell / Nm_{car} v_{x,i}} - 1)$$

$$\Delta t_2 - \Delta t_1 = \frac{Nm_{car}}{\lambda}(e^{\lambda 2\ell / Nm_{car} v_{x,i}} - e^{\lambda \ell / Nm_{car} v_{x,i}}).$$

The amount of grain added to the second car is therefore

$$\Delta m_2 = \lambda(\Delta t_2 - \Delta t_1) = Nm_{car}(e^{\lambda 2\ell / Nm_{car} v_{x,i}} - e^{\lambda \ell / Nm_{car} v_{x,i}}).$$

For the Nth car, we have

$$\Delta t_N - \Delta t_{N-1} = \frac{Nm_{car}}{\lambda}(e^{\lambda N\ell / Nm_{car} v_{x,i}} - e^{\lambda(N-1)\ell / Nm_{car} v_{x,i}})$$

$$\Delta m_N = \lambda(\Delta t_N - \Delta t_{N-1})$$

$$= Nm_{car}(e^{\lambda N\ell / Nm_{car} v_{x,i}} - e^{\lambda(N-1)\ell / Nm_{car} v_{x,i}}).$$

The ratio of the amount of grain in the last car to the amount in the first car is therefore

$$\frac{\Delta m_N}{\Delta m_1} = \frac{(e^{\lambda N\ell / Nm_{car} v_{x,i}} - e^{\lambda(N-1)\ell / Nm_{car} v_{x,i}})}{e^{\lambda \ell / Nm_{car} v_{x,i}} - 1} = e^{\lambda(N-1)\ell / Nm_{car} v_{x,i}} \checkmark \tag{3}$$

This expression answers both of your boss's questions. It gives the ratio of the amounts of grain in the last and first cars and shows that the variables affecting the amount in each car are the number N of cars in the train, length ℓ of each car, loading rate λ, inertia m_{car} of an empty car, and the initial x component of the train's velocity $v_{x,i}$. $\checkmark$

❹ **EVALUATE RESULT** We assumed that the tracks are straight and horizontal, the cars are identical when empty, and the effects of friction are negligibly small. None of these assumptions is unreasonable in this context. We assumed that the grain falls vertically and that it enters the cars at a constant rate. Given that the chute is of a fixed size and shape, it is not unreasonable to assume a constant loading rate, at least over time intervals of a few minutes—more long than enough to get several cars past the chute. This assumption is reasonable but not mandatory. If the loading rate were a controlled variable, however, it could be adjusted to match the speed of the train and your boss would have no problem.

We see from Eq. 2 that the loading time interval increases as expected for cars farther and farther from car 1, as Δx increases. Note also that Δt goes to zero in Eq. 2 as $\Delta x \to 0$ (because it takes no time to move no distance).

The ratio of Eq. 3 grows as the number N of cars increases. This is to be expected because eventually the train slows to a crawl. Let's look at some data. Suppose $N = 4$ cars, $\ell = 15$ m, $m_{car} = 2.7 \times 10^4$ kg, $v_{x,i} = 1.0$ m/s, and $\lambda = 1.5 \times 10^3$ kg/s. Substituting these numbers into Eq. 3 yields a ratio of 1.9; that is, the last car has almost twice the grain of the first car! Even with 3 cars, the third car will contain 50% more grain than the first. It looks like the company should control the loading rate to match the train velocity. You may have a promotion coming when you suggest this and provide the time-distance relationship needed to adjust the loading rate.

Guided Problem 4.8 Rocket speed

Suppose a rocket that has a combined rocket + fuel inertia m_i starts from rest and then expels fuel at a rate dm/dt. The speed of the fuel as it exits the rocket at any instant t is the difference between the forward speed of the rocket at this instant and the constant nozzle speed of the ejected fuel v_{fuel} (that is, v_{fuel} is the speed with which the fuel is ejected when the rocket is at rest). Use conservation of momentum to show that, once enough fuel has been expelled to reduce the combined rocket + fuel inertia to m_f, the change in the rocket's speed $\Delta v_{rocket} = v_{rocket,f} - v_{rocket,i}$ is

$$v_{rocket,f} - v_{rocket,i} = v_{fuel} \ln \frac{m_i}{m_f}.$$

Ignore any gravity effects. (This classic rocketry formula was first worked out by Russian engineer K. Tsiolkovskii in 1897.)

❶ GETTING STARTED

1. As in Worked Problem 4.7, this is a process with continuous changes in inertia and velocity of parts of a system. How can the approach of Worked Problem 4.7 be modified to suit these circumstances? Consider analyzing a short time interval dt and then integrating.

2. Draw one sketch at arbitrary instant t showing the rocket moving at velocity v_{rocket}. The rocket has inertia $m + dm$, where dm represents a tiny amount of fuel still on board but about to be expelled. This is the initial instant. Then draw a second sketch showing the tiny amount of fuel and the rocket as separate objects at instant $t + dt$, a short time later.

3. The rocket moves at a slightly higher velocity $(v_{rocket} + dv_{rocket})$ in the second sketch. What is the velocity of the fuel element dm in the second sketch?

❷ DEVISE PLAN

4. Notice that there are no numbers here but that all of the symbols except v_{rocket} (and dv_{rocket}) can be considered known quantities.

5. Write the equation that compares the momentum at instant t with the momentum at instant $t + dt$.

6. Examine your equation for each situation—with the fuel of inertia dm on board and with it expelled. Can some terms be canceled?

7. Separate variables to get all v terms on the left side of the equation and all m terms on the right side, then integrate. When you integrate, what sign is associated with dm?

❸ EXECUTE PLAN

❹ EVALUATE RESULT

8. Does the expression you obtain behave as you might expect? For example, does the velocity increase with time?

9. Make sure any assumptions you made are not unreasonable. Were any assumptions needed about the mechanism of fuel ejection or about the type of fuel?

Questions and Problems

For instructor-assigned homework, go to MasteringPhysics® (MP)

Dots indicate difficulty level of problems: • = easy, •• = intermediate, ••• = hard; CR = context-rich problem.

4.1 Friction

1. Two identical hockey pucks slide over the same rough surface. The time interval needed for puck 2 to stop is twice that needed for puck 1 to stop. What explanation can there be for this? •

2. Figure P4.2 shows the velocity of a block of wood as a function of time. The block is sliding over a horizontal surface. Describe the physical processes that led to this graph. •

Figure P4.2 v_x (m/s)

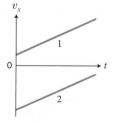

3. The velocity-versus-time graph in Figure P4.3 shows the motion of two different objects sliding across a horizontal surface. Could the change in the x component of velocity with time be attributed to friction in each case? ••

Figure P4.3 v_x

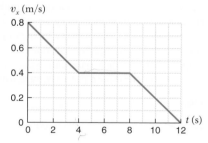

4. Consider the two velocity-versus-time graphs shown in Figure P4.4. Are the motions represented by these curves best described as similar or as different? Is the effect of friction on the motion plausibly more pronounced in one case? ••

Figure P4.4

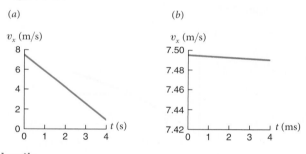

(a) v_x (m/s)

(b) v_x (m/s)

4.2 Inertia

5. Two objects collide on a low-friction track. Object 1 experiences a magnitude of change in velocity $|\Delta \vec{v}_1| = 3$ m/s, while object 2 has $|\Delta \vec{v}_2| = 1$ m/s. How do the inertias of these two objects compare? •

6. In a collision experiment, the ratio of the velocity change between two carts of equal inertia is found to equal 1. What happens to this ratio if the experiment is repeated in the following conditions? (*a*) The inertia of each cart is doubled, and the same initial velocities are used. (*b*) The original value of inertia is used, and the initial velocities of the carts are doubled. •

7. Two carts of equal inertia are moving in opposite directions toward each other. Figure P4.7 represents the positions of the carts until they collide. Sketch the positions of the carts after the collision. •

Figure P4.7 x (mm)

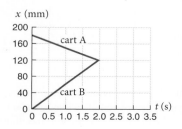

8. Cart 1 initially at rest is struck by cart 2, which has twice the inertia of cart 1. Figure P4.8 shows the velocity of cart 2 as a function of time. Complete the graph by adding the velocity of cart 1 during the same time. ••

Figure P4.8

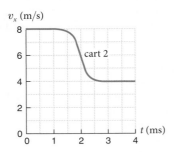

9. Figure P4.9 is the position-versus-time graph for a collision, along a straight line, between two identical amusement-park bumper cars A and B. The inertias of the passengers are different. (a) Which post-collision solid line is a continuation of the dotted line for car A? Which is a continuation of the dashed line for car B? (b) Which car contains the passenger having the greater inertia? ••

Figure P4.9

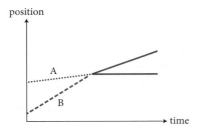

10. Figure P4.10 shows the $v_x(t)$ curves for two carts, A and B, that collide on a low-friction track. What is the ratio of their inertias? ••

Figure P4.10

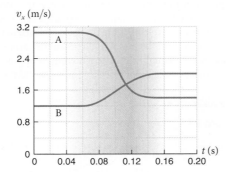

11. Figure P4.11 is a velocity-versus-time graph for two objects, A and B, before and after they collide. Object B is initially at rest. What factor relates the inertias of objects A and B? ••

Figure P4.11

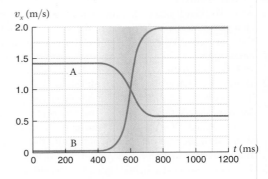

4.3 What determines inertia?

12. Which has greater inertia: a 1-quart milk carton filled with feathers or the same carton filled with buckshot? •
13. A jeweler pounds a small ingot of gold into a thin sheet. What happens to the inertia of the gold? •
14. Five different objects are formed from a constant volume V of different materials, and the objects are placed on various surfaces. Rank the following objects in order of increasing inertia: a cubic lead block on ice, a cubic plastic block on ice, a sphere of lead on ice, a pyramid of plastic on concrete, and a pyramid of lead on concrete. •
15. Which has greater inertia: a bottle full of water or the same bottle after the water has been drunk? Why is this so when the volume of the bottle does not change? ••
16. Does the inertia of a bicycle tire change when you add air to the tire? ••
17. The velocity-versus-time graphs in Figure P4.17 all depict the motion of carts that have the same size and shape. The different graphs show motion on different tracks: a smooth icy track, a dusty unpolished track, and a rough damaged track. In one case wooden carts are used, in one case plastic cars are used, and in one case one cart of each material is used. The carts may initially be moving in opposite directions, in the same direction, or one may be at rest. For each graph, label the track type, cart type, and initial motion. If you cannot determine a property, state why. ••

Figure P4.17

(a)

(b)

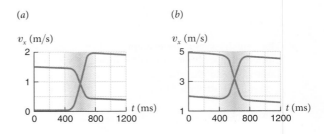

(c)

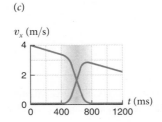

4.4 Systems

18. The label on a bag of cookies lists the number of cookies in the bag, the serving size, and the number of calories per serving. Is each of these quantities intensive or extensive? ●

19. You are riding a bus and thinking about the number of passengers on board. (*a*) Is the number of passengers an extensive or intensive quantity? (*b*) Draw a system diagram to help account for the number of passengers. Where is the system boundary? (*c*) Does the number of passengers remain constant? (*d*) Does the number of passengers remain constant as the bus follows its route while its doors stay closed? ●

20. A quart is a unit of volume. (*a*) If you have 1 quart of water in one container and 1 quart of water in another container and you pour all the water from both into a larger container, what volume of water do you end up with? (*b*) Suppose you have a 1-quart container filled with marbles and a large pail containing 1 quart of water. When you dump the marbles into the pail, is the final volume the same as in part *a*? Why or why not? (*c*) Compare the inertias in each case before and after mixing. ●●

21. Figure P4.21 shows a person on a truck throwing a ball to a friend on the ground. In how many ways can you divide these things into a system and an environment? ●●

Figure P4.21

4.5 Inertial standard

22. If two carts collide on a low-friction track, it is possible for both carts to slow down. Yet Eq. 4.1 $\left(\frac{m_u}{m_s} = -\frac{\Delta v_{sx}}{\Delta v_{ux}}\right)$ seems to say that, because inertia is always positive, if one velocity increases, the other must decrease. Explain the apparent discrepancy. ●

23. Cart A, of inertia 1.0 kg, is initially at rest on a low-friction track; cart B, of unknown inertia, has an initial velocity of $+3.0$ m/s in your coordinate system. After the two carts collide, the final velocities are $\vec{v}_A = +2.0$ m/s and $\vec{v}_B = -3.0$ m/s. What is the inertia of cart B? ●

24. A 2.0-kg cart collides with a 1.0-kg cart that is initially at rest on a low-friction track. After the collision, the 1.0-kg cart moves to the right at 0.40 m/s and the 2.0-kg cart moves to the right at 0.30 m/s. What was the initial velocity of the 2.0-kg cart? ●●

25. A 1-kg standard cart collides with a 5.0-kg cart initially at rest on a low-friction track. After the collision, the standard cart is at rest and the 5.0-kg cart has a velocity of 0.20 m/s to the left. What was the initial velocity of the standard cart? ●●

26. Figure P4.26 is the position-versus-time graph for a collision between two carts on a low-friction track. Cart 1 has an inertia of 1.0 kg; cart 2 has an inertia of 4.0 kg. (*a*) What are the initial and final velocities of each cart? (*b*) What is the change in the velocity of each cart? (*c*) Do the values you calculated in part *b* satisfy Eq. 4.1? (*d*) Draw a velocity-versus-time graph for this collision. (*e*) Does cart 1 have a nonzero

acceleration? If so, when, and what is the sign of the acceleration? (*f*) Does cart 2 have a nonzero acceleration? If so, when, and what is the sign of the acceleration? ●●

Figure P4.26

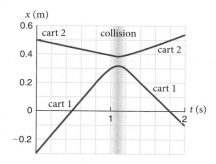

27. The bumper boats at your local theme park each have an inertia of 90 kg. In boat 1 are a man of unknown inertia, a 45-kg woman, and a 3.0-kg dog. In boat 2 are an 80-kg father, a 50-kg mother, and a 30-kg son. Boat 1 collides at 1.5 m/s with boat 2, which is initially at rest. Two seconds after the collision, boat 2 has moved 2.3 m and boat 1 has moved 0.26 m in the opposite direction. (*a*) Taking the initial direction of motion of boat 1 as positive, what is the velocity of each boat after the collision? (*b*) What is the change in velocity of each boat? (*c*) What is the man's inertia? (*d*) If the collision took 0.50 s, what was each boat's average acceleration during the collision? ●●●

28. A 1-kg standard cart collides with a cart A of unknown inertia. Both carts appear to be rolling with significant wheel friction because their velocities change with time as shown in Figure P4.28. (*a*) What are the carts' velocities at $t = 0$, $t = 5.0$ s, $t = 6.0$ s, and $t = 10$ s? (*b*) When not colliding, are the carts speeding up or slowing down? (*c*) Is the acceleration of each cart the same before and after the collision? (*d*) What is the inertia of cart A? ●●●

Figure P4.28

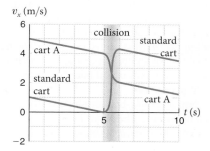

4.6 Momentum

29. Which has greater momentum: a 0.14-kg baseball pitched at 45 m/s or a 0.012-kg bullet fired at 480 m/s? ●

30. The student next to you says, "Momentum is inertia times velocity. That means momentum is proportional to inertia, and so things with more inertia have more momentum." What would you say to the student to clarify the matter? ●

31. Two identical cars traveling at 20 m/s both slow to a stop. The driver of car A applies the brakes hard, so that the car comes to a stop in 3.0 s, while the driver of car B brakes gently and comes to a stop in 7.0 s. Which car has the greater change in momentum? •

32. A bicyclist coasting along a road at speed v collides with a grasshopper flying in the opposite direction at the same speed. The grasshopper sticks to the bicyclist's helmet. Make a sketch showing (a) the two velocity vectors and the two momentum vectors before the collision, (b) the velocity and momentum vectors after the collision, and (c) the velocity change and the momentum change for the bicyclist and for the grasshopper. ••

33. Is it possible for the momentum of a system consisting of two carts on a low-friction track to be zero even if both carts are moving? ••

34. A 1.0-kg standard cart A collides with a 0.10-kg cart B. The x component of the velocity of cart A is $+0.60$ m/s before the collision and $+0.50$ m/s after the collision. Cart B is initially traveling toward cart A at 0.40 m/s, and after the collision the x component of its velocity is $+0.60$ m/s. What are (a) the x component of the change in the momentum of A, (b) the x component of the change in the momentum of B, and (c) the sum of these two x components of the changes in momentum? ••

35. You're rolling solid rubber balls on the kitchen floor. Ball 1 has a density of 1.00×10^3 kg/m^3 and a radius of 25.0 mm. Ball 2 has an unknown density and a radius of 40.0 mm and is initially at rest. You roll ball 1 at an initial speed of 3.00 m/s, and the two balls collide head-on. Ball 1 reverses direction and comes back to you at 2.00 m/s, and after the collision, the speed of ball 2 is 1.00 m/s. With the positive x axis in the direction of ball 1's initial motion, what are (a) the magnitudes of the initial and final momenta of ball 1, (b) the initial and final momenta of ball 2, and (c) the density of ball 2? ••

36. What is the magnitude of the momentum change of two gallons of water (inertia about 7.3 kg) as it comes to a stop in a bathtub into which it is poured from a height of 2.0 m? ••

37. From what height would a car have to fall in order for the magnitude of its momentum to equal the magnitude of its momentum when it is moving on a highway at 30 m/s? ••

38. (a) Write an expression relating the average acceleration, Δp, and Δt for an object of constant inertia m. (b) Given your result in part a, what can you say about the magnitude of the acceleration of an object coming to rest after being dropped on a soft bed versus a hard floor? (c) Discuss the rationale for airbags in cars. •••

39. You drop a 0.15-kg ball to the floor from a height of 2.0 m, and it bounces to a height of 1.6 m. What is the magnitude of the change in its momentum as a result of the bounce? •••

4.7 Isolated systems

40. When a rifle is fired, it *recoils*, kicking backward into the shooter's shoulder. Why? •

41. A 4.0-kg rifle fires a 10-g bullet at 800 m/s. With what speed does the rifle recoil (move backward toward the shooter's shoulder)? •

42. What is the magnitude of the momentum of the system that consists of all the molecules making up the air in your dorm room? ••

43. A car collides with a telephone pole. (a) Does the car alone form an isolated system? (b) The car and the pole? (c) The car, the pole, and Earth? ••

44. Start walking from a standstill. Can you consider your body an isolated system? ••

45. In a pairs skating competition, a 75-kg male skater moving at 4.0 m/s collides (gently) with his stationary, 50-kg female partner and raises her in a lift. Neither of them makes a horizontal push at the instant of the pickup. What is the speed of the pair after the collision? (Hint: The vertical motion of the female has no bearing on the question.) ••

46. A girl wearing ice skates stands in a skating rink and throws her backpack to a bench just off the ice. (a) If her skates are aligned parallel to the direction of the throw, what happens to her as a result? (b) When the backpack lands on the bench, it stops. Does the skater stop at the same instant? If not, where does the momentum of the skater-backpack system go? ••

47. A cart moving on a low-friction track has a momentum of 6 kg·m/s to the right. At the end of the track is a wall. (a) What is the momentum of the system that consists of the cart and the wall? (b) After the cart collides with the wall, the cart's momentum is 6 kg·m/s to the left. What is the wall's momentum after impact? (c) Is the system defined in part a an isolated system? Why doesn't the wall move? ••

4.8 Conservation of momentum

48. What does conservation of momentum say about the motion of a single cart that doesn't collide with anything as it moves on a low-friction track? •

49. The three carts described in the table collide simultaneously. What is the momentum of the system of three carts (a) before the collision and (b) after the collision? (c) Is the system of three carts isolated during the collision, according to these data? •

Cart	Momentum p_x (kg · m/s)	
	Before collision	After collision
1	$+6.0$	-4.0
2	-2.0	$+2.0$
3	-3.0	$+3.0$

50. Two carts collide on a low-friction track. Cart 1 has an initial momentum of $+10$ kg·m/s $\hat{\imath}$ and a final momentum of -2.0 kg·m/s $\hat{\imath}$. If cart 2 has a final momentum of -6.0 kg·m/s $\hat{\imath}$, what was its initial momentum? •

51. Two carts A and B collide on a low-friction track. Measurements show that their initial and final momenta are $\vec{p}_{A,i} = +10$ kg·m/s$\hat{\imath}$, $\vec{p}_{A,f} = +2.0$ kg·m/s $\hat{\imath}$, $\vec{p}_{B,i} = -4.0$ kg·m/s $\hat{\imath}$, and $\vec{p}_{B,f} = +4.0$ kg·m/s$\hat{\imath}$. (a) What is the change in the momentum of each cart during the collision? (b) What is the sum of these changes in momenta? (c) Is this collision consistent with conservation of momentum? Why or why not? ••

52. Two carts A and B collide on a low-friction track. Measurements show that their initial and final momenta are $\vec{p}_{A,i} = +3.0\ \text{kg} \cdot \text{m/s}\,\hat{\imath}$, $\vec{p}_{A,f} = +1.0\ \text{kg} \cdot \text{m/s}\,\hat{\imath}$, $\vec{p}_{B,i} = +2.0\ \text{kg} \cdot \text{m/s}\,\hat{\imath}$, and $\vec{p}_{B,f} = -6.0\ \text{kg} \cdot \text{m/s}\,\hat{\imath}$. (a) What is the change in the momentum of each cart during the collision? (b) What is the sum of these changes in momenta? (c) Is this collision consistent with the physics you know? Why or why not? ●●

53. A moving object collides with an object at rest. (a) Is it possible for both objects to be at rest after the collision? (b) Is it possible for just one object to be at rest after the collision? If so, which one? Ignore any effects due to friction. ●●

54. Estimate the magnitude of impulse that you impart to a nail (and to whatever the nail is buried in) when you hit it with a hammer. ●●

55. A 1200-kg automobile traveling at 15 m/s collides head-on with a 1600-kg automobile traveling at 10 m/s in the opposite direction. (a) Is it possible to predict the velocities of the cars after the collision? (b) Is it possible to predict the value that any pertinent physical quantity has immediately after the collision? ●●

56. A 2.0-kg cart and a 3.0-kg cart collide on a low-friction track. The 3.0-kg cart is initially moving at 1.0 m/s to the right, but after the collision it is moving at 5.0 m/s to the right. After the collision, the 2.0-kg cart is moving to the right at 3.0 m/s. (a) What was the 2.0-kg cart's initial velocity? (b) What would your answer be if, other quantities being unchanged, the 3.0-kg cart were initially moving to the left at 1.0 m/s? ●●

57. Warmly dressed in several layers of clothing, you are standing at the center of a frozen pond. There is not enough friction to permit you to walk to the edge of the pond. How can you save yourself? (Never mind how you got to the center of the pond.) ●●

58. Two male moose charge at each other with the same speed and meet on a icy patch of tundra. As they collide, their antlers lock together and they slide together with one-third of their original speed. What is the ratio of their inertias? ●●

59. An 80-kg physicist and a friend are ice-skating. The physicist, distracted, collides from behind at 7.0 m/s with his friend, who is skating at 5.0 m/s in the same direction. After the collision, the physicist continues in the same direction at 4.0 m/s, but his friend is now moving (still in the same direction) at 8.0 m/s. What is her inertia? ●●

60. You are driving your 1000-kg car at a velocity of $(20\ \text{m/s})\hat{\imath}$ when a 9.0-g bug splatters on your windshield. Before the collision, the bug was traveling at a velocity of $(-2.0\ \text{m/s})\hat{\imath}$. Before the collision, what were (a) your car's momentum and (b) the bug's momentum? (c) What is the change in velocity of the car due to its encounter with the bug? ●●

61. Sledding down a hill, you are traveling at 10 m/s when you reach the bottom. You (inertia 70 kg) then move across horizontal snow toward a 200-kg boulder but jump off the sled (inertia 5.0 kg) the instant before it hits the boulder. The boulder is sitting on *very* slick ice and moves freely when the sled hits it. The sled bounces back, moving at 6.0 m/s. (a) At what speed does the boulder move after the sled hits it? (b) If you stayed on the sled, what would your momentum be just before you hit the boulder? (c) Suppose you and the sled hit the boulder and continue forward after the collision at 2.0 m/s. Would the boulder's after-collision speed be higher or lower than the speed you calculated in part a? ●●

62. A fire hose sprays water against a burning building. The stream of water has cross-sectional area A and density ρ and moves with speed v toward the building. Assume the water splatters against the building without reflecting. What magnitude of impulse does the stream of water deliver during the time interval Δt? ●●

63. Playing pool, you send the cue ball head-on (that is, along the line joining the centers of the two balls) into the stationary 8 ball, and the cue ball stops as a result of the collision. (a) Describe the momentum of each ball before and after the collision. (b) Show that your answer to part a is consistent with Eq. 4.18, $\Delta \vec{p} = \vec{J}$. (Note: Pool balls have identical inertias.) ●●

64. Three identical carts on a low-friction track have putty on their ends so that they stick together when they collide. In Figure P4.64a, two carts already stuck together and moving with speed v are about to collide with the third cart initially at rest. In Figure P4.64b, the single cart moving with the same speed v is about to collide with the two carts already stuck together and initially at rest. In which case is the final velocity of the trio of carts greater? ●●

Figure P4.64

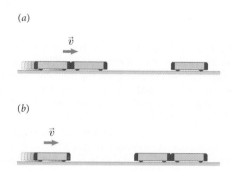

65. A load of coal is dropped from a bunker into a railroad hopper car of inertia 3.0×10^4 kg coasting at 0.50 m/s on a level track. The car's speed is 0.30 m/s after the coal falls. What is the inertia of the load of coal? ●●

66. Three identical carts on a low-friction track have putty on their ends so that they stick together when they collide. In Figure P4.66a, two carts already stuck together and moving at speed v collide with the single cart initially at rest. In Figure P4.66b, there are two collisions: First the center cart, moving at speed v, collides with the cart on the right (and they stick together). Then the left cart, moving at speed v, collides with those two carts. In which case is the final speed of the three carts greater? ●●

Figure P4.66

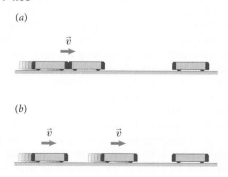

67. Figure P4.67 is a momentum-versus-time graph for a collision between two carts. Carts 1 and 2 have inertias of 1.0 kg and 3.0 kg, respectively. What are the x components of the initial and final velocities of (a) cart 1 and (b) cart 2? (c) What are the momentum changes $\Delta \vec{p}_1$ and $\Delta \vec{p}_2$? (d) Do the values you calculated in part c satisfy Eq. 4.18, $\Delta \vec{p} = \vec{J}$? (e) Draw the velocity-versus-time graph for this collision. (Hint: You can do this by making some simple changes to Figure P4.67.) •••

Figure P4.67

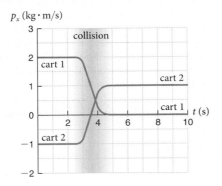

68. Carts A and B collide on a low-friction track. Rank, from largest to smallest, the following four collisions according to the magnitude of the change in the momentum of cart B, which has twice the inertia of cart A. (a) A initially moving right at 1.0 m/s, B initially stationary; stick together on impact. (b) A initially stationary, B initially moving left at 1.0 m/s; stick together on impact. (c) A initially moving right at 1.0 m/s; after impact, A moving left at 0.33 m/s, B moving right at 0.67 m/s. (d) A initially moving right at 1.0 m/s, B initially moving left at 1.0 m/s; stick together on impact. •••

69. A red cart and a green cart, both traveling at speed v on a low-friction track, approach each other head-on. The red cart is moving to the right; the green cart, whose inertia is three times that of the red cart, is moving to the left. A black cart that has a spring on its left end and putty on its right end sits on the track between the two moving carts. The moving carts hit the black cart at the same instant. After the collision, the red cart moves left at speed v, and the green and black carts move together at speed $v/5$. What is the inertia of the black cart, expressed as a multiple of the inertia of the red cart? •••

70. A 1.0-kg standard cart collides on a low-friction track with cart A. The standard cart has an initial x component of velocity of $+0.40$ m/s, and cart A is initially at rest. After the collision the x component of velocity of the standard cart is $+0.20$ m/s and the x component of velocity of cart A is $+0.60$ m/s. After the collision, cart A continues to the end of the track and rebounds with its speed unchanged. Before the carts collide again, you drop a lump of putty onto cart A, where it sticks. After the second collision, the x component of velocity of the standard cart is -0.20 m/s and the x component of velocity of cart A is $+0.40$ m/s. What is the inertia of the putty? •••

71. A 400-kg shipboard cannon fires a 20-kg ball at 60 m/s. The cannon's resulting recoil speed across the deck is regarded as excessive. How much inertia (in the form of sandbags, say) must be added to the cannon if its recoil speed must be reduced to 2.0 m/s? •••

72. You point a tennis-ball serving machine straight up and put a piece of plastic wrap across the top of the barrel, placing a 10-g marble on top of the barrel. When a 58-g tennis ball in the machine is fired, it sends the marble into the air. The marble rises 200 m above the barrel mouth, and the tennis ball rises 54 m. Neglecting the plastic wrap, what is the ball's launch speed? •••

Additional Problems

73. A golf ball has one-tenth the inertia and five times the speed of a baseball. What is the ratio of the magnitudes of their momenta? •

74. A 30,000-kg trailer truck approaches a one-lane bridge at 2.2 m/s; a 2400-kg minivan approaches the bridge from the other direction at 30 m/s. Which vehicle has a momentum of greater magnitude? •

75. You and a friend are bowling. She rolls her 4.5-kg ball at 10 m/s; you roll your ball at 8.0 m/s. What inertia must your ball have if its momentum is to be the same as hers? •

76. Two identical carts collide in three experiments on a low-friction track. In each case, cart A initially has velocity $\vec{v}$ and cart B is at rest. Sketch the velocity-versus-time graph for each cart (a) if the carts stick together after the collision, (b) if cart A is stationary after the collision, and (c) if cart A has velocity $-\vec{v}/10$ after the collision. ••

77. Draw diagrams that show the initial and final velocity vectors and the initial and final momentum vectors when a rapidly moving golf ball hits (a) a golf ball at rest and (b) a basketball at rest. In each case, assume that the golf ball moves along the line connecting the centers of the two balls. ••

78. The position of a certain airplane during takeoff is given by $x = \frac{1}{2}bt^2$, where $b = 2.0$ m/s^2 and $t = 0$ corresponds to the instant at which the airplane's brakes are released at position $x = 0$. The empty but fueled airplane has an inertia of 35,000 kg, and the 150 people on board have an average inertia of 65 kg. What is the magnitude of the airplane's momentum 15 s after the brakes are released? ••

79. A bullet of inertia m_{bullet} is fired horizontally at a speed $v_{bullet,i}$ into a stationary wooden block of inertia m_{block} lying on a low-friction surface. The bullet passes through the block and emerges with a speed $v_{bullet,f}$. Determine the final speed of the block in terms of the given quantities. ••

80. The speed of a bullet can be measured by firing it at a wooden cart initially at rest and measuring the speed of the cart with the bullet embedded in it. Figure P4.80 shows a 12-g bullet fired at a 4.0-kg cart. After the collision, the cart rolls at 1.8 m/s. What is the bullet's speed before it strikes the cart? ••

Figure P4.80

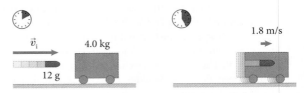

81. The World War II–era rocket launcher called the bazooka was essentially a tube open at both ends. On the basis of momentum considerations, how is the firing of a bazooka different from the firing of a cannon, which is a tube open at one end only? ••

82. In a football game, a 95-kg player carrying the ball can run the 50-m dash in 5.5 s. Two opponents are available to stop him with a head-on tackle. One opponent has an inertia of 110 kg and a 50-m time of 6.6 s; the other has an inertia of 90 kg and a 50-m time of 5.1 s. Which has the better chance of stopping the 95-kg player, and why? ●●

83. In the days before rocketry, some people argued that rocket engines would not work in space because there is no atmosphere for the exhaust to push against. Even today, some people think that a rocket requires a launch pad to push against in order to lift off. Refute such arguments with an argument based on momentum considerations. ●●

84. In some collisions, the velocity of one participant changes little while that of the other changes a lot, as Figure P4.84 illustrates. (a) In which direction (positive or negative) are the objects moving before the collision? (b) After the collision? (c) What is the ratio of the inertia of the larger object to the inertia of the smaller object? (d) Does friction play an important role in this collision? ●●

Figure P4.84

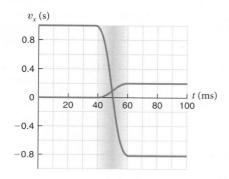

85. In a head-on collision between two cars of different inertias, you might prefer to be the driver of the car that has the greater inertia, at least if you considered momentum only. Why? ●●

86. A single-stage rocket is in deep space coasting at $v_{rocket,i} = 2.0 \times 10^3$ m/s. It fires its engine, which has an exhaust speed of $v_{exhaust} = 1.0 \times 10^3$ m/s. What is the rocket's speed $v_{rocket,f}$ after it has lost one-third of its inertia by exhausting burned fuel? Assume that the fuel is ejected all at once. ●●

87. In the process of moving out of your house, you are dropping stuff out a second-floor window to a friend 4.0 m below. You are about to drop a 6.0-kg stereo speaker when you begin to worry that catching anything that has a momentum greater than 50 kg·m/s might be harmful. ●●● CR

88. You are in mission control in Florida, supervising a probe on its way to Mars. The most recent report indicates that the probe will soon need to fire a quick burst of its booster rocket to increase its speed by 5.2 m/s. You know that the spent fuel is ejected at a speed $v_{fuel} = 800$ m/s $- v_i$, where v_i is the initial speed of the probe before firing, and that 10% of the probe's inertia is currently fuel. Your main worry is whether enough fuel is left for landing maneuvers. ●●● CR

89. You have just launched your new hot-air balloon and are hovering 30.0 m above the ground. As you are checking equipment, the burner fails. You know that if you do nothing, the balloon will descend with a constant acceleration of 1.50 m/s², and that the balloon's basket will be seriously damaged if it hits the ground with a momentum of 2850 kg·m/s or more. The combined inertia of the balloon, basket, and all equipment and fuel is 195 kg, not counting yourself (85.0 kg) and ten 5.00-kg sandbags carried as ballast. You make some quick calculations and take action to save the day. ●●● CR

90. A single gas molecule of inertia m is trapped in a box and travels back and forth with constant speed v between opposite walls A and B a distance ℓ apart. At each collision with a wall, the molecule reverses direction without changing speed. Write algebraic expressions for (a) the magnitude of the change in momentum of the molecule as it collides with wall B, (b) the amount of time that elapses between collisions with wall B; (c) the number of collisions per second the molecule makes with wall B, and (d) the change in momentum undergone by wall B, per second, as a result of these collisions. ●●●

91. Consider a two-stage rocket made up of two engine stages, each of inertia m when empty, and a payload of inertia m. Stages 1 and 2 each contain fuel of inertia m, so that the rocket's inertia before any fuel is spent is $5m$. Each stage exhausts fuel at a speed $v_{fuel} = v_{ex} - v_i$, where v_i is the speed of the rocket at the instant the fuel is spent. The rocket is initially at rest in deep space. Stage 1 fires, ejects its fuel all at once, and then detaches from the remainder of the rocket. The same process is repeated with stage 2. (a) What is the final speed of the payload? (Hint: Determine the speed of the rocket after stage 1 fires. Then, because stage 1 is detached from the rocket, *redefine the system* to include only the payload and stage 2, and determine the speed of the rocket after stage 2 fires.) (b) Now consider another rocket of inertia $5m$ but with only a single engine stage, of inertia $2m$, carrying fuel of inertia $2m$. What is the final speed of the payload in this case? (c) Which design yields a higher payload speed: two stage or single stage? Why? ●●●

Answers to Review Questions

1. No one has yet developed a completely frictionless surface, but there are many ways to minimize friction: smooth or polish the surfaces of contact; apply a slippery substance, such as grease or oil; place the object on rollers or mount it on wheels with very good bearings; and float the object on a cushion of air, as is done with low-friction tracks.

2. The changes in velocity are equal in magnitude but of opposite sign.

3. The magnitude of the velocity change of A is half that of B because the ratio of the magnitudes of velocity changes is the inverse of the ratio of the inertias. The two velocity changes are in opposite directions.

4. The inertias are identical because shape has no effect on an object's inertia. The inertia of an object is determined *entirely* by the type of material of which the object is made and by the amount of that material contained in the object.

5. Inertia depends on the material, so they are not identical. Experience suggests that the iron object has greater inertia.

6. A system is an object or group of objects that we can separate in our minds from the surrounding environment.

7. An extensive quantity depends on the extent (size) of the system that you choose; an intensive quantity does not.

8. Input, output, creation, and destruction can change the value.

9. It means that the extensive quantity cannot be created or destroyed. Therefore only two processes can change the value of the quantity in a system: input and output.

10. Fasten the baseball to one standard cart and arrange a collision between this cart and the other standard cart. Measure the change in velocity of each object; then use the fact that the ratio of these changes in velocity is the inverse of the ratio of the inertias of the two objects. Subtract 1 kg from the inertia of the cart with the baseball to obtain the baseball's inertia.

11. No. Inertia is always positive. If one of the objects involved in a collision had negative inertia, *both* objects would have an increase in velocity. Such collisions have never been observed.

12. Their inertias are equal.

13. The bumblebee has greater momentum. The train has zero velocity, and so it has zero momentum. The train has much greater inertia than the bumblebee, however.

14. Even though the inertia of the baseball is greater than the inertia of the bullet, their momenta (and the results of the collisions) can be comparable if the speed of the bullet is proportionately larger than the speed of the baseball.

15. No. Only scalars can be negative, and momentum is a vector. However, the x component of momentum can be negative if the momentum points in the negative x direction.

16. For purposes of this chapter, an interaction is two objects acting on each other such that at least one of them is accelerated.

17. Internal interactions are between two objects in a system, whereas external interactions are between an object inside the system and an object outside the system. The distinction is important because external interactions can change the momentum of a system, while internal interactions cannot.

18. An isolated system is one for which there are no external interactions. Because the momentum of an isolated system does not change with time, we can use information about the momentum at one instant to determine the momentum at a later instant.

19. The statements are not equivalent, but *a* depends on *b*. Statement *b* means that momentum cannot be created or destroyed. Statement *a* applies to only an isolated system, and it applies *because b* is true. By definition, there are no transfers of momentum across the boundary of an isolated system, and so none of the four mechanisms that might change the system's momentum—input, output, creation, and destruction—are available.

20. Conservation of momentum wouldn't be affected in the least. The 1-kg standard is a convention only.

21. *Conservation of momentum* is a verbal statement of the fact that momentum can be neither created nor destroyed. The *momentum law* is a mathematical statement given by Eq. 4.18, $\Delta \vec{p} = \vec{J}$, saying that any change in momentum is due to transfers of momentum. Because it contains no terms for the creation or destruction of momentum, the momentum law embodies the conservation of momentum.

22. Impulse, $\vec{J}$, represents the transfer of momentum between the environment and a system. It is a vector with the same units as momentum, kg · m/s, and, in the nomenclature of Section 4.4, it accounts for the "input" and "output" of the system. Impulse is related to the change in momentum of a system by Eq. 4.18: $\vec{J} = \Delta \vec{p}$.

Answers to Guided Problems

Guided Problem 4.2 8.2×10^{-2} m/s

Guided Problem 4.4 *c*, ejecting both at once to attain the highest speed for the shuttle:

$$v_{\text{f}} = \frac{(m_1 + m_2)}{m_{\text{shuttle}}} v$$

Guided Problem 4.6 After the first impact:

$$v_{\text{target}} = \frac{m_{\text{bullet}}}{(m_{\text{target}} + m_{\text{bullet}})} v_{\text{muzzle}};$$

after the second impact: $v_{\text{target}} = \dfrac{2m_{\text{bullet}}}{(m_{\text{target}} + 2m_{\text{bullet}})} v_{\text{muzzle}}$

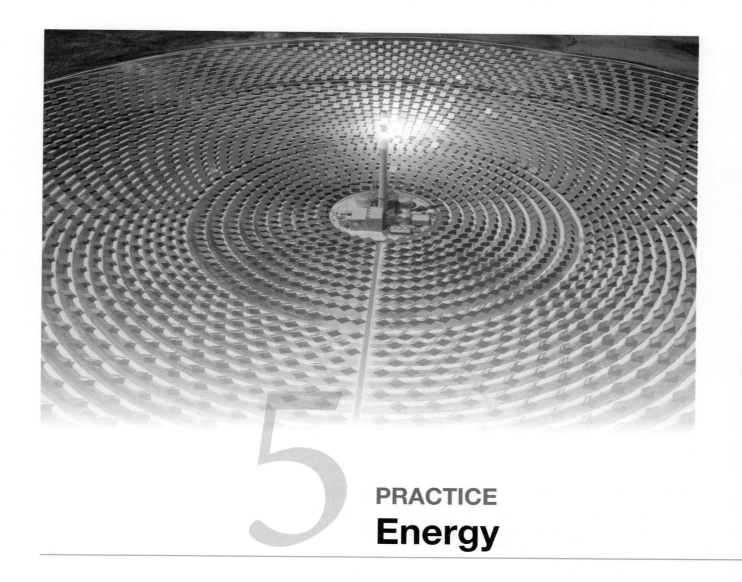

5

PRACTICE
Energy

Chapter Summary 70
Review Questions 71
Developing a Feel 72
Worked and Guided Problems 73
Questions and Problems 79
Answers to Review Questions 85
Answers to Guided Problems 85

Chapter Summary

Kinetic energy (Sections 5.2, 5.5)

Concepts The **kinetic energy** of an object is the energy associated with its motion. Kinetic energy is a positive scalar quantity and is independent of the direction of motion.

Quantitative Tools The **kinetic energy** K of an object of inertia m moving at speed v is

$$K = \tfrac{1}{2}mv^2. \qquad (5.12)$$

The SI unit of kinetic energy is the **joule** (J):

$$1\,\text{J} = 1\,\text{kg} \cdot \text{m}^2/\text{s}^2.$$

Relative velocity, states, and internal energy (Sections 5.1, 5.3, 5.5)

Concepts In a collision between two objects, the velocity of one object relative to the velocity of the other object is the **relative velocity** $\vec{v}_{12}$. The magnitude of the relative velocity is the **relative speed** v_{12}.

The **state** of an object is its condition as specified by some complete set of physical parameters. Energy associated with the object's state but not with its motion is called the **internal energy** of the object.

We can consider a system of two colliding objects to be isolated during the collision. Therefore the momentum of the system remains constant during all the collisions we study.

Quantitative Tools The **relative velocity** $\vec{v}_{12}$ of object 2 relative to object 1 is

$$\vec{v}_{12} \equiv \vec{v}_2 - \vec{v}_1. \qquad (5.1)$$

The **relative speed** v_{12} of object 2 relative to object 1 is the magnitude of $\vec{v}_{12}$:

$$v_{12} = |\vec{v}_2 - \vec{v}_1|.$$

Because momentum is a conserved quantity, the momentum of a system remains constant during a collision:

$$p_{x,\text{i}} = p_{x,\text{f}}.$$

Types of collisions (Sections 5.5, 5.6, 5.8)

Concepts The **coefficient of restitution** e for a collision is a positive, unitless quantity that tells how much of the initial relative speed is restored after the collision.

For an **elastic collision,** the relative speed is the same before and after the collision, and the coefficient of restitution is equal to 1. The collision is **reversible,** and the kinetic energy of the system made up of the colliding objects is constant.

For an **inelastic collision,** the relative speed after the collision is less than it was before the collision. The coefficient of restitution is between 0 and 1, and the collision is **irreversible.** The kinetic energy of the objects changes during the collision, but the energy of the system does not change. If the objects stick together, the final relative speed is zero; the collision is **totally inelastic,** and the coefficient of restitution is 0.

For an **explosive separation,** kinetic energy is gained during the collision and the coefficient of restitution is greater than 1.

Quantitative Tools The **coefficient of restitution** e is

$$e = \frac{v_{12\text{f}}}{v_{12\text{i}}} = -\frac{v_{2x,\text{f}} - v_{1x,\text{f}}}{v_{2x,\text{i}} - v_{1x,\text{i}}}. \qquad (5.18, 5.19)$$

For an **elastic collision,**

$$v_{12\text{i}} = v_{12\text{f}} \qquad (5.3)$$

$$K_\text{i} = K_\text{f} \qquad (5.14)$$

$$e = 1.$$

For an **inelastic collision,**

$$v_{12\text{f}} < v_{12\text{i}}$$

$$K_\text{f} < K_\text{i}$$

$$0 < e < 1.$$

For a **totally inelastic collision,**

$$v_{12\text{f}} = 0 \qquad (5.16)$$

$$e = 0.$$

For an **explosive separation,**

$$v_{12\text{f}} > v_{12\text{i}}$$

$$K_\text{f} > K_\text{i}$$

$$e > 1.$$

Conservation of energy (Sections 5.3, 5.4, 5.7)

Concepts The energy of any system is the sum of the kinetic energies and internal energies of all the objects that make up the system.

The law of **conservation of energy** states that energy can be transferred from one object to another or converted from one form to another, but it cannot be destroyed or created.

A **closed system** is one in which no energy is transferred in or out. The energy of such a system remains constant.

Quantitative Tools The energy of a system is

$$E = K + E_{int}. \tag{5.21}$$

The law of conservation of energy requires the energy of a **closed system** to be constant:

$$E_i = E_f. \tag{5.22}$$

Review Questions

Answers to these questions can be found at the end of this chapter.

5.1 Classification of collisions

1. What is relative velocity, and how does it differ from relative speed?
2. Explain the order and meaning of the subscripts in the relative velocity symbol $\vec{v}_{12}$.
3. What is the main differentiating characteristic of (*a*) elastic collisions, (*b*) inelastic collisions, and (*c*) totally inelastic collisions?

5.2 Kinetic energy

4. What is kinetic energy?
5. What is the sign of the kinetic energy of an object that is moving in the negative *x* direction?
6. Is it possible for an object's kinetic energy to be negative or zero?

5.3 Internal energy

7. What is meant by the *state of an object*?
8. What is the difference between a reversible process and an irreversible process?
9. In an inelastic collision, what is the relationship between the system's kinetic energy and internal energy?
10. How do you calculate the change in internal energy in an inelastic collision?

5.4 Closed systems

11. What is a closed system? Is it the same as an isolated system?
12. What is energy conversion, and how is it different from energy transfer?

5.5 Elastic collisions

13. How does the proof that kinetic energy remains constant in an elastic collision (Eqs. 5.4 through 5.13) depend on the elastic nature of the collision?

14. Explain why the relative velocity of two objects changes sign in an elastic collision.

5.6 Inelastic collisions

15. What is the definition of coefficient of restitution?
16. What is the numerical value of the coefficient of restitution for an elastic collision and for a totally inelastic collision?
17. Explain why Eq. 5.19 has a minus sign in front of the fraction.
18. In an elastic collision between a pair of objects where you know the inertias and initial velocities, you can use the fact that the momentum and kinetic energy of the system remain constant to determine the motion of both objects afterward—two equations, two unknowns. In an inelastic collision, however, the kinetic energy does not remain constant. Can you predict the final velocities of both objects in this case, given the inertias and initial velocities?

5.7 Conservation of energy

19. Are you creating energy when you throw a baseball?
20. Can there be a change in the physical state of a system if there is no change in its kinetic energy?
21. Without direct knowledge of how to compute internal energy values, how can we determine the change in the internal energy of a closed system?

5.8 Explosive separations

22. Where does the kinetic energy increase come from in an explosive separation?
23. Is it possible for an explosive separation to be elastic?

Developing a Feel

Make an order-of-magnitude estimate of each of the following quantities. Letters in parentheses refer to hints below. Use them as needed to guide your thinking.

1. The relative velocity of an airliner cruising east seen from a train speeding west (K, R)
2. The relative speed of the Moon in orbit as seen from Earth (E, W)
3. The kinetic energy of a tennis ball as it crosses the net after service (P, I)
4. The kinetic energy of a small car driving along the freeway (G, L)
5. The kinetic energy of an airliner moving at cruising speed (S, K)
6. The speed of the center of a tennis racquet just before the ball is served (P, H, C, X)
7. The change in the kinetic energy of a golf ball hit for a long drive from the tee (A, F, O, T)
8. The chemical energy released during the explosive separation of a professional fireworks shell (N, D, Z)
9. The kinetic energy converted to internal energy when a switching engine couples with a rail car initially at rest (V, J, Y)
10. The amount of useful energy your car can obtain from 1 gallon of gasoline at freeway speed (B, G, M, U, Q)

Hints

A. What distance does the golf ball travel in the air?
B. While cruising on a freeway, how much speed does a car lose in 5 s if it is shifted into neutral?
C. What is the coefficient of restitution for the ball-racquet collision?
D. What is the maximum radius of the exploding sphere?
E. What time interval is needed for one Moon orbit?
F. For how long a time interval is the golf ball in the air?
G. What is the inertia of a typical small car?
H. Considering the inertia of the ball relative to the inertia of the arm plus racquet, how much does the speed of the racquet change during the collision?
I. What is the inertia of a tennis ball?
J. What is the rail car's inertia?
K. What is the speed of a cruising airliner?
L. What is the speed of the car on the freeway?
M. How much kinetic energy is lost by a coasting car in 5 s?
N. What is the inertia of the fireworks payload?
O. What is the speed of a golf ball leaving the tee?
P. What is the speed of a tennis ball after service?
Q. What is the fuel consumption of a typical car on the freeway?
R. What is the speed of a speeding train?
S. What is the inertia of an airliner?

T. What is the inertia of a golf ball?
U. How many kilometers does a car travel in 5 s at freeway speed?
V. What is the switching engine's inertia?
W. What is the radius of the Moon's orbit?
X How does the velocity of the ball relative to that of the racquet compare before and after the collision?
Y. What is the engine's speed before coupling?
Z. How long a time interval is required for the sphere to expand?

Key (all values approximate)

A. 1×10^2 m; B. 4 m/s; C. $e \approx 1$; D. 3×10^1 m; E. 1 month, or 3×10^1 days; F. 3 s; G. 1×10^3 kg; H. very little, because the inertia ratio is large; I. 6×10^{-2} kg; J. 2×10^4 kg; K. 2×10^2 m/s; L. 3×10^1 m/s; M. 1×10^5 J; N. 2 kg; O. 4×10^1 m/s; P. 4×10^1 m/s; Q. 0.1 L/km; R. 3×10^1 m/s in the United States; much faster in some other countries; S. 1×10^5 kg; T. 5×10^{-2} kg; U. 0.2 km; V. 8×10^4 kg; W. 4×10^8 m; X. relative speed remains roughly the same, but the direction of the velocity reverses; Y. 0.4 m/s; Z. 1 s

Worked and Guided Problems

Procedure: Choosing a closed system

When we analyze energy changes, it is convenient to choose a system for which no energy is transferred to or from the system (a closed system). To do so, follow this procedure:

1. Make a sketch showing the initial and final conditions of the objects under consideration.
2. Identify all the changes in state or motion that occur during the time interval of interest.
3. Choose a system that includes all the objects undergoing these changes in state or motion. Draw a dashed

line around the objects in your chosen system to represent the system boundary. Write "closed" near the system boundary to remind yourself that no energy is transferred to or from the system.

4. Verify that nothing in the surroundings of the system undergoes a change in motion or state that is related to what happens inside the system.

Once you have selected a closed system, you know that its energy remains constant.

These examples involve material from this chapter but are not associated with any particular section. Some examples are worked out in detail; others you should work out by following the guidelines provided.

Worked Problem 5.1 Lofty arrow

You shoot a 0.12-kg arrow vertically upward at 40 m/s. Calculate the arrow's kinetic energy (*a*) at the start of its flight, immediately after it leaves the bow, and (*b*) when it reaches half of its maximum height. (*c*) Use energy arguments to estimate the position of the arrow when it has half its original speed.

❶ GETTING STARTED This is a kinematics problem with an energy twist. We can use the kinematics we learned in Chapter 3 first to calculate the velocity (and thus the speed) of the arrow at any position for parts *a* and *b* and then to calculate the position at any speed for part *c*. In addition, because we now know the connection between speed and kinetic energy, we can connect position and kinetic energy. The arrow in motion is not a closed system because of the gravitational interaction, but the arrow + Earth system is closed. A system diagram would include Earth and the arrow but would not provide much additional information. Instead, taking the kinematics approach, we draw a motion diagram (Figure WG5.1).

Figure WG5.1

❷ DEVISE PLAN The definition of kinetic energy should be sufficient to tackle part *a* because both the inertia and the initial velocity are known. Parts *b* and *c* are more involved. In free fall, acceleration is downward and constant. With our choice in Figure WG5.1

to have the positive *x* axis pointing upward, we have $a_x = -g$. The relationship between the *x* components of velocity and position that does not explicitly involve time is Eq. 3.13:

$$v_{x,f}^2 = v_{x,i}^2 + 2a_x(x_f - x_i).$$

This equation may be applied between any two points in the motion, and so we can use it to determine either the arrow's height at a specified speed or its speed at a specified height. This plus the definition of kinetic energy should take care of parts *b* and *c*.

❸ EXECUTE PLAN

(*a*) $\qquad K_{start} = \frac{1}{2}mv_{start}^2$

$\qquad\qquad = \frac{1}{2}(0.12 \text{ kg})(40 \text{ m/s})^2 = 96 \text{ J.} ✔$

(*b*) Before we can determine the value of *K* at the instant the arrow reaches half its maximum height, we must obtain the distance to the top of the flight. We know the initial position (0) and the *x* components of the initial velocity (+40 m/s), the final velocity (0; why?), and the acceleration (−9.8 m/s²; why minus?), and so we can write

$$v_{x,\text{top}}^2 = v_{x,\text{start}}^2 + 2a_x(x_{\text{top}} - x_{\text{start}})$$

$$x_{\text{top}} = x_{\text{start}} + \frac{v_{x,\text{top}}^2 - v_{x,\text{start}}^2}{2a_x} = 0 + \frac{(0)^2 - (40 \text{ m/s})^2}{2(-9.8 \text{ m/s}^2)}$$

$$= 82 \text{ m}.$$

We now obtain the squared velocity at half this maximum height, $x_{\text{half}} = x_{\text{top}}/2$:

$$v_{x,\text{half}}^2 = v_{x,\text{start}}^2 + 2a_x(x_{\text{half}} - x_{\text{start}})$$

$$= (+40 \text{ m/s})^2 + 2(-9.8 \text{m/s}^2)(41 \text{ m} - 0) = 796 \text{ m}^2/\text{s}^2.$$

PRACTICE

The arrow's speed at half its maximum height is therefore

$$v_{\text{half}} = |\sqrt{v_{x,\text{half}}^2}| = |\sqrt{796 \text{ m}^2/\text{s}^2}| = 28 \text{ m/s}.$$

(Notice that this speed is *not* half the initial speed.) The kinetic energy at this point is thus

$$K_{\text{half}} = \tfrac{1}{2}mv_{\text{half}}^2 = \tfrac{1}{2}(0.12 \text{ kg})(796 \text{ m}^2/\text{s}^2) = 48 \text{ J}. ✔$$

This is half the initial kinetic energy. Having gained half its maximum height, half the kinetic energy has been converted to internal energy of the arrow + Earth system.

(*c*) We know from part *b* that at half its maximum height the arrow's speed is 28 m/s. This means that the position where the speed is 20 m/s, half the initial value, must be above the halfway point. Let's make a guess. Because the velocity in $K = \tfrac{1}{2}mv^2$ is squared, reducing *v* to half its initial value reduces *K* to a quarter of its initial value. We know that $v = 0$ at the top of the arrow's trajectory (see Section 3.3). From part *b* we know that when the arrow is at half its maximum height, half of its kinetic energy has been converted to internal energy. Perhaps at three-quarters of the distance to the top of the flight, three-quarters of the kinetic energy has been converted to internal energy, as shown in the energy bars of Figure WG5.2. Thus, we estimate that the height at which the velocity is +20 m/s, half the original velocity, is about

$$(0.75)(82 \text{ m}) = 61 \text{ m} ✔$$

above the starting position.

Figure WG5.2

❹ **EVALUATE RESULT** Comparing the answers to parts *a* and *b* shows that the arrow's speed decreases as the arrow rises, as expected. A maximum distance of 82 m is a reasonable height for an arrow, consistent with the 40-m/s initial speed. We assumed that the arrow was in free fall so that we could use the free-fall value for acceleration, and this is a reasonable assumption for an arrow on its upward flight. The rest of the solution follows from kinematics and the definition of kinetic energy.

In part *c*, we were told to use energy arguments to determine a position. If we had not been restricted that way, we could have used our kinematics equation to obtain the position at which the arrow was moving at +20 m/s, and so let's do that now as a check:

$$v_{x,(c)}^2 = v_{x,\text{start}}^2 + 2a(x_{(c)} - x_{\text{start}})$$

$$x_{(c)} = x_{\text{start}} + \frac{v_{x,(c)}^2 - v_{x,\text{start}}^2}{2a} = 0 + \frac{(20 \text{ m/s})^2 - (40 \text{ m/s})^2}{2(-9.8 \text{ m/s}^2)}$$

$$= 61 \text{ m},$$

in nice agreement with our energy-based guess.

Guided Problem 5.2 Throwing a punch

A boxer delivers a hard blow to the chin of his opponent. The inertia of the boxer's hand (with glove) and forearm is 3.0 kg, and the inertia of the opponent's head is 6.5 kg. You learned in neurobiology class about 10 J of extra internal energy will render an opponent wobbly-kneed, and you guess that about half of the converted energy will end up in the opponent's head. Assuming a coefficient of restitution $e = 0.20$, with what speed does the boxer's fist have to contact the opponent's head in order to deliver the punch?

❶ **GETTING STARTED**

1. Do the boxer's hand, glove, and forearm form an isolated or closed system? What if the opponent's head is included?
2. What kind of diagram might be useful?

❷ **DEVISE PLAN**

3. Using your diagram as a guide, write the relevant equation(s).
4. Make sure you have enough information to solve for the requested answer.
5. Does the opponent's head remain stationary as the punch is delivered?

❸ **EXECUTE PLAN**

6. What value should you use for the converted energy?

❹ **EVALUATE RESULT**

Worked Problem 5.3 Poor parking

A car traveling at 3.0 m/s whizzes into an empty parking space. At the same instant, a truck that has an inertia 50% greater than that of the car is using the empty parking space to take a shortcut through the parking lot. Both vehicles are coasting (in other words, both have zero acceleration), and just before they hit head-on, the truck is moving at 4.0 m/s. (*a*) If during the collision 75% of the initial kinetic energy of the car-truck system is converted to internal energy, what are the final velocities of the two vehicles? (*b*) What is the coefficient of restitution for the collision?

❶ GETTING STARTED Our first step is to draw a system diagram and then add the information about initial velocities (Figure WG5.3). For the short duration of the collision, the car and truck form an isolated system. Because we are told that a given percentage of kinetic energy is converted, we can also assume that the system is closed. We are not given inertia values but are told only that the truck's inertia is 50% greater than that of the car. We hope we do not need the actual values, but we put the information we have in the diagram to help us think clearly about the situation.

Figure WG5.3

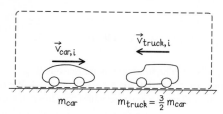

m_{car} $m_{truck} = \frac{3}{2} m_{car}$

We are asked to determine the two "final velocities," which we assume means "final velocities immediately after the collision" so that any interaction with the environment can be neglected. This information is contained in the momenta of the vehicles. Because kinetic energy is lost, we know that the collision is inelastic.

❷ DEVISE PLAN The momentum of the two-vehicle system after the collision is the same as the momentum before the collision. Thus we should draw a diagram representing the initial and final momenta to help plan our approach (Figure WG5.4). To get the directions of the final motion, we shall have to deal with the signs of the momenta with reference to an axis. Let us choose to point our positive x axis toward the left and include it in our diagram. We have drawn both final momenta as being in the positive x direction because the truck has the greater initial momentum.

Figure WG5.4

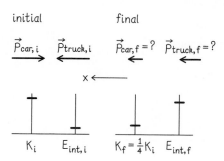

Along the x axis we chose, we have $v_{car\,x,i} = -v_{car,i}$ and $v_{truck\,x,i} = +v_{truck,i}$, where $v_{car,i} = 3.0$ m/s and $v_{truck,i} = 4.0$ m/s. From this, we know that the x components of the initial momenta are $p_{car\,x,i} = m_{car}\,v_{car\,x,i}$ and $p_{truck\,x,i} = m_{truck}\,v_{truck\,x,i}$. We also know that, from immediately before to immediately after the collision, the momentum of our system is constant. From this information, we can write

$$p_{car\,x,f} + p_{truck\,x,f} = p_{car\,x,i} + p_{truck\,x,i}.$$

We know only the two quantities on the right, however, and so we need more information. The energy of the system does not change under our assumptions, but the final kinetic energy is reduced substantially as kinetic energy is converted to internal energy. We illustrate this with energy bars sketched on the system diagram, and we write the information in equation form:

$$\Delta E = 0, \quad \text{but} \quad K_f = \tfrac{1}{4}K_i.$$

Using this kinetic energy equation to obtain the final velocities, we can calculate the coefficient of restitution from the expression

$$e = -\frac{v_{car\,x,f} - v_{truck\,x,f}}{v_{car\,x,i} - v_{truck\,x,i}}.$$

❸ EXECUTE PLAN (*a*) Conservation of momentum gives us

$$m_{car}v_{car\,x,f} + m_{truck}v_{truck\,x,f} = m_{car}v_{car\,x,i} + m_{truck}v_{truck\,x,i}.$$

That the truck's inertia is given in terms of the car's inertia allows us to get rid of the m_{truck} factors and then simplify further by dividing through by m_{car}:

$$m_{car}v_{car\,x,f} + \tfrac{3}{2}m_{car}v_{truck\,x,f} = m_{car}v_{car\,x,i} + \tfrac{3}{2}m_{car}v_{truck\,x,i}$$

$$v_{car\,x,f} + \tfrac{3}{2}v_{truck\,x,f} = v_{car\,x,i} + \tfrac{3}{2}v_{truck\,x,i}. \quad (1)$$

The final velocities are unknown, and so we have one equation with two unknowns, which means we do not yet have enough information to solve the problem. We need one more equation, so we turn to our energy equation:

$$K_f = \tfrac{1}{4}K_i$$

$$\tfrac{1}{2}m_{car}v_{car\,x,f}^2 + \tfrac{1}{2}m_{truck}v_{truck\,x,f}^2$$

$$= \tfrac{1}{4}\left(\tfrac{1}{2}m_{car}v_{car\,x,i}^2 + \tfrac{1}{2}m_{truck}v_{truck\,x,i}^2\right)$$

$$\tfrac{1}{2}m_{car}v_{car\,x,f}^2 + \tfrac{1}{2}\left(\tfrac{3}{2}m_{car}\right)v_{truck\,x,f}^2$$

$$= \tfrac{1}{4}\left[\tfrac{1}{2}m_{car}v_{car\,x,i}^2 + \tfrac{1}{2}\left(\tfrac{3}{2}m_{car}\right)v_{truck\,x,i}^2\right].$$

Multiplying through by 4 and canceling m_{car}, we get

$$2v_{car\ x,f}^2 + 3v_{truck\ x,f}^2 = \tfrac{1}{2}v_{car\ x,i}^2 + \tfrac{3}{4}v_{truck\ x,i}^2. \qquad (2)$$

This result combined with Eq. 1 gives us two equations containing two unknowns. To solve them, we express $v_{car\ x,f}$ in Eq. 1 in terms of $v_{truck\ x,f}$:

$$v_{car\ x,f} = v_{car,x,i} + \tfrac{3}{2}v_{truck\ x,i} - \tfrac{3}{2}v_{truck\ x,f}$$

$$= (-v_{car,i}) + \tfrac{3}{2}(+v_{truck,i}) - \tfrac{3}{2}v_{truck\ x,f}$$

$$= (-3.0\ \text{m/s}) + \tfrac{3}{2}(+4.0\ \text{m/s}) - \tfrac{3}{2}v_{truck\ x,f}$$

$$= +3.0\ \text{m/s} - \tfrac{3}{2}v_{truck\ x,f}. \qquad (3)$$

Now we use this result in Eq. 2 and solve for $v_{truck\ x,f}$:

$$2v_{car\ x,f}^2 + 3v_{truck\ x,f}^2 = \tfrac{1}{2}v_{car\ x,i}^2 + \tfrac{3}{4}v_{truck\ x,i}^2$$

$$2(+3.0\ \text{m/s} - \tfrac{3}{2}v_{truck\ x,f})^2 + 3v_{truck\ x,f}^2$$

$$= \tfrac{1}{2}(-3.0\ \text{m/s})^2 + \tfrac{3}{4}(+4.0\ \text{m/s})^2$$

$$(18\ \text{m}^2/\text{s}^2) - (18\ \text{m/s})v_{truck\ x,f} + \tfrac{18}{4}v_{truck\ x,f}^2 + 3v_{truck\ x,f}^2$$

$$= \frac{9.0\ \text{m}^2/\text{s}^2}{2} + (12\ \text{m}^2/\text{s}^2)$$

$$(18\ \text{m}^2/\text{s}^2) - (18\ \text{m/s})v_{truck\ x,f} + \tfrac{30}{4}v_{truck\ x,f}^2$$

$$= \frac{33\ \text{m}^2/\text{s}^2}{2}$$

$$5.0v_{truck\ x,f}^2 - (12\ \text{m/s})v_{truck\ x,f} + (1.0\ \text{m}^2/\text{s}^2)$$

$$= 0.$$

Solving this quadratic equation yields

$$v_{truck\ x,f} = +2.3\ \text{m/s} \quad \text{or} \quad +0.086\ \text{m/s}.$$

Substituting these values into Eq. 3 and solving for $v_{car\ x,f}$ gives

$$v_{truck\ x,f} = +2.3\ \text{m/s}, \ v_{car\ x,f} = -0.45\ \text{m/s}$$

or $\qquad v_{truck\ x,f} = +0.086\ \text{m/s}, \ v_{car\ x,f} = +2.9\ \text{m/s}.$

We must choose the result that corresponds to the physical situation. The first result says that the truck keeps moving in the positive x direction and the car keeps moving in the negative x direction, implying that they pass through each other, a physically impossible result. The second solution has the truck continuing but the car switching direction, which is plausible. Consequently, with our having assigned leftward as the positive x direction, the physically correct velocities are

$$v_{truck\ x,f} = +0.086\ \text{m/s} \quad (0.086\ \text{m/s to the left}) ✔$$

$$v_{car\ x,f} = +2.9\ \text{m/s} \quad (2.9\ \text{m/s to the left}). ✔$$

(b) The coefficient of restitution is

$$e = -\frac{v_{car\ x,f} - v_{truck\ x,f}}{v_{car\ x,i} - v_{truck\ x,i}}$$

$$= -\frac{+2.9\ \text{m/s} - (+0.086\ \text{m/s})}{-3.0\ \text{m/s} - (+4.0\ \text{m/s})} = +\frac{2.8}{7.0} = 0.40. ✔$$

❹ EVALUATE RESULT The final speeds are reasonable given the initial speeds (which are a bit high for a parking space!). We made sure the velocity directions are realistic when we chose the appropriate root of the quadratic equation. The kinetic energy of the system does decrease substantially, as expected, with the truck moving very slowly after the collision. Because the magnitude of the change in momentum for the two vehicles is equal but the inertias are not, the truck should have the smaller change in velocity. Its velocity change is $(+0.086\ \text{m/s}) - (+4.0\ \text{m/s}) = -3.9\ \text{m/s}$, while the car's is $(+2.9\ \text{m/s}) - (-3.0\ \text{m/s}) = 5.9\ \text{m/s}$, just as we expected. The coefficient of restitution is reassuringly less than 1 for this inelastic collision.

Guided Problem 5.4 Burrowing bullet

A gun that has a muzzle velocity of 600 m/s is used to fire a 12.0-g bullet horizontally into a 4.00-kg block of wood initially at rest. The bullet passes completely through the block, with negligible loss of inertia to either object. After the collision, the block slides at 1.20 m/s in the direction of the bullet's motion. What is the change in internal energy in the bullet-block system? Ignore any effects due to friction between the block and the surface over which it slides.

❶ GETTING STARTED

1. Which objects should you include in your system? Draw a system diagram for these objects.
2. Which conservation laws apply here: momentum, energy, or both? How could you represent what happens to these quantities graphically?

❷ DEVISE PLAN

3. Write equations relating the initial and final values. Does any equation have a single unknown (and hence can be solved immediately)?
4. What is the relationship between the change in internal energy and the change in kinetic energy?

❸ EXECUTE PLAN

❹ EVALUATE RESULT

Worked Problem 5.5 Explosive beat

You decide to build a *ballistocardiograph* that works on the following principle: A patient lies flat on a slab that floats on a cushion of air so that it can move freely in the horizontal direction. When the heart pumps blood horizontally in one direction, the slab and patient move in the opposite direction. For a resting patient, the heart pumps blood preferentially toward the head. The resulting recoil speed can be measured and can be correlated with the medical information you are after, which is the heart's ability to pump blood. You anticipate having patients as large as 1.0×10^2 kg. You know that a normal heart, each time it pumps, converts 2.0 mJ of chemical energy to kinetic energy and moves about 50 g of blood. You also know you can buy velocity sensors that can detect the slab's speed to a sensitivity of 1.0×10^{-5} m/s, and you want the measured speed to be comfortably higher than this—say, by a factor of 10. What is the highest practical value for the slab's inertia if its maximum speed with a patient lying on it is to be 1.0×10^{-4} m/s? Assume that any friction experienced by the slab sliding over the air cushion is negligible and that the slab and patient move as a single unit.

❶ GETTING STARTED The heart is converting internal energy to the kinetic energy of the blood moving in one direction and the kinetic energy of the slab and patient moving in the opposite direction. This is an explosive event and a signal that we should probably use conservation of momentum in our analysis, which means our first step is to choose an isolated system and draw a system diagram for it (Figure WG5.5a). The "initial" diagram is at an instant between heartbeats, when the patient, the slab, and the small amount of blood about to be pumped can be thought of as motionless. The initial system momentum is therefore zero. We approximate that all the pumped blood moves toward the patient's head. After we add the unknown final velocity information, Figure WG5.5(b) shows the notation we shall use. We suspect we may need a coordinate system, and so we arbitrarily choose our positive x axis pointing to the left. The "final" diagram shows the small volume of blood moving toward the head and the patient and slab recoiling. We suspect that the amount of internal energy converted to kinetic energy will give us information about the speeds.

Figure WG5.5

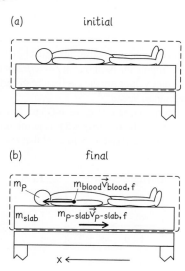

(a)   initial

(b)   final

❷ DEVISE PLAN Conservation of momentum gives us, for the isolated system of blood, patient, and slab,

$$\vec{p}_f = \vec{p}_i$$

$$m_{blood}\vec{v}_{blood,f} + (m_{patient} + m_{slab})\vec{v}_{patient+slab,f} = 0. \quad (1)$$

We want to know the value of m_{slab} that will generate the minimum detectable value for $|\vec{v}_{patient+slab,f}| = 1.0 \times 10^{-4}$ m/s. Because we have two unknowns, m_{slab} and $\vec{v}_{blood,f}$, we need another equation. Knowing that the change in the internal energy is $\Delta E_{int} = -2.0$ mJ $= -2.0 \times 10^{-3}$ J, we should be able to use conservation of energy for this closed system to get another equation:

$$\Delta K + \Delta E_{int} = 0$$

$$K_f - K_i = -\Delta E_{int}$$

$$\left[\tfrac{1}{2}m_{blood}v^2_{blood\,x,f} + \tfrac{1}{2}(m_{patient} + m_{slab})v^2_{patient+slab\,x,f}\right] - 0$$

$$= -\Delta E_{int} \quad (2)$$

This gives us two equations with everything known except two values: m_{slab} and $v_{blood\,x,f}$. We should first use Eq. 1 to express $v_{blood\,x,f}$ in terms of m_{slab} and then substitute into Eq. 2.

❸ EXECUTE PLAN Notice that the patient and slab move together as one object. It is convenient to use the subscript ps to replace the unwieldy patient+slab. This will require that we solve first for the combined inertia of patient and slab and then subtract the known maximum inertia of the patient to obtain the desired inertia of the slab. With this substitution, Eq. 1 gives

$$m_{blood}v_{blood\,x,f} = -m_{ps}v_{ps\,x,f}$$

$$v_{blood\,x,f} = -\frac{m_{ps}}{m_{blood}}v_{ps\,x,f}.$$

Then Eq. 2 gives

$$\tfrac{1}{2}m_{blood}v^2_{blood\,x,f} + \tfrac{1}{2}m_{ps}v^2_{ps\,x,f} = -\Delta E_{int}$$

$$\tfrac{1}{2}m_{blood}\left(-\frac{m_{ps}}{m_{blood}}v_{ps\,x,f}\right)^2 + \tfrac{1}{2}m_{ps}v^2_{ps\,x,f} = -\Delta E_{int}$$

$$\tfrac{1}{2}\frac{m^2_{ps}}{m_{blood}}v^2_{ps\,x,f} + \tfrac{1}{2}m_{ps}v^2_{ps\,x,f} = -\Delta E_{int}$$

$$\left(\tfrac{1}{2}\frac{v^2_{ps\,x,f}}{m_{blood}}\right)m^2_{ps} + (\tfrac{1}{2}v^2_{ps,x,f})m_{ps} + \Delta E_{int} = 0. \quad (3)$$

Because two coefficients involve the same quantity, we divide by that quantity:

$$\frac{1}{m_{blood}}m^2_{ps} + m_{ps} + \frac{2\Delta E_{int}}{v^2_{ps\,x,f}} = 0.$$

We solve this quadratic equation for m_{ps} using the quadratic formula:

$$m_{ps} = \frac{-1 \pm \sqrt{(1)^2 - 4\left(\dfrac{1}{m_{blood}}\right)\left(\dfrac{2\Delta E_{int}}{v^2_{ps\,x,f}}\right)}}{2\left(\dfrac{1}{m_{blood}}\right)}.$$

Substituting the known numerical values gives

$$m_{ps} = \frac{-1 \pm \sqrt{1 - 4\left[\dfrac{1}{5.0 \times 10^{-2} \text{ kg}}\right]\dfrac{2(-2.0 \times 10^{-3} \text{ J})}{(1.0 \times 10^{-4} \text{ m/s})^2}}}{2\left[\dfrac{1}{5.0 \times 10^{-2} \text{ kg}}\right]},$$

which simplifies to

$$m_{ps} = \frac{-1 \pm \sqrt{1 + 3.2 \times 10^7}}{4.0 \times 10^1 \text{ kg}^{-1}}.$$

The positive solution to this equation is $m_{ps} = 1.4 \times 10^2$ kg. We now can obtain the maximum design value for the slab's inertia:

$$m_{slab} = m_{ps} - m_{patient} = 140 \text{ kg} - 100 \text{ kg} = 40 \text{ kg}. ✔$$

Guided Problem 5.6 Useful approximations

In Worked Problem 5.5, we had to solve a quadratic equation (Eq. 3). Solving such equations can be a bother when you wish to make a quick calculation. The procedure used in Worked Problem 5.5 can be bypassed in explosive separations if the inertia of one object is much greater than that of the other(s). Consider the explosive separation that breaks an object initially at rest into two fragments, one of which has much greater inertia than the other: $m_1 \ll m_2$. In the explosive separation, some of the internal energy of the system is converted to kinetic energy in the fragments. Show that the following are good approximations for the division of energy between the fragments:

$$\tfrac{1}{2} m_2 v_2^2 \approx \frac{m_1}{m_2} \Delta K \tag{A}$$

$$\tfrac{1}{2} m_1 v_1^2 \approx \Delta K, \tag{B}$$

where v_1 is the speed of the fragment of inertia m_1 and v_2 is the speed of the fragment of inertia m_2. Equation B tells us that almost all of the internal energy converted to kinetic energy goes to the fragment that has less inertia. Only a small fraction m_1/m_2 of the internal energy goes to the fragment that has greater inertia.

① **GETTING STARTED**
1. Is the approach of Worked Problem 5.5 relevant?
2. Draw a system diagram for the initial and final situations. It may be necessary to use velocities rather than speeds when dealing with momentum.

② **DEVISE PLAN**
3. Is it possible to follow the procedure of Worked Problem 5.5 using the simpler notation m_1, v_1 and m_2, v_2?
4. In your new version of Eq. 3 of Worked Problem 5.5, isolate the kinetic energy of the fragment with greater inertia, $\tfrac{1}{2} m_2 v_2^2$.
5. Which is true: $m_1/m_2 \gg 1$ or $m_1/m_2 \ll 1$?

③ **EXECUTE PLAN**
6. Remember to approximate, especially if you are adding or subtracting a small number to or from a very large number.
7. You should now be able to do the step that gives Eq. A.
8. To derive Eq. B, repeat the derivation of Worked Problem 5.5 but this time in terms of the speed v_1 of the object with less inertia m_1.

④ **EVALUATE RESULT**
9. Use the numerical values given in Worked Problem 5.5 in Eq. A to estimate the maximum practical inertia for the slab. Is the value you calculate consistent with the numerical result obtained in Worked Problem 5.5?

④ **EVALUATE RESULT** This is not an especially low inertia, and so the answer suggests that such a device might be feasible. Looking at the math confirms that this is a maximum inertia for the slab because a larger slab would require either less patient inertia or a smaller value of v_{ps}. The assumption of an isolated system is not unreasonable, given the air-cushion design. The assumption that all 50 g of blood moves toward the head is an oversimplification, but it is consistent with the problem statement, and in the absence of more accurate blood flow information we have no better alternative.

(Bonus task: You should convince yourself that the slab moves back and forth as the heart beats rather than moving farther and farther in one direction.)

Worked Problem 5.7 Designer bowling

A 7.5-kg bowling ball rolls down the lane toward the 1.5-kg head pin at 5.0 m/s. The bowling alley owner wants to know how fast that pin can go flying, so that the end of the lane can be reinforced properly. You mumble confidently about coefficient of restitution, inertia, and maximum speed, and as the boss leaves she mentions that she wants this information yesterday. Ignore friction between the alley floor and the ball.

① **GETTING STARTED** What have we got to work with? Well, for the short time interval of the collision, the ball and pin form an isolated system whose momentum must remain constant. The system may or may not be closed, but it seems probable that the less energy transferred out of the ball-pin system, the more likely the pin will attain its maximum speed. There is no obvious source of energy to be transferred into the system, so assuming a closed system should be a good approximation. There is also no obvious mechanism for an explosive

separation, so the coefficient of restitution must be between 1 and 0. We could simply compute the final velocity in terms of an unknown e and then try different values for the coefficient of restitution to see which yields the highest speed.

We draw a system diagram (Figure WG5.6), as usual, with initial and final pictures to help clarify what is going on and allow us to assign symbols for the variables and establish the direction of a positive x axis.

Figure WG5.6

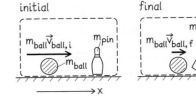

② DEVISE PLAN Conservation of momentum tells us that

$$\vec{p}_f = \vec{p}_i$$

$$m_{ball}v_{ball\,x,f} + m_{pin}v_{pin\,x,f} = m_{ball}v_{ball\,x,i} + m_{pin}v_{pin\,x,i}, \quad (1)$$

where we assume that the original motion of the ball is in the positive x direction. We know both inertias and both initial velocities, taking $v_{pin\,x,i}$ to be 0. Thus we have two unknowns, $v_{ball\,x,f}$ and $v_{pin\,x,f}$, but we also have a second equation to use:

$$e = \frac{v_{rel,f}}{v_{rel,i}} = -\frac{v_{ball\,x,f} - v_{pin\,x,f}}{v_{ball\,x,i} - v_{pin\,x,i}} = \text{trial value} \quad (2)$$

Thus, assuming a known trial value for e, we have two equations and two unknowns, the x components of the final velocities.

③ EXECUTE PLAN A logical step toward solution is to manipulate Eq. 2 to look more like Eq. 1 (with all the final values on the left and all the initial values on the right) and then look for a way to eliminate an unknown:

$$m_{ball}v_{ball\,x,f} + m_{pin}v_{pin\,x,f} = m_{ball}v_{ball\,x,i} + m_{pin}v_{pin\,x,i} \quad (1)$$

$$v_{ball\,x,f} - v_{pin\,x,f} = -e(v_{ball\,x,i} - v_{pin\,x,i}) \quad (2)$$

Noting that there are different signs on the two final x components of pin velocity, but the same sign on the two final x components of ball velocity, we plan to subtract Eq. 2 from Eq. 1 to eliminate $v_{ball\,x,f}$. To make this work, we multiply Eq. 2 by m_{ball}:

$$m_{ball}v_{ball\,x,f} - m_{ball}v_{pin\,x,f} = -em_{ball}(v_{ball\,x,i} - v_{pin\,x,i})$$

and subtract this equation from Eq. 1:

$$0 + (m_{ball} + m_{pin})v_{pin\,x,f}$$

$$= m_{ball}(1 + e)v_{ball\,x,i} + (m_{pin} - em_{ball})v_{pin\,x,i}$$

$$v_{pin\,x,f} = \frac{m_{ball}(1 + e)v_{ball\,x,i} + (m_{pin} - em_{ball})v_{pin\,x,i}}{(m_{ball} + m_{pin})}.$$

Noting that the initial x component of the pin velocity is zero, we obtain

$$v_{pin\,x,f} = \frac{m_{ball}(1 + e)v_{ball\,x,i} + 0}{(m_{ball} + m_{pin})} = \frac{m_{ball}v_{ball\,x,i}}{(m_{ball} + m_{pin})}(1 + e)$$

$$= \frac{7.5\ \text{kg}\,(5.0\ \text{m/s})}{(7.5\ \text{kg} + 1.5\ \text{kg})}(1 + e).$$

The speed we seek is the magnitude of this x component of velocity.

It is now clear that the larger the value of e, the higher the final speed of the pin. Because the largest reasonable value of e is 1, we compute the maximum speed as

$$v_{pin,f} = |v_{pin\,x,f}| = |(4.17\ \text{m/s})(1 + 1)| = 8.34\ \text{m/s} = 8.3\ \text{m/s}. \checkmark$$

The owner should design for this speed.

④ EVALUATE RESULT The speed we obtain for the pin is almost 70% faster than the speed of the incoming ball, but that is reasonable given that the ball has greater inertia and keeps rolling in the same direction. We therefore expect the pin to recoil faster than the incoming ball. Because the ball-pin system is closed, we can also check our answer using energy conservation. For an elastic collision, the kinetic energy is constant and so

$$\tfrac{1}{2}m_{ball}v_{ball,f}^2 + \tfrac{1}{2}m_{pin}v_{pin,f}^2 = \tfrac{1}{2}m_{ball}v_{ball,i}^2 + \tfrac{1}{2}m_{pin}v_{pin,i}^2$$

$$\tfrac{1}{2}m_{pin}v_{pin,f}^2 = \tfrac{1}{2}m_{ball}v_{ball,i}^2 - \tfrac{1}{2}m_{ball}v_{ball,f}^2 + \tfrac{1}{2}m_{pin}v_{pin,i}^2.$$

There are still two unknown final speeds, so we compute one of them using Eq. (2).

$$v_{ball\,x,f} = v_{pin\,x,f} - e(v_{ball\,x,i} - v_{pin\,x,i})$$

$$v_{ball\,x,f} = +8.34\ \text{m/s} - 1(+5.0\ \text{m/s} - 0)$$

$$v_{ball\,x,f} = +3.34\ \text{m/s}, \quad v_{ball,f} = 3.3\ \text{m/s}$$

Using the values computed above for $e = 1$, we have

$$\tfrac{1}{2}(1.5\ \text{kg})v_{pin,f}^2$$

$$= \tfrac{1}{2}(7.5\ \text{kg})(5.0\ \text{m/s})^2 - \tfrac{1}{2}(7.5\ \text{kg})(3.34\ \text{m/s})^2 + 0$$

$$v_{pin,f}^2 = 69.2\ \text{m}^2/\text{s}^2$$

$$v_{pin,f} = 8.3\ \text{m/s},$$

which is the result we obtained.

Guided Problem 5.8 Fast service

Champion table-tennis players can swing the paddle at speeds of 20 m/s. How fast is the ball going after one of these serves?

① GETTING STARTED
1. Choose a system and draw a system diagram.
2. Is your system isolated? Closed? Which conservation laws are appropriate to describe it?
3. Draw before and after pictures, and label all known and unknown quantities. It may seem at first like you don't have enough information, but plunge ahead anyway.

② DEVISE PLAN
4. What is the initial velocity of the ball?
5. Based on what you expect the properties of a Ping-Pong ball to be, which type of collision is this likely to be: elastic, inelastic, or totally inelastic?

6. No inertia values are given. Assuming the hand and forearm are part of the paddle, how does the inertia of the ball compare with the inertia of the paddle? How will the paddle's final velocity after the collision compare with its initial velocity before the collision?
7. Write the relative velocity equation. Do you need another equation?

③ EXECUTE PLAN

④ EVALUATE RESULT
8. How does the final speed of the ball compare with the initial speed of the paddle?
9. Which type of collision did you assume? Is that assumption justified (or at least not unreasonable)? How would your result change if you modified the coefficient of restitution a bit?

Questions and Problems

Dots indicate difficulty level of problems: • = easy, •• = intermediate, ••• = hard; CR = context-rich problem.

5.1 Classification of collisions

1. (*a*) While driving a car at 25 m/s, you pass a truck traveling in the same direction at 22 m/s. If you assign the direction in which the two vehicles are moving as the positive *x* direction of a coordinate system, what is the truck's velocity relative to you? (*b*) Now a motorcycle passes you at 29 m/s. What is its velocity relative to you? •

2. Suppose you have an isolated system in which two objects about to collide have equal and opposite momenta. If the collision is totally inelastic, what can you say about the motion after the collision? •

3. Figure P5.3 shows velocity-versus-time graphs for two situations in which a pair of objects collide. For each situation, decide whether the collision is elastic, inelastic, or totally inelastic. ••

Figure P5.3

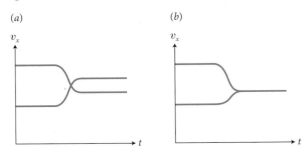

4. When you throw a tennis ball against a wall with some initial speed, is it possible for the ball to bounce back with a higher speed? ••

5.2 Kinetic energy

5. What does doubling an object's velocity do to its momentum and to its kinetic energy? •

6. As a result of an elastic collision between carts 1 and 2, the kinetic energy of cart 1 doubles. Does the kinetic energy of cart 2 change? If so, by what amount? •

7. A common energy unit used in food chemistry is the *food calorie* (1 Cal = 1000 cal = 4186 J). What must be the inertia of a person who is moving at a walking pace of 1.0 m/s and has kinetic energy numerically equal to the food energy in a jelly doughnut (300 Cal)? •

8. (*a*) Write an expression for the kinetic energy of an object in terms of its momentum p and inertia m. (*b*) If kinetic energy can be written in terms of momentum, how can the kinetic energy of a system change in an inelastic collision even though the law of momentum conservation forces the system's momentum to remain unchanged? ••

9. If two objects A and B have the same momentum but A has four times the kinetic energy of B, what is the ratio of their inertias? ••

10. If two objects A and B have the same kinetic energy but A has four times the momentum of B, what is the ratio of their inertias? ••

11. Object X has great inertia, and object Y has much less inertia. If the objects have the same momentum, which has more kinetic energy? If they have the same kinetic energy, which has greater momentum? ••

12. Draw a graph of kinetic energy versus distance fallen for a brick that is falling vertically from the top of a tall building. ••

5.3 Internal energy

13. You see a book on a table and give the book a push. It slides across the table and then comes to rest. Describe any changes in the physical state of the book-table system. •

14. Which of the following interactions are reversible: (*a*) a collision between two billiard balls, (*b*) a hand tossing a coin in the air, (*c*) a collision between hockey players on ice, (*d*) the firing of a cannon, (*e*) the lighting of a match? •

15. When a loaded cannon is fired, which physical variables are affected? •

16. A small bullet is fired into a large piece of wood. After the bullet penetrates the wood, the assembly moves as one unit along a low-friction track in the direction of travel of the bullet. (*a*) After the bullet is stuck in the piece of wood, is the momentum of the wood (not including the bullet) *greater than, equal to,* or *less than* the initial momentum of the bullet? State any assumptions explicitly. (*b*) Is the combined kinetic energy of the bullet and the wood after the bullet is stuck in the wood *greater than, equal to,* or *less than* the initial kinetic energy of the bullet? ••

17. A block slides with speed v across a low-friction, horizontal surface and collides with and sticks to an identical block at rest. The combination then strikes a spring attached to a fixed wall, bouncing off elastically. (*a*) Choose a system and discuss whether it is isolated. (*b*) Sketch energy bar diagrams for the initial state, the state after the first collision, the state during the collision with the spring, and the state after the final collision. (*c*) Discuss which types of internal energy are involved during this entire sequence of events. (*d*) Determine the final speed of the combined blocks. ••

5.4 Closed systems

18. Is it possible for a change of physical state to occur in an isolated system? In a closed system? •

19. Explain what is going on in terms of energy when your brakes overheat as you use them continuously coasting down a steep hill on a bike or in a car. •

20. Imagine making two springy devices, each made up of a dozen or so metal blocks loosely connected by springs, and then making the two collide head-on. Do you expect the collision to be elastic, inelastic, or totally inelastic? If the collision is not elastic, where does the kinetic energy go? ••

21. Two cars come to a stop from the same initial speed, one braking gently and the other braking hard. Which car converts more kinetic energy to internal energy (for example, thermal energy in the brakes)? ••

22. In the past, cars were built to be fairly rigid in collisions. Now "crumple zones" (areas designed to crumple or deform during a collision) are deliberately engineered. Why do such zones make a car safer in a collision? ••

5.5 Elastic collisions

23. A ball bounces elastically from the floor. It has *x* component of velocity $+v$ just before it hits the floor. What is the magnitude of its change in momentum in the collision? What is the change in its kinetic energy? Are your two answers consistent with each other? •

24. A 1200-kg car initially at rest undergoes constant acceleration for 8.8 s, reaching a speed of 10 m/s. It then collides with a stationary car that has a perfectly elastic spring bumper. What is the final kinetic energy of the two-car system? •

25. An elastic collision takes place between a 0.080-kg toy car moving at 10×10^2 m/h and a 0.016-kg toy car moving at 20×10^9 mm/yr. What is the kinetic energy of the system in joules? •

26. You want to give a third-grader a common example that illustrates how much energy a joule is. Show that a 0.1-kg ring of keys falling out of your pocket develops about 1 J of kinetic energy before hitting the ground. Can you think of another example? ••

27. You have an inertia of 52 kg and are standing at rest on an iced-over pond in your skates. Suddenly, your 60-kg brother skates in from the right with x component of velocity -5.0 m/s and collides elastically with you. (a) What is the siblings' relative speed after the collision? (b) If your brother's final x component of velocity is -0.36 m/s, what is your final velocity? (c) Is your answer to part b consistent with your answer to part a? (d) What is the change in kinetic energy of the system of you and your brother? ••

28. Two carts, of inertias m_1 and m_2, collide head-on on a low-friction track. Before the collision, which is elastic, cart 1 is moving to the right at 10 m/s and cart 2 is at rest. After the collision, cart 1 is moving to the left at 5 m/s. What are the speed and direction of motion of cart 2 after the collision? If $m_2 = 6$ kg, what is the value of m_1? ••

29. Show that in an elastic collision between two objects of inertias m_1 and m_2, with initial x components of velocity $v_{1i} > 0$ and $v_{2i} = 0$, the final x components of velocity are

$$v_{1x,f} = \left(\frac{m_1 - m_2}{m_1 + m_2} \right) v_{1x,i}$$

$$v_{2x,f} = \left(\frac{2m_1}{m_1 + m_2} \right) v_{1x,i}.$$

Discuss the cases $m_1 \ll m_2$, $m_1 = m_2$, and $m_1 \gg m_2$. Using everyday objects, give an example of each of these three cases. ••

30. Consider two hockey pucks identical in every respect except that one is black and the other is white. The black puck is initially at rest on the ice. A player shoots the white puck directly at the black one with velocity $\vec{v}_{white}$. The white puck hits the black puck elastically dead center. What are the magnitude and direction of the velocity of each puck after the collision? ••

31. The device shown in Figure P5.31 is called a Newton's cradle. Any collision between the balls in the device in Figure P5.31 is elastic. (a) If one end ball is pulled back and let go, as shown, would it be a violation of conservation of momentum if two balls at the other end bounced up at half the first ball's speed? (b) Would the result in part a be a violation of conservation of energy? (c) Show that the *only* result consistent with both conservation laws is that for n balls raised and let go, n balls rise on the other side. ••

Figure P5.31

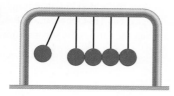

32. Figure P5.32 shows a pattern for the balls in a Newton's cradle just before collision. The arrows indicate the directions of motion of the balls. All the balls have the same size and inertia, and they collide elastically. Assume that all the collisions happen at the same instant. Sketch the pattern of the balls immediately after the collision. ••

Figure P5.32

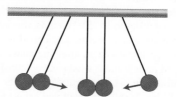

33. A cart of inertia m_1 and velocity $\vec{v}_{1i}$ collides elastically with a cart of inertia m_2 initially moving at velocity $-0.5\vec{v}_{1i}$. What is the speed of each cart after the collision? •••

34. Two carts are initially moving to the right on a low-friction track, with cart 1 behind cart 2. Cart 1 has a speed twice that of cart 2 and so moves up and rear-ends cart 2, which has twice the inertia of cart 1. What is the speed of each cart right after the collision if the collision is elastic? •••

5.6 Inelastic collisions

35. Cart A, with inertia 0.400 kg, is moving on a low-friction track at a speed of 2.2 m/s. It collides with cart B, which is initially at rest. After the collision, cart A continues moving in its original direction at 1.0 m/s, and cart B starts moving in the same direction at 3.0 m/s. Is this collision elastic, inelastic, or totally inelastic? What is the inertia of cart B? •

36. A physics student driving a 1200-kg car runs into the rear of a 2000-kg car stopped at a red light. From the fact that the joined vehicles skidded forward 4.0 m, the investigating police officer calculates that the speed after the collision was 6.6 m/s. Based on this information, calculate the student's speed before impact. •

37. A basketball is thrown horizontally at a heavy door that is free to move. For the two cases in which the ball is (a) well inflated and (b) flat, sketch a system diagram of the ball and the door, showing momentum vectors and energy bars just before and just after the collision. Make your picture the view from above. •

38. Draw a velocity-versus-time graph for a collision between a 2-kg object moving at an initial speed of 4 m/s and a 6-kg object initially at rest, if the coefficient of restitution is 0.25. ••

39. Consider four identical cars. Two of them collide head-on when both are traveling at 34 m/s. The other two collide head-on when both are traveling at 25 m/s. What is the ratio of the amount of kinetic energy converted to internal energy in the two collisions? ••

40. Take the common case where an object of inertia m_1 collides totally inelastically with a stationary object of inertia m_2. Show that the fraction of kinetic energy converted in the collision is $m_2/(m_1 + m_2)$. Comment on the amount of energy converted for the cases $m_2 \gg m_1$ and $m_2 \ll m_1$. ••

41. Show that when a moving object collides with an identical stationary object, the ratio of the final and initial kinetic energies is related to the coefficient of restitution e by

$$\frac{K_f}{K_i} = \tfrac{1}{2}(1 + e^2). \; \bullet\bullet$$

42. A 2000-kg truck is sitting at rest (in neutral) when it is rear-ended by a 1000-kg car going 25 m/s. After the collision, the two vehicles stick together. (a) What is the final speed of the car-truck combination? (b) What is the kinetic energy of the two-vehicle system before the collision? (c) What is the kinetic energy of the system after the collision? (d) Based on the results of parts b and c, what can you conclude about which type of collision this is? (e) Calculate the coefficient of restitution for this collision. Is this the result you expect for e? $\bullet\bullet$

43. Which head-on collision between a small car and a large truck causes a larger conversion of kinetic energy: one in which their initial momenta are the same magnitude, or one in which their initial kinetic energies are the same? Assume the same kinetic energy for the two-vehicle system in both cases. $\bullet\bullet\bullet$

44. You shoot a 0.0050-kg bullet into a 2.0-kg wooden block at rest on a horizontal surface (Figure P5.44). After hitting dead center on a hard knot that runs through the block horizontally, the bullet pushes out the knot. It takes the bullet 1.0 ms to travel through the block, and as it does so, it experiences an x component of acceleration of -4.9×10^5 m/s². After the bullet pushes the knot out, the knot and bullet together have an x component of velocity of $+10$ m/s. The knot carries 10% of the original inertia of the block. (a) What is the initial velocity of the bullet? (b) Using conservation of momentum, compute the final velocity of the block after the collision. (c) Calculate the initial and final kinetic energies of the block-knot-bullet system. Does the kinetic energy of the system change during the collision? (d) Can you calculate the coefficient of restitution for this collision? Which type of collision is it? $\bullet\bullet\bullet$

Figure P5.44

5.7 Conservation of energy

45. An experienced bartender knows just the right initial speed of a glass of beer to get it to come to a stop in front any customer sitting along the bar. Say the initial speed needed to move a glass all the way to the end of the bar is v_{end}. In terms of v_{end}, with what speed should an identical glass of beer be released if it is to stop at a customer sitting three-quarters of the way down the bar? (The kinetic energy converted to thermal energy because of friction is proportional to the distance the glass skids.) $\bullet$

46. A car goes into a skid and gradually comes to a stop, accelerating at a constant rate. At the midpoint of the skid, how much of its kinetic energy has it lost? $\bullet$

47. You roll a 0.250-kg wooden croquet ball toward a 0.050-kg golf ball at rest. (a) If the wooden ball travels at 5.0 m/s before the impact with the golf ball and then at 4.0 m/s after the impact, what is the speed of the golf ball after impact? (b) Is the collision elastic? (c) What is the change in internal energy of the two-ball system? $\bullet\bullet$

48. A 1200-kg car is backing out of a parking space at 5.0 m/s. The unobservant driver of a 1800-kg pickup truck is coasting through the parking lot at a speed of 3.0 m/s and runs straight into the rear bumper of the car. (a) What is the change in internal energy of the two-vehicle system if the velocity of the pickup is 1.5 m/s backward after they collide? (b) What is the coefficient of restitution for this collision? $\bullet\bullet$

49. Explain how a police officer at the scene of a car crash can judge whether or not you were speeding when you rear-ended a stopped car, based solely on the length of the skid marks after the collision. Is the speed the officer estimates for your speed just before the collision faster than, equal to, or slower than your actual speed? $\bullet\bullet$

50. A wagon is coasting along a level sidewalk at 5.00 m/s. Its wheels have very good bearings. You are standing on a low wall, and you drop vertically into the wagon as it passes by. The wagon has an inertia of 100 kg, and your inertia is 50.0 kg. (a) Use conservation of momentum to determine the speed of the wagon after you are in it. (b) Use conservation of energy to determine that speed. (c) After comparing your answers in parts a and b, explain which method is correct. $\bullet\bullet$

51. The extinction of the dinosaurs is, according to one theory, attributed to a collision between Earth and an asteroid about 10 km in diameter. Assume that the asteroid had about the same density as Earth. Estimate the energy released by such an impact in megatons of TNT, where 1 megaton $= 4.2 \times 10^{15}$ J. $\bullet\bullet\bullet$

5.8 Explosive separations

52. Can you tell from the coefficient of restitution whether a collision has added kinetic energy to a system, taken some away, or left the system's kinetic energy unchanged? $\bullet$

53. If the video of an explosive separation (like the firing of a gun) is run backward, the event looks like a totally inelastic collision, with two or more originally separate pieces all sticking together. How are the coefficient of restitution for an explosive separation and that for a totally inelastic collision related to each other? $\bullet$

54. At the peak of its vertical flight, a fireworks shell explodes into two pieces. Piece 1 has three times the speed of piece 2. What is the ratio of the inertias of the pieces? What is the ratio of their kinetic energies? $\bullet\bullet$

55. A 52-kg ice skater (this value includes her body, her clothing, and several 1.0-kg snowballs she is carrying) is at rest on the ice. She throws a snowball to the right at 10 m/s. (a) What is her speed after the throw? Is her velocity to the left or to the right? (b) Calculate the coefficient of restitution for this event. She next throws a second snowball but this time at a speed of 20 m/s to the left. (c) What is her speed after this throw? Is her velocity to the left or to the right? (d) Calculate the coefficient of restitution for this event. (e) Calculate the change in kinetic energy in this second event. Where does the added kinetic energy come from? (f) If one food calorie equals 4184 J, how many food calories does the skater burn when she throws the second snowball? $\bullet\bullet$

56. A mysterious crate has shown up at your place of work, Firecracker Company, and you are told to measure its inertia. It is too heavy to lift, but it rolls smoothly on casters. Getting an inspiration, you lightly tape a 0.60-kg iron block to the side of the crate, slide a firecracker between the crate and the block, and light the fuse. When the firecracker explodes, the block goes one way and the crate rolls the other way. You measure the crate's speed to be 0.055 m/s by timing how long it takes to cross floor tiles. You look up the specifications of the firecracker and find that it releases 9.0 J of energy. That's all you need, and you quickly calculate the inertia of the crate. What is that inertia? ••

57. A two-stage rocket is traveling at 4000 m/s before the stages separate. The 3000-kg first stage is pushed away from the second stage with an explosive charge, after which the first stage continues to travel in the same direction at a speed of 2500 m/s. (*a*) How fast, and in what direction, is the 1500-kg second stage traveling after separation? (*b*) How much energy is released by the explosive separation of the two stages? ••

58. A system consists of a 4.0-kg cart and a 1.0-kg cart attached to each other by a compressed spring. Initially, the system is at rest on a low-friction track. When the spring is released, an explosive separation occurs at the expense of the internal energy of the compressed spring. If the change in the spring's internal energy during the separation is 1.0 kJ, what is the speed of each cart right after the separation? ••

59. A space shuttle of inertia *m* is attached to a booster rocket that has an inertia nine times greater. This system is moving at a speed of 800 m/s in outer space, as seen by observers in a nearby space station. Then explosive bolts are detonated, separating the shuttle from the rocket and thrusting the shuttle forward at a speed of 100 m/s relative to the rocket. What are the velocities of the rocket and the shuttle right after the explosion? ••

60. A uranium-238 atom can break up into a thorium-234 atom and a particle called an *alpha particle, α*-4. The numbers indicate the inertias of the atoms and the alpha particle in *atomic mass units* (1 amu = 1.66×10^{-27} kg). When a uranium atom initially at rest breaks up, the thorium atom is observed to recoil with an *x* component of velocity of -2.5×10^5 m/s. How much of the uranium atom's internal energy is released in the breakup? ••

61. A gun with inertia 5.0 kg fires a 10-g bullet at a stationary target located 1.0 km away. After the bullet leaves the gun, its speed decreases (constant *x* component of acceleration $a = -1.0$ m/s^2) so that the bullet hits the target at 299 m/s. If the direction in which the bullet moves is along a positive *x* coordinate axis, what is the recoil velocity of the gun? ••

62. An assembled system consists of cart A of inertia m_A, cart B of inertia m_B, and a spring of negligible inertia, clamped together so that the fully compressed spring is aligned between the front end of cart B and the back end of cart A. The internal energy of the system, stored in the compressed spring, is E_{spring}. The system is put on a low-friction track and given a small shove so that it moves to the right at speed v_i, with cart A in front. Once the system is moving, the clamp is released (by remote control, so that the motion of the system as a whole is not affected in any way). As the spring expands, the carts move apart. What are their final speeds? •••

63. Consider the same initial system of spring and carts A and B as in Problem 62. After the system is set in motion with common speed v_i, the clamp is again released. What is the speed of cart A at the instant when the spring is still partially compressed and stores one-fourth of its initial internal energy E_{spring}? •••

64. A toy water-rocket consists of an elongated plastic tank with a nozzle at one end (Figure P5.64). A hand pump is used to get pressurized air into the tank. If there is nothing but air in the tank, the performance is lackluster. If some water is put in the tank before it is pumped up with air, however, the flight is dramatically better. Why? •••

Figure P5.64

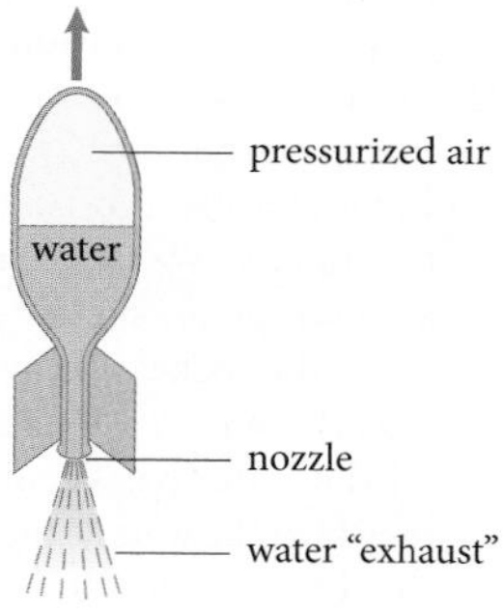

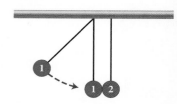

pressurized air

water

nozzle

water "exhaust"

65. A fictional subatomic particle, the solarino, is initially at rest. It then explodes into three particles: two lightons and a heavyon. The inertia of a heavyon is twice the inertia of a lighton. In the explosive separation, the three particles can be kicked out in either direction along a straight line, but the internal energy of the solarino is always entirely converted to the kinetic energy of the three particles. (*a*) Draw the paths of the three particles for the case in which the heavyon has the greatest possible speed. (*b*) Draw the paths of the particles for the case in which one of the lightons has the greatest possible speed. •••

Additional Problems

66. Which has more kinetic energy: a 0.14-kg baseball traveling at 45 m/s or a 0.012-kg bullet traveling at 480 m/s? •

67. Why is it easier to hit a home run from a pitched fastball than from a ball tossed in the air by the batter? •

68. You and a friend are playing catch with a medicine ball. The game has gotten a little unfriendly, and you'd like to knock your friend down with the next throw. To achieve this, should you suggest that your friend catch the ball with his hands or bounce it elastically back to you with his fists? ••

69. Your uncle works at the railroad freight yard, and he has asked you to watch the yard hands couple 20,000-kg boxcars together. As he leaves to get a sandwich, he tells you, "Whatever you do, don't let those cars bang together too hard. The couplings will take only 10,000 J on a hit before they shatter." You ask how hard is too hard, but your uncle has already ducked out. How fast can one moving car bump into a stationary car without the coupling shattering? ••• CR

70. Two solid spheres hung by thin threads from a horizontal support (Figure P5.70) are initially in contact with each other. Sphere 1 has inertia $m_1 = 0.050$ kg, and sphere 2 has inertia $m_2 = 0.10$ kg. When pulled to the left and released, sphere 1 collides elastically with sphere 2. At the instant just before the collision takes place, sphere 1 has kinetic energy $K_1 = 0.098$ J. (*a*) What is the velocity of sphere 1 right before the collision? (*b*) What is the kinetic energy of the system before the collision? (*c*) What is the velocity of each sphere after the collision? (*d*) From part *c*, calculate the kinetic energy after the collision. Does the value you get equal the result of part *b*? Explain why or why not. (*e*) Calculate the coefficient of restitution of the collision. Is this the result you expect? ••

Figure P5.70

PRACTICE

71. Repeat Problem 70 for two spheres made of modeling clay. All the data remain the same, but this time the two spheres stick together upon collision. ••

72. In the toy water-rocket of Problem 64, the pressurized air can store internal energy. Assume you have a rocket that expels water at a speed of 15.0 m/s and the opening through which the water leaves is 10.0 mm in diameter. (*a*) Using Figure P5.72 as a template, sketch energy bar graphs for three instants: when the rocket is fully pressurized but has not been launched, after the launch when half the water has been expelled, and after all the water has been expelled. (*b*) What is the approximate inertia of the water expelled in the first second of flight? (*c*) Does the rocket accelerate at a constant rate? ••

Figure P5.72

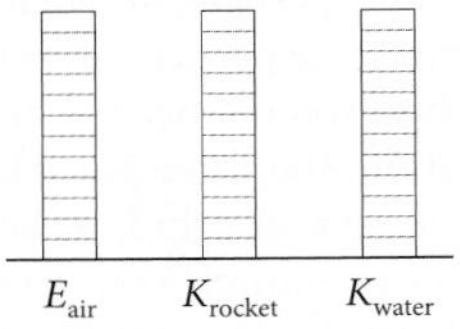

E_{air} K_{rocket} K_{water}

73. Suppose that, as it slides across the floor of a basketball arena, a hockey puck loses kinetic energy because of friction. The amount of energy lost is directly proportional to the distance the puck slides. Sketch a graph of kinetic energy versus time during this slide and then verify with a calculation. •••

74. While bowling one day, you begin to wonder about the inertia of a bowling pin. You know that balls and pins generally make elastic collisions and that your 6.5-kg bowling ball continues to move forward after a collision, even when you hit several pins to make a strike. As it happens, you have just managed to knock down nine of the ten pins, leaving only one pin upright at the end of the lane. You ask your friend to make a video of your next shot so that you can later carefully observe the initial and final velocities of your ball. Sure enough, you hit the remaining pin head-on (make the spare), and after a little video analysis you decide that your ball lost about 40% of its initial speed when it hit the pin. ••• CR

75. For a system of two identical cars, A and B, of the same inertia *m* but moving at two different velocities, v_A and v_B, show that the kinetic energy of the two-car system can be expressed as the sum of two terms: the kinetic energy of a double car moving with one-half the sum of their velocities, plus the kinetic energy of a double car moving with one-half the difference of their velocities:

$$K_{sum} = \tfrac{1}{2}(2m)\left[\frac{v_A + v_B}{2}\right]^2$$

$$K_{dif} = \tfrac{1}{2}(2m)\left[\frac{v_A - v_B}{2}\right]^2.$$

Interpret the meaning of each of these terms. •••

76. A 1000-kg car traveling with an *x* component of velocity of +20 m/s collides head-on with a 1500-kg light truck traveling with an *x* component of velocity of −10 m/s. (*a*) If 10% of the system's kinetic energy is converted to internal energy during the collision, what are the final speeds of the car and truck? (*b*) If the car had instead rear-ended the same truck moving with an *x* component of velocity of +10 m/s, how would your answer to part *a* change? •••

77. Your little brother is outside squirting a basketball with the stream of water from a garden hose. As the ball rolls across the yard, you begin to wonder what you would need to know in order to determine the rate at which momentum is transferred from the water to the ball. It should not be hard to measure both the radius *r* of the hose and the flow rate *Q* of water out of the hose in cubic meters per second. The density of water is certainly known, but you aren't sure whether the collision is elastic or inelastic. ••• CR

78. Two hockey players push off from a clinch, recoiling from rest. Jean-Claude has an inertia that is 50% greater than Pierre's inertia. After they push off and move across the ice in opposite directions, they lose kinetic energy at the same rate until they come to a stop. (*a*) Who travels farther? (*b*) How much farther? •••

Answers to Review Questions

1. Relative velocity is the difference between the velocity of a reference object and the velocity of an object being studied. We are free to choose which object is the object of study and which is the reference object, but we must distinguish which is which because relative velocity is a vector. Relative speed is the magnitude of the relative velocity and is the same regardless of which object is the object of study and which is the reference object.

2. The last subscript represents the object whose velocity we are studying. The first subscript represents the object we are measuring relative to (the reference object).

3. (*a*) The relative speed does not change. (*b*) The relative speed changes. (*c*) The final relative speed is zero (the objects stick together).

4. The kinetic energy of an object is the energy associated with the motion of the object. It can be computed as one-half the product of the object's inertia and the square of its speed.

5. The kinetic energy of a moving object is always positive, regardless of the direction in which the object is moving.

6. Kinetic energy can never be negative. It is zero if an object is stationary.

7. The *state* of an object is its condition as specified by the complete set of the physical variables that describe the object, such as shape and temperature.

8. In a reversible process, there is no permanent change in the objects' physical states; in an irreversible process, there is a permanent change in the physical states.

9. The internal energy increases by an amount equal to the decrease in the kinetic energy so that the energy in the system is unchanged.

10. We have no direct way of computing changes in internal energy at this point, but in the absence of energy transfer across the system boundary, the change in internal energy is the negative of the change in kinetic energy that takes place during the collision. Thus determining the change in kinetic energy and taking its negative is one way to determine the internal energy change.

11. A closed system is one in which no energy is transferred across its boundary. This is not the same as an isolated system (one in which no momentum is transferred across its boundary).

12. Energy conversion is energy changing from one form to another, such as from chemical energy to thermal energy; it may, but need not, involve more than one object. Energy transfer is the exchange of energy from one object to another; it may, but need not, involve the energy converting from one form to another.

13. The relative speeds are the same before and after the collision, a fact expressed in Eq. 5.4. This is true only in elastic collisions.

14. Imagine that you are one object in the collision. Before the collision, the distance between you and the other object is decreasing, but after the collision, the distance between you and the other object is increasing. So, the relative velocity between you and the other object changes sign.

15. The coefficient of restitution is the ratio of the relative speed of two objects after their collision to their relative speed before the collision.

16. Elastic: $e = 1$; totally inelastic: $e = 0$.

17. The minus sign is needed to make e positive. Because relative velocity changes sign after a collision, the fraction $v_{12\,x,f}/v_{12\,x,i}$ is always negative, and so the minus sign in front of the fraction makes e always positive.

18. No. The final velocities cannot be predicted unless additional information is available—the value of the coefficient of restitution, for instance, or the change in internal energy.

19. No. You are merely converting some of the internal energy of your muscles to the kinetic energy of the ball and your hand.

20. Yes. Internal energy may be transferred from one object in the system to another or may be converted from one form to another (say, chemical to thermal), which means a change in the physical state.

21. Conservation of energy (Eq. 5.23) states that the energy of a closed system remains constant. Thus if we can compute ΔK for the system, ΔE_{int} must be its equal and opposite value, as in Eq. 5.24.

22. Some of the internal energy of the exploding objects is converted to kinetic energy.

23. No. Because the relative speed of the objects increases (from zero before the separation to a nonzero value after), $e = 1$ can never be true. In another approach, you can reason that the kinetic energy of the system changes in an explosive separation. Because the kinetic energy does not change in elastic collisions, no explosive separation can be elastic.

Answers to Guided Problems

Guided Problem 5.2 5.0 m/s
Guided Problem 5.4 1.92×10^3 J
Guided Problem 5.8 40 m/s

6
PRACTICE
Principle of Relativity

Chapter Summary 87

Review Questions 89

Developing a Feel 90

Worked and Guided Problems 91

Questions and Problems 97

Answers to Review Questions 104

Answers to Guided Problems 104

PRACTICE

Chapter Summary

Inertial reference frames (Sections 6.1–6.4, 6.6)

Concepts A **reference frame** is a combination of a reference axis that defines a direction in space and a reference point that defines the origin from which motion is measured. The **Earth reference frame** is a reference frame at rest relative to Earth.

An **inertial reference frame** is one in which the **law of inertia** holds, which means a reference frame in which an isolated object remains either at rest or in motion at constant velocity.

The **principle of relativity** states that the laws of the universe are the same in all inertial reference frames.

The **zero-momentum reference frame** for a system of objects is the reference frame in which the momentum of the system is zero. The velocity of this reference frame is equal to the velocity of the center of mass of the system.

Quantitative Tools The position $\vec{r}_{cm}$ of the **center of mass** of a system of objects of inertias $m_1, m_2, \ldots$ located at positions $\vec{r}_1, \vec{r}_2, \ldots$ is

$$\vec{r}_{cm} \equiv \frac{m_1\vec{r}_1 + m_2\vec{r}_2 + \cdots}{m_1 + m_2 + \cdots}. \tag{6.24}$$

The **center-of-mass velocity** of a system of objects is

$$\vec{v}_{cm} \equiv \frac{d\vec{r}_{cm}}{dt} = \frac{m_1\vec{v}_1 + m_2\vec{v}_2 + \cdots}{m_1 + m_2 + \cdots}. \tag{6.26}$$

This is also the velocity of the **zero-momentum reference frame** for this system.

Relative motion (Section 6.5)

Concepts The **Galilean transformation equations** allow us to transform quantities measured in one inertial reference frame into their values in another inertial reference frame when the two reference frames are moving at constant relative velocity. The transformation equations tell us that *time intervals*, *lengths*, and *accelerations* are the same in any two inertial reference frames moving at constant relative velocity.

Quantitative Tools The **Galilean transformation equations** relate the time t and position $\vec{r}_e$ for an event (e) measured in an inertial reference frame A to these quantities measured for the event in any other inertial reference frame B. If the reference frames have a constant relative velocity $\vec{v}_{AB}$ and have their origins coinciding at $t = 0$, the transformation equations are

$$t_B = t_A = t \tag{6.4}$$

and

$$\vec{r}_{Be} = \vec{r}_{Ae} - \vec{v}_{AB}t_e. \tag{6.5}$$

As a consequence of these equations, the velocity $\vec{v}_{Ao}$ of an object (o) measured in an inertial reference frame A is related to the object's velocity measured in any other inertial reference frame B by

$$\vec{v}_{Ao} = \vec{v}_{AB} + \vec{v}_{Bo}. \tag{6.14}$$

The transformation equations also give the relationship between accelerations measured in any two inertial reference frames A and B:

$$\vec{a}_{Ao} \equiv \vec{a}_{Bo}. \tag{6.11}$$

Convertible kinetic energy (Section 6.7)

Concepts The **translational kinetic energy** of a system is the kinetic energy associated with the motion of its center of mass. For an isolated system, this kinetic energy is **nonconvertible** because it cannot be converted to internal energy (if it were, the momentum of the system would not be constant).

The **convertible kinetic energy** of an isolated system is the portion of the system's

Quantitative Tools The **translational (nonconvertible) kinetic energy** K_{cm} of a system is

$$K_{cm} \equiv \tfrac{1}{2}mv_{cm}^2, \tag{6.32}$$

where m is the system's inertia and v_{cm} is the speed of its center of mass.

The **convertible kinetic energy** of a two-particle system is

$$K_{conv} = \tfrac{1}{2}\mu v_{12}^2, \tag{6.40}$$

kinetic energy that can be converted to internal energy. This energy is the same in all inertial reference frames.

The **kinetic energy** of a system can be split into a convertible part K_{conv} and a nonconvertible part. The nonconvertible part is the system's translational kinetic energy K_{cm}.

where μ is the **reduced inertia** (or *reduced mass*) of the system, given by

$$\mu \equiv \frac{m_1 m_2}{m_1 + m_2}. \tag{6.39}$$

The kinetic energy of a system is

$$K = K_{cm} + K_{conv}. \tag{6.35}$$

Conservation laws and relativity (Section 6.8)

Concepts Changes in the momentum and energy of a system are the same in any two inertial reference frames.

Quantitative Tools If A and B are any two reference frames moving at constant velocity relative to each other, then the changes in the momentum and energy of any system are

$$\Delta \vec{p}_{A\,sys} = \Delta \vec{p}_{B\,sys} \tag{6.47}$$

$$\Delta K_B + \Delta E_{B\,int} = \Delta K_A + \Delta E_{A\,int}. \tag{6.56}$$

Review Questions

Answers to these questions can be found at the end of this chapter.

6.1 Relativity of motion

1. A truck moves at constant velocity on a freeway. Describe how the truck's motion is seen (*a*) by an observer in a car moving with the same constant velocity as the truck and (*b*) by an observer in a car moving at any other constant velocity.
2. On a breezy day, riding a bicycle into the wind can feel like riding into a windstorm. Riding in the direction of the wind on the same breezy day, however, you can sometimes feel no wind at all.

6.2 Inertial reference frames

3. During a given time interval, an observer in inertial reference frame 1 measures an object's change in momentum to be $\Delta \vec{p} = 0$. What is the value of $\Delta \vec{p}$ measured during the same time interval by an observer in inertial reference frame 2, which is moving east at 10 m/s relative to reference frame 1?
4. What is the law of inertia, and of what use is it?
5. An observer sees an isolated object undergo a change in momentum. Is the observer in an inertial reference frame?
6. (*a*) If an object's acceleration is zero in one inertial reference frame, is it zero in all other inertial reference frames? (*b*) If an object is at rest in one inertial reference frame, is it at rest in all other inertial reference frames?

6.3 Principle of relativity

7. State and explain the principle of relativity.
8. Is there any experiment you can do in your own reference frame to determine whether your reference frame is moving at a constant velocity? If yes, can you determine the value of that velocity?
9. If an object is moving forward in one inertial reference frame, it is always possible to find another inertial reference frame in which it is moving backward. If the object is accelerating forward in the first reference frame, which way is it accelerating in the second reference frame?
10. Someone says to you, "Momentum isn't a conserved quantity! All I have to do is change to a different inertial reference frame and the momentum of the system I'm looking at is different from what it was in my first reference frame." How should you respond?
11. Two observers witness an elastic collision from different inertial reference frames. What do they disagree on and what do they agree on regarding kinetic energy? Regarding momentum?

6.4 Zero-momentum reference frame

12. What is the zero-momentum reference frame for a system? How can you use it to solve a problem if the only velocity data you have for the system were measured in a different inertial reference frame?

13. (*a*) Compare the initial and final values of the magnitude of the momentum of a particle involved in an elastic collision when viewed from the zero-momentum reference frame. (*b*) What happens to the direction of the particle's momentum?

6.5 Galilean relativity

14. What are the Galilean transformation equations (Eqs. 6.4 and 6.5) used for?
15. Suppose observers C and D both measure a train's velocity but from different inertial reference frames. C reports velocity $\vec{v}_{Ct}$, and D reports $\vec{v}_{Dt}$. What other information must C have in order to determine whether or not D's measurement is accurate?

6.6 Center of mass

16. How does an object's inertia differ when measured in two different inertial reference frames?
17. How does the position of a system's center of mass measured using one reference point differ from the position measured using some other reference point?
18. In an isolated system that contains moving parts, how does the system's center of mass move?
19. What is the momentum of a system of objects as measured by someone moving along with the center of mass of the system?

6.7 Convertible kinetic energy

20. Is the convertible kinetic energy of a system of objects the same in all inertial reference frames?
21. Which part of a system's kinetic energy changes when an observer switches from one inertial reference frame to another: the convertible part or the nonconvertible part?
22. Can all the initial kinetic energy of two colliding objects be converted to internal energy?
23. (*a*) What is the maximum amount of kinetic energy that can be converted to internal energy in a collision? (*b*) In which type of collision is this maximum conversion possible?

6.8 Conservation laws and relativity

24. (*a*) If the change in momentum of a system is zero in one inertial reference frame, is it zero in all other inertial reference frames? (*b*) If the change in momentum of a system has a specific magnitude and direction in one inertial reference frame, does it have the same magnitude and direction in all other inertial reference frames?
25. (*a*) If the change in energy of a system is zero in one inertial reference frame, is it zero in all inertial reference frames? (*b*) If the change in energy of a system has a specific magnitude in one inertial reference frame, does it have the same magnitude in all other inertial reference frames?

Developing a Feel

Make an order-of-magnitude estimate of each of the following quantities. Letters in parentheses refer to hints below. Use them as needed to guide your thinking.

1. The velocity of the zero-momentum reference frame relative to the Earth reference frame for a head-on collision between a bug and an 18-wheel truck traveling at highway speed (Z, X, R)
2. The velocity of the zero-momentum reference frame relative to the Earth reference frame for a collision between a parked, fully loaded furniture delivery truck and a car traveling on a city street (Z, M, C)
3. The velocity of the zero-momentum reference frame relative to the Earth reference frame for a collision in which a large bird, while flying, is hit from behind by a bullet (E, I, V, P)
4. The reduced inertia of the Earth-Moon system (AA, J, Y)
5. The location of the center of mass of a five-car, one-locomotive train (BB, G, L)

6. The location of the center of mass of your body when you are standing (S, K, T, W)
7. The velocity of the center of mass of a system of 16 billiard balls (including the cue ball) just after the first break shot (A, F, O, U)
8. The kinetic energy converted to internal energy during a head-on collision between a bug and an 18-wheel truck traveling at highway speed (Z, X, D, R)
9. The maximum convertible kinetic energy for a collision between a parked, fully loaded furniture delivery truck and a car traveling on a city street (Z, M, AA, N, C)
10. The maximum convertible kinetic energy for a high-school wrestler slammed to the mat (B, H, Q)

Hints

A. Is the system of balls an isolated system during the break-shot collision?
B. What is an appropriate inertia for the wrestling mat?
C. What is a typical speed of a car driving on a city street?
D. What is the inertia of a bug?
E. What is the inertia of a large bird?
F. What is the speed of the cue ball approaching the 15 other balls?
G. What is the ratio of the inertia of a train car to the inertia of a locomotive?
H. With what speed does the wrestler hit the mat?
I. What is the inertia of a bullet?
J. What is the inertia of Earth?
K. Is your density roughly uniform?
L. How do the lengths of a locomotive and a train car compare?
M. How do the inertias of the two vehicles compare?
N. What is the reduced inertia of the car-truck system?
O. How does the center-of-mass velocity of the billiard balls change during the collision?
P. What is the speed of a bullet in flight?
Q. What is the reduced inertia of the system?
R. What is a typical highway speed of an 18-wheel truck?
S. How tall are you?
T. What is your basic shape when standing?
U. What is the center-of-mass speed of the billiard balls before the collision?

V. What is the speed of a large bird in flight?
W. Where is the center of mass of a uniform cylinder?
X. How do the inertias of the bug and 18-wheel truck compare?
Y. What is the inertia of the Moon?
Z. How is the velocity of the zero-momentum reference frame related to individual velocities and inertias?
AA. How is the reduced inertia related to individual inertias?
BB. How is the center of mass related to individual inertias?

Key (all values approximate)

A. yes, with the cue ball in the system; B. the mat is supported by Earth, so the appropriate inertia is huge; C. 1×10^1 m/s; D. a large bug: 5×10^{-4} kg; E. 2 kg; F. 1×10^1 m/s; G. 1:10; H. 4 m/s; I. 2×10^{-2} kg; J. 6×10^{24} kg; K. yes, L. they are about equal; M. the truck has about five times the inertia of the car; N. 5/6 of car inertia, so 1×10^3 kg; O. it does not change; P. 4×10^2 m/s; Q. about the same as the wrestler's inertia—say, 8×10^1 kg; R. 3×10^1 m/s; S. 2 m; T. cylindrical; U. from Eq. 6.26, $(10 \text{ m/s})/16 = 0.6$ m/s; V. 6 m/s; W. halfway along its length; X. the bug has almost zero inertia relative to the inertia of the 18-wheel truck; Y. 7×10^{22} kg; Z. sum of products of individual inertias and velocities divided by sum of inertias, Eq. 6.26; AA. product divided by sum, Eq. 6.39; BB. sum of products of individual inertias and positions divided by sum of inertias, Eq. 6.24

Worked and Guided Problems

Procedure: Applying Galilean relativity

In problems dealing with more than one reference frame, you need to keep track not only of objects, but also of reference frames. For this reason, each quantity is labeled with two subscripts. The first subscript denotes the observer; the second denotes the object of interest. For example, if we have an observer on a train and also a car somewhere on the ground but in sight of the train, then $\vec{a}_{Tc}$ is the train observer's measurement of the acceleration of the car. Once you understand this notation and a few basic operations, working with relative quantities is easy and straightforward.

observer A's measurement of velocity of car:

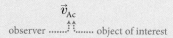

$$\vec{v}_{Ac}$$

observer ⋯⋯⋮⋮⋯⋯ object of interest

1. Begin by listing the quantities given in the problem, using this double-subscript notation.
2. Write the quantities you need to determine in the same notation.

3. Use subscript cancellation (Eq. 6.13) to write an equation for each quantity you need to determine, keeping the first and the last subscripts on each side the same. For example, in a problem where you need to determine $\vec{v}_{Tc}$ involving a moving observer B, write

$$\vec{v}_{Tc} = \vec{v}_{TB} + \vec{v}_{Bc}.$$

4. If needed, use subscript reversal (Eq. 6.15) to eliminate any unknowns.
5. Use the kinematics relationships from Chapters 2 and 3 to solve for any remaining unknowns, making sure you stay in one reference frame.

You can use this procedure and the subscript operations for any of the three basic kinematic quantities (position, velocity, and acceleration).

These examples involve material from this chapter but are not associated with any particular section. Some examples are worked out in detail; others you should work out by following the guidelines provided.

Worked Problem 6.1 Running a train

On a train carrying the university team to a track meet, a sprinter in one of the cars practices his starts by running down the aisle. In the reference frame of the train, the sprinter starts from rest and runs toward the rear of the car. After 2.0 s, he has accelerated to a speed of 10 m/s. If the train is moving at a constant 30 m/s relative to the Earth reference frame, what does an observer standing alongside the tracks measure for the sprinter's initial velocity, final velocity, and acceleration?

❶ **GETTING STARTED** We must translate information from the train reference frame T to the Earth reference frame E, which suggests Galilean transformations for velocity and acceleration. As usual, we begin by making a sketch, paying attention to the directions of the vectors (Figure WG6.1).

Figure WG6.1

initial

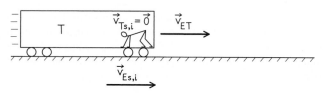

final

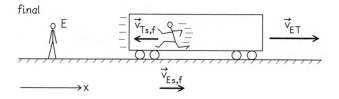

Because the train does not accelerate, T is an inertial reference frame, as is the Earth reference frame E. We are given enough information to calculate the sprinter's initial velocity $\vec{v}_{Ts,i}$, final velocity $\vec{v}_{Ts,f}$, and acceleration $\vec{a}_{Ts}$ measured by an observer at rest in the train reference frame, and we need to determine what the trackside observer measures for these three quantities, which we denote as $\vec{v}_{Es,i}$, $\vec{v}_{Es,f}$, and $\vec{a}_{Es}$ because they are being measured in the Earth reference frame.

❷ **DEVISE PLAN** As Figure WG6.1 shows, we define the positive x direction as the direction in which the train is moving. The train is therefore moving with a velocity that has x component $v_{ETx} = +30\,\text{m/s}$ as measured by the observer in the Earth reference frame. (And to a passenger on the train, the trackside observer moves along the x axis with a velocity that has x component $v_{TEx} = -30\,\text{m/s}$.)

We need to transform velocity information from the T reference frame to the E reference frame, and so we use Eq. 6.14 with the subscripts appropriate to this problem:

$$\vec{v}_{Es} = \vec{v}_{ET} + \vec{v}_{Ts}. \tag{1}$$

For the sprinter's acceleration, we note from Eq. 6.11 that the two inertial reference frames give the same result:

$$\vec{a}_{Es} = \vec{a}_{Ts}.$$

We can use Eq. 1 for both the sprinter's initial velocity and his final velocity measured by the trackside observer. To calculate the acceleration this observer measures, we do not have enough information to take the time derivative of the velocity because we know the sprinter's velocity at only two instants. Thus all we can do is either compute his average acceleration or assume his acceleration is constant. In either case, we have for the acceleration measured by the observer in the Earth reference frame

$$\vec{a}_{\text{Es}} = \frac{\Delta \vec{v}_{\text{Es}}}{\Delta t}.$$

❸ **EXECUTE PLAN** We first rewrite Eq. 1 in terms of the x components:

$$v_{\text{Es}x} = v_{\text{ET}x} + v_{\text{Ts}x}.$$

For the x component of the initial velocity we have

$$v_{\text{Es}\,x,\text{i}} = (+30 \text{ m/s}) + (0) = +30 \text{ m/s}, ✔$$

and for the x component of the final velocity we have

$$v_{\text{Es}\,x,\text{f}} = (+30 \text{ m/s}) + (-10 \text{ m/s}) = +20 \text{ m/s}. ✔$$

The x component of the acceleration is then

$$a_{\text{Es}x} = \frac{\Delta v_{\text{Es}x}}{\Delta t} = \frac{(+20 \text{ m/s}) - (+30 \text{ m/s})}{2.0 \text{ s}} = -5.0 \text{ m/s}^2. ✔$$

❹ **EVALUATE RESULT** The positive signs for the x components of the velocities mean that the observer in the Earth reference frame sees the sprinter moving in the positive x direction even though the sprinter is running toward the rear of the car. One way to think about this is to imagine that both the sprinter and the rear wall of the car are headed in the positive x direction but that the speed at which the rear wall moves in this direction is higher than the speed at which the sprinter moves in this direction. The minus sign for acceleration in our coordinate system means that the acceleration is directed toward the rear of the car.

We can check for consistency by calculating the x component of the acceleration with the information measured in the train reference frame:

$$a_{\text{Ts}x} = \frac{\Delta v_{\text{Ts}x}}{\Delta t} = \frac{(-10 \text{ m/s}) - (0)}{2.0 \text{ s}} = -5.0 \text{ m/s}^2.$$

This value agrees with the value obtained using data from the Earth reference frame. The assumption of constant acceleration is not unreasonable, given the limited data at hand. Whether or not the sprinter maintains constant acceleration, the value we obtained is his average acceleration for the 2.0-s interval.

Guided Problem 6.2 Safe passage

You are driving at 35 m/s on the highway when you come up behind a truck traveling at 25 m/s in the same direction. You'd like to have 20 m of space between the front of your car and the back of the truck when you move into the passing lane and 20 m of space between the back of your car and the front of the truck when you switch back into your original lane. If the truck is 10 m long, how many seconds do you need in the passing lane if you keep your speed constant at 35 m/s?

❶ **GETTING STARTED**
1. Draw a sketch showing the velocities of the truck and car, with arrows to indicate magnitudes and directions. In what reference frame is each velocity in your sketch measured?
2. How does the passing look to the truck's driver? Which is the simpler way to work this problem: from the point of view of the truck's reference frame or from the point of view of the Earth reference frame?

3. Redo your sketch in the appropriate reference frame, if necessary, depending on your answer to question 2.
4. What should you assume for the length of the car?

❷ **DEVISE PLAN**
5. How is time interval related to displacement for motion at constant speed? In the reference frame you have chosen, which vehicle is undergoing a displacement?
6. Translate your sketch into algebraic equations that contain the desired time interval.

❸ **EXECUTE PLAN**

❹ **EVALUATE RESULT**
7. Is the size of your answer reasonable? Consider that most drivers do not allow such margins in actual practice.
8. Does the length of the car matter?

Worked Problem 6.3 Stacked pennies

You have ten stacks of pennies lined up as in Figure WG6.2—one penny in the first stack, two in the second stack, three in the third, and so on. Where along this row is the center of mass of the system consisting of all the pennies?

Figure WG6.2

❶ **GETTING STARTED** We suspect that the center of mass of the system is located somewhere in one of the middle stacks, and we

know there is a formula to calculate the location involving inertias and positions (Eq. 6.24). We are not told any inertias or dimensions, but we can assume that all the pennies have identical inertias and dimensions. We will use symbols for the inertia and diameter of each penny and answer the question based on that.

❷ **DEVISE PLAN** Let us call the inertia of each penny m and the diameter of each penny d. (Using a variable to identify an unknown is a better practice than assigning a real number, both because using variables is a more general approach and because numbers can confuse the issue when we derive a final expression.) For the origin of our coordinate system, we arbitrarily choose the center of the leftmost penny, indicated by the black dot in Figure WG6.2.

The lack of numerical values means our answer will be in terms of m and d (unless these values cancel in the algebra). We number the stacks from left to right: $n = 1$ for the leftmost stack, $n = 2$ for the stack to its right, and so on. If we call the inertia of any one stack m_n, the position of the center of mass is

$$x_{cm} = \frac{\sum m_n x_n}{\sum m_n}.$$

We have to figure out the inertia m_n of each stack and its distance x_n from our origin. Let us choose to the right as the positive direction of the x axis. We know that there is one more penny on each stack as we go from left to right and that the position increases by a distance $x = d$ as we move rightward from one stack to the next. We need appropriate equations for m_n and x_n as functions of m, d, and n.

❸ EXECUTE PLAN We note that the number of pennies in a stack equals the stack number n. This means $m_n = nm$. We see that the $n = 1$ stack (containing one penny) is located at $x_1 = 0$ and that the center of each of the other stacks is located a distance $x_n = (n - 1)d$ from the origin. With ten stacks, Eq. 6.24 becomes

$$x_{cm} = \frac{\sum m_n x_n}{\sum m_n} = \frac{\sum_{n=1}^{10} (nm)[(n-1)d]}{\sum_{n=1}^{10} nm} = \frac{md \sum_{n=1}^{10} n(n-1)}{m \sum_{n=1}^{10} n}$$

$$= d \frac{\frac{1}{3}(10-1)(10)(10+1)}{\frac{1}{2}(10)(10+1)} = \frac{2}{3}(10-1)d = 6d. \checkmark$$

This result places the center of mass of the system at the center of the seventh stack, six stacks to the right of the first one. If it is not obvious that the seventh stack is at $x = 6d$, start with your finger on the origin in Figure WG6.2 and count d's to the seventh stack. Note that the center-of-mass location is independent of the inertia m of the pennies.

❹ EVALUATE RESULT The center is somewhere near the middle of this distribution, which is what we expect. Notice that there are $10 + 9 + 8 = 27$ pennies to the right of the seventh stack, but only $1 + 2 + 3 + 4 + 5 + 6 = 21$ pennies to the left of it. However, the pennies to the right of stack 7 are closer to the calculated center of mass, which means this is not an unreasonable result.

Guided Problem 6.4 Balancing a wheel

You have just had a new set of tires put on your car, and now the mechanic has to balance the wheels. For one of the wheels, the center of mass once the tire is installed is located 1.0 mm from the center of the axle. The mechanic corrects for this by crimping a small lead bar on the wheel rim 200 mm away from the axle. With this added inertia, the center of mass is at the center of the axle. If the wheel with the tire installed has an inertia of 10 kg, what is the inertia of the lead bar?

❶ GETTING STARTED

1. Draw a diagram that shows the information given, and add an x axis so that you can specify locations. Where is a convenient place to put the origin of your coordinate system?

❷ DEVISE PLAN

2. The goal is to restore the center of mass to the center of the axle so that $x_{cm} = x_{axle}$. What is the value of the x coordinate of the axle with your chosen origin?

3. Write the equation for the location of the center of mass of the balanced wheel, using the labels you put in your drawing. Convince yourself that you now have everything you need to solve this equation.

❸ EXECUTE PLAN

❹ EVALUATE RESULT

4. Should the numerical value you obtained for the lead bar inertia be negative or positive?

5. Does your calculated inertia value seem reasonable?

Worked Problem 6.5 Fore!

The 0.100-kg head of a golf club is moving at 45 m/s when it strikes a stationary 0.050-kg golf ball. (a) What is the kinetic energy of this two-object system in the Earth reference frame? (b) In the zero-momentum reference frame, what are the velocities of the club and ball just before they collide? (c) What is their relative velocity in the zero-momentum reference frame? (d) How much of the kinetic energy of the system can be converted to internal energy? (e) What is the translational kinetic energy of the system in the zero-momentum reference frame and in the Earth reference frame?

❶ GETTING STARTED Let us make two drawings, one in each reference frame, partly to establish labels for various quantities and partly to define an x axis (Figure WG6.3). We label the Earth reference frame E, the zero-momentum reference frame Z, and use c for club variables and b for ball variables. Notice that Figure WG6.3 represents the initial situation; it is not necessary to draw the final situation because none of the unknowns requires analysis of the situation after the collision.

Figure WG6.3

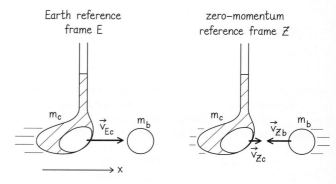

In drawing our diagram for the zero-momentum reference frame, we must be sure to show that the relative velocity measured in this reference frame is the same as the value measured in the Earth reference frame. The momenta of the club and ball should add up to zero in our zero-momentum reference frame, so the club's velocity arrow should be shorter than the ball's velocity arrow because the club's inertia is greater than that of the ball.

❷ DEVISE PLAN

(a) The kinetic energy of the system is the sum of the kinetic energies of the objects.

(b) We first use the fact that v_{EZx}, the value an observer in the Earth reference frame measures for the x component of the velocity of the zero-momentum reference frame, is the same as v_{Ecmx}, the value an observer in the Earth reference frame measures for the x component of the velocity of the center of mass of the system, which is given by Eq. 6.26:

$$v_{Ecmx} = \frac{m_c v_{Ecx} + m_b v_{Ebx}}{m_c + m_b}.$$

We know all the quantities on the right side of this equation, and so we can calculate v_{Ecmx}. Next we use this Earth-measured center-of-mass velocity in Eq. 6.14 to calculate the x components of the velocities of the club and ball measured in the zero-momentum reference frame, using the equality $v_{ZEx} = -v_{EZx} = -v_{Ecmx}$:

$$v_{Zbx} = v_{ZEx} + v_{Ebx} = -v_{Ecmx} + v_{Ebx}$$
$$v_{Zcx} = v_{ZEx} + v_{Ecx} = -v_{Ecmx} + v_{Ecx}.$$

(c) The relative velocity is the same in both reference frames and so can be evaluated in either.

(d) This result gives us enough information to compute the kinetic energy in the zero-momentum reference frame as well as the convertible kinetic energy, which is the kinetic energy of the ball and club in the zero-momentum reference frame.

(e) We need to measure the translational kinetic energy K_{cm} of the system in both reference frames, using the expression

$$K_{cm} = \tfrac{1}{2}(m_c + m_b)(v_{cm})^2$$

in each one.

❸ EXECUTE PLAN

(a) The kinetic energy of the club-ball system measured in the Earth reference frame is

$$K_E = \tfrac{1}{2}m_c v_{Ecx}^2 + \tfrac{1}{2}m_b v_{Ebx}^2$$
$$= \tfrac{1}{2}(0.100\,\text{kg})(45\,\text{m/s})^2 + \tfrac{1}{2}(0.050\,\text{kg})(0)^2$$
$$= 1.0 \times 10^2\,\text{J}. ✔$$

(b) The value an observer in the Earth reference frame measures for the x component of the velocity of the zero-momentum reference frame is the same as the value that Earth observer measures for the x component of the velocity of the center of mass of the system:

$$v_{Ecmx} = \frac{m_c v_{Ecx} + m_b v_{Ebx}}{m_c + m_b} = \frac{(0.100\,\text{kg})(+45\,\text{m/s})}{0.100\,\text{kg} + 0.050\,\text{kg}} = +30\,\text{m/s}.$$

The x components of the velocities of club and ball measured by an observer in the zero-momentum reference frame are therefore

$$v_{Zbx} = -v_{Ecmx} + v_{Ebx} = -30\,\text{m/s} + 0 = -30\,\text{m/s} ✔$$
$$v_{Zcx} = -v_{Ecmx} + v_{Ecx} = -30\,\text{m/s} + 45\,\text{m/s} = +15\,\text{m/s}. ✔$$

(c) We get the x component of relative velocity in the zero-momentum reference frame from our transformed velocities in part b:

$$v_{Zbcx} = v_{Zcx} - v_{Zbx} = (+15\,\text{m/s}) - (-30\,\text{m/s}) = +45\,\text{m/s}. ✔$$

(d) The convertible kinetic energy is the kinetic energy of the system measured in the zero-momentum reference frame:

$$K_{conv} = K_Z = \tfrac{1}{2}m_c(v_{Zcx})^2 + \tfrac{1}{2}m_b(v_{Zbx})^2$$
$$= \tfrac{1}{2}(0.100\,\text{kg})(+15\,\text{m/s})^2 + \tfrac{1}{2}(0.050\,\text{kg})(-30\,\text{m/s})^2$$
$$= 34\,\text{J}. ✔$$

(e) The translational kinetic energy of this system in the zero-momentum reference frame is easy to calculate because the system's center of mass is at rest in this reference frame:

$$K_{Zcm} = \tfrac{1}{2}(m_c + m_b)(v_{Zcmx})^2 = \tfrac{1}{2}(0.150\,\text{kg})(0)^2 = 0. ✔$$

The system has no translational kinetic energy in the zero-momentum reference frame because the system's center of mass is not translating in that reference frame! In the Earth reference frame, the translational kinetic energy is

$$K_{Ecm} = \tfrac{1}{2}(m_c + m_b)(v_{Ecmx})^2$$
$$= \tfrac{1}{2}(0.150\,\text{kg})(30\,\text{m/s})^2 = 68\,\text{J}. ✔$$

❹ EVALUATE RESULT The signs and magnitudes of the x components of the velocities in the zero-momentum reference frame match our expectations shown in Figure WG6.3. The x component of the relative velocity should be the same in all reference frames. We computed it to be $v_{Zbcx} = 45\,\text{m/s}$ in the zero-momentum reference frame. It should be the same in the Earth reference frame:

$$v_{Ebcx} = v_{Ecx} - v_{Ebx} = +45\,\text{m/s} - 0 = +45\,\text{m/s}.$$

We can check our calculation of convertible energy by using the definition of convertible kinetic energy with our Earth values:

$$K_{conv} = K_E - K_{Ecm} = 101.3\,\text{J} - 67.5\,\text{J} = 34\,\text{J}.$$

This energy is the amount that could be converted to internal energy in the collision, and it should be (and is) the same in both reference frames. This does not mean that all of it is converted, however. Because of the design and construction of the golf ball, we expect the collision to be nearly elastic, which means that little of the system's kinetic energy is converted to internal energy.

Guided Problem 6.6 Crunch!

Compare the energy lost in a totally inelastic collision between a moving car and a stationary car of the same inertia with the energy lost in a totally inelastic collision between a moving car and a bridge abutment.

❶ GETTING STARTED

1. Even if you think they might not be needed, it is good practice to make sketches of the two crashes. What scalar and/or vector variables should you include? It is generally good to include any variables that you use in your equations. (You can add these symbols to your sketch as you go along.)
2. What amount of kinetic energy is convertible (for instance, in deforming a car and making noise)?

❷ DEVISE PLAN

3. Because the collision is totally inelastic in both cases, what fraction of the available convertible energy is converted to internal energy?
4. Write the equation for convertible kinetic energy (Eq. 6.40). What quantity do you need to determine in this equation?

5. Which variable in this equation is different in the two cases? What is its value in each case?
6. Suppose in each case the moving car has inertia m_{moving} and what it runs into has inertia m_{rest}. Write an expression for the reduced inertia in this general case.

❸ EXECUTE PLAN

7. Are there too many unknowns? Perhaps if you make a ratio of the converted energies in the two cases some variables will cancel.

❹ EVALUATE RESULT

8. Based on the result, which would you rather collide with: a parked car or a concrete wall? Do your results agree with your common sense?
9. Were any assumptions needed? If so, check that they are not unreasonable.

Worked Problem 6.7 Hot reactors

Fast-moving neutrons in the core of a nuclear fission reactor are slowed down by collisions with carbon nuclei in the graphite moderator in the core. The ratio of the inertia of a carbon nucleus to the inertia of a neutron is about 12:1. If a neutron moving at 2.0×10^7 m/s strikes a stationary carbon nucleus head-on in an elastic collision, what are the velocities of the two particles after the collision?

❶ GETTING STARTED Figure WG6.4 is a drawing that displays the information given. Because we may need either velocity or momentum equations, it is convenient to show the neutron's initial velocity with a labeled vector arrow and to indicate the inertias m and $12m$.

Figure WG6.4

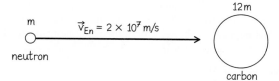

We could answer the question by using conservation of momentum and kinetic energy as we did in Chapter 5, but the algebra might be intimidating. This is because we are looking at the collision from the reference frame in which the carbon nucleus is at rest (the Earth reference frame E), our analysis will be easier if we move to the zero-momentum reference frame Z.

❷ DEVISE PLAN We anticipate a Galilean transformation to the zero-momentum reference frame. We can use Eq. 6.26 to get the velocity of the zero-momentum reference frame measured by an observer in the Earth reference frame:

$$\vec{v}_{EZ} = \vec{v}_{cm} = \frac{m_n \vec{v}_n + m_c \vec{v}_c}{m_n + m_c}.$$

Because the momenta sum to zero in the zero-momentum reference frame, the collision looks like what is shown in Figure WG6.5, where we have included an x axis so that we can assign signs to our vectors. Because this is the zero-momentum reference frame, the momentum in the positive x direction must cancel the momentum in the negative x direction. Therefore we know that, because of the large differences in inertias, the speed of the carbon nucleus must be much lower than the speed of the neutron in order to have $|m_n \vec{v}_n| = |m_c \vec{v}_c|$.

Figure WG6.5

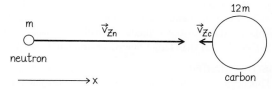

We can compute the initial velocity of each particle in the zero-momentum reference frame using Eq. 6.14 with the appropriate subscripts:

$$\vec{v}_{Zn} = \vec{v}_{ZE} + \vec{v}_{En} = -\vec{v}_{EZ} + \vec{v}_{En} = -\vec{v}_{cm} + \vec{v}_{En}$$

$$\vec{v}_{Zc} = \vec{v}_{ZE} + \vec{v}_{Ec} = -\vec{v}_{EZ} + \vec{v}_{Ec} = -\vec{v}_{cm} + \vec{v}_{Ec}.$$

Because this is an elastic collision, the final relative velocity in the x direction is the negative of the initial value. This means that in the zero-momentum reference frame the momentum of each particle changes direction by 180°. We should have enough information to determine the final velocity of each particle after the collision. We then do another transformation using Eq. 6.14 to determine the final velocities measured in the Earth reference frame.

❸ **EXECUTE PLAN** We first compute the transformation to the zero-momentum reference frame by determining the x component of the velocity of the center of mass measured in the Earth reference frame:

$$v_{cmx} = \frac{m(+v_n) + 12m(0)}{m + 12m} = +\tfrac{1}{13}v_n.$$

The x components of the initial velocities of the two particles in the zero-momentum reference frame are then

$$v_{Znx,i} = -v_{cmx} + v_{Enx,i} = -\tfrac{1}{13}v_n + v_n = +\tfrac{12}{13}v_n$$

$$v_{Zcx,i} = -v_{cmx} + v_{Ecx,i} = -\tfrac{1}{13}v_n + 0 = -\tfrac{1}{13}v_n.$$

The negative sign for $v_{Zcx,i}$ means that the carbon nucleus is moving in the negative x direction. In the zero-momentum reference frame, the initial momenta are equal in magnitude and opposite in direction, and the same is true after the collision, when the momenta have both changed direction by 180° (Figure WG6.6).

Figure WG6.6

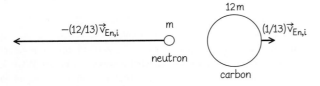

To answer the posed question, we must transform back to the Earth reference frame, again using Eq. 6.14:

$$v_{Enx,f} = v_{EZx} + v_{Znx,f} = +\tfrac{1}{13}v_n + \left(-\tfrac{12}{13}v_n\right) = -\tfrac{11}{13}v_n$$

$$v_{Ecx,f} = v_{EZx} + v_{Zcx,f} = +\tfrac{1}{13}v_n + \left(+\tfrac{1}{13}v_n\right) = +\tfrac{2}{13}v_n.$$

The neutron reverses direction and speeds away from the carbon nucleus at a velocity of

$$-\tfrac{11}{13}(2.0 \times 10^7 \text{ m/s}) = -1.7 \times 10^7 \text{ m/s}, ✔$$

and the carbon nucleus moves in the positive x direction at a velocity of

$$+\tfrac{2}{13}(2.0 \times 10^7 \text{ m/s}) = +3.1 \times 10^6 \text{ m/s}. ✔$$

❹ **EVALUATE RESULT** The neutron has lost only 15% of its initial speed, and the post-collision speed of the carbon nucleus is much lower than that of the neutron. This is what we expect for a collision between, say, a golf ball and a basketball, whose inertias differ by a factor of ten or so.

Guided Problem 6.8 Slowing neutrons

In a nuclear reactor, a few uranium nuclei spontaneously split into the nuclei of lighter elements plus several fast neutrons. These neutrons are slowed by the *moderator*, which makes them more likely to be absorbed by other uranium nuclei. That causes them to split and increases the rate of the nuclear reaction. As noted in Worked Problem 6.7, this is usually achieved through elastic collisions between the neutrons and atomic nuclei. The moderator used to slow the neutrons is made of small atoms (like hydrogen and carbon) rather than larger ones (like iron and lead). Why?

❶ **GETTING STARTED**

1. This collision problem is similar to Worked Problem 6.7, but the important question is: How is it different?
2. What is the desired outcome of the collision between the neutrons and the atoms in the moderator? What physical quantity is associated with this goal?

❷ **DEVISE PLAN**

3. How can you modify the Worked Problem 6.7 analysis to allow a variable inertia m_T for the target nucleus?

4. Do you need any additional information to do the analysis?
5. Determine the neutron's velocity after the collision in terms of m_n and m_T.

❸ **EXECUTE PLAN**

6. What is the smallest value possible for the ratio m_T/m_n? (Hint: A hydrogen nucleus "target" is generally a single proton.) What happens to the velocity of the recoiling neutron when this ratio has its smallest value?
7. What is the largest value of m_T one might reasonably have? What happens to the velocity of the recoiling neutron when m_T has this largest value?

❹ **EVALUATE RESULT**

8. Compare your results with your expectations for a collision between two everyday objects whose inertias are in the same ratio as the inertias of the neutron and target.

Questions and Problems

For instructor-assigned homework, go to MasteringPhysics®   (MP)

Dots indicate difficulty level of problems: • = *easy,* •• = *intermediate,* ••• = *hard;* CR = *context-rich problem.*

6.1 Relativity of motion

1. Moving sidewalks are commonplace at large airports, but you must use caution getting on and off them. What should you do to make the transition easy when stepping onto one of these sidewalks? When stepping off? •

2. A flying bug hits the helmet of a coasting bicyclist. (What kind of collision do you expect this to be?) Draw the momentum vectors for the bug and the bicyclist before and after the collision (*a*) in the reference frame in which the bug is initially at rest and (*b*) in the reference frame in which the bicyclist is initially at rest. •

3. The data in the table were recorded for two race cars, A and B, driving down a straight stretch of track. Determine the average velocity of car B as measured by an observer in car A. ••

Time (s)	Car A position (m)	Car B position (m)
1	40	20
2	65	50
3	90	80
4	115	110
5	140	140
6	165	170
7	190	200
8	215	230
9	240	260
10	265	290

4. An observer in the Earth reference frame observes cart A and cart B moving toward each other with the same speed until they collide and come to rest. Consider this event as seen by observer A moving along with cart A. Assume that this observer matches cart A's initial velocity but does not collide with cart B (that is, observer A continues along at the same velocity that cart A had initially). Would the observer in the Earth reference frame and observer A obtain the same value for (*a*) the initial velocity of cart A, (*b*) the final velocity of cart B after the collision, and (*c*) the relative velocity of the two carts after the collision? ••

5. You are sitting, facing rearward, in the bed of a pickup truck and hop off while the truck is moving. When you land, do you move toward or away from the truck (*a*) from the point of view of a person standing alongside the road and (*b*) from the point of view of a person sitting in the truck bed with you? ••

6. Two objects, A and B, of equal inertia approach each other with relative velocity $\vec{v}_{AB}$ and collide elastically. For each object, draw a velocity-versus-time graph for the interval starting a few seconds before the collision and ending a few seconds after the collision (*a*) from the reference frame in which A is initially at rest, (*b*) from the reference frame in which B is initially at rest, and (*c*) from the zero-momentum reference frame. ••

6.2 Inertial reference frames

7. A woman standing beside a road sees a car accelerate from rest to 30 m/s. Describe the car's motion as seen by the driver of a truck traveling in the same direction as the car at a constant 30 m/s. What if the truck is moving in the opposite direction? Is the direction of acceleration of the car affected by the (constant) velocity of the truck driver relative to Earth? •

8. On a long bus ride, you walk from your seat to the back of the bus to use the restroom. If the bus is driving at 100 km/h, and you walk at 2.0 m/s from your seat to the restroom, how quickly are you moving relative to the ground? •

9. A pickup truck has several empty soda cans loose in the bed. Why do the cans roll forward in the bed when the truck slows down? •

10. You drop your keys in a high-speed elevator going up at a constant speed. Do the keys accelerate faster toward the elevator floor than they would (*a*) if the elevator were not moving? (*b*) if the elevator were accelerating downward? ••

11. Train A, m_A = 150,000 kg, is traveling west at 60 km/h. Train B, m_B = 100,000 kg, behind train A on the same track, is traveling west at 88 km/h and so is gaining on train A. Because the engineer driving B fails to slow down, B runs into the back of A. The two trains stick together and then move as a single unit after the collision. What is the momentum of each train before the collision and after the collision according to (*a*) an observer standing alongside the tracks and (*b*) an observer in a westbound automobile traveling at 100 km/h on a road that parallels the tracks? ••

12. A 1000-kg car traveling east at 50 km/h passes over the top of a hill and hits a 3000-kg truck stalled in the middle of the lane. The impact causes the truck to roll eastward at 15 km/h. (*a*) What is the coefficient of restitution for the collision? During the collision, how much kinetic energy is converted to internal energy (*b*) in the Earth reference frame and (*c*) in a reference frame moving at 5 km/h eastward along the road? ••

13. You are riding a 450-kg horse at 14.4 km/h east along a desert road. You have inertia equal to 60.0 kg. A police officer driving past (whom you know and who knows your inertia and the horse's inertia) measures your speed relative to the police car and calculates your kinetic energy to be 16.32 kJ. What possible speed(s) could the police car have been driving at the instant the officer measured your speed? •••

6.3 Principle of relativity

14. Two cars collide head-on on a busy street. An observer standing on the street witnesses the accident and calculates how much of the cars' initial kinetic energy went into deforming the cars upon collision, E_{def}. A police officer was driving next to one of the cars at a matching speed when the cars collided. Would this officer calculate a higher value of E_{def}, a lower value of E_{def}, or the same value of E_{def} if he used the velocities of the cars measured in his reference frame? •

15. At an airport, two business partners both walk at 1.5 m/s from the gate to the main terminal, one on a moving sidewalk and the other on the floor next to it. The partner on the moving sidewalk gets to the end in 60 s, and the partner on the floor reaches the end of the sidewalk in 90 s. What is the speed of the sidewalk in the Earth reference frame? •

16. Do the laws of the universe change if the clocks in two inertial reference frames, A and B, are not synchronized (in other words, if $t_A = t_B + \Delta\tau$, where $\Delta\tau$ is some constant time interval)? ●●

17. You place a 0.10-kg sonic ranger on a low-friction track in front of a 0.50-kg cart to measure the cart's velocity in the Earth reference frame, which turns out to be $+(1.0 \text{ m/s})\hat{\imath}$. You are distracted, the cart hits the ranger in a totally inelastic collision, and the two objects then move forward together. A friend is running toward the cart with a velocity of $-(3.0 \text{ m/s})\hat{\imath}$ with her sonic ranger on and pointed at the cart. (a) What is the momentum of the cart, with the ranger stuck to it, in your friend's reference frame? (b) What velocity does her sonic ranger measure for the cart after the collision? ●●

18. At the roller rink, two 20-kg girls accelerate toward each other until they are each moving at 2.0 m/s in the Earth reference frame. They then collide stomach-to-stomach, grab on to each other, and fall to the floor. Calculate the momentum of each girl before the collision and after the collision (a) in the Earth reference frame and (b) in the reference frame of a mother who started skating with one of the girls and then continued on after the collision without changing speed. (c) The fathers decide the collision looks like fun and give it a try. Explain, in terms of energy, why they don't enjoy it nearly as much as their daughters do. ●●

19. Carts A and B are identical and are moving toward each other on a track. The speed of cart A is v, while the speed of cart B is $2v$. In the Earth reference frame, the system of the two carts has kinetic energy K. Is there any other reference frame in which the two-cart system has the same kinetic energy K? If so, describe this reference frame. Otherwise, explain why no such reference frame exists. ●●●

6.4 Zero-momentum reference frame

20. Two identical cars approach each other head-on while traveling at the same speed. What does an observer in the Earth reference frame measure for the speed of the zero-momentum reference frame of this two-car system? Suppose the same cars are traveling in the same direction. Now what does an observer in the Earth reference frame measure for the speed of the zero-momentum reference frame? ●

21. Is the kinetic energy of a system zero when measured from the zero-momentum reference frame for the system? ●

22. Two identical particles, A and B, collide elastically. In the zero-momentum reference frame, what can you say about the ratios of the final and initial speeds, $v_{A,f}/v_{A,i}$ and $v_{B,f}/v_{B,i}$? ●

23. Figure P6.23 shows three identical carts placed on rails (so that they can slide easily left and right) and connected to one another by springs. There are two simple, symmetrical ways for this system to vibrate in the zero-momentum reference frame. One way is the sweeping-X pattern shown in the figure. Draw the other way. ●●

Figure P6.23

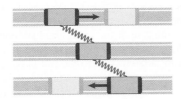

24. Consider two objects, A and B, of inertias m_A and $m_B \gg m_A$. If the two are moving at constant velocities with $\vec{v}_A \neq \vec{v}_B$, is the velocity of the zero-momentum reference frame closer to $\vec{v}_A$ or to $\vec{v}_B$? ●●

25. A 4000-kg dump truck is parked on a hill. The parking brake fails, and the truck rolls down the hill. It then coasts briefly along a flat stretch of road at 36 km/h before hitting a stationary 1000-kg car. The car sticks to the grill of the truck, and the two vehicles continue moving forward. (a) What is the speed of the combined unit immediately after the collision? (b) What is the momentum of the combined unit in the reference frame of a jogger approaching the collision at 2.0 m/s and moving in the direction opposite the direction of the velocity of the moving unit? (c) How fast, and in which direction, must the jogger be running in order for the momentum of the combined unit to be zero in his reference frame? ●●

26. Object A has ten times the inertia of object B. They approach each other with relative velocity $\vec{v}_{AB}$ and collide elastically. For each object, draw a velocity-versus-time graph for the interval starting a few seconds before the collision and ending a few seconds after the collision (a) from the reference frame in which A is initially at rest, (b) from the reference frame in which B is initially at rest, and (c) from the zero-momentum reference frame. ●●●

27. In an elastic collision between a lightweight object and a heavy object, which one carries away more of the kinetic energy? Does the answer depend on the initial speeds? (Hint: Begin with the zero-momentum reference frame for clarity.) ●●●

6.5 Galilean relativity

28. When two identical objects traveling at the same speed collide head-on, an observer standing in the Earth reference frame sees both objects changing direction. Do observers in every inertial reference frame that is moving relative to the Earth reference frame also see both objects changing direction? ●

29. You toss a ball into the air and note the time interval between the ball leaving your hand and reaching its highest position. While you are doing this, a construction worker being lifted on a hydraulic platform at constant speed also notes the time interval needed for the ball to reach its highest position. Is the time interval reported by the worker *longer, shorter,* or *the same as* the interval you report? ●●

30. A student runs an experiment with two carts on a low-friction track. As measured in the Earth reference frame, cart 1 ($m = 0.36$ kg) moves from left to right at 1.0 m/s as the student walks along next to it at the same velocity. (a) What velocity $\vec{v}_{E2,i}$ in the Earth reference frame must cart 2 ($m = 0.12$ kg) have before the collision if, in the student's reference frame, cart 2 comes to rest right after the collision and cart 1 reverses direction and travels from right to left at 0.33 m/s? (b) What does the student measure for the momentum of the two-cart system? (c) What does a person standing in the Earth reference frame measure for the momentum of each cart before the collision? ●●

31. Notice that in *Principles* Figure 6.12 the time intervals of the interaction are all the same, regardless of the reference frame shown. Is this true in all reference frames? Why or why not? ●●

32. In a three-car crash, car A bumps into the back end of car B, which then goes forward and bumps into the back end of car C. Is the distance that car B moves between the collisions the same in all inertial reference frames? ••

33. Riding up an escalator while staying on the same step for the whole ride takes 30 s. Walking up the same escalator takes 20 s. How long does it take to walk down the up escalator? ••

34. A woman is on a train leaving the station at 4.0 m/s, while a friend waving goodbye runs alongside the car she's in. (*a*) If the friend is running at 6.0 m/s and moving in the same direction as the train, how fast must the woman walk, and in which direction, to keep up with him? (*b*) Once the train has reached a speed of 10 m/s, how fast must the woman walk, and in which direction, to keep up with her friend? ••

35. Airline pilots who fly round trips know that their round-trip travel time increases if there is any wind. To see this, suppose that an airliner cruises at speed v relative to the air. (*a*) For a flight whose one-way distance is d, write an expression for the interval Δt_{calm} needed for a round trip on a windless day. Ignore any time spent on the ground, and assume that the airliner flies at cruising speed for essentially the whole trip. (*b*) Now assume there is a wind of speed w. It doesn't matter which way the wind is blowing; all that matters is that it is a head wind in one direction and a tail wind in the opposite direction. Write an expression for the time interval Δt_{wind} needed for a round trip on the day this wind is blowing. Then show that your expression reduces to

$$\Delta t_{wind} \approx \Delta t_{calm}\left[1 + \left(\frac{w}{v}\right)^2\right]$$

provided that $w \ll v$. (For the last step, you may need to know that $(1 \pm z)^b \approx 1 \pm bz$ when $z \ll 1$.) •••

6.6 Center of mass

36. The inertia of an object is m measured when the object is at rest in the Earth reference frame. According to Galilean relativity, what is its inertia measured by an observer moving past the object with a constant velocity $\vec{v}$ in the positive x direction? What is the object's momentum according to this observer? •

37. At what distance from the center of Earth is the center of mass of the Earth-Moon system? •

38. (*a*) Determine the location of the center of mass of the system shown in Figure P6.38. All three disks are made of sheet metal of the same material, and the diameters are 1.0 m, 2.0 m, and 3.0 m. (*b*) Repeat the calculation for three solid spheres all made of the same metal and having the same diameters as in part *a*. ••

Figure P6.38

39. A boy and a girl are resting on separate rafts 10 m apart in calm water when the girl notices a small beach toy floating midway between the rafts. The girl and her raft have twice the inertia of the boy and his raft. The rafts are connected by a rope 12 m long, so she decides to pull on the rope, drawing the rafts together until she can reach the toy. Which raft gets to the toy first? How much distance is there between the two rafts when the first one reaches the toy? ••

40. The two cubes in Figure P6.40 have different inertias. The cubes are connected to each other by a spring, and a hammer strikes them in the two ways, (*a*) and (*b*), shown in the figure. Assuming that the same impulse is transferred from the hammer, does the center-of-mass motion after the collision depend on which cube is struck first? ••

Figure P6.40

(*a*)   (*b*)

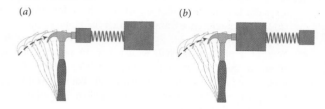

41. Determine the position of the center of mass of the baton shown in Figure P6.41, taking the origin of your coordinate axis to be (*a*) the center of the larger ball, (*b*) the center of the smaller ball, and (*c*) a point 1.0 m to the left of the larger ball. How much calculation was required for each of the three parts of this problem? ••

Figure P6.41

0.20 kg   0.10 kg   0.10 kg

|←—— 1.0 m ——→|←—— 1.0 m ——→|

42. The empty cubical box shown in Figure P6.42 has no top face; that is, the box is made up of only five square faces. If all five faces have the same inertia, at what height above the bottom of the box is the center of mass? ••

Figure P6.42

43. How can you tell from the motion of the center of mass of an isolated system whether the reference frame from which the motion is measured is inertial? ••

44. Determine the center of mass of a pool cue whose diameter decreases smoothly from 40 mm to 10 mm over its 1.40-m length (Figure P6.44). Assume that the cue is made from solid wood, with no hidden weights inside. (Hint: See Appendix D for the center-of-mass computation for extended objects. You will find it easier to do the integral for a complete cone. The pool cue is a truncated cone—that is, a cone with its conical tip removed. Slicing off a piece is like adding negative inertia.) •••

Figure P6.44

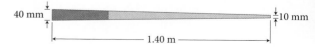

40 mm   10 mm

|←————————— 1.40 m —————————→|

6.7 Convertible kinetic energy

45. An object of inertia m_1 collides totally inelastically with a stationary object of inertia m_2. Plot the fraction of the kinetic energy lost as a function of m_2/m_1 in the range $m_2/m_1 = 0$ to $m_2/m_1 = 4$, and discuss what happens as m_2/m_1 approaches infinity. •

46. A 3.0-g particle is moving toward a stationary 7.0-g particle at 3.0 m/s. What percentage of the original kinetic energy is convertible to internal energy? •

47. Think of a system of two objects of different inertias $m_1 < m_2$. How much larger than m_1 can the reduced inertia μ of the pair be? How much smaller than m_1 can it be? •

48. (a) Is there a reference frame in which the kinetic energy of a system is a minimum? If so, what is this reference frame? (b) Is there a reference frame in which the kinetic energy of a system is a maximum? If so, what is this reference frame? ••

49. You hit a pitched baseball with a bat. In which reference frame is the translational (nonconvertible) kinetic energy greater: the reference frame in which the bat is at rest before the collision or the reference frame in which the ball is at rest before the collision? ••

50. What is the ratio K_1/K_2 when K_1 is the kinetic energy converted to internal energy when two cars each initially traveling at 88 km/h collide head-on and K_2 is the kinetic energy converted to internal energy when a car moving at 88 km/h hits a stationary car? Assume that both cars have the same inertia and that the collisions are totally inelastic. (Hint: Determine how much of the energy is required to move the center of mass in each case.) ••

51. A 0.075-kg disk initially at rest in the Earth reference frame is free to move parallel to a horizontal bar through a hole at the disk's center. The disk is struck face-on by a 0.050-kg paintball traveling at 11 m/s, as illustrated in Figure P6.51. (a) What is the initial relative velocity of the disk and the paintball? (b) What is their reduced inertia? (c) If all the paint stays on the disk, what percentage of the original kinetic energy is converted to internal energy? (d) What is the change in kinetic energy measured by someone riding a bicycle at the velocity of the paintball before the collision? (e) Confirm your answer to part d by calculating the final and initial kinetic energies in the cyclist's reference frame. ••

Figure P6.51

52. Take the common case where a moving object of inertia m_{moving} collides with a stationary object of inertia m_{rest}. (a) Show that the fraction of kinetic energy not convertible in the collision is $m_{\text{moving}}/(m_{\text{moving}} + m_{\text{rest}})$. Interpret for the case where $m_{\text{rest}} \gg m_{\text{moving}}$ and for the case where $m_{\text{rest}} \ll m_{\text{moving}}$. (b) Does the value of the fraction depend on the elasticity of the collision? (c) Why is this energy not convertible? ••

53. Ball 1 is moving toward you at 10 m/s, and you decide to throw ball 2 at it to make it reverse its velocity. The balls collide head-on, and the coefficient of restitution for the collision is 0.90. (a) If ball 1 has an inertia of 0.500 kg and ball 2 has an inertia of 0.600 kg, how fast must ball 2 be traveling in order to reverse the velocity of ball 1? (b) What is the initial relative velocity of the two balls? (c) What is their reduced inertia? (d) What percentage of the original kinetic energy is convertible? (e) What are the final velocities of the two balls immediately after the collision? ••

54. Ball 1 from Problem 53 is moving away from you at 5.0 m/s, and you decide to throw ball 2 at it to make it go faster. Once again the balls collide head-on, and the coefficient of restitution for the collision is 0.90. (a) Given the inertias in Problem 53, how fast must ball 2 be traveling in order to double the velocity of ball 1? (b) What is the initial relative velocity of the two balls? (c) What is their reduced inertia? (d) What percentage of the original kinetic energy is convertible? (e) What are the final velocities of the two balls immediately after the collision? ••

55. If the relative velocities are the same in both cases, which has more convertible kinetic energy: a collision between two objects each of inertia m or a collision between an object of inertia m_1 and an object of inertia $m_2 < m_1$ with $m_1 + m_2 = 2m$? ••

56. After coming down a slope, a 60-kg skier is coasting northward on a level, snowy surface at a constant 15 m/s. Her 5.0-kg cat, initially running southward at 3.8 m/s, leaps into her arms, and she catches it. (a) Determine the amount of kinetic energy converted to internal energy in the Earth reference frame. (b) What is the velocity, measured in the Earth reference frame, of an inertial reference frame in which the cat's kinetic energy does not change? ••

57. You toss a 0.40-kg ball at 9.0 m/s to a 14-kg dog standing on an iced-over pond. The dog catches the ball and begins to slide on the ice. (a) Measured from the Earth reference frame, what is the velocity of the dog immediately after he catches the ball? (b) Measured from the Earth reference frame, what is the velocity of an inertial reference frame in which the ball's kinetic energy does not change? (c) Measured from the Earth reference frame, how much of the original kinetic energy of the system is convertible? (d) Measured from the reference frame described in part b, how much of the original kinetic energy of the system is convertible? ••

58. Instead of defining a reduced inertia μ to characterize the convertible kinetic energy of a system, we could define a *reduced velocity* v_{red} as follows: For a system of two particles, one of inertia m_1 and velocity v_1 and the other of inertia m_2 and velocity v_2, the kinetic energy is

$$K = K_{\text{cm}} + K_{\text{conv}} = \tfrac{1}{2}(m_1 + m_2)v_{\text{cm}}^2 + \tfrac{1}{2}(m_1 + m_2)v_{\text{red}}^2.$$

Derive an expression for the reduced velocity in terms of the given variables. (Just as the center-of-mass velocity is a weighted mean of the individual velocities, the reduced velocity is a weighted difference between the individual velocities.) ••••

6.8 Conservation laws and relativity

59. In *Principles* you learned about the zero-momentum reference frame. You may be curious whether there is such a thing as a zero-energy reference frame. Does a zero-kinetic energy reference frame always exist, never exist, or sometimes exist? •

60. After a totally inelastic collision, the kinetic energy of an isolated system composed of two objects is zero. What was the momentum of the system in the same reference frame before the collision? •

61. A mother penguin and her chick are on a flat, icy surface. The mother is lying at rest 0.50 m from the edge of the water. The chick, which has one-fourth of its mother's inertia, is sliding, collides with her inelastically, and bounces back at one-eighth of its original speed. The mother wakes up as she hits the water 0.40 s later. (*a*) How fast was the chick going when it hit her? (*b*) To a penguin paddling directly toward them at 1.0 m/s, what is the mother's momentum before and after the collision in terms of m_{mother}? ••

62. A 50-kg ice skater moves across the ice at a constant speed of 2.0 m/s. She is caught by her 70-kg partner, and then the pair continues to glide together. He is at rest when he catches her, and immediately afterward they both coast. (*a*) What is their velocity just after the catch? (*b*) A skater follows the girl at a constant speed of 1.0 m/s. What is the girl's change in momentum in the reference frame of the skater? ••

63. In an inertial reference frame F, an orange of inertia m_{orange} and velocity $\vec{v}_{orange}$ collides totally inelastically with an apple of inertia m_{apple} initially at rest. (*a*) When this collision is viewed from an inertial reference frame G, the kinetic energy of the orange is unchanged by the collision. What does an observer in the F reference frame measure for the velocity of the G reference frame? (*b*) For the case where $m_{orange} = m_{apple}$, sketch the velocity vectors of the two objects in each reference frame before and after the collision. ••

64. A 50-kg meteorite moving at 1000 m/s strikes Earth. Assume the velocity is along the line joining Earth's center of mass and the meteor's center of mass. (*a*) Calculate the amount of kinetic energy converted to internal energy measured from the Earth reference frame. (*b*) What is the gain in Earth's kinetic energy? (*c*) Why does the ground around where the meteor strikes get so hot? ••

65. A 0.20-kg softball is traveling at a velocity of 20 m/s to the east relative to Earth. It collides head-on with a 0.40-kg rubber ball traveling at a velocity of 10 m/s to the west. (*a*) If the system's kinetic energy, as measured from the Earth reference frame, decreases by 20% because of the collision, what are the final velocities of the balls? (*b*) What change in internal energy has occurred? (*c*) An observer watches this collision from a reference frame moving at a velocity of 15 m/s to the east relative to the Earth reference frame. What changes in kinetic and internal energies does this observer measure? (*d*) What changes in kinetic and internal energies would be measured by an observer in a reference frame traveling at 20 m/s east relative to the Earth reference frame? ••

66. Derive an expression showing that when an elastic collision between two objects is viewed from the zero-momentum reference frame, the direction of the momentum of each object is reversed and the magnitude of the change in momentum of each object is twice the magnitude of that object's initial momentum. •••

Additional Problems

67. Just as a car passes a school crossing guard, a child throws a toy from the back seat of the car toward his sister in the front seat. The toy is thrown at a speed of 2.0 m/s relative to the car, and the car is traveling at 10 m/s relative to Earth. Draw the velocity vectors for the toy, the car, and the crossing guard from the reference frame of (*a*) the car, (*b*) the guard, and (*c*) the toy. •

68. A 20-kg child is sliding on an icy surface toward her mother at 3.0 m/s. Her 68-kg mother starts toward her at 2.0 m/s, intending to catch her. What percentage of the original kinetic energy is convertible? •

69. An 80-kg man is walking at 2.0 m/s. A 10-kg dog is running at five times that speed in the same direction. At what speed and in what direction relative to the man would you have to be jogging in order for the dog to have the same momentum as the man in your reference frame? •

70. One rider is in a descending elevator that is accelerating to a stop. Another is in an elevator that is accelerating upward from rest. With their eyes closed, can the riders tell which of the two elevators they are in? ••

71. Suppose you are in a noninertial reference frame observing two isolated objects of inertias m_1 and m_2. What can you say about (*a*) the apparent acceleration of each object and (*b*) the apparent change in momentum of each object? ••

72. A bug of inertia m_B collides with the windshield of a Mack truck of inertia $m_T \gg m_B$ at an instant when the relative speed of the two is v_{BT}. (*a*) Express the system momentum in the truck's reference frame, then transform that expression to the bug's reference frame, and in so doing remove $m_B v_{BT}$ from the expression. (Remember, in the bug's reference frame, the bug is initially at rest and the truck is moving.) (*b*) Now express the system momentum in the bug's reference frame, then transform that expression to the truck's reference frame, and in so doing remove $m_T v_{BT}$ from the expression. (*c*) Is there something wrong here? How can we change the momentum by a small amount $m_B v_{BT}$ doing the transformation one way and by a large amount $m_T v_{BT}$ doing the transformation the other way? ••

73. A 0.045-kg golf ball moving at 50 m/s (measured in the Earth reference frame) collides inelastically with a 1.8-kg, heavy-duty plastic flowerpot sitting on a windowsill. The coefficient of restitution for the collision is 0.50. Calculate the final velocities of the ball and the pot. Do this problem by first transforming to the zero-momentum reference frame, where the collision is much simpler, and then transforming back to the Earth reference frame. (Assume that all the action happens in one dimension and that the friction between pot and sill is insignificant.) ••

74. Asteroid A1, $m_{A1} = 3.60 \times 10^6$ kg, and asteroid A2, $m_{A2} = 1.20 \times 10^6$ kg, collide head-on in space. Approximate (rather poorly) the collision as being elastic. Observers

PRACTICE

watch the event from two space platforms. An observer on platform Q measures the speed of A1 to be 528 m/s and the speed of A2 to be 315 m/s, and sees A2 heading directly toward A1. The second platform, platform Z, is at rest in the zero-momentum reference frame of the asteroids. (*a*) What is the combined momentum of the asteroids measured from platform Q? (*b*) What is the velocity of platform Z relative to platform Q? (*c*) Draw velocity, momentum, and kinetic energy graphs for the asteroids as viewed from platform Z, using graphs like those in Figure P6.74. Assume the collision begins at $t = 40$ s and lasts for 20 s. (*d*) Add a curve for the momentum of the system to the momentum graph you constructed in part *c*. (*e*) What happens to the kinetic energy when its curve bends? (*f*) On your kinetic energy graph from part *c*, show what an observer on platform Z sees for the convertible and nonconvertible kinetic energies of the asteroids. (*g*) Repeat parts *c*–*f* as seen from platform Q. ••

Figure P6.74

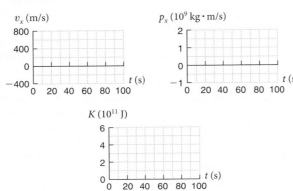

75. Assume that the asteroids in Problem 74 collide totally inelastically. Answer the same questions posed in that problem for this situation. ••

76. A 1500-kg van is coasting to a stoplight at 15 m/s. A 1000-kg car behind the van is doing the same thing at 25 m/s and crashes into the rear of the van. The bumpers collide with a coefficient of restitution of 0.70. (*a*) What are the final velocities of the two vehicles? (*b*) How much kinetic energy is converted to internal energy in this collision? (*c*) How much energy is converted if the bumpers lock together in a totally inelastic collision? ••

77. A 0.30-kg cart traveling along a low-friction track at 2.0 m/s relative to Earth collides with a 0.50-kg cart traveling in the same direction at 1.0 m/s relative to Earth. If the system's kinetic energy measured in the Earth reference frame increases by 30% as a result of the collision, what are the final velocities of the carts in this reference frame? ••

78. The transformation between position and time measurements in an inertial reference frame I and position and time measurements in a constantly accelerating (noninertial) reference frame N is given by

$$t_N = t_I = t$$

$$\vec{r}_N = \vec{r}_I - \vec{v}_{IN}t - \tfrac{1}{2}\vec{a}_{IN}t^2,$$

where $\vec{v}_{IN}$ is the velocity of the noninertial reference frame at $t = 0$ measured from the inertial reference frame and $\vec{a}_{IN}$ is the acceleration of the noninertial reference frame measured from the inertial reference frame. Derive by differentiation the transformation laws, giving v_N and a_N in terms of v_I and a_I. •••

79. The medallion shown in Figure P6.79 has been made by cutting a small circular piece out of a larger circular disk. The diameter of the original disk is twice the diameter of the hole, and the thickness of the disk is uniform. You begin to wonder about the location of the center of mass of this object, but the calculation seems impossibly difficult. Then you visualize two circular pieces coming together in superposition. ••• CR

Figure P6.79

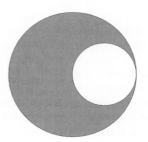

80. As an avid biker, you've come to realize that your top speed is really an air speed, not a ground speed, because air resistance is very noticeable when you ride fast. You like to ride at top speed for exactly an hour every day, and you know that on a calm day, you can ride 20 km before having to turn around and come back the same way. Today, though, you have a 20-km/h head wind for the first leg of the trip (which of course is a tailwind on the way back). You must be back home in an hour, so you look at your map. ••• CR

81. An uncoupled 20,000-kg railroad hopper car coasts along a track at 2.0 m/s. As it passes a grain chute, the chute opens (Figure P6.81), grain fills the car at a rate of 4000 kg/s for 5.0 s, and the chute closes. (*a*) How fast is the car going after the chute closes? (*b*) How far does the car travel during the 5.0 s in which grain is being added? (This distance is the length of the trail of grain in the car.) (*c*) Where is the center of mass of the grain-car system at the end of the 5.0 s? (*d*) Where is the center of mass of the trail of grain at the end of the 5.0 s? (*e*) Why doesn't the center of mass of the trail of grain coincide with the point midway between the two ends of the trail? •••

Figure P6.81

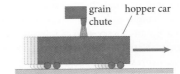

82. A skateboarder coasting at 5.0 m/s decides to coast into a friend going 4.0 m/s on rollerblades in the same direction. Unfortunately, the rollerblader stops right before the collision, too quickly for the skateboarder to react. The rollerblader turns around as he stops and pushes his arms out to ward off the collision. The rollerblader and his gear have a combined inertia of 70 kg, and the skateboarder and his gear have a combined inertia of 79 kg. (*a*) If the skateboarder ends up moving backward at 2.0 m/s, what is the coefficient of restitution for the collision? (*b*) What are the changes in kinetic and internal energies measured by an observer standing alongside the collision? (*c*) What are the changes in kinetic and internal energies measured by an observer in the zero-momentum reference frame for the skateboarder-rollerblader system? (*d*) If the rollerblader had not turned around and stopped, so that the skateboarder was able to grab him and keep coasting, what amount of kinetic energy would have been lost in the collision? ●●●

Answers to Review Questions

1. (*a*) The observer in the car moving with the same velocity as the truck sees the truck as not moving. (*b*) The observer in the car moving at any constant velocity different from the truck velocity sees the truck as moving at some constant velocity that is different from the observer's velocity.

2. In a reference frame attached to the wind, the air is at rest. When you ride into the wind, your velocity measured in this reference frame is higher than your velocity measured in the Earth reference frame. The higher your speed measured in the wind reference frame, the more you feel like you are riding through a storm. When you are moving in the direction of the wind and your speed happens to be the same as the wind speed, you sense the air as being at rest; in other words, you feel no wind at all.

3. The observer in reference frame 2 also measures $\Delta \vec{p} = 0$ because the measured value is zero in *any* inertial reference frame.

4. The law of inertia: In an inertial reference frame, any isolated object at rest remains at rest and any isolated object in motion keeps moving with constant velocity. This law is useful in determining whether a reference frame is inertial: You observe as many isolated objects as possible, and if any of them violates the law, you are observing from a noninertial reference frame. If no objects violate the law, you are observing from an inertial reference frame.

5. The observer is in a noninertial reference frame because the law of inertia does not hold.

6. (*a*) Yes. The acceleration is the same in all inertial reference frames. (*b*) No. An observer in some other inertial reference frame can measure a nonzero velocity for the object. The only thing you can say about this measured nonzero velocity is that it is constant.

7. The principle of relativity: The laws of the universe are the same in all inertial reference frames. This means that it is not necessary to invent different laws to explain results obtained by observers moving at any constant velocity with respect to each other; one set of laws is universally valid.

8. No, because one inertial reference frame is physically indistinguishable from any other inertial reference frame. You can determine whether or not your reference frame is inertial, but you can only measure its velocity relative to the velocity of other inertial reference frames.

9. Because the second reference frame is inertial, the object's acceleration in it is the same as the acceleration in the first reference frame.

10. Conservation of momentum doesn't mean that the momentum of the system must have the same numerical value in all inertial reference frames. What it does mean is that momentum can be neither created nor destroyed, so that any changes in system momentum must be accounted for in terms of impulse (momentum input or output). Because changes in system momentum are identical in all inertial reference frames, momentum conservation is equally valid in all of them.

11. They disagree on the kinetic energy of each object and on the amount of kinetic energy in the system, but they agree that the kinetic energy of the system does not change as a result of the collision. They also disagree about the momentum of each object and the system momentum, but they agree that the momentum of the system does not change.

12. The zero-momentum reference frame is the inertial reference frame in which the momentum of a system is zero. It can be identified by computing the system momentum in some inertial reference frame. If the system momentum is zero in your chosen reference frame, you have the zero-momentum reference frame for the system. If your computed momentum is not zero, compute the velocity of the zero-momentum reference frame by dividing the system momentum you calculated in your chosen reference frame by the system inertia. Subtract this velocity from each object's velocity to obtain the velocity of each object in the zero-momentum reference frame.

13. (*a*) The magnitude does not change. (*b*) The direction reverses.

14. They give us a means to relate measurements made in one inertial reference frame to those made in another.

15. C must know D's velocity $\vec{v}_{CD}$ measured in C's reference frame. If D's measurement is accurate, then the three quantities are related to one another as in Eq. 6.14: $\vec{v}_{Ct} = \vec{v}_{CD} + \vec{v}_{Dt}$.

16. The inertia is the same in *all* inertial reference frames because inertias are determined from Eq. 4.2, $m_o = -(\Delta v_{sx}/\Delta v_{ox})m_s$, and changes in velocity are identical in different inertial reference frames.

17. There is no difference. The numerical value of the *x* component of position may change with the specific reference point (origin) chosen, but the actual position of the center of mass of a system is a property of the system that is independent of the reference point.

18. The system momentum is constant because the system is isolated. The center of mass moves with the momentum of the system, which means it moves with constant momentum. (Put another way, the center of mass is at rest in the zero-momentum reference frame for the system.) If the inertia of the system is also unchanging, the center of mass moves at constant velocity.

19. The momentum is zero because the center-of-mass velocity is that of the zero-momentum reference frame. (Compare *Principles* Eqs. 6.23 and 6.26.)

20. Yes. This was established for a particular collision in the *Principles* volume, but it is also logically required that energy that appears in the form of thermal energy or crumpled fenders must not depend on the reference frame of the inertial observer.

21. The nonconvertible kinetic energy changes, which is the part associated with the motion of the center of mass.

22. Not in general. The translational kinetic energy of the system's center of mass is nonconvertible energy, and so it cannot be converted to internal energy. If you happen to be using the zero-momentum reference frame, though, this translational kinetic energy is zero, so in this case all of the kinetic energy of the system is convertible and hence can be converted.

23. (*a*) The maximum amount is all of the convertible kinetic energy: $\frac{1}{2}\mu v_{rel,i}^2$. (*b*) This maximum conversion happens in totally inelastic collisions because $e = 0$.

24. (*a*) Yes, because "zero change in momentum" means the system is isolated, and a system that is isolated in one inertial reference frame is isolated in all inertial reference frames. (*b*) Yes. The change in the momentum of a system is always the same in any inertial reference frame, even though the momentum magnitudes and directions of objects in the system are not always the same, and even though the magnitude and direction of the system momentum may vary from one reference frame to another.

25. (*a*) Yes, because $\Delta E = 0$ means the system is closed, and a system that is closed in one inertial reference frame is closed in all inertial reference frames. (*b*) Yes. The change in energy of a system is the same in all inertial reference frames even though the energy changes of the objects in the system are not always the same, and even though the energy of the system may vary from one reference frame to another.

Answers to Guided Problems

Guided Problem 6.2 5.0 s plus 0.1 s for each meter length of the car

Guided Problem 6.4 5×10^{-2} kg

Guided Problem 6.6 K_{conv} hitting abutment $\approx 2K_{conv}$ hitting parked car

Guided Problem 6.8 Small nuclei make the best targets because the recoil velocity of the neutron gets lower as the target inertia decreases.

PRACTICE
Interactions

Chapter Summary   106

Review Questions   108

Developing a Feel   109

Worked and Guided Problems   110

Questions and Problems   117

Answers to Review Questions   122

Answers to Guided Problems   123

Chapter Summary

The basics of interactions (Sections 7.1, 7.5, 7.6, 7.7)

Concepts An **interaction** is an event that produces either a physical change or a change in motion. A *repulsive interaction* causes the interacting objects to accelerate away from each other, and an *attractive interaction* causes them to accelerate toward each other.

The **interaction range** is the distance over which an interaction is appreciable. A long-range interaction has an infinite range; a short-range interaction has a finite range.

A **field** is a model used to visualize interactions between objects. According to this model, each object that takes part in an interaction produces a field in the space surrounding itself, and the fields mediate the interaction between the objects.

A **fundamental interaction** is one that cannot be explained in terms of other interactions. The four known fundamental interactions are the **gravitational interaction** (a long-range attractive interaction between objects that have *mass*), the **electromagnetic interaction** (a long-range interaction between objects that have *electrical charge*; this interaction can be either attractive or repulsive), the **weak interaction** (a short-range repulsive interaction between subatomic particles), and the **strong interaction** (a short-range interaction between *quarks,* the building blocks of protons, neutrons, and certain other subatomic particles; this interaction can be either attractive or repulsive).

Quantitative Tools If two objects of inertias m_1 and m_2 interact, the ratio of the x components of their accelerations is

$$\frac{a_{1x}}{a_{2x}} = -\frac{m_2}{m_1}. \tag{7.6}$$

Potential energy (Sections 7.2, 7.8, 7.9)

Concepts **Potential energy** is a coherent form of internal energy associated with reversible changes in the *configuration* of an object or system. Potential energy can be converted entirely to kinetic energy.

Gravitational potential energy is the potential energy associated with the relative position of objects that are interacting gravitationally.

Elastic potential energy is the potential energy associated with the reversible deformation of objects.

Changes in potential energy are *independent of path*. This means that the change in an object's potential energy as the object moves from a position x_1 to any other position x_2 depends *only* on x_1 and x_2, and *not* on the path the object takes in moving from x_1 to x_2.

Quantitative Tools The potential energy U of a system of two interacting objects can always be written in the form

$$U = U(x), \tag{7.12}$$

where $U(x)$ is a unique function of a position variable x that quantifies the configuration of the system.

Near Earth's surface, if the vertical coordinate of an object of inertia m changes by Δx, the gravitational potential energy U^G of the Earth-object system changes by

$$\Delta U^G = mg\Delta x. \tag{7.19}$$

Energy dissipation during interactions (Sections 7.3, 7.4, 7.8, 7.10

Concepts All energy can be divided into two fundamental classes: energy associated with motion (kinetic energy) and energy associated with the configuration of interacting objects (potential energy). Each class of energy comes in two forms: *coherent* and *incoherent*. Energy is coherent if it involves ordered motion or configuration; it is incoherent if it involves random motion or configuration. For example, the kinetic energy of a moving object is coherent because all of its atoms move in the same way, whereas the thermal energy of an object is incoherent because the atoms move randomly.

The **mechanical energy** E_{mech} of a system is the sum of the system's coherent energies (kinetic energy K and potential energy U).

The **thermal energy** E_{th} of a system is the incoherent internal energy associated with the random motion of atoms that make up the objects in the system. Thermal energy cannot be entirely converted to coherent energy.

Source energy E_s is incoherent energy (such as chemical, nuclear, solar, and stored solar energy) used to produce other forms of energy.

Dissipative interactions are irreversible interactions that involve changes in thermal energy.

Nondissipative interactions are reversible interactions that convert kinetic energy to potential energy, and vice versa.

Quantitative Tools

$$E_{mech} = K + U. \tag{7.9}$$

During a dissipative interaction, the sum of the changes in all forms of energy in a closed system is zero:

$$\Delta K + \Delta U + \Delta E_s + \Delta E_{th} = 0. \tag{7.28}$$

During a nondissipative interaction, the mechanical energy of a closed system does not change:

$$\Delta E_{mech} = \Delta K + \Delta U = 0. \tag{7.8}$$

Review Questions

Answers to these questions can be found at the end of this chapter.

7.1 The effects of interactions

1. A pool ball collides head-on and elastically with a second pool ball initially at rest. Do any of these properties of the system made up of the two balls change during the interaction: (*a*) momentum, (*b*) kinetic energy, (*c*) the sum of all forms of energy in the system? (Assume no significant interaction between the pool table and the balls.)

2. In an elastic collision between two objects, does their relative speed stay the same during the interaction?

3. Consider a system made up of two objects that collide with each other. Describe the difference between what happens to the system's kinetic energy when the collision is (*a*) totally inelastic, (*b*) inelastic, and (*c*) elastic.

4. Summarize the characteristics of an interaction.

7.2 Potential energy

5. Describe the difference between kinetic and potential energy.

7.3 Energy dissipation

6. Give an example of an interaction that converts kinetic energy to thermal energy.

7. How does energy of motion differ from energy of configuration? How does coherent energy differ from incoherent energy?

7.4 Source energy

8. What is the ultimate source of virtually all the energy we use?

9. What distinguishes dissipative interactions from nondissipative ones?

10. Think of several interactions that occur in everyday life. For each interaction, describe which category of energy is converted and to which category it is converted.

7.5 Interaction range

11. Describe the two models physicists use to represent interactions between objects without requiring direct touching.

7.6 Fundamental interactions

12. Which of the four fundamental interactions do we commonly observe in everyday life? Why not all of them?

13. Which fundamental interaction exerts the most control (*a*) in chemical processes and (*b*) in biological processes?

14. The strength of the gravitational interaction is minuscule compared with the strength of the electromagnetic interaction. Yet we can study the interactions of most ordinary objects without considering electromagnetic interactions, while it is essential that we include gravitational interactions. Give a reason why this is so.

7.7 Interactions and accelerations

15. How are the acceleration and inertia of an object 1 related to the acceleration and inertia of an object 2 when the objects collide (*a*) elastically and (*b*) inelastically?

16. List any assumptions made in deriving the relationship between ratios of acceleration and inertias for two interacting objects.

7.8 Nondissipative interactions

17. Explain why mechanical energy remains constant when nondissipative interactions take place in a closed system.

18. Explain how we know that, in a closed system in which all interactions are nondissipative, potential energy is a function of position only.

19. An object subject to only nondissipative interactions moves from point A to point B and then back to A. How does the object's initial kinetic energy compare with its final kinetic energy?

7.9 Potential energy near Earth's surface

20. In Eq. 7.19, is Δx a horizontal or vertical displacement? Does it matter?

21. For an object placed at a given position relative to the ground, is the value of the object's gravitational potential energy either always positive or always negative?

7.10 Dissipative interactions

22. A car collides totally inelastically with a parked car. In the (isolated) system made up of the two cars, is all the initial kinetic energy dissipated during the collision?

23. In a collision between two objects, how much of the convertible kinetic energy is converted to thermal energy, and how much is temporarily stored as potential energy during the collision (*a*) when $e = 1$ and (*b*) when $e = 0$?

Developing a Feel

Make an order-of-magnitude estimate of each of the following quantities. Letters in parentheses refer to hints below. Use them as needed to guide your thinking.

1. The energy dissipated when you clap your hands once (S, K, A)
2. The energy dissipated when you throw a tennis ball against a wall (E, X, D)
3. The thermal energy released by burning three logs in a fireplace (I, P, AA)
4. The upward acceleration of Earth you cause by jumping off a chair (G, L, U)
5. The acceleration of a rifle when a bullet is fired from it (T, Z, Q)
6. The energy supplied by your muscles when you climb a flight of stairs (H, C, Y, L)
7. The energy supplied by an escalator when it carries 30 people up one flight (L, Y, C, F)
8. The maximum potential energy stored when two protons collide head-on if, before the collision, each is moving at 3×10^7 m/s (J, W)
9. The maximum elastic potential energy stored when you perform a bungee jump (L, R, O, Y, W)
10. The gravitational potential energy stored as a 30-story building is constructed (M, V, N, B, Y)

Hints

A. What is the coefficient of restitution for the collision when you clap your hands?
B. What is the inertia of a 30-story building?
C. Does the speed change on the way up?
D. What is the coefficient of restitution?
E. What is the inertia of a tennis ball?
F. What is the height between floors of a building?
G. What is the inertia of Earth?
H. What is the vertical height of a flight of stairs?
I. How much energy is released by breaking one chemical bond?
J. What is the inertia of a proton?
K. What is the inertia of one hand and forearm?
L. What is the inertia of a typical person?
M. What is the height of a 30-story building?
N. What fraction of the volume of a 30-story building is structural concrete and steel?
O. What is the maximum height difference from bottom to top of one bungee jump?
P. What is the size of an atom?
Q. Over what length does the bullet accelerate along the rifle barrel?
R. Is the inertia of the bungee cord negligible in this situation?
S. What is the speed of each hand just before the two touch each other?

T. What is the ratio $m_{\text{rifle}}/m_{\text{bullet}}$?
U. What is your acceleration in free fall?
V. Where is the vertical location of this building's center of mass?
W. What is the minimum kinetic energy in the zero-momentum reference frame?
X. What is the speed of the ball just before it hits the wall?
Y. How does the gravitational potential energy of the system change?
Z. With what speed does the bullet leave the rifle?
AA. How many atoms are there in a fireplace log?

Key (all values approximate)

A. 0; B. 1×10^7 kg; C. generally no; D. 0.7; E. 0.1 kg; F. 6 m; G. 6×10^{24} kg; H. 3 m; I. 1×10^{-19} J; J. 2×10^{-27} kg; K. 2 kg; L. 7×10^1 kg; M. 1×10^2 m; N. 1/8; O. 1×10^2 m; P. 1×10^{-10} m; Q. 1 m; R. yes; S. 1 m/s; T. 4×10^2, or 400:1; U. 1×10^1 m/s²; V. less than halfway up—say, 4×10^1 m above ground; W. 0, because in this reference frame there must be at least one instant at which $v_{\text{rel}} = 0$; X. 2×10^1 m/s; Y. increases as object rises, decreases as object falls; Z. 4×10^2 m/s; AA. 1×10^{27} atoms

Worked and Guided Problems

These examples involve material from this chapter but are not associated with any particular section.
Some examples are worked out in detail; others you should work out by following the guidelines provided.

Worked Problem 7.1 Watch out below!

A student standing at the top of a wall throws a ball straight down, giving it speed v_i as it leaves her hand. Use energy methods to determine the speed v_f at which the ball strikes the pavement a distance h below her hand. Express your answer in terms of v_i and h.

❶ GETTING STARTED We define our system as being the ball and Earth so that the gravitational interaction can be expressed in terms of potential energy. The initial system diagram displays the given information, and the final system diagram shows the sought quantity, the final speed v_f (Figure WG7.1).

Figure WG7.1

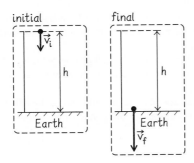

The change in Earth's kinetic energy while the ball is falling is negligible. We know that the ball speeds up because of the acceleration due to gravity, and we are required to use energy methods to determine by how much. If we let the inertia of the ball be m, we know that the ball starts out with initial kinetic energy $\frac{1}{2}mv_i^2$. As the ball travels downward, the system's gravitational potential energy decreases as this energy gets converted to kinetic energy by the gravitational interaction. Ignoring any dissipative interactions (like air resistance), we can assume that the system's mechanical energy (kinetic plus potential) does not change.

❷ DEVISE PLAN Because we have assumed that the mechanical energy of the system is constant, we can set the final mechanical energy equal to the initial value:

$$K_f + U_f = K_i + U_i. \tag{1}$$

We should include the kinetic energy of each object in our system, but we have assumed that Earth's kinetic energy does not change, so the initial and final Earth kinetic energies will simply cancel and need not be explicitly included.

Each gravitational potential energy term depends on the height, and we know neither the initial nor final height. However, if we collect terms, then only the height difference will be needed, and we do know that. We also know the initial kinetic energy term for the ball, which leaves only one unknown: the final kinetic energy of

the ball. Because the kinetic energy formula is $\frac{1}{2}mv^2$, we can determine the final speed if we know the final kinetic energy. The ball's inertia m is not given, so we must be sure our answer does not contain this unknown quantity.

❸ EXECUTE PLAN Using the expression for gravitational potential energy near Earth's surface, $U^G = mgx$, we write Eq. 1 in the form

$$\frac{1}{2}mv_f^2 + mgx_f = \frac{1}{2}mv_i^2 + mgx_i.$$

Isolating the final kinetic energy, we obtain

$$\frac{1}{2}mv_f^2 = \frac{1}{2}mv_i^2 + mg(x_i - x_f). \tag{2}$$

The difference between the initial and final positions should be the height h, but to ensure that we make no error, we add specific values for x_i and x_f to our drawing (Figure WG7.2). The x axis has to point up because that was an assumption made in the derivation of $U^G = mgx$. We know that the zero level of the potential energy is an arbitrary choice, and so we choose the position at which the ball was released. With this choice, $x_f = -h$ because the ball's final position is a distance h in the negative x direction. We can now solve Eq. 2 for v_f:

$$\frac{1}{2}mv_f^2 = \frac{1}{2}mv_i^2 + mg[0 - (-h)]$$

$$mv_f^2 = mv_i^2 + 2mgh$$

$$v_f = \sqrt{v_i^2 + 2gh}. ✔$$

Figure WG7.2

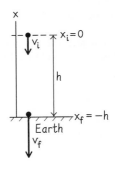

❹ EVALUATE RESULT Because the distance h is a magnitude (as is g) and v_i^2 is always positive, the quantity under the square-root sign is positive. This is reassuring because the final speed must be a real number, not an imaginary one. We expect v_f to increase if either v_i or h increases, and our expression predicts this behavior. We suspect that the actual final speed of a real ball would be lower than this because of air resistance, but our result should be fairly accurate if the ball is more like a baseball than a Ping-Pong ball and if the wall is not too tall.

Guided Problem 7.2 Safe play

A soft rubber ball is held at a height h above the floor and then released. The coefficient of restitution for the ball-floor collision is 0.25. To what height does the ball rise on its first bounce, expressed as a fraction of h?

❶ GETTING STARTED

1. Begin by choosing your system and drawing a system diagram that includes the given information and the quantity you seek (labeled with a question mark).

❷ DEVISE PLAN

2. What is the numerical value of the ball's speed at the highest point the ball reaches on its first bounce?

3. What kinds of energy conversions take place during the fall, in the ball's collision with the floor, and in the bounce?

❸ EXECUTE PLAN

4. With what speed does the ball strike the floor?

5. What is the ball's speed immediately after it bounces off the floor?

6. How much kinetic energy is dissipated in the ball-floor collision?

7. What is the mechanical energy of the system as the ball leaves the floor? What is the mechanical energy at the ball's highest bounce distance above the floor?

❹ EVALUATE RESULT

Worked Problem 7.3 Moving explosion

Two 1.0-kg carts coupled together are initially moving to the right at 2.0 m/s on a horizontal, low-friction track. The coupling contains an explosive charge that is ignited remotely, releasing 18 J of energy. Half of the released energy is dissipated into incoherent energy, such as noise, thermal energy, and damage to the carts. (*a*) What are the velocities of the carts just after the explosive separation? (*b*) Draw a bar diagram showing the energy distribution before and after the explosive separation for the system composed of the two carts and the explosive coupling.

❶ GETTING STARTED We are told that the carts are coupled together and initially moving to the right at 2.0 m/s. Given that half of the energy released by the explosive charge is dissipated, we conclude that the other half goes into increasing the kinetic energy of the system. We assume that the rear car, which we can call cart 2, is pushed to the left by the collision (slowing it down) and that the leading cart (call it cart 1) is pushed to the right (speeding it up). Figure WG7.3 shows the system diagram of this moving explosive separation based on these assumptions, with the positive x direction chosen to the right. (Although we expect $\vec{v}_{2f}$ to be in the negative x direction, we do not know for sure. The solution will tell us whether our assumption is right.)

Figure WG7.3

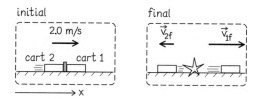

Because this is an explosive separation, we expect to use conservation of momentum. We are also given energy information, which we must incorporate into our analysis.

❷ DEVISE PLAN It is often easier to analyze collisions in the zero-momentum reference frame. We are given information in the Earth reference frame, however, so we need to work with Galilean transformations. In the zero-momentum reference frame Z, the system

momentum is always zero, and so we know it is zero both before and after the explosive separation:

$$m_1 v_{Z1x,i} + m_2 v_{Z2x,i} = (m_1 + m_2)v_{Zx,i} = 0 \tag{1}$$

$$m_1 v_{Z1x,f} + m_2 v_{Z2x,f} = 0 \tag{2}$$

where $v_{Zx,i}$ represents the initial x component of the velocity of the system in the zero-momentum reference frame. Because the inertias are not zero, we conclude from Eq. 1 that

$$v_{Z1x,i} = v_{Z2x,i} = v_{Zx,i} = 0.$$

Because we have two unknowns, $v_{Z1x,f}$ and $v_{Z2x,f}$, we need another independent equation, and so we turn to the energy information. The energy released is energy initially locked up in the chemical bonds of the explosive charge, which means it is a type of source energy. We are given how much of this source energy is released and how much is dissipated to incoherent energy (an amount that is the same in any reference frame). For simplicity, we combine all of the released incoherent energy (noise, deformation, and so on) and label it as thermal energy. Because the sum of all the forms of energy in the system is constant in this explosive separation, the energetics in the Z reference frame are described by

$$\Delta K_Z + \Delta E_{Zs} + \Delta E_{Zth} = 0. \tag{3}$$

Equations 1–3 should give us all the information we need to calculate the final velocities in the zero-momentum reference frame. Then we do a Galilean transformation to obtain the velocities in the Earth reference frame.

❸ EXECUTE PLAN

(*a*) In the Earth reference frame, the x component of the initial velocity of the system's center of mass is given by Eq. 6.26:

$$v_{cmx} = \frac{m_1 v_{1x,i} + m_2 v_{2x,i}}{m_1 + m_2}.$$

Because $m_1 = m_2 = m$, this expression simplifies to

$$v_{cmx} = \frac{mv_{1x,i} + mv_{2x,i}}{m + m} = \frac{v_{1x,i} + v_{2x,i}}{2}$$

$$= \frac{(+2.0 \text{ m/s}) + (+2.0 \text{ m/s})}{2} = +2.0 \text{ m/s}.$$

In the zero-momentum reference frame the final momentum (like the initial momentum) must be zero, and so from Eq. 2 we have

$$m_1 v_{Z1x,f} + m_2 v_{Z2x,f} = mv_{Z1x,f} + mv_{Z2x,f} = 0$$

$$v_{Z1x,f} = -v_{Z2x,f}. \tag{4}$$

Let us use E to represent the quantity of source energy released, $E = 18$ J. Because the source energy decreases as the chemical bonds in the explosive charge break, the source energy term in Eq. 3 becomes $\Delta E_{Zs} = -E$. Because half of the released source energy is dissipated to thermal energy, the thermal energy term in Eq. 3 is $\Delta E_{Zth} = +\frac{1}{2}E$. Equation 3 then becomes

$$\Delta K_Z + \Delta E_{Zs} + \Delta E_{Zth} = 0$$

$$\left[\left(\tfrac{1}{2} m_1 v_{Z1x,f}^2 + \tfrac{1}{2} m_2 v_{Z2x,f}^2 \right) - \left(\tfrac{1}{2} m_1 v_{Z1x,i}^2 + \tfrac{1}{2} m_2 v_{Z2x,i}^2 \right) \right]$$

$$+ (-E) + (\tfrac{1}{2}E) = 0$$

$$\left(\tfrac{1}{2} m v_{Z1x,f}^2 + \tfrac{1}{2} m v_{Z2x,f}^2 \right) - \left(\tfrac{1}{2} m(0)^2 + \tfrac{1}{2} m_2(0)^2 \right) = +\tfrac{1}{2}E.$$

Next we substitute from Eq. 4 to get

$$\tfrac{1}{2} m(-v_{Z2x,f})^2 + \tfrac{1}{2} m v_{Z2x,f}^2 = \tfrac{1}{2}E$$

$$v_{Z2x,f} = +\sqrt{\frac{E}{2m}} \quad \text{or} \quad -\sqrt{\frac{E}{2m}}.$$

The positive root implies that cart 2 passes through cart 1 in the zero-momentum reference frame instead of being pushed away, and so we must discard this root as physically impossible. With the negative root, we have

$$v_{Z2x,f} = -\sqrt{\frac{E}{2m}}$$

$$v_{Z1x,f} = -v_{Z2x,f} = -\left(-\sqrt{\frac{E}{2m}} \right) = +\sqrt{\frac{E}{2m}}.$$

In a procedure analogous to that used in Worked Problem 6.5, the Galilean transformation to the Earth reference frame is

$$v_{E1x,f} = v_{Z1x,f} + v_{cmx} = +\sqrt{\frac{E}{2m}} + v_{cmx}$$

$$= +\sqrt{\frac{18 \text{ J}}{(2)(1.0 \text{ kg})}} + 2.0 \text{ m/s} = +3.0 \text{ m/s} + 2.0 \text{ m/s}$$

$$= +5.0 \text{ m/s} ✔$$

and

$$v_{E2x,f} = v_{Z2x,f} + v_{cmx} = -\sqrt{\frac{E}{2m}} + v_{cmx}$$

$$= -3.0 \text{ m/s} + 2.0 \text{ m/s}$$

$$= -1.0 \text{ m/s}. ✔$$

Because to the right is the positive x direction in our coordinate system, cart 1 moves to the right after the explosive separation and cart 2 moves to the left, just as we assumed when we began our solution.

(b) The energy diagrams should contain a bar for each energy category: kinetic, potential, source, and thermal (Figure WG7.4). We know that $E_{s,i} = E = 18$ J, $E_{s,f} = 0$, $E_{th,i} = 0$, and $E_{th,f} = \frac{1}{2}E = 9.0$ J. We leave both U bars empty because there is no interaction in this system to temporarily store potential energy. The initial and final kinetic energies are

$$K_i = \tfrac{1}{2} m_1 (v_{1i})^2 + \tfrac{1}{2} m_2 (v_{2i})^2$$

$$= \tfrac{1}{2}(1.0 \text{ kg})(2.0 \text{ m/s})^2 + \tfrac{1}{2}(1.0 \text{ kg})(2.0 \text{ m/s})^2 = 4.0 \text{ J}$$

$$K_f = \tfrac{1}{2} m_1 (v_{1f})^2 + \tfrac{1}{2} m_2 (v_{2f})^2$$

$$= \tfrac{1}{2}(1.0 \text{ kg})(5.0 \text{ m/s})^2 + \tfrac{1}{2}(1.0 \text{ kg})(1.0 \text{ m/s})^2 = 13 \text{ J}.$$

These values yield the bar diagram shown in Figure WG7.4. ✔

Figure WG7.4

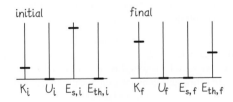

initial

final

$K_i \quad U_i \quad E_{s,i} \quad E_{th,i}$ $\qquad$ $K_f \quad U_f \quad E_{s,f} \quad E_{th,f}$

❹ **EVALUATE RESULT** We can check our results in several ways. It is reassuring that we obtained a center-of-mass velocity of $v_{cmx} = +2.0$ m/s because the two carts are initially locked together and moving at that velocity.

Let us make several other checks, although you normally might not do all of them. We used conservation of momentum in the zero-momentum reference frame. We can verify that this law is satisfied in the Earth reference frame with our calculated numbers:

$$p_{x,i} = m_1 v_{1x,i} + m_2 v_{2x,i}$$

$$= (1.0 \text{ kg})(+2.0 \text{ m/s}) + (1.0 \text{ kg})(+2.0 \text{ m/s})$$

$$= +4.0 \text{ kg} \cdot \text{m/s}$$

$$p_{x,f} = m_1 v_{1x,f} + m_2 v_{2x,f}$$

$$= (1.0 \text{ kg})(+5.0 \text{ m/s}) + (1.0 \text{ kg})(-1.0 \text{ m/s})$$

$$= +4.0 \text{ kg} \cdot \text{m/s}.$$

We can verify that the system's energy remains constant by calculating its change in the Earth reference frame:

$$\Delta K + \Delta E_s + \Delta E_{th} = (13 \text{ J} - 4.0 \text{ J}) + (0 - 18 \text{ J}) + (9.0 \text{ J} - 0)$$

$$= 0.$$

We can also check consistency in the kinetic energy:

$$K = \tfrac{1}{2}(m_1 + m_2)(v_{cmx})^2 + K_{conv}$$

$$= \tfrac{1}{2}(m_1 + m_2)(v_{cmx})^2 + \tfrac{1}{2}\left(\frac{m_1 m_2}{m_1 + m_2}\right)(v_{2x} - v_{1x})^2,$$

where for K_{conv} we have substituted the expression given in Eq. 6.38. Because the inertias are equal, we can simplify to

$$K_f = \tfrac{1}{2}(m + m)(v_{cmx})^2 + \tfrac{1}{2}\frac{m^2}{m + m}(v_{2x,f} - v_{1x,f})^2$$

$$= m(v_{cm})^2 + \tfrac{1}{4}m(v_{2x,f} - v_{1x,f})^2$$

$$= (1.0\ \text{kg})(2.0\ \text{m/s})^2 + \tfrac{1}{4}(1.0\ \text{kg})\left[(-1.0\ \text{m/s}) - (+5.0\ \text{m/s})\right]^2$$

$$= 4.0\ \text{J} + 9.0\ \text{J} = 13\ \text{J}$$

in agreement with our calculation for the bar diagram in part *b*.

Guided Problem 7.4 An unequal match

On a low-friction track, two carts coupled together are initially moving to the right at 2.0 m/s. The rear cart has an inertia of 2.0 kg, and the lead cart's inertia is one-third of that. An explosive charge attached to the coupling is ignited remotely, releasing 27 J of energy. One-third of that energy is dissipated to incoherent energy, such as noise, thermal energy, and damage to the carts. Use the Earth reference frame for the following tasks. (*a*) Determine the velocities of the carts just after the explosive separation. (*b*) Draw a bar diagram showing the energy distribution before and after the explosive separation.

❶ GETTING STARTED

1. How is this problem similar to Worked Problem 7.3? How does it differ from Worked Problem 7.3?
2. What is a good choice for your system?

3. Draw a system diagram showing all relevant information.
4. Which system quantities remain constant during the explosive separation?

❷ DEVISE PLAN

5. Write the equations describing the conservation laws for the quantities named in question 4, as viewed in the Earth reference frame.
6. How many unknowns do you have? Do you have a sufficient number of equations to solve for them?

❸ EXECUTE PLAN

❹ EVALUATE RESULT

7. Check your numerical answer for consistency in at least two ways.

Worked Problem 7.5 Exploding rocket

A rocket is made up of an unpowered payload of inertia $m_{payload}$ and a powered booster connected to the payload by an explosive coupler. The rocket is fired straight up, uses all its fuel, and reaches the peak of its trajectory at height h_1 above the ground. The coupler is detonated at the peak of the trajectory. The quantity of energy released by the explosion is $E > 0$, one-quarter of which is dissipated to incoherent energy. The explosion separates the booster from the payload, and the latter shoots straight up along its original line of motion. Immediately after the separation, the inertia of the booster is five times that of the payload. What maximum height h_2 above the ground does the payload reach after the explosion? Assume that all the incoherent energy is in the form of thermal energy, and report your answer in terms of h_1, E, and $m_{payload}$.

❶ GETTING STARTED This problem involves three stages of motion that are best analyzed separately. First there is the trip from ground to h_1, which needs no analysis because h_1 is given information. The remaining two stages are the explosive separation and the post-separation motions of the booster and payload under the influence of Earth's gravitational field.

First we choose an isolated and closed system and draw a system diagram. A good choice is the system composed of booster, explosive coupler, payload, and Earth. In our diagram (Figure WG7.5), we show the initial location of the rocket (the ground), the rocket at height h_1 just before separation, the booster and payload at h_1 just after separation, and the payload at its maximum height h_2. Because we are considering gravitational potential energy, we choose to point the x axis vertically upward, with the origin at ground level.

The rocket is at the top of its trajectory when the explosive separation occurs, which means that its velocity at the instant of

separation is zero. So, we already have complete information about the momentum and energy of the objects in our system at height h_1, and the initial journey from ground to h_1 need not be analyzed! Because the payload travels straight up after the separation, conservation of momentum tells us that the booster must be thrown downward by the explosion. As the payload moves up, its interaction with Earth's gravitational field causes it to slow down as its kinetic energy gets converted to gravitational potential energy until its velocity is zero at the final maximum height we seek. This last portion of the motion, from h_1 to h_2, is simply projectile motion of the payload, which we can analyze with the techniques of Chapter 3 or with energy methods, as we wish.

Figure WG7.5

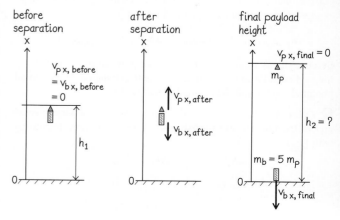

❷ DEVISE PLAN Because the height h_2 the payload reaches is determined by the payload velocity right after separation, we first examine the explosion. In the explosion, an amount E of source energy stored in the coupler is converted to other forms ($\Delta E_s = -E$), with one-quarter dissipated as thermal energy ($\Delta E_{th} = +E/4$). Because the booster and payload do not have a chance to move much during the explosion, they are at essentially the same location just before and just after the explosion. Thus we can consider the gravitational potential energy to remain unchanged during the explosion, $\Delta U^G = 0$. Therefore, the expression for the changes in energy for the rocket-Earth system during the explosion can be written as

$$\Delta K_{\text{rocket-Earth}} + \Delta E_s + \Delta E_{th} = 0. \qquad (1)$$

Because the rocket has zero speed and consequently zero kinetic energy just before the explosion, we have

$$\Delta K_{\text{rocket-Earth}} = K_{\text{after}} - K_{\text{before}} = K_{\text{after}} - 0.$$

Therefore Eq. 1 gives us an expression for the kinetic energy just after the explosive separation:

$$K_{\text{after}} + \Delta E_s + \Delta E_{th} = 0$$
$$K_{\text{after}} = -\Delta E_s - \Delta E_{th}.$$

We must then determine how much of this kinetic energy is associated with the payload because

$$\Delta K_{\text{rocket-Earth}} = K_{\text{after}} = K_{\text{payload, after}} + K_{\text{booster, after}}. \qquad (2)$$

To determine how much kinetic energy is associated with the payload, we consider the explosive separation of the booster and payload separately. During the explosion we can consider the payload-booster system to be isolated and so its momentum remains constant, or $\Delta \vec{p}_{\text{payload}} + \Delta \vec{p}_{\text{booster}} = 0$. We can then use this relation to figure out the relationship between the velocities of the payload and the booster just after the explosive separation. Given the payload velocity just after the separation, we can determine the kinetic energy of the payload. After the separation, the mechanical energy in the payload-Earth system should remain constant because now there is no longer any source energy being dissipated to incoherent forms. For the mechanical energy of the payload-Earth system, we therefore can say

$$\Delta K_{\text{payload-Earth}} + \Delta U^G_{\text{payload-Earth}} = 0. \qquad (3)$$

Finally, we realize that Earth's inertia is much greater than that of any rocket. This means that only a vanishingly small fraction of the gravitational potential energy goes into changing Earth's kinetic energy (see Developing a Feel 4), which implies

$$\Delta K_{\text{payload-Earth}} = \Delta K_{\text{payload}} + \Delta K_{\text{Earth}} = \Delta K_{\text{payload}} + 0.$$

Because this is so, the only kinetic energy we need to consider is the payload's. From it we can determine the change in gravitational potential energy and the payload's final height h_2.

❷ EXECUTE PLAN We first determine the sum of the kinetic energies of the booster and payload right after separation, beginning with Eq. 1:

$$\Delta K_{\text{rocket-Earth}} + \Delta E_s + \Delta E_{th} = 0$$
$$\Delta K_{\text{rocket-Earth}} + (-E) + (+E/4) = 0$$
$$\Delta K_{\text{rocket-Earth}} = \tfrac{3}{4}E.$$

Substituting from Eq. 2 gives

$$\Delta K_{\text{rocket-Earth}} = K_{\text{payload, after}} + K_{\text{booster, after}} = \tfrac{3}{4}E$$
$$\tfrac{1}{2} m_{\text{payload}} v^2_{\text{payload, after}} + \tfrac{1}{2} m_{\text{booster}} v^2_{\text{booster, after}} = \tfrac{3}{4}E. \qquad (4)$$

To determine how this kinetic energy is distributed between payload and booster, we use the fact that the momentum of the payload-booster system is constant during the explosion to determine the relationship of their velocities just after the explosive separation:

$$p_{\text{payload } x, \text{ after}} + p_{\text{booster } x, \text{ after}} = p_{\text{payload } x, \text{ before}} + p_{\text{booster } x, \text{ before}}$$

$$m_{\text{payload}}(+v_{\text{payload, after}}) + m_{\text{booster}}(-v_{\text{booster, after}}) = 0 + 0$$

$$v_{\text{booster, after}} = \frac{m_{\text{payload}}}{m_{\text{booster}}} v_{\text{payload, after}},$$

where we have used the velocity directions shown in Figure WG7.5. Substituting this result into Eq. 4 and using the fact that, after separation, $m_{\text{booster}} = 5 m_{\text{payload}}$, we obtain

$$\tfrac{1}{2} m_{\text{payload}} v^2_{\text{payload, after}} + \tfrac{1}{2} m_{\text{booster}} \left(\frac{m_{\text{payload}}}{m_{\text{booster}}} v_{\text{payload, after}} \right)^2 = \tfrac{3}{4}E$$

$$\left(\tfrac{1}{2} m_{\text{payload}} v^2_{\text{payload, after}} \right) \left(1 + \frac{m_{\text{payload}}}{m_{\text{booster}}} \right) = \tfrac{3}{4}E$$

$$\left(\tfrac{1}{2} m_{\text{payload}} v^2_{\text{payload, after}} \right) \left(1 + \frac{m_{\text{payload}}}{5 m_{\text{payload}}} \right) = \tfrac{3}{4}E$$

$$\left(\tfrac{1}{2} m_{\text{payload}} v^2_{\text{payload, after}} \right) \left(\tfrac{6}{5} \right) = \tfrac{3}{4}E$$

$$\tfrac{1}{2} m_{\text{payload}} v^2_{\text{payload, after}} = \tfrac{15}{24}E = \tfrac{5}{8}E.$$

Now we are ready to determine the quantity we are asked for—h_2, the maximum height reached by the payload after separation. We use Eq. 3 (the mechanical energy for the payload-Earth system remains constant), remembering that the subscript final means values at the maximum payload height h_2 and after means values just after separation (note that, for our choice of origin for the x axis, x_{final} and x_{after} are equivalent to h_2 and h_1, respectively):

$$\Delta K_{\text{payload}} + \Delta U^G_{\text{payload-Earth}} = 0$$

$$\left[\tfrac{1}{2} m_{\text{payload}} v^2_{\text{payload, final}} - \tfrac{1}{2} m_{\text{payload}} v^2_{\text{payload, after}} \right]$$
$$+ \left[m_{\text{payload}} g x_{\text{final}} - m_{\text{payload}} g x_{\text{after}} \right] = 0$$

$$\left[\tfrac{1}{2} m_{\text{payload}} (0)^2 - \tfrac{5}{8}E \right] + \left[m_{\text{payload}} g(+h_2) - m_{\text{payload}} g(+h_1) \right] = 0$$

$$m_{\text{payload}} g h_2 = m_{\text{payload}} g h_1 + \tfrac{5}{8}E$$

$$h_2 = h_1 + \tfrac{5}{8} \frac{E}{m_{\text{payload}} g}. ✔$$

PRACTICE

③ EVALUATE RESULT If E is increased, we expect the payload to go higher, which is what our algebraic result predicts. If $m_{payload}$ is increased, we expect the opposite, and this, too, is consistent with our result. Further, the solution suggests that the amount of energy released in the explosive separation, E, appears to be about double the energy required to lift the payload through a height difference

$h_2 - h_1$, so the order of magnitude of the result is correct. The assumption $m_{rocket} \ll m_{Earth}$ is reasonable.

One caveat is that we have used an expression for gravitational potential energy that is valid only near Earth's surface; our solution is applicable to only low-flying rockets, a situation we will correct in Chapter 13.

Guided Problem 7.6 Booster impact

A toy rocket comprises an unpowered, 1.0-kg payload and a powered booster connected by an explosive coupler. The rocket is fired straight up and reaches a speed of 90 m/s as the last of its fuel is exhausted 100 m above the ground. At this instant, the coupler is detonated, and the two parts separate from each other. The energy released by the explosive separation is 3000 J, with 1000 J of it converted to incoherent energy. Immediately after the explosive separation, the booster has an inertia of 2.0 kg. Use energy and momentum methods to calculate the speed at which the booster strikes the ground.

① GETTING STARTED

1. How is this problem similar to Worked Problem 7.5? How does it differ in information and in specific task?
2. Choose your system (or systems) and draw a version of Figure WG7.5 that contains the information supplied in this problem, using a question mark to identify the quantity you need to calculate.

② DEVELOP PLAN

3. Write an expression for the kinetic energy of the payload-booster system right after the separation.
4. Which law can you use to establish a relationship between $\vec{v}_{payload}$ and $\vec{v}_{booster}$ just after the separation?
5. What is the booster's x coordinate just after the separation, and what is it just as the booster hits the ground?
6. What is the booster's change in potential energy? How is this change related to its change in kinetic energy?

③ EXECUTE PLAN

④ EVALUATE RESULT

Worked Problem 7.7 Somewhat springy potential energy

A 4.0-kg block takes part in only one interaction (with a fixed object) that causes potential energy to be stored in the block-object system as the block moves along an x axis. At $t = 0$, the block is at $x_0 = +2.0$ m and moving in the negative x direction at 2.0 m/s. The amount of potential energy stored depends on the block's position along the axis and is given by $U(x) = ax^2 + bx + c$, where $a = +1.0$ J/m^2, $b = -2.0$ J/m, and $c = -1.0$ J. (a) Draw a diagram of this potential energy as a function of x, and show the value of the block's mechanical energy on your diagram. (b) What is the block's speed after it has traveled 3.0 m from x_0? (c) At what value of x is the block's velocity zero? (d) Describe what the block does after it passes through that value of x.

① GETTING STARTED The block and fixed object form a closed, nondissipative system because the only interaction is associated with a potential energy. We can get the potential energy and kinetic energy when the block is at one position and thus determine its mechanical energy at that position. Knowing that the mechanical energy remains constant during the motion allows us to sketch the graph for part a. We then need only to determine the potential energy when the block is at any other position in order to obtain the block's kinetic energy and hence its speed there. This covers parts b and c, and hopefully also provides insight about part d.

② DEVISE PLAN We are given the speed $v_0 = 2.0$ m/s at $x_0 = +2.0$ m, from which we can calculate the kinetic energy $K_0 = K(x_0) = \frac{1}{2}mv_0^2$ and the potential energy $U_0 = U(x_0)$. No sources of energy are involved and no energy is dissipated during the motion, and so the mechanical energy remains constant. Consequently we know that for any position x,

$$E_{mech} = K(x_0) + U(x_0) = K(x) + U(x).$$

Knowing either $K(x)$ or $U(x)$ at any x allows us to calculate the other value:

$$K(x) = E_{mech} - U(x)$$
$$U(x) = E_{mech} - K(x).$$

After the block travels 3.0 m in the negative x direction from $x_0 = +2.0$ m, it is at $x_b = -1.0$ m, where we use the subscript b to remind ourselves that this value is needed to solve part b. We can now readily calculate its kinetic energy at that position.

At the instant when the block's velocity is zero at x_c (subscript c because this position is needed to solve part c), its kinetic energy is zero: $K(x_c) = 0$. We then get the potential energy from $U(x_c) = E_{mech} - K(x_c) = E_{mech} - 0$, and calculate the value of x_c that gives the block that amount of potential energy.

A similar procedure determines the block's motion thereafter, for part d.

③ EXECUTE PLAN

(a) We can substitute values of x into the $U(x)$ equation to produce the graph. However, to help us in our thinking (and review some tools for more complex problems), it is good to find out whether there are any extrema (maxima and/or minima) in the potential energy curve by determining the value of x at which the derivative of U with respect to x is zero:

$$\frac{d}{dx}U(x) = \frac{d}{dx}(ax^2 + bx + c) = 2ax + b$$

$$\frac{dU}{dx} = 0 = 2ax + b = 2(+1.0 \text{ J/m}^2)x + (-2.0 \text{ J/m}).$$

This equation has a single solution, $x = +1.0$ m, and at that value of x the potential energy is

$$U(+1.0 \text{ m}) = (1.0 \text{ J/m}^2)(+1.0 \text{ m})^2 - (2.0 \text{ J/m})(+1.0 \text{ m}) - 1.0 \text{ J}$$

$$= -2.0 \text{ J}.$$

Because the second derivative of the potential energy with respect to x is positive ($+2.0 \text{ J/m}^2$) everywhere, the potential energy curve is concave up. The one extremum is therefore a minimum. To pin down the shape of the curve further, we compute the value of U at a few other values of x: $U(-2.0 \text{ m}) = +7.0$ J, $U(-1.0 \text{ m}) = +2.0$ J, and $U(0) = -1.0$ J. Symmetry gives us the shape on the other side of the minimum, and we end up with the graph shown in Figure WG7.6.

Figure WG7.6

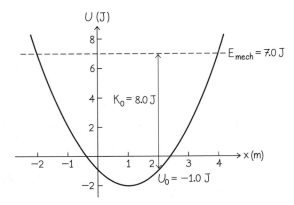

The graph shows that U decreases as the block travels in the negative x direction. Because $E_{\text{mech}} = U + K$ stays constant, decreasing U means increasing K, and increasing K means increasing v; in other words, the block speeds up as it moves from $x = +2.0$ m to $x = +1.0$ m. At $x = +1.0$ m, U has its lowest value, and so K must have its highest value. As the block continues to the left of $x = +1.0$ m, it slows down because U increases and so K and v must decrease.

The kinetic energy at $x_0 = +2.0$ m is

$$K_0 = \tfrac{1}{2}mv_0^2 = \tfrac{1}{2}(4.0 \text{ kg})(2.0 \text{ m/s})^2 = 8.0 \text{ J}.$$

The mechanical energy when the block is at $x_0 = +2.0$ m is then

$$E_{\text{mech}} = \tfrac{1}{2}mv_0^2 + U_0 = (8.0 \text{ J}) + (-1.0 \text{ J}) = 7.0 \text{ J}.$$

Because the system's mechanical energy is constant, this is also the value of E_{mech} at any other value of x. We include a dashed horizontal line on the graph to illustrate this constant value of E_{mech}. ✔

(b) After the block has traveled 3.0 m from x_0, moving in the negative x direction, it is at $x_b = -1.0$ m. We can determine its speed at this position once we get the kinetic energy:

$$K_b = K(x_b) = E_{\text{mech}} - U(x_b) = 7.0 \text{ J} - (+2.0 \text{ J}) = +5.0 \text{ J}$$

$$\tfrac{1}{2}mv_b^2 = K_b$$

$$v_b = \sqrt{\frac{2K_b}{m}} = \sqrt{\frac{2(5.0 \text{ J})}{4.0 \text{ kg}}} = 1.6 \text{ m/s.} \ ✔$$

(c) We determine x_c, the position at which the block's velocity is zero, by using the fact that its kinetic energy is also zero at this position:

$$K_c + U_c = E_{\text{mech}}$$

$$0 + (ax^2 + bx + c) = E_{\text{mech}}$$

$$(1.0 \text{ J/m}^2)x_c^2 + (-2.0 \text{ J/m})x_c + (-1.0 \text{ J}) = 7.0 \text{ J}$$

$$x_c^2 - (2.0 \text{ m})x_c - (8.0 \text{ m}^2) = 0.$$

This expression can be factored, giving

$$[x_c + (2.0 \text{ m})][x_c - (4.0 \text{ m})] = 0.$$

The solutions to this equation are $x_c = -2.0$ m and $x_c = +4.0$ m, which agree with the values we read from Figure WG7.6 (the two values of x at which the curve intersects the $E_{\text{mech}} = 7.0$ J dashed line). Because the block is moving in the negative x direction, the solution we want here is $x_c = -2.0$ m. ✔

(d) At $x_c = -2.0$ m, the block has zero velocity because its kinetic energy is zero. Its acceleration is not zero at this location; indeed, its acceleration is always in the direction of decreasing potential energy. For the block at $x_c = -2.0$ m, this means directed along the positive x axis. Because the direction of motion reverses here, this position is called a *turning point* of the motion.

As the block moves in the positive x direction, it picks up speed as the potential energy decreases along the curve in Figure WG7.6, until it reaches $x = +1.0$ m. After it moves through this position of minimum potential energy, the block slows down as kinetic energy is converted to potential energy. The block eventually again has zero velocity at the other value we obtained in part c, $x_c = +4$ m, which is the second turning point of the motion. The block then again moves in the direction of decreasing potential energy, which now is to the left. ✔

❹ **EVALUATE RESULT** We can check whether our calculated answers are consistent with our graph. The kinetic energy K_b at $x_b = -1.0$ m in the graph (the amount of energy that must be added to the potential energy at this point to reach the E_{mech} line) is 5.0 J, which is what we calculated in part b. We can see from where the E_{mech} line crosses the potential energy curve that the left turning point is at $x = -2.0$ m as we calculated.

Guided Problem 7.8 Over the hill

A particle takes part in just one interaction with a fixed object. The interaction causes potential energy to be stored in the particle-object system as the particle moves along an x axis. The particle is released from rest at $x_0 = -3.0$ m, and the amount of potential energy stored (in joules) is given by $U(x) = ax + bx^2 + cx^3$, where $a = +12$ J/m, $b = +3.0$ J/m^2, and $c = -2.0$ J/m^3. (a) Make a graph of the potential energy that also displays the particle's mechanical energy. (b) In which direction does the particle move initially? (c) Describe the particle's motion after it leaves x_0. (d) At which positions along the curve you drew in part a is the particle speeding up, and at which positions is it slowing down? (e) What is the particle's kinetic energy at $x = -1.0$ m, $x = +1.0$ m, and $x = +3.0$ m?

① GETTING STARTED

1. Which portions of Worked Problem 7.7 are relevant?
2. How can you use calculus to determine the extrema of the potential energy function?

3. What happens to the amount of potential energy stored as x gets larger in the positive direction and as x gets larger in the negative direction?

② DEVISE PLAN

4. Calculate the potential energy at a few values of x. [Hint: What values are useful for part e?] What are the values of the potential energy at the extrema?
5. What is the kinetic energy of the particle at the instant it is released? How can you use this information to determine the (constant) mechanical energy of the system?
6. What determines in which direction the particle tends to move once it is released?

③ EXECUTE PLAN

7. Where is the potential energy increasing, and where is it decreasing?

④ EVALUATE RESULT

Questions and Problems

Dots indicate difficulty level of problems: • = easy, •• = intermediate, ••• = hard; CR = context-rich problem.

7.1 The effects of interactions

1. A piece of salami is squashed between two pieces of bread. Is there one interaction between the salami and bread or more than one? •
2. Explain why you cannot cause Earth to accelerate very much, even if you jump up and down on it. •
3. Sketch curves, to the same scale, of kinetic energy versus time for an elastic collision between cart A, originally moving with speed v, and cart B, originally at rest, when (a) $m_B = m_A$, (b) $m_B = 2m_A$, and (c) $m_B = m_A/2$. ••
4. Two objects of inertias m_1 and m_2 start from rest and then interact with each other (assume neither is interacting with any other object). (a) What is the ratio of their x components of velocity at any instant? (b) What is the ratio of their kinetic energies at any instant? ••
5. Two toy cars ($m_1 = 0.200$ kg and $m_2 = 0.250$ kg) are held together rear to rear with a compressed spring between them. When they are released, the cars are free to roll away from the ends of the spring. If you measure the acceleration of the 0.200-kg car to be 2.25 m/s^2 to the right, what is the acceleration of the other car? ••

7.2 Potential Energy

6. You are at the swimming pool and want to make a splash. You climb the ladder to the high diving board, take a big bounce, and produce a prodigious splash. Describe the changes in kinetic and potential energy involved in this motion, beginning when you start to climb and ending when you hit the water. •
7. You throw a stone straight up and then catch it as it falls. Draw the energy bar diagrams for the stone-Earth system (a) just after the stone leaves your hand, (b) at the top of the stone's path, and (c) just before the stone is back in your hand. •

8. A ring is attached at the center of the underside of a trampoline. A sneaky teenager crawls under the trampoline and uses the ring to pull the trampoline slowly down while his 75-kg mother is sleeping on it. When he releases the trampoline, she is launched upward. As she passes through the position at which she was before her son stretched the trampoline, her speed is 3.0 m/s. How much elastic potential energy did the son add to the trampoline by pulling it down? Assume the interaction is nondissipative. ••
9. An elevator might be raised and lowered directly by a winch, as in Figure P7.9a, or the mechanism might include a counterweight that moves up when the elevator moves down, as in Figure P7.9b. In which case is the change in the gravitational potential energy of the Earth-elevator-winch system lower? ••

Figure P7.9

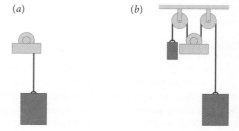

(a)　　　　　　(b)

10. *Principles* Figures 7.7 and 7.25 suggest that potential energy in a nondissipative interaction builds up and then diminishes. Is this always the case? ••
11. Two blocks are held together with a compressed spring between them on the surface of a slippery table. One block has three times the inertia of the other. When the blocks are released, the spring pushes them away from each other. What is the ratio of their kinetic energies after the release? •••

7.3 Energy dissipation

12. A trampoline's springs are what allow you to bounce, and if you bounce long enough, the springs get warm. Using *Principles* Figure 7.10 as a guide, classify the energy in the springs that accounts for (a) the bouncing and (b) the heating up. (c) Where does this energy come from? •

13. You throw a baseball straight up, and because of air resistance some of the ball's energy is converted to thermal energy during the whole trip. Which takes longer: the trip up or the trip down? •

14. An arrow is shot into a hollow pipe resting on a horizontal table and flies out the other end. While the arrow travels in the pipe, its feathers brush against the walls of the pipe. (a) Which type of collision is the arrow-pipe interaction: elastic, inelastic, or totally inelastic? (b) Is there an instant when the velocity of the arrow relative to the pipe is necessarily zero? (c) Describe the energy conversions in the pipe-arrow system. ••

15. (a) When you stretch a Slinky spring a moderate amount and then release it, it returns to its original state, and all the potential energy stored by the stretching is converted to another form. If you stretch too much, however, you can permanently bend the spring. Where does the stored potential energy go in this case? (b) If you use pliers to carefully straighten the bent spring so that it looks the same as before, do you expect to recover the energy dissipated when you bent it? ••

16. Mattresses are often a linked set of coil springs. Using *Principles* Figure 7.10 as a guide, classify the energy that you give the mattress in each of these cases: (a) You slide the mattress from one side of a room to the other side. (b) You jump on the mattress several times. (c) You run your fingers along the tops of the coils, so that they jiggle around. (d) By jumping hard, you manage to snap some of the links between the coils. ••

17. Make a graph of the kinetic energy, potential energy, and incoherent energy versus time for (a) a moving billiard ball colliding with a stationary billiard ball, and (b) a moving ball of cookie dough colliding with, and sticking to, an identical but stationary ball of cookie dough. ••

18. When a steel ball and a soft rubber ball are dropped from the same height onto a concrete surface, the steel ball bounces higher. When they are dropped onto packed sand, however, the rubber ball bounces higher. Explain the difference. •••

7.4 Source energy

19. Choose an appropriate closed system and draw a bar diagram representing the energy conversions and transfers that occur during each process of *Principles* Checkpoint 7.9: (a) a ball launching as the compressed spring it sits on expands, (b) a ball released from some height and falling to the ground, (c) a coasting bicycle slowing down, (d) a car accelerating on a highway. •

20. When an 800-kg compact car accelerates from rest to 27 m/s, it consumes 0.0606 L of gasoline, and 1.0 L of gasoline contains approximately 3.2×10^7 J of energy. What is the efficiency of the car (in other words, what percentage of the energy available from the fuel is delivered to the wheels)? ••

21. (a) Figure P7.21 shows initial and final energy bars for a closed system. Adjust the height of any two energy bars in the final version, on the right, so that the diagram does not violate what we know about energy. (There are numerous ways for you to do this.) (b) Describe an interaction that would generate an energy bar diagram that matches your version of Figure P7.21. ••

Figure P7.21

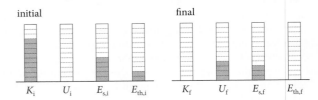

22. Two objects initially moving at velocities that have x components v_{1x} and v_{2x} collide totally inelastically. (a) Show that the ratio of the kinetic energy dissipated in the interaction is

$$\frac{\Delta K_1}{\Delta K_2} = -\frac{v_{cmx} + v_{1x,i}}{v_{cmx} + v_{2x,i}}.$$

(b) What does the minus sign in the expression in part a mean? ••

7.5 Interaction range

23. If, on the atomic scale, "contact" between two objects is in fact not *physical* contact, does that have implications for the contact between atoms in a piece of copper? •

24. Look again at *Principles* Checkpoint 7.10. Suppose you and a friend stand within an arm's length of each other and throw a ball back and forth. Describe a method to model an attractive interaction based on throwing the ball. ••

25. Look again at *Principles* Checkpoint 7.10. (a) When you throw a ball to a friend, at what instant does your momentum change: at the beginning, middle, or end of the ball's flight? (b) At which of these three instants does your friend's momentum change? ••

26. Consider the interactions involved when (a) you use a TV remote control to change the channel, and (b) you use your cell phone to text your friend. Classify each interaction as long range or short range. ••

7.6 Fundamental interactions

27. Suppose that, instead of two types of electrically charged particles in the universe, there was only one type. What would you expect to happen to the structure of the universe if the particles produced (a) an attractive interaction and (b) a repulsive interaction? •

28. Explain why friction is not considered a fundamental interaction. •

29. Suppose that the range of the strong interaction were suddenly increased by 20 orders of magnitude, with all other interactions unchanged. Describe the changes that might occur in the structure of our world. ••

30. Picture two objects, A and B, taking part in some interaction involving gauge particles. If object A radiates gauge particles uniformly in all directions, how does the strength of the interaction depend on the distance from A to B? Assume that the gauge particles do not interact with one another and that the strength of the interaction is proportional to the number of gauge particles encountered by B. •••

7.7 Interactions and accelerations

31. On a low-friction track, a 0.66-kg cart initially going at 1.85 m/s to the right collides with a cart of unknown inertia initially going at 2.17 m/s to the left. After the collision, the 0.66-kg cart is going at 1.32 m/s to the left, and the cart of unknown inertia is going at 3.22 m/s to the right. The collision takes 0.010 s. (*a*) What is the unknown inertia? (*b*) What is the average acceleration of each cart? (*c*) Use your answer to part *a* to check your answer to part *b*. •

32. Two children on ice skates pull toward each other on a rope held taut between them. The inertia of one child is 30 kg, and the inertia of the other is 25 kg. (*a*) If at one instant the 30-kg child is accelerating at 1.0 m/s^2 to the left, what is the acceleration of the 25-kg child at that instant? (*b*) What happens to each acceleration if the 25-kg child stops pulling and just hangs on? •

33. A 0.010-kg bullet is fired from a 5.0-kg gun with a muzzle velocity of 250 m/s. (*a*) While the bullet is traveling in the barrel, what is the ratio of the acceleration of the gun to the acceleration of the bullet as a function of time? (*b*) What is the speed of the bullet relative to the ground at the instant the bullet leaves the barrel? (*c*) What is the speed of the gun relative to the ground at that instant? ••

34. A 1500-kg car going at 6.32 m/s collides with a 3000-kg truck at rest. If the collision is totally inelastic and takes place over an interval of 0.203 s, what is the average acceleration of each vehicle? Are the ratio of accelerations and the ratio of inertias related the way you expect them to be? ••

35. A glob of putty thrown to the right bounces off a 0.500-kg cart initially at rest on a low-friction track. The collision takes 0.15 s, and the coefficient of restitution is 0.64. The cart has a final velocity of 1.0 m/s to the right, and the glob has a final velocity of 0.834 m/s to the left. (*a*) What is the inertia of the glob? (*b*) What is the average acceleration of each object during the collision? ••

36. An 8.20-kg object is sliding across the ice at 2.34 m/s. An internal explosion occurs, splitting the object into two equal chunks and adding 16 J of kinetic energy to the system. What is the average acceleration of the two chunks if the explosive separation takes place over a 0.16-s time interval? ••

37. A railroad car that has an inertia of 5.04×10^4 kg is moving to the right at 4.25 m/s when it collides and couples with three identical railroad cars that are already coupled together and moving in the same direction at 2.09 m/s. If the acceleration of the single car is 5.45 m/s^2 to the left while it is changing velocity, how long does the interaction take? ••

38. A 90-kg halfback on a football team runs head-on into a 120-kg opponent at an instant when neither has his feet on the ground. The halfback is initially going west at 10 m/s, and his opponent is initially going east at 4.37 m/s. The collision is totally inelastic. (*a*) If the collision takes 0.207 s, what is the acceleration of each player? (*b*) How much kinetic energy is converted to incoherent energy during the collision? ••

39. You are watching a hockey game on your digital video recorder. Your team's goalie is at rest when he catches a 0.16-kg puck moving straight toward him. The announcer says it was a 40-m/s slap shot, based on radar gun data. Normally the goalie would be braced, but in this case he is resting on parallel skate blades so that there is negligible friction to impede his subsequent motion. The collision is too fast to observe quantitatively, but by running the video in slow motion, you can

tell that the goalie's glove is 150 mm from his body when the puck strikes and is right against his chest after the successful catch. It is also clear that the stunned goalie (holding the puck) drifts backward at 71 mm/s for a second or two after the collision is over. (*a*) Compute the average acceleration of the puck and of the goalie during the collision. (*b*) Determine the inertia of the goalie (including his equipment). (*c*) How much kinetic energy is converted to thermal energy in the glove? Check your answers by solving in two ways when possible. •••

40. Two objects, one red and one yellow, start from rest and interact with each other along an *x* axis. (*a*) If $m_{red} < m_{yellow}$, what are the ratios of their changes in momentum, velocity, and kinetic energy? (*b*) Compare these ratios with the ratio of their accelerations. (*c*) What does your answer to part *a* tell you about the relative changes of these quantities for a ball and Earth when you drop the ball? •••

7.8 Nondissipative interactions

41. Draw a bar diagram representing the various kinds of energy for the cart-spring system in *Principles* Figure 7.26*b* when the end of the cart attached to the spring is at locations x_1, x_2, and x_3. •

42. The potential energy of an interaction is given by $U(x) = ax^2$, where $a = +6.4$ J/m^2. (*a*) If the initial speed of a 0.82-kg object in this system is 2.23 m/s at $x = 0$, how far does the object travel before it reaches a speed of $v = 0$? (*b*) Does your answer in part *a* depend on whether the object is traveling in the positive or negative *x* direction? •

43. Suppose we have a block mounted on a spring and define the zero point for the elastic potential energy of the system to be the relaxed position (meaning the block is positioned such that the spring is neither compressed nor stretched). What is the appropriate sign of the elastic potential energy when the block is repositioned so that the spring is (*a*) stretched and (*b*) compressed? (Hint: Think about what happens to the elastic potential energy when the stretched or compressed spring is released.) ••

44. A 0.36-kg cart and a 0.12-kg cart are held together with a compressed spring between them. When they are released, the 0.36-kg cart moves at 1.1 m/s to the right. How much elastic potential energy was stored in the spring before the release? ••

45. A 0.530-kg cart moving at 0.922 m/s to the right collides elastically with a 0.25-kg cart initially at rest. The 0.25-kg cart then moves off rapidly and compresses a spring before the 0.530-kg cart can catch it again. At the instant the 0.25-kg cart comes to rest, what is the energy stored in the spring? ••

46. A low-friction track is set up with a spring on the right end. In the middle of the track, a 2.0-kg cart moving to the right at 5.0 m/s overtakes a 3.0-kg cart moving to the right at 2.0 m/s. After the collision, the 2.0-kg cart is moving at 1.7 m/s. (*a*) What is the coefficient of restitution of the collision? (*b*) How much energy is stored in the spring when it is at its maximum compression after the collision? ••

47. For a certain interaction, the potential energy in joules is given by $U(x) = ax + bx^2$, where $a = +4.0$ J/m and $b = -2.0$ J/m^2. If, at some instant, a 10-kg object taking part in this interaction has an *x* component of velocity of -3.0 m/s and a position $x = +2.0$ m, what is its speed (*a*) at $x = +1.0$ m and (*b*) at $x = -1.0$ m? ••

PRACTICE

48. You release a block from the top of a long, slippery inclined plane of length ℓ that makes an angle θ with the horizontal. The magnitude of the block's acceleration is $g \sin \theta$ (see Section 3.7). (*a*) For an *x* axis pointing down the incline and having its origin at the release position, derive an expression for the potential energy of the block-Earth system as a function of *x*. (*b*) How would your expression change if you put the origin of the *x* axis at the bottom of the incline? (*c*) Does the location of the origin make any difference in the dynamics of the situation? (*d*) Use the expression you derived in part *a* to determine the speed of the block at the bottom of the incline. Does your speed value agree with the result obtained by a kinematic calculation? ●●●

49. On a low-friction track, a 0.36-kg cart initially moving to the right at 2.05 m/s collides elastically with a 0.12-kg cart initially moving to the left at 0.13 m/s. The 0.12-kg cart bounces off the 0.36-kg cart and then compresses a spring attached to the right end of the track. (*a*) At the instant of maximum compression of the spring, how much elastic potential energy is stored in the spring? (*b*) If the spring then returns all of this energy to the cart, and the two carts again collide, what is the final velocity of each cart? ●●●

7.9 Potential energy near Earth's surface

50. A 70-kg woman walks to the top of the Empire State Building, 380 m above street level. What is the change in the gravitational potential energy of the system comprising the woman and Earth? ●

51. At graduation, you toss your mortarboard cap straight up at some initial speed *v*. How fast is it moving when it comes back down to your hands? How do you know? ●

52. An office worker has his lunch on the 53rd floor of an office building. In which case is the gravitational potential energy of the lunch-Earth system greatest: (*a*) the worker carried the lunch up to the 53rd floor in the elevator, (*b*) he carried it up the stairs to the 53rd floor, or (*c*) he got a ride in a helicopter to the top of the 68-floor building and then carried the lunch down in the elevator to the 53rd floor? ●

53. A 0.0135-kg bullet is fired from ground level directly upward. If its initial speed is 300 m/s, how high does it go? (Neglect air resistance.) ●

54. A typical loaded commercial jet airplane has an inertia of 2.1×10^5 kg. (*a*) Based on the time intervals involved, which do you expect to require more energy: getting the plane up to cruising speed or getting it up to cruising altitude? (*b*) How much energy does it take to get the plane to a cruising speed of 270 m/s? (Ignore wind drag.) (*c*) How much energy does it take to get the plane to a cruising altitude of 10,400 m if it travels to that altitude at its (constant) cruising speed? ●●

55. Two 2.4-kg blocks are connected by a string draped over the edge of a slippery table, so that one block is on the table and the other is just hanging off the edge. A restraint holds the block on the table in place, and the string is 0.50 m long. After the restraint is removed, what speed does the pair of blocks have at the instant the upper one is pulled off the table? ●●

56. A 30-kg child jumps to the ground from the top of a fence that is 2.0 m high. You analyze the problem using upward as the positive *x* direction. (*a*) Taking $x = 0$ to be at the bottom of the fence, what are the initial potential energy of the child-Earth system and the change in the system kinetic energy during the jump? (*b*) Repeat part *a* for $x = 0$ at the top of the fence. ●●

57. You lean out of your dorm window, which is 12 m above the ground, and toss a 0.12-kg ball up to a friend at a window 11 m above you. (*a*) What is the slowest initial speed at which you can throw the ball so that he just catches it? (*b*) If you throw the ball at this speed and your friend misses it, what is its kinetic energy the instant before it hits the ground? (*c*) Suppose you throw the ball directly downward at the speed calculated in part *a*. What is its kinetic energy just before it hits the ground? ●●

58. Two blocks are hung by a string draped over a pulley, a 1.0-kg block on the left and a 3.0-kg block on the right. The two blocks start out at rest and at the same height. (*a*) What is the change in the gravitational potential energy of the system of blocks and Earth when the 3.0-kg block has dropped 0.53 m? (*b*) What is the change in the kinetic energy of the system between release and this instant? (*c*) What is the velocity of the 1.0-kg block at this instant? ●●

59. You drop a rubber ball from a height of 3.0 m. It bounces off a concrete surface to a height of 2.7 m. (*a*) What is the coefficient of restitution for this collision? (*b*) You want to get the ball to bounce upward to a height of 7.3 m. From the same starting point, how fast must you throw the ball, and in which direction? ●●

60. You throw a 0.52-kg target upward at 15 m/s. When it is at a height of 10 m above the launch position and moving downward, it is struck by a 0.323-kg arrow going 25 m/s upward. Assume the interaction is instantaneous. (*a*) What is the velocity of the target and arrow immediately after the collision? (*b*) What is the speed of the combination right before it strikes the ground? ●●

61. A uniform chain of inertia *m* and length ℓ is lying on a slippery table. When one quarter of its length hangs over the edge, the chain begins to slip off. How fast is it moving when the last bit of it slips off the table? (Ignore friction.) ●●●

62. A uniform chain of inertia *m* and length ℓ is lying on a slippery table. When just the tip hangs over the edge, the chain begins to slip off. (Ignore friction.) Calculate the speed of the chain as a function of time. (Hint: $\int \frac{dx}{v} = \int dt$. What's the connection between *x* and *v*?) ●●●

63. Two blocks are connected by a string draped over a small pulley. The smaller block has half the inertia of the larger block. For which configuration, (*a*) or (*b*), in Figure P7.63 are the speeds of the blocks greatest after the larger block has descended a distance *d*? Ignore friction between the table and the sliding block, and check your answer with a computation. ●●●

Figure P7.63

(a)   *(b)*

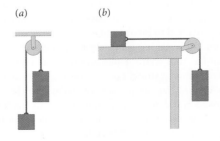

7.10 Dissipative interactions

64. Gravity-powered roller coasters have a motorized chain assembly that hauls the cars up to the top of the first hill. No additional source energy is supplied for the rest of the trip. What is the maximum possible height for all subsequent hills? •

65. You drop a ball from rest from a window 12 m above the ground, and just before it hits the ground its speed is recorded to be 14.6 m/s. What percentage of the ball's kinetic energy is dissipated due to air resistance? •

66. A 5.3-kg steel ball is dropped from a height of 20 m into a box of sand and sinks 0.20 m into the sand before stopping. How much energy is dissipated through the interaction with the sand? ••

67. Two cars collide inelastically on a city street. For the two-car system, which of the following are the same in any inertial reference frame: (*a*) the kinetic energy, (*b*) the momentum, (*c*) the amount of energy dissipated, (*d*) the momentum exchanged? ••

68. A 0.70-kg basketball dropped on a hardwood floor rises back up to 65% of its original height. (*a*) If the basketball is dropped from a height of 1.5 m, how much energy is dissipated in the first bounce? (*b*) How much energy is dissipated in the fourth bounce? (*c*) To which type of incoherent energy is the dissipated energy converted? ••

69. An 80-kg man standing on a frozen lake tosses a 0.500-kg football to his dog. (*a*) If the ball leaves his hands at 15 m/s relative to Earth, what minimum source energy did the man supply? (*b*) His 20-kg dog, standing still, catches the ball and slides with it. How much energy is dissipated in the catch? ••

70. A 2.2-kg measuring instrument is mounted on a balloon by your scientific team for atmospheric studies. At the top of its flight, the instrument is released from the balloon and falls most of the way back to Earth before a parachute opens. You are told that the magnitude of the acceleration at any instant t before the parachute opens is given by $|a| = ge^{-t/\tau}$, where g is the acceleration due to gravity, e is the base of natural logarithms, and τ is a time constant that depends on the shape of the instrument and in this case is 5.68 s. Your primary concern is how much the instrument heats up as it falls, due to air resistance. At what rate, in joules per second, is energy dissipated before the parachute opens? Express your answer as a function of t, where $t = 0$ at release. (Hint: Integrate the acceleration to calculate speed and displacement.) •••

71. A nursery rhyme goes "Humpty Dumpty sat on a wall, Humpty Dumpty had a great fall. All the king's horses and all the king's men couldn't put Humpty Dumpty together again." (*a*) Describe the energy of the Earth-HD system before, during (including the instant just before impact), and after the fall. Use bar diagrams showing K, U, E_s, and E_{th} to illustrate your description. (*b*) In terms of the energy of the Earth-HD system, explain why all the king's horses and all the king's men could not put HD together again. (*c*) Now assume that things had turned out differently for HD. The king's men were able to set up a net so that he fell into the net and bounced back up and once again sat on the wall. Draw energy bar diagrams for this situation, showing the distribution of energy during the fall, at the instant HD arrives at the net (choose the instant at which the net with HD in it is stretched to the lowest position), during the ascent, and when he is back sitting on the wall. •••

72. A small block of wood of inertia m_{block} is released from rest a distance h above the ground, directly above your head. You decide to shoot it with your pellet gun, which fires a pellet of inertia m_{pellet}. After the block has fallen a distance d, the pellet hits it and becomes embedded in it, kicking it upward. At the instant of impact, the pellet is moving at speed v_{pellet}. (*a*) To what maximum height h_{max} does the pellet-block system rise above the ground? Express your answer in terms of the given variables and known values, such as g. (*b*) How much energy is dissipated in the collision? •••

Additional Problems

73. Is the collision between two billiard balls a nondissipative interaction? (Hint: What does the fact that you can hear the collision tell you?) •

74. Draw before and after energy bars for the collision shown in *Principles* Figure 6.8*a* and 6.8*b*. •

75. The observation deck of the Washington Monument is 152 m above the ground. How many king-sized candy bars, each supplying 1.3 MJ of source energy, must an 80-kg man consume to replenish the energy he expends in climbing the 825 stairs up to the observation deck? •

76. You drop a ball from some height. At which position do you expect its speed to be one-half the speed at which it eventually hits the ground: higher than the midpoint of the path, at the midpoint, or lower than the midpoint? Does air resistance change your answer? ••

77. Is it impossible to convert incoherent energy to coherent energy? Can you think of an example where thermal energy can (at least partially) be converted to coherent energy? ••

78. People unenlightened about physics sometimes fire a gun straight up into the air to celebrate big events. By discussing the forms of energy at each stage of the process, explain why this is a bad idea. ••

79. Two 1.20-kg carts on a low-friction track are rigged with magnets so that the carts repel each other when they approach each other. One cart has an initial velocity of +0.323 m/s, and the other has an initial velocity of −0.147 m/s. What is the maximum energy stored in the magnetic field during the collision, assuming that the carts never touch each other? ••

80. Wind turns a windmill that stores energy in a battery. Assume that the windmill turns at a constant rate while the battery is charging. The battery is then used to run a toy car that moves at a constant speed until the battery goes dead. Draw bar diagrams representing a sequence of instants that depict the various conversions of energy from one form to another in this situation. For each conversion, name the system you are describing. ••

81. Not looking where you are going, you and your bike collide at 12 m/s into the back of a car stopped at a red light. The car does not have its brakes applied and so is jolted forward. The driver immediately leaps out crying, "Whiplash!" Facing a day in court, you have to determine the acceleration of the car as a result of the collision. You note that you and your bike (combined inertia 80 kg) came to a complete halt in the collision and that the rim of your front wheel was pushed all the way to the center hub. The diameter of the bike wheel before the crash was 0.75 m, and a reference book tells you that the inertia of the car is 1800 kg. ••• CR

82. An amusement park commissions your company to design a roller coaster. Your company specializes in an approach in which the cars are accelerated on a horizontal section of track by magnetic motors rather than being hauled up to the top of a hill and released. The park officials would like the riders to be accelerated via your horizontal design and then sent straight up a vertical length of track 66.4 m tall. (After the cars pass this high point, the ride is the regular ups and downs of a conventional roller coaster.) Your calculations tell you that 212 motors will get the loaded cars up to the necessary speed at an acceleration of 0.85g, but you wonder whether or not the park has enough horizontal distance available for the acceleration device. ●●● CR

83. A 1.00-kg cart has attached to its front end a device that explodes when it hits anything, releasing a quantity of energy E. This cart is moving to the right with speed v when it collides head-on with a 2.00-kg cart traveling to the left at the same speed v. The explosive goes off when the carts hit, causing them to rebound from each other. If one-quarter of the explosive energy is dissipated into the incoherent energy of noise and deformation of the carts, what is the final velocity of each cart? Solve in the zero-momentum reference frame and then check your result by solving in the Earth reference frame. ●●●

Answers to Review Questions

1. (*a*) The momentum remains constant. (*b*) The kinetic energy dips briefly as some or all of it is converted to internal energy during the collision. (*c*) The sum of all forms of energy in the system remains constant.
2. No. The relative speed must be nonzero before the collision, but it changes to zero at some instant during the interaction.
3. In all cases a portion of the kinetic energy of the system is converted to internal energy. In part *a*, none of this converted energy reappears as kinetic energy after the collision. In part *b*, some of this energy reappears as kinetic. In part *c*, all of it reappears as kinetic energy after the collision.
4. An interaction involves two objects and results in a change of motion, a change of physical state, or both. If a change of motion is involved, the ratio of the *x* components of the accelerations of the objects is equal to the negative of the inverse ratio of their inertias. Both the momentum and kinetic energy of each object generally change as a result of the interaction, but if the system is isolated, its momentum does not change. If the system is closed, its energy does not change but its kinetic energy changes during the interaction as some portion is converted to internal energy. When the interaction is an elastic collision, all of the converted kinetic energy ultimately reappears as kinetic energy after the interaction.
5. Kinetic energy is the energy any moving object has because of its motion. Potential energy is energy stored in a system due to reversible changes in the configuration state of the system.
6. Many answers are possible. Two possibilities are bending (coherent kinetic energy) a paper clip back and forth, which heats it up (incoherent thermal energy), and rubbing your hands together to warm them.
7. Energy of motion is energy due to the motion of objects or molecules in a system. Energy of configuration is energy associated with the locations of objects or molecules in a system. Coherent energy is associated with either motions or configurations that are organized, so that individual objects or molecules in the system either all move with a common velocity or are all displaced by a small amount along a common direction. Incoherent energy is associated with either motions or configurations that are disorganized, so that objects or molecules in the system move randomly or are displaced randomly.
8. Solar radiation generated by nuclear reactions in the Sun is the ultimate source.
9. Nondissipative interactions cause reversible changes; dissipative interactions cause irreversible changes.
10. Many answers are possible. In the interaction between a car and the gasoline in its fuel tank, source energy (chemical energy of the gasoline) is converted to thermal energy, to the car's kinetic energy, and, if the car is climbing a hill, to gravitational potential energy. When you rub your hands together to warm them, chemical energy

in the molecules that make up the muscles you use to move your hands is converted to kinetic energy, which in turn is converted to thermal energy by friction.
11. In the field model, every object is surrounded by a field, and it is with this field that other objects interact. In the gauge-particle model, two objects interact by exchanging fundamental particles called gauge particles.
12. Only the gravitational and electromagnetic interactions have a long enough range to be noticed on the macroscopic scale. The range of the strong and weak interactions is too short to cause changes that we can observe on the macroscopic scale.
13. (*a*) The electromagnetic interaction is most important. (*b*) Again, the electromagnetic interaction is most important.
14. Although the electromagnetic interaction is much stronger, it can be either attractive or repulsive depending on the signs of the electrically charged particles involved in the interaction. Because there are equal numbers of the two types of particles in any macroscopic object, the particles tend to arrange themselves to produce electrical neutrality and to minimize the effect of the interaction. In contrast, the gravitational interaction is always attractive, and no cancellation diminishes its cumulative effect. Because we live on the surface of Earth, the very large effect of the gravitational interaction between macroscopic objects and Earth is generally not negligible.
15. (*a*) $a_{1x}/a_{2x} = -m_2/m_1$. (*b*) Again $a_{1x}/a_{2x} = -m_2/m_1$ because *Principles* Eq. 7.6 holds for all interactions.
16. Two assumptions were made—that the system of interacting objects is isolated, and the inertias of the objects are not changed by the interaction.
17. For a closed system, energy must not change, so the changes in the four categories of energy—kinetic, potential, source, and thermal—must cancel. Both source energy and thermal energy always involve dissipation, so if all interactions are nondissipative, only kinetic and potential energy changes are possible, and they must add to zero. Mechanical energy is the sum of kinetic and potential energies and hence remains constant in such situations.
18. Nondissipative interactions cause only reversible changes. A movie recording the object(s) in such a system must appear the same when run either forward or backward. That means the kinetic energy of each object must have the same value at a given position regardless of whether the object is moving forward or backward at that position (or not moving at all). But mechanical energy remains constant in such systems and is equal to the sum of kinetic and potential energies. If the kinetic energy must be expressible as a function of position, then so must the potential energy.
19. Because of the reversibility of changes due to nondissipative interactions, the initial and final kinetic energies must be equal at the common position (point A).

20. Vertical. Yes, it does matter because changes in gravitational potential energy are zero for horizontal displacements.
21. No, the gravitational potential energy can be either positive or negative, depending on which position you choose for your $U^G = 0$ position.
22. No, only the convertible part of the system's energy is converted. The inconvertible part remains as system kinetic energy. Some final nonzero velocity (and hence final nonzero kinetic energy) is needed to keep the system's momentum constant.
23. (a) Equation 7.26, $\Delta E_{th} = -\Delta K = \frac{1}{2}\mu v_{12i}^2(1 - e^2)$, tells us that none of the convertible energy is converted to thermal energy (which means the collision is elastic). All of it is temporarily stored as potential energy. (b) Equation 7.26 tells us that all of the convertible kinetic energy is ultimately converted to thermal energy, but all or part of it may still be temporarily stored as potential energy.

Answers to Guided Problems

Guided Problem 7.2 $h/16$
Guided Problem 7.4 (a) rear $= -0.12$ m/s, lead $= +8.4$ m/s;

(b)

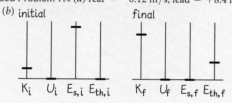

Guided Problem 7.6 $v_f = 78$ m/s
Guided Problem 7.8

(a)

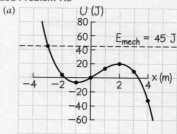

x	U(x)
-3	45
-2	4
-1	-7
0	0
1	13
2	20
3	9
4	-32

(b) toward positive x; (c) accelerates toward positive x, then accelerates toward negative x between $x = -1.0$ m and $x = +2.0$ m, then accelerates toward positive x again; (d) speeding up from $x = -3.0$ m to $x = -1.0$ m and from $x = +2.0$ m to large x values, slowing down between $x = -1.0$ m and $x = +2.0$ m; (e) $K(x = -1.0$ m$) = +52$ J, $K(x = +1.0$ m$) = +32$ J, $K(x = +3.0$ m$) = +36$ J

8

PRACTICE
Force

Chapter Summary   125

Review Questions   127

Developing a Feel   128

Worked and Guided Problems   129

Questions and Problems   136

Answers to Review Questions   144

Answers to Guided Problems   144

PRACTICE

Chapter Summary

Characteristics of forces (Sections 8.1, 8.2, 8.3, 8.7)

Concepts When an object participates in one interaction only, the **force** exerted on the object is given by the time rate of change in the object's momentum.

A **contact force** is a force that one object exerts on another object only when the two objects are in physical contact.

A **field force** is a force (such as gravity) that one object exerts on another object without the requirement that the two objects be in physical contact.

When two objects interact, the forces they exert on each other form an **interaction pair,** and these forces have equal magnitudes but opposite directions.

Quantitative Tools The SI unit of force is the **newton** (N):

$$1\ \text{N} = 1\ \text{kg} \cdot \text{m/s}^2. \tag{8.9}$$

For an interaction pair of forces,

$$\vec{F}_{12} = -\vec{F}_{21}. \tag{8.15}$$

Some important forces (Sections 8.6, 8.8, 8.9)

Concepts A taut, flexible object (such as a spring, rope, or thread), when subjected to equal and opposite *tensile forces* applied at either end, experiences along its length a *stress* called **tension.** If the object is very light, the tension is the same everywhere in it.

Hooke's law relates the force that a spring exerts on a load to the distance the spring is stretched (or compressed) from its relaxed position.

Quantitative Tools The x component of the force of gravity exerted on an object of inertia m near Earth's surface is

$$F^G_{\text{Eo}x} = -mg, \tag{8.17}$$

where the minus sign indicates that the force is directed downward.

Hooke's law: If a spring is stretched (or compressed) by a *small* distance $x - x_0$ from its unstretched length x_0, the x component of the force it exerts on the load is

$$(F_{\text{by spring on load}})_x = -k(x - x_0), \tag{8.20}$$

where k is called the **spring constant** of the spring.

Effects of force (Sections 8.1, 8.4, 8.5, 8.7)

Concepts The vector sum of the forces exerted on an object is equal to the time rate of change of the momentum of the object.

The **equation of motion** for an object relates the object's acceleration to the vector sum of the forces exerted on it.

Newton's laws of motion describe the effects forces have on the motion of objects.

If an object is at rest or moving with constant velocity, it is in **translational equilibrium**. In this case, the vector sum of the forces exerted on it is equal to zero.

A **free-body diagram** for an object is a sketch representing the object by a dot and showing all the forces exerted *on* it.

Quantitative Tools Vector sum of forces exerted on an object:

$$\sum \vec{F} \equiv \frac{d\vec{p}}{dt}. \tag{8.4}$$

Equation of motion:

$$\vec{a} = \frac{\sum \vec{F}}{m}. \tag{8.7}$$

If the inertia m of an object is constant, Newton's second law is usually written as

$$\sum \vec{F} = m\vec{a}. \tag{8.6}$$

For two interacting objects, 1 and 2, Newton's third law of motion states that

$$\vec{F}_{12} = -\vec{F}_{21}. \tag{8.15}$$

Impulse (Section 8.10)

Concepts The **impulse** $\vec{J}$ produced by forces exerted on an object is the product of the vector sum of the forces and the time interval over which the forces are exerted. The impulse delivered to the object is also equal to the change in its momentum.

Quantitative Tools For a constant force,

$$\Delta\vec{p} = \vec{J} = \left(\Sigma\vec{F}\right)\Delta t. \tag{8.24, 8.25}$$

For a time-varying force,

$$\Delta\vec{p} = \vec{J} = \int_{t_i}^{t_f} \Sigma\vec{F}(t)\, dt. \tag{8.26}$$

System of interacting objects (Sections 8.11, 8.12)

Concepts The center of mass of a system of objects accelerates as though all the objects were located at the center of mass and the external force were applied at that location.

Quantitative Tools Acceleration of a system of objects:

$$\vec{a}_{cm} = \frac{\Sigma\vec{F}_{ext}}{m}. \tag{8.45}$$

Review Questions

Answers to these questions can be found at the end of this chapter.

8.1 Momentum and force

1. Does a force exerted on an object always increase the object's speed? If yes, explain why. If no, provide an example in which an exerted force does not increase the object's speed.
2. What are the magnitude and direction of the vector sum of the forces exerted on a car traveling on a straight downward slope of a highway at a constant 100 km/h?
3. What is the relationship between momentum and force?

8.2 The reciprocity of forces

4. Is there any circumstance in which an object 1 may exert a force on an object 2 without an equal, reciprocal force being exerted by object 2 on object 1?
5. A cup filled with coffee sits on a table. Do the downward force the cup exerts on the table and the downward force the coffee exerts on the bottom of the cup form an interaction pair?

8.3 Identifying forces

6. What is the principal difference between a contact force and a field force?
7. Characterize each of the following as a contact force or a field force: (*a*) the force that causes a book sliding across a polished floor to eventually slow down; (*b*) the force that causes a wine glass to fall downward after it has been knocked off a table; (*c*) the force that causes one magnet to repel another; (*d*) the force exerted by the wind on a sailboat; (*e*) the force that causes an object attached to one end of a stretched rubber band to move toward the object attached to the other end.

8.4 Translational equilibrium

8. Can an object that is in translational equilibrium have any forces exerted on it?
9. How can you determine, by observing an object, whether or not it is in translational equilibrium?

8.5 Free-body diagrams

10. Describe the three main elements in a free-body diagram.
11. Give at least two reasons it is important to draw a free-body diagram to analyze physical situations.
12. Why should an arrow representing the vector sum of the forces exerted on an object *not* appear in the free-body diagram for the object?

8.6 Springs and tension

13. How is the change in the length of a spring related (*a*) to the magnitude F_{os} of the force exerted by an object on the spring and (*b*) to the magnitude F_{so} of the force exerted by the spring on the object?
14. What is an elastic force?
15. Under what condition(s) is a pulling force exerted by a hand on one end of a rope, spring, or thread transmitted undiminished to a block attached to the other end?

8.7 Equation of motion

16. Knowing all the forces exerted on an object gives you direct information about which aspect of the object's motion?
17. What condition allows you to convert the relationship $\sum \vec{F} = d\vec{p}/dt$ to the relationship $\sum \vec{F} = m\vec{a}$?
18. State Newton's three laws of motion in your own words.

8.8 Force of gravity

19. (*a*) Which gravitational force exerted by Earth has the larger magnitude: that exerted on an apple or that exerted on a sack of apples? (*b*) How do you reconcile your answer to part *a* with the observation that, when they are in free fall, the apple and the sack of apples accelerate at the same rate?
20. An object at rest has zero acceleration. Does this mean that Eq. 8.17 ($F_{Eox}^G = -mg$) does not accurately represent the force of gravity exerted on the object?

8.9 Hooke's law

21. How does the numerical value of the spring constant k for a spring under compression compare with the value of k for the same spring when it is stretched?
22. Equation 8.20 relates the force exerted by a spring to the displacement of the spring's free end: $(F_{\text{by spring on load } x}) = -k(x - x_0)$. What is the meaning of the symbol x_0?
23. What is the physical significance of the slope of the curve in *Principles* Figure 8.18, a graph that plots the displacement of the free end of a spring as a function of the x component of the force exerted on the spring?

8.10 Impulse

24. State the impulse equation in words.
25. Suppose that one or more of the forces exerted on an object is not constant, but that you do have information about how each force exerted on the object varies as a function of time during the time interval of interest. How can this information be used to compute the impulse?

8.11 Systems of two interacting objects

26. What distinguishes an internal force from an external force?
27. Explain what is special about the motion of the center of mass of a system made up of two interacting objects when some external force is exerted on the system.

8.12 Systems of many interacting objects

28. What is the numerical value of the vector sum of all internal forces exerted in a system composed of many objects?
29. Explain why the motion of rigid objects is usually easier to analyze than the motion of deformable objects.

Developing a Feel

Make an order-of-magnitude estimate of each of the following quantities. Letters in parentheses refer to hints below. Use them as needed to guide your thinking.

1. The magnitude of the vector sum of forces needed to give a bus the acceleration seen in a sports car that can go from 0 to 60 mi/h in 5 s (K, O)
2. The magnitude of the average force exerted by a bat on a baseball hit straight toward the pitcher (E, W, J, U)
3. The magnitude of the vector sum of forces exerted on an airliner accelerating for takeoff (P, I, G)
4. The magnitude of the gravitational force exerted by Earth on an elephant (S)
5. The magnitude of the upward force exerted by the air on an airliner flying horizontally at constant velocity (A, P, V)
6. Over and above the force required to stand on the floor, the magnitude of the extra force exerted by a professional basketball player on the floor in order to leap for a slam dunk (H, C, X, R, L)
7. The spring constant of the spring in a ballpoint pen (F, T)
8. The spring constant of an automobile spring (D, Z)
9. The magnitude of the impulse required to get a minivan up to freeway speed (N, Y)
10. The magnitude of the impulse required to stop a speeding train in 60 s (B, M, Q)

Hints

A. Is the airliner in translational equilibrium?
B. What is the speed of a speeding train?
C. What has happened to the basketball player's kinetic energy when he is at the top of the jump?
D. By what distance does a car fender drop when an adult of average inertia sits on the car?
E. What is the baseball's incoming velocity?
F. Through what distance is the spring compressed when you click the pen?
G. What is the speed just as the airliner's wheels lift off the ground?
H. How does Newton's third law of motion apply to this situation?
I. How far along the runway does the airliner move before the wheels lift off the ground?
J. What is the baseball's outgoing velocity?
K. What is the magnitude of the acceleration of a sports car?
L. Over what time interval is the force exerted?
M. What is the inertia of a train?
N. What is freeway speed?
O. How does the inertia of a bus compare with the inertia of a sports car?
P. What is the inertia of an airliner?
Q. What portion of the momentum must change in the time interval?
R. What is the inertia of a professional basketball player?

S. What is the inertia of an elephant?
T. What is the magnitude of the force you must exert to click the pen?
U. For what time interval is the ball in contact with the bat?
V. What other force is exerted vertically on the airliner?
W. What is the inertia of a baseball?
X. What is the height of the basketball player's jump?
Y. What is the inertia of a minivan?
Z. What is the inertia of an average adult?

Key (all values approximate)

A. yes; B. 3×10^1 m/s; C. it has all been converted to gravitational potential energy; D. 4×10^1 mm; E. 4×10^1 m/s toward batter; F. 5 mm; G. 9×10^1 m/s; H. it tells us that $F_{\text{by player on floor}} = F_{\text{by floor on player}}$; I. 1 km; J. a bit faster than the incoming velocity—say, 5×10^1 m/s away from batter; K. 5 m/s²; L. 0.4 s; M. 5×10^6 kg; N. 3×10^1 m/s; O. $m_{\text{bus}} \approx 10 m_{\text{car}}$; P. 1×10^5 kg; Q. the entire momentum, 2×10^8 kg·m/s, must go to zero; R. 9×10^1 kg; S. 6×10^3 kg; T. 3 N; U. 2×10^{-2} s; V. the gravitational force; W. 0.2 kg; X. each part of the body rises 0.7 m; Y. 2×10^3 kg; Z. 8×10^1 kg

Worked and Guided Problems

Procedure: Drawing free-body diagrams

1. Draw a center-of-mass symbol (a circle with a cross) to indicate the object you wish to consider*—this object is your system. Pretend the object is by itself in empty space (hence the name *free body*). If you need to consider more than one object in order to solve a problem, draw a separate free-body diagram for each.

2. List all the items in the object's environment that are in contact with the object. These are the items that exert *contact forces* on the object. *Do not add these items to your drawing*! If you do, you run the risk of confusing the forces exerted *on* the object with those exerted *by* the object.

3. Identify all the forces exerted *on* the object by objects in its environment. (For now, omit from consideration any force not exerted along the object's line of motion.) In general, you should consider (*a*) the *gravitational field force* exerted by Earth on the object and (*b*) the *contact force* exerted by each item listed in step 2.

4. Draw an arrow to represent each force identified in step 3. Point the arrow in the direction in which the force is exerted and place the tail at the center of mass. If possible, draw the lengths of the arrows so that they reflect the relative magnitudes of the forces. Finally, label each arrow in the form

$$\vec{F}^{\text{type}}_{\text{by on}},$$

where "type" is a single letter identifying the origin of the force (c for contact force, G for gravitational

force), "by" is a single letter identifying the object exerting the force, and "on" is a single letter identifying the object subjected to that force (this object is the one represented by the center of mass you drew in step 1).

5. Verify that all forces you have drawn are exerted **on** and not **by** the object under consideration. Because the first letter of the subscript on $\vec{F}$ represents the object exerting the force and the second letter represents the object on which the force is exerted, every force label in your free-body diagram should have the same second letter in its subscript.

6. Draw a vector representing the object's acceleration *next to* the center of mass that represents the object. Check that the vector sum of your force vectors points in the direction of the acceleration. If you cannot make your forces add up to give you an acceleration in the correct direction, verify that you drew the correct forces in step 4. If the object is not accelerating (that is, if it is in translational equilibrium), write $\vec{a} = 0$ and make sure your force arrows add up to zero. If you do not know the direction of acceleration, choose a tentative direction for the acceleration.

7. Draw a reference axis. If the object is accelerating, it is often convenient to point the positive x axis in the direction of the object's acceleration.

When your diagram is complete it should contain only the center-of-mass symbol, the forces exerted *on* the object (with their tails at the center of mass), an axis, and an indication of the acceleration of the object. Do not add anything else to your diagram.

---

* Representing, say, a car by a single point might seem like an over-simplification, but because we are interested only in the motion of the car as a whole, any details (like its shape or orientation) do not matter. Any unnecessary details you add to your drawing only distract from the issue at hand.

These examples involve material from this chapter but are not associated with any particular section.
Some examples are worked out in detail; others you should work out by following the guidelines provided.

Worked Problem 8.1 Moving a block

A forklift is attempting to pull a 1400-kg block of granite across a warehouse floor with a rope, aided by a push from a worker (Figure WG8.1). The tension in the rope is 2500 N, and the worker exerts a horizontal push of 200 N. The block slides across the floor at a constant speed of 1.5 m/s. What is the magnitude of the frictional force exerted on the block?

Figure WG8.1

❶ GETTING STARTED We are asked to calculate the magnitude of a frictional force. We note that, in addition to the frictional force, there are two other horizontal contact forces exerted on the block: the pulling force exerted by the rope and the pushing force exerted by the worker. A relationship that involves force and motion is needed, so we will likely need *Principles* Eq. 8.6, $\sum \vec{F} = m\vec{a}$, Newton's second law when inertia is constant. A free-body diagram (Figure WG8.2) helps us organize the information. The directions of the pulling and pushing forces must be the same, so we arbitrarily draw those to the right in the figure. The frictional force opposes slipping between the block and the floor surface. If we were to stop pushing and pulling, friction should cause the block to slow and stop, so friction should be directed to the left, opposite the velocity of the block. We note that this velocity is constant, which means the acceleration is zero.

Figure WG8.2

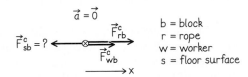

$\vec{a} = \vec{0}$

$\vec{F}^c_{sb} = ?$ $\vec{F}^c_{rb}$ $\vec{F}^c_{wb}$

b = block
r = rope
w = worker
s = floor surface

❷ DEVISE PLAN We know the tension in the rope and the magnitude of the worker's pushing force. We know that the acceleration is zero, and we know the direction of the frictional force. We arbitrarily choose the positive x direction to be the direction of motion (to the right in Figure WG8.2). The three forces are horizontal, so we need only their x components to relate their values:

$$\sum F_x = 0. \qquad (1)$$

One equation is sufficient to solve for the one unknown frictional force.

❸ EXECUTE PLAN Using Figure WG8.2 as a guide, we express the left side of Newton's second law as the sum of the x components of the individual forces. By comparing the directions of the forces with the positive x direction, we see that the x components of the pushing and pulling forces are positive and that of the frictional force is negative. If we substitute the magnitude of the forces, Eq. 1 becomes

$$F^c_{rbx} + F^c_{wbx} + F^c_{sbx} = 0$$
$$+F^c_{rb} + F^c_{wb} + (-F^c_{sb}) = 0 \qquad (2)$$
$$F^c_{sb} = F^c_{rb} + F^c_{wb} = 2500 \text{ N} + 200 \text{ N} = 2700 \text{ N}. ✔$$

❹ EVALUATE RESULT If we substitute this answer into Eq. 2, the signs in the equation are consistent with the force directions shown in Figure WG8.2. No assumptions were required to solve this problem, nor did we require all the numerical values that were given.

Guided Problem 8.2 Car springs

As an automotive engineer, you have been assigned the task of designing a set of springs for a car. If each of the four springs must compress a distance of 30 mm when the 640-kg car body rests on them, what value of the spring constant k should each spring have?

❶ GETTING STARTED
1. What interactions does the car body take part in?
2. Draw a free-body diagram of the car body.

❷ DEVISE PLAN
3. After the springs stop bouncing, what is the body's acceleration? Add that information to your free-body diagram.
4. Decide on an appropriate direction for the positive x axis and indicate it on the diagram.

5. Using your free-body diagram as a guide, write the component version of the appropriate force equation.

❸ EXECUTE PLAN
6. Solve for the force exerted by each spring on the body.
7. How is this force related to the spring constant?

❹ EVALUATE RESULT
8. Examine your expression for k. If the inertia of the car were increased, what would happen to k? Is this reasonable?
9. Did you make any assumptions?

Worked Problem 8.3 Truckin' on

A tractor is pulling two trailers. Trailer 1 has an inertia of 4000 kg, and trailer 2 has an inertia of 6000 kg. The tractor pulls its load with a force of 2.5 kN when it starts from rest. (*a*) What is the magnitude of the tractor's acceleration? (*b*) What is the tension in the coupling between the two trailers?

❶ GETTING STARTED Figure WG8.3 identifies the objects involved: tractor t, trailer 1, and trailer 2. Because two objects are being pulled, we may need to draw a free-body diagram for each one. What shall we put in our free-body diagrams? The motion is horizontal, so we concern ourselves only with horizontal forces.

Figure WG8.3

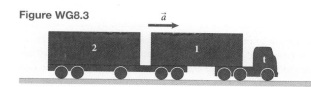

$\vec{a}$

The only force exerted on trailer 2 is the forward pull exerted by the coupling with trailer 1. The forces exerted on trailer 1 are the 2.5-kN force exerted by the tractor pulling forward and the force exerted by the coupling with trailer 2. We ignore air resistance because the truck starts from rest and so is moving at low speed. We also assume the wheels are well greased so that bearing friction does not impede the motion. As long as the tractor and two trailers move together, they must have a common acceleration, and it is in the direction in which the tractor pulls (which we make the positive x direction). We can therefore use the same acceleration symbol a for the tractor and both trailers, and our free-body diagrams are as shown in Figure WG8.4.

Figure WG8.4

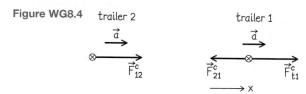

trailer 2   trailer 1

② **DEVISE PLAN** Because this problem deals with acceleration and force, we start with Newton's second law. With two free-body diagrams we obtain two equations, one from each diagram:

$$\sum F_{1x} = m_1 a_{1x} = m_1 a_x$$

$$\sum F_{2x} = m_2 a_{2x} = m_2 a_x.$$

The unknown common acceleration $a = |\vec{a}| = |a_x|$ and the unknown coupling tension $\mathcal{T} = |\vec{F}^{\text{tensile}}| = |\vec{F}_{12}^{c}| = |\vec{F}_{21}^{c}|$ will appear in both equations. We know the magnitude of the tractor's pulling force: $F_{t1}^{c} = 2.5$ kN. Two equations with two unknowns are all we need.

③ **EXECUTE PLAN**
(*a*) Our second-law equations give us

$$\sum F_{1x} = F_{21x}^{c} + F_{t1x}^{c} = m_1 a_x = -F^{\text{tensile}} + F^{\text{pull}} = m_1(+a)$$

$$\sum F_{2x} = F_{12x}^{c} = m_2 a_x = +F^{\text{tensile}} = m_2(+a). \tag{1}$$

Adding these two equations eliminates the tensile force and allows us to solve for the acceleration of each object:

$$m_2 a + m_1 a = (+F^{\text{tensile}}) + (-F^{\text{tensile}}) + (F^{\text{pull}})$$

$$(m_1 + m_2)a = F^{\text{pull}}$$

$$a = \frac{F^{\text{pull}}}{m_1 + m_2} = \frac{2.5 \times 10^3 \text{ N}}{4000 \text{ kg} + 6000 \text{ kg}} = 0.25 \text{ m/s}^2. \checkmark \tag{2}$$

(*b*) We compute the tension $\mathcal{T} = |\vec{F}^{\text{tensile}}|$ in the coupling by substituting the expression we derived for a in Eq. 2 into Eq. 1:

$$\mathcal{T} = m_2\left(\frac{F^{\text{pull}}}{m_1 + m_2}\right) \tag{3}$$

$$\mathcal{T} = \frac{m_2}{m_1 + m_2}F^{\text{pull}} = \frac{6000 \text{ kg}}{4000 \text{ kg} + 6000 \text{ kg}}(2.5 \times 10^3 \text{ N})$$

$$= 1.5 \times 10^3 \text{ N}. \checkmark$$

④ **EVALUATE RESULT** Notice from Eq. 2 that the acceleration is the force exerted by the tractor divided by the inertia of the trailers, which is what we would expect if we had lumped the two trailers together as one object to get the acceleration.

The magnitude of the tension we obtained for the coupling, 1.5×10^3 N, is numerically equal to the magnitude of the tensile force required to accelerate just trailer 2 and should be less than the magnitude of the force exerted by the tractor, 2.5×10^3 N, which has to get both trailers to accelerate at the same rate.

Our algebraic expression for tension in the coupling (Eq. 3) states that the tension is some fraction of the 2.5-kN force exerted by the tractor, as expected. It also predicts that as the inertia of trailer 1 decreases to zero, the tension in the coupling is just the force the tractor pulls with, which is to be expected if trailer 1 becomes just a connector. As m_2 goes to zero, the tension also goes to zero because no force is needed to accelerate a trailer of zero inertia.

We did ignore friction and air resistance, which is reasonable at low speed.

Guided Problem 8.4 Rubber-band leash

A child drags a toy dog around at constant speed, using a rubber band as a leash. If the spring constant of the rubber band is 20 N/m and the frictional force between the toy and the floor is 0.734 N, how much does the rubber band stretch?

① **GETTING STARTED**

1. What physical quantity is appropriate to represent the unknown you seek?
2. Which object, the rubber band or the toy, should be the subject of a free-body diagram?
3. Don't forget to include an x axis on your free-body diagram to indicate the positive direction.

② **DEVISE PLAN**

4. What relates the tensile force exerted on the object in your free-body diagram to the distance the rubber band stretches?

5. How does the number of equations you have compare with the number of unknowns?

③ **EXECUTE PLAN**

6. Solve your equations to obtain an algebraic expression for the desired unknown.
7. Substitute the values given.

④ **EVALUATE RESULT**

8. Does your algebraic expression make sense as you imagine a larger or smaller frictional force? What about a larger or smaller spring constant?
9. Did you make any simplifying assumptions that would significantly affect your answer?

PRACTICE

Worked problem 8.5 Retarding a mine cart

In Figure WG8.5, a hanging block suspended by a rope hung over a pulley is used to accelerate a loaded mine cart along rail tracks to the edge of a cliff, where the load is dumped. To help reduce its acceleration as it nears the edge, the cart is attached to a horizontal spring of spring constant k, and the other end of the spring is attached to a wall. Ignoring friction and the small amount of stretch in the rope, determine the acceleration of the cart when the spring is stretched a distance d beyond its relaxed length.

Figure WG8.5

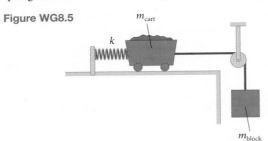

❶ **GETTING STARTED** The rope tends to accelerate the cart to the right in Figure WG8.5, but the stretched spring tends to accelerate it to the left. If the spring is stretched only a little, the force it exerts on the cart is small, and the pull of the rope causes the cart to accelerate to the right. If the spring is stretched a lot, the magnitude of the force it exerts on the cart may be greater than the magnitude of the force exerted by the rope, accelerating the cart to the left.

The block is subject to both the upward tug of the rope and the downward tug of gravity, and so we do not know which way it is accelerating. However, unless the rope goes slack, stretches, or breaks, the magnitude of the acceleration of the cart and the block must be the same. This common magnitude of acceleration is horizontal for the cart and vertical for the block, and the directions of these accelerations must allow the rope to remain taut and of constant length.

Even though the cart moves horizontally and the block moves vertically, each moves in only one dimension, which means this is just two one-dimensional problems coupled by the tension in the rope.

❷ **DEVISE PLAN** We begin with a free-body diagram of the cart because it is the cart's acceleration we are asked to calculate (Figure WG8.6). The cart is subject to two horizontal forces, one exerted by the rope and one by the spring. We know the value of neither, and so, counting the acceleration we seek, there are three unknowns!

The expression we must obtain for the acceleration involves the symbols given in the problem statement. We will be able to substitute for the unknown spring force using $F^c_{scx} = -k(x - x_0) = -kd$, where x is the position of the cart end of the spring at any instant when the spring is either stretched or compressed and x_0 is the position of the cart end of the spring when the spring is relaxed. Because we have no obvious substitution for the force exerted by the rope, we need another equation, which suggests that we draw a free-body diagram of the other object, the block [Figure WG8.6]. We show the acceleration in both free-body diagrams but add a

Figure WG8.6

$\longrightarrow x$

$\vec{F}^c_{sc} \longleftarrow$ cart $\longrightarrow \vec{F}^c_{rc}$
m_c
$\vec{a} = ?$

$\vec{F}^c_{rb}$

block $\otimes$
m_b $\downarrow \vec{a} = ?$

$\vec{F}^G_{Eb}$

b = block
c = cart
r = rope
s = spring
E = Earth

question mark to indicate that we are unsure of its direction. However, we know that if the block accelerates downward, the cart accelerates to the right and if the block accelerates upward, the cart accelerates to the left. Thus both accelerations are either along the positive x directions we have chosen (to the right and downward) or along the negative x directions. This allows us to use a single symbol, a_x, for the x component of the acceleration of either object.

We are now prepared to analyze this problem using forces, generating equations of motion for each free-body diagram. Hooke's law (*Principles* Eq. 8.20) is also involved because it allows us to substitute kd for the unknown force magnitude F^c_{sc}. Reciprocity of forces allows us to write $|\vec{F}^c_{cr}| = |\vec{F}^c_{rc}| = |\vec{F}^c_{rb}| = |\vec{F}^c_{br}| = \mathcal{T}$, where $\mathcal{T}$ is the tension in the rope (a scalar). Note that the middle equality, $|\vec{F}^c_{rc}| = |\vec{F}^c_{rb}|$, is due to the ability of a rope to transmit force undiminished from one end to the other and so depends on the rope having a negligibly small inertia.

The free-body diagram for the block allows us to relate the tension (magnitude of the tensile force) in the rope to the known variables m_b and g and to the unknown acceleration $\vec{a}$ or, equivalently, its x component a_x.

❸ **EXECUTE PLAN** The sum of the x components of all forces exerted on the cart is

$$\sum F_{cx} = F^c_{scx} + F^c_{rcx} = m_c a_x.$$

Because we don't know the direction of the acceleration, we choose not to substitute a sign and magnitude for that quantity. However, we can substitute for the force components, using Figure WG8.6 as a guide:

$$-F^c_{sc} + F^c_{rc} = m_c a_x$$
$$-kd + \mathcal{T} = m_c a_x. \qquad (1)$$

Repeating the process using Figure WG8.6 for the block, we obtain

$$\sum F_{bx} = F^G_{Ebx} + F^c_{rbx} = m_b a_x$$
$$+F^G_{Eb} - F^c_{rb} = +m_b a_x$$
$$+m_b g - \mathcal{T} = m_b a_x. \qquad (2)$$

We combine Eqs. 1 and 2 to eliminate the unknown tension.

$$+m_b a_x + m_c a_x = +m_b g - \mathcal{T} - kd + \mathcal{T}$$
$$(m_b + m_c)a_x = m_b g - kd$$

$$a_x = \frac{m_b g - kd}{m_b + m_c}. \checkmark$$

This expression for the x component of acceleration can be either positive or negative, which means that the acceleration of the cart may be either to the right ($a_x > 0$) or to the left ($a_x < 0$).

❸ **EVALUATE RESULT** Our expression for a_x tells us that the acceleration of the cart can be to the left or to the right, depending on the size of kd relative to the magnitude $m_b g$. If either the stretch distance d or the spring constant k is small enough to make $kd < m_b g$, our expression gives a positive value for a_x and the cart accelerates in the positive x direction. If either d or k is large enough to make $kd > m_b g$, we get a negative value for a_x and the cart accelerates in the negative x direction. The acceleration becomes very small if the inertia of the cart is very large, as expected.

Guided Problem 8.6 Atwood machine

A simple device for fully or partially counterbalancing objects, the Atwood machine, consists of two blocks connected by a cable draped across a stationary pulley that rotates freely. The pulley serves only to change the direction of the cable without increasing or decreasing the tension in it. For the simple Atwood machine depicted in Figure WG8.7, what are (a) the magnitude of acceleration of each block and (b) the tension in the cable after the blocks are released? Assume the blocks are released from rest, and ignore any friction in the pulley.

Figure WG8.7

5.0 kg   7.0 kg

❶ **GETTING STARTED**

1. Draw a free-body diagram for each block.
2. In which direction should you draw the acceleration vector in each diagram? What is an appropriate direction for the x axis in each diagram?
3. Are the accelerations of the two blocks related to each other?

❷ **DEVISE PLAN**

4. How many equations are needed to solve this problem?
5. For each free-body diagram you can write one equation, paying attention to the signs for each term.
6. Can you use Newton's third law to eliminate any unknowns?

❸ **EXECUTE PLAN**

7. Isolate the unknowns, and obtain expressions for the acceleration and tension.
8. Substitute the values given, and compute numerical values for the unknowns.

❹ **EVALUATE RESULT**

9. Do all magnitudes come out positive? If not, determine the source of your error.
10. Examine your expression for a. If the inertia of either block is increased or decreased, what happens to the value of a? Is this reasonable?
11. Examine your expression for $\mathcal{T}$. If the inertia of either block is increased or decreased, what happens to the value of $\mathcal{T}$? Is this reasonable?
12. Compare the numerical values you obtained to known values, such as the acceleration in free fall and the gravitational force on each block. Are the results reasonable?

Worked Problem 8.7 Pulley power

The block-and-tackle system in Figures WG8.8a and WG8.8b is composed of two pulleys connected by a looped rope, with the upper pulley suspended from a fixed support attached to the ceiling. Starting at the point where the person holds the rope, the rope goes over the upper pulley (rope C), goes under the lower pulley (rope A), is attached to the upper pulley (rope B), which is linked to the ceiling by the support. The object to be lifted from the ground to the platform is hung from the lower pulley. If the pulleys have negligible inertia and good bearings, no energy is lost to turning the pulleys or to friction. If the inertia of the rope is also negligible, the tension is the same everywhere along the rope.

Assume that a block of inertia m_b is the object being lifted and that each pulley has negligible inertia, $m_p \ll m_b$. (a) With what force magnitude must a person pull on rope segment C to give the block an upward acceleration of magnitude a? (b) As the block moves, what is the tension $\mathcal{T}_{support}$ in the support that attaches the upper pulley to the ceiling? (c) Suppose the free end of the rope falls off the upper pulley and the person cannot reach the ceiling to re-wrap the rope. He can climb to the platform and use the lower pulley and the rope (with the fixed end of the rope still attached to the ceiling through the support). He positions himself above the block and lower pulley (Figure WG8.8c) and lifts the combination by pulling straight up on the free end of the rope. If he wants to give the block an upward acceleration of magnitude a, how does the magnitude of the force he must exert compare with the force magnitude determined in part a? (d) Now suppose he discards the pulleys, ties the rope around the block (Figure WG8.8d), and lifts it with the same acceleration $\vec{a}$ by pulling upward on the rope. Compare the force he must exert in this case with your answers to parts a and c.

❶ **GETTING STARTED** (a) The lower pulley and the block are raised (as a single unit) by rope segments A and B. The problem statement allows us to assume that the tension $\mathcal{T}$ in the rope is transmitted undiminished throughout the rope, so that this tension equals the magnitude of the pulling force exerted by the worker on segment C, which is what we must calculate in part a. Let's call this tension in the three rope segments $\mathcal{T}_{segment}$ to distinguish it from $\mathcal{T}_{support}$, the tension in the ceiling support we need in part b.

(b) To determine the tension $\mathcal{T}_{support}$ in the support attaching the system to the ceiling, we use the fact that in an object under tension, the opposing forces creating the tension have equal magnitudes. The two tensile forces here are $\vec{F}_{cs}^c$, the upward force exerted by the ceiling on the support, and $\vec{F}_{ps}^c$, the downward force

Figure WG8.8

(a), (b)   (c)   (d)

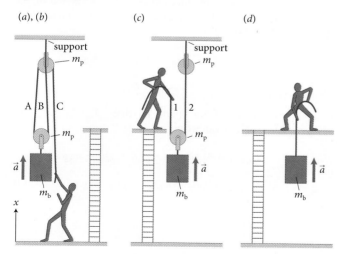

exerted by the upper pulley on the support. If we know either force magnitude, we know $\mathcal{T}_{\text{support}}$. Because we know nothing about the force exerted by the ceiling, we'll have to work with the force exerted by the pulley on the support.

(c), (d) The free-body diagram for the lower pulley-block system used with Newton's second law gives us what we need to answer part c. A similar approach should apply to the block alone in part d.

❷ **DEVISE PLAN** (a) Let us call the combination of the lower pulley and block our *load* and say that it has an upward acceleration of magnitude a. Rope segments A and B exert on the load upward forces of magnitudes $F_{A\ell}^c$ and $F_{B\ell}^c$, and Earth exerts a gravitational force of magnitude $m_\ell g = (m_p + m_b)g$. We need to draw the free-body diagram for the load and then apply Newton's second law to get the magnitude of the downward pulling force the worker exerts on segment C. This magnitude equals the common tension $\mathcal{T}_{\text{segment}}$ in the rope segments: $\mathcal{T}_{\text{segment}} = F_{wC}^c$. The free-body diagram for the load is shown in Figure WG8.9.

Figure WG8.9

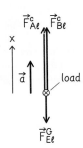

(b) We need to determine the force $\vec{F}_{ps}^c$ exerted by the upper pulley on the support because the magnitude of this force equals the tension $\mathcal{T}_{\text{support}}$ in the support. From the reciprocity of forces we know that $\vec{F}_{ps}^c$ has the same magnitude as the force $\vec{F}_{sp}^c$ exerted by the support on the pulley. So, if we calculate the magnitude F_{sp}^c, we have $\mathcal{T}_{\text{support}}$. The support exerts an upward force of magnitude F_{sp}^c on the upper pulley. The three rope segments exert on the upper pulley downward forces of magnitude $F_{Ap}^c = F_{Bp}^c = F_{Cp}^c = \mathcal{T}_{\text{segment}}$. Because this pulley does not rise or fall, its acceleration in the vertical direction is zero, which is information we'll need when we work with the second-law equation.

(c) The approach we used in part a should work for this part also, but now we need concern ourselves only with the lower pulley plus block. To avoid confusion, we label the two rope segments 1 and 2 for this case (Figure WG8.8c). The worker exerts on segment 1 an upward force of magnitude F_{w1}^c, and this magnitude is again equal to the tension in either rope segment.

(d) The same approach should apply, this time focusing on only the block. Here we use F_{wr}^c for the magnitude of the force exerted by the worker on the (only) rope.

❸ **EXECUTE PLAN** (a) Summing the x components of forces exerted on the load gives us

$$\sum F_{\ell x} = m_\ell a_x$$

$$F_{A\ell x}^c + F_{B\ell x}^c + F_{E\ell x}^G = (m_p + m_b)(+a). \tag{1}$$

Because the tensile forces in the three rope segments all have the same magnitude and because this magnitude is equal to the magnitude F_{wC}^c of the pulling force the worker must exert on rope segment C, we have

$$|F_{wCx}^c| = |F_{A\ell x}^c| = |F_{B\ell x}^c|.$$

Using Figure WG8.9 as a guide, we can substitute F_{wC}^c for the first two terms in Eq. 1, along with a value for the gravitational force component:

$$F_{wC}^c + F_{wC}^c - m_\ell g = m_\ell a$$

$$2F_{wC}^c = m_\ell a + m_\ell g = (m_\ell)(a + g)$$

$$F_{wC}^c = \tfrac{1}{2}(m_p + m_b)(a + g). ✔ \tag{2}$$

Remember, this force magnitude F_{wC}^c is equal to the tension $\mathcal{T}_{\text{segment}}$ in the rope segments. We'll need this information in our solution for part b.

(b) To get the tension in the support, our plan is to calculate F_{sp}^c, the magnitude of the force exerted by the support on the upper pulley, because this magnitude is equal to $\mathcal{T}_{\text{support}}$. The free-body diagram for the upper pulley is shown in Figure WG8.10. Summing force components gives

$$\sum F_{px} = m_p a_{px}$$

$$F_{spx}^c + F_{Apx}^c + F_{Bpx}^c + F_{Cpx}^c + F_{Epx}^G = m_p(0).$$

Figure WG8.10

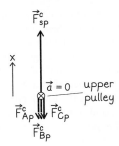

As noted when we devised our plan for part b, $F_{Ap}^c = F_{Bp}^c = F_{Cp}^c = \mathcal{T}_{\text{segment}}$, where $\mathcal{T}_{\text{segment}}$ is the tension in any rope segment. Substituting $\mathcal{T}_{\text{segment}}$ for the terms that represent forces exerted by the three rope segments and taking into account the direction in which these forces are exerted give us

$$F_{sp}^c - \mathcal{T}_{\text{segment}} - \mathcal{T}_{\text{segment}} - \mathcal{T}_{\text{segment}} - m_p g = 0$$

$$F_{sp}^c = 3\mathcal{T}_{\text{segment}} + m_p g.$$

Because we know from part a that $\mathcal{T}_{\text{segment}} = F_{wC}^c = \tfrac{1}{2}(m_p + m_b)(a + g)$, we can write

$$F_{sp}^c = \mathcal{T}_{\text{support}} = 3\left[\tfrac{1}{2}(m_p + m_b)(a + g)\right] + m_p g$$

$$\mathcal{T}_{\text{support}} = \tfrac{3}{2}(m_p + m_b)(a + g) + m_p g. ✔ \tag{3}$$

(*c*) The physical situation and free-body diagram for the load are illustrated in Figure WG8.11: Rope segments 1 and 2 pull upward on the load with a common tension equal to the force magnitude F_{w1}^c exerted by the worker on rope segment 1: $F_{w1}^c = |F_{1\ell x}^c| = |F_{2\ell x}^c|$. The free-body diagram is just like the diagram for the load in part *a*, and so the mathematical solution is the same here:

$$\Sigma F_{\ell x} = m_\ell a_x$$

$$F_{1\ell x}^c + F_{2\ell x}^c + F_{E\ell x}^G = m_\ell(+a)$$

$$F_{w1}^c + F_{w1}^c - m_\ell g = m_\ell a$$

$$F_{w1}^c = \tfrac{1}{2}(m_p + m_b)(a + g). ✔$$

Figure WG8.11

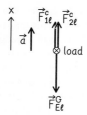

With only one pulley, the person must pull upward with the same magnitude of force he exerts downward on the rope in part *a*.

(*d*) When both pulleys are removed and the person pulls straight up on the load (which now consists of only the block), the free-body diagram has only two forces as shown in Figure WG8.12. The required force magnitude F_{wr}^c is

$$\Sigma F_{\ell x} = F_{r\ell x}^c + F_{E\ell x}^G = m_b a_x$$

$$F_{wr}^c - m_b g = m_b(+a)$$

$$F_{wr}^c = m_b(a + g).$$

Figure WG8.12

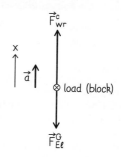

We see that if $m_p \ll m_b$, we have

$$F_{wC}^c = F_{w1}^c = \tfrac{1}{2}F_{wr}^c = \tfrac{1}{2}m_b(a + g). \tag{4}$$

Twice as much force is needed to lift the block without using pulleys. ✔

❶ **EVALUATE RESULT** Equation 4 shows the mechanical advantage of the block and tackle: This simple system of a rope and two pulleys allows you to raise the load with approximately half the effort of a direct lift! Equation 4 also reveals that the advantage of using two pulleys, with one fixed to the ceiling, must be in something other than the required force. The advantage is one of direction: You should find it easier to pull down (rather than up) to lift a heavy object.

The $+a$ term in Eq. 2 implies that you have to pull harder if you wish the load to have a greater acceleration upward, which is exactly what we expect. Equation 3 implies that the more inertia we have to lift against gravity, the larger the support force has to be, which is reasonable.

We did ignore any effort to turn the pulleys, a topic we will return to in Chapter 11. Real pulleys are often of negligible inertia and nearly frictionless, so the conditions of this problem are reasonable. We also approximated the three rope segments to be vertical even though Figure WG8.8*a* shows segments B and C as slightly angled. Because they are nearly vertical, our answer should be fairly accurate.

Guided Problem 8.8 Support force revisited

Redo part *b* of Worked Problem 8.7. This time, though, calculate $\mathcal{T}_{support}$ by using *Principles* Eq. 8.44 in the form $\Sigma F_{ext\,x} = ma_{cm\,x}$, where the system comprises the two pulleys and the block.

❶ **GETTING STARTED**

1. Sketch a system diagram and list the inertias and accelerations of the individual parts of the system.
2. What external forces are exerted on the objects in the system? (Hint: Why is it hard to raise the block?)
3. Are the forces exerted by rope segments A, B, and C internal or external forces? In other words, do they represent interactions between objects in the system or interactions between the system and objects external to the system?

❷ **DEVISE PLAN**

4. Draw a free-body diagram with an appropriate *x* axis for your system.

5. In the second-law equation, $\Sigma F_{ext\,x} = ma_{cm\,x}$, what is m?
6. What is the acceleration of the system's center of mass, given that two of the objects in the system are accelerating and one is not?
7. Write the sum of the external forces exerted on the system, based on your free-body diagram, and determine the sign and magnitude of each force.

❸ **EXECUTE PLAN**

8. Solve for the tension in the support $\mathcal{T}_{support}$.

❹ **EVALUATE RESULT**

9. Does your expression agree with the one obtained in part *b* of Worked Problem 8.7? If it does not, you might not have correctly identified all the external forces exerted by Earth (an external object) on the system components.

Questions and Problems

For instructor-assigned homework, go to MasteringPhysics® **MP**

Dots indicate difficulty level of problems: • = *easy*, •• = *intermediate*, ••• = *hard*; CR = *context-rich problem.*

8.1 Momentum and force

1. A truck is traveling at a constant 25 m/s when a motorcycle speeds past at a constant 40 m/s. On which vehicle, if either, is the magnitude of the vector sum of forces exerted on the vehicle greater? •

2. You wish to move a crate. (*a*) Is there a minimum value for the magnitude of the pushing force you have to exert to set the crate in motion? (*b*) Is there a minimum value for the magnitude of the vector sum of forces that must be exerted on the crate to set it in motion? •

3. Sketch the $\sum F_x(t)$ curve for the vector sum of forces exerted on the object whose *x* component of momentum is plotted in Figure P8.3. Curve sections are numbered for reference. ••

Figure P8.3

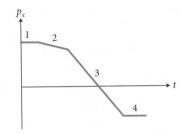

4. The *x* component of the vector sum of forces exerted on a cart moving on a low-friction track is measured in the laboratory with the aid of a spring-loaded sensor and a monitoring computer. From the data in the table, sketch the $\sum F_x(t)$ curve of the cart and use your graph to estimate the change in the cart's momentum. ••

F (N)	t (s)
0.00	0.00
0.15	0.020
0.32	0.040
0.50	0.060
0.57	0.080
0.37	0.10
0.21	0.12
0.11	0.14
0.070	0.16
0.040	0.18

5. If a single known force is exerted on an object for a certain time interval, you know the change in the object's momentum. However, you don't know its change in kinetic energy. Why not? •••

6. Consider a coal mining cart being loaded with coal as it moves along a horizontal track. Suppose that a tractor exerts a constant horizontal force on this cart. Is there any situation in which the cart-coal system would not accelerate? •••

8.2 The reciprocity of forces

7. A worker pushes a box in a factory. In each case, decide which force has the greater magnitude: the force exerted by the worker on the box or the force exerted by the box on the worker. (*a*) The box is heavy and does not move no matter how hard the worker pushes. (*b*) Some contents are removed, and now when pushed the box slides across the floor at constant speed. (*c*) The worker pushes harder, and the box accelerates. •

8. In many science fiction movies, someone is portrayed moving something with her or his mind—an amazing feat called *telekinesis*. What does your knowledge of forces predict about the effect of a telekinetic event on the gifted person? •

9. Which of the following can exert a force: (*a*) a hammer, (*b*) a nail, (*c*) a person, (*d*) a chair, (*e*) a ship, (*f*) water, (*g*) Earth? ••

10. During a tennis volley, a ball that arrives at a player at 40 m/s is struck by the racquet and returned at 40 m/s. The other player, realizing that the ball is out of bounds, catches it in her hand. Assuming the time interval of contact is the same in both cases, compare the force exerted by the first player's racquet on the ball with the force exerted by the second player's hand on the ball. ••

8.3 Identifying forces

11. A pitcher throws a fastball toward home plate. (*a*) When the ball is halfway to the plate, is the pitcher's push still exerted on the ball? (*b*) What forces, if any, are exerted on the ball at the halfway location? •

12. A car's acceleration forward must be due to a force exerted on the car, and the only thing that can push forward on the car is the road (in contact with the tires). So, what purpose does the engine have? •

13. A baby is bouncing in a seat fastened by an elastic cord to a hook on the ceiling. At the top of his bounce, his velocity is instantaneously zero. Is the vector sum of forces exerted on him at that instant zero? Why or why not? ••

14. (*a*) Can an object ever exert both a contact force and a field force on another object? (*b*) Can an object ever exert a contact force on something without exerting a field force on that same object? ••

8.4 Translational equilibrium

15. You push on a refrigerator, but it doesn't move. Explain why not. •

16. You are discussing physics homework with a friend while the two of you are riding your bikes. Distracted, you collide with a parked car. The car doesn't move. Your friend says, "Of course not! The force of the collision wasn't enough to overcome the inertia of the car." How do you respond? •

17. What interaction does a bathroom (spring) scale actually measure? ••

18. A table sits on the floor, and a vase containing water and flowers sits on top of the table. All objects are at rest. Identify all forces exerted on the table, and describe how these forces are related. ••

8.5 Free-body diagrams

19. You are in a stationary elevator, so that the contact force exerted by the floor on you is equal in magnitude to the force of gravity exerted on you. When the elevator starts downward, which force changes? What happens to its magnitude? •

20. When you are standing motionless on the ground, your feet exert a force on Earth. Why doesn't Earth move away from you? ••

21. You push on a crate, and it starts to move but you don't. Draw a free-body diagram for you and one for the crate. Then use the diagrams your knowledge about forces and interactions to explain why the crate moves but you don't. ••

22. The force of air resistance on any object falling through the air increases with the object's speed. What happens to the acceleration of a raindrop as it falls from a cloud? ••

23. You are carrying your 3-year-old niece on your shoulders in an elevator cabin that is moving upward with velocity 1.5 m/s and constant downward acceleration of 1.0 m/s² (Figure P8.23). Your inertia is 50 kg, your niece's inertia is 20 kg, and the cabin's inertia is 100 kg. (a) Draw free-body diagrams for the elevator cabin, your niece, and yourself, labeling the forces using the notation introduced in *Principles* Section 8.5. Make sure the lengths of the force vectors are appropriate to the magnitudes of the forces. (b) State the magnitudes of as many forces as possible, and identify all the forces that form interaction pairs. Explain briefly how you obtained these magnitudes. •••

Figure P8.23

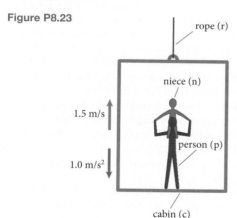

8.6 Springs and tension

24. A delivery person in an elevator is holding a package by an elastic cord. (Don't ask why.) (a) What happens to the length of the cord when the elevator accelerates upward? Draw the free-body diagram for the package in this case. (b) What happens to the cord length when the elevator slows to a stop after its ascent? Draw the free-body diagram for the package in this case. •

25. Suppose a rope that is hanging by one end from the ceiling has a large inertia. Is the tension in the rope uniform throughout the length? ••

26. Walking beside a pasture, you and a fellow student see a farmer pulling a mule with a rope and getting nowhere. Your friend says, "The force with which the mule is pulling on the rope has the same magnitude as the force with which the farmer is pulling on the rope, but the two forces point in opposite directions. Because the two forces cancel, the tension in the rope is zero." How do you respond? ••

27. In a game of tug of war, the blue team is beating the red team, and the red team is accelerating toward the mud pit between the two teams. Which has the greater magnitude, if either: the force the blue team exerts on the rope or the force the red team exerts on the rope? Assume that the gravitational force exerted on the rope is negligible compared with the forces exerted by the two teams on the rope. ••

28. The standard coupling between railroad cars must be capable of withstanding the maximum tensile force exerted on any coupling in a given train. (a) If a locomotive is pulling ten cars and speeding up, on which coupling is the greatest tensile force exerted? (b) Is this tensile force due to stretching or compression? (c) If the locomotive is slowing the train down, on which coupling is the greatest tensile force exerted? (d) Is this tensile force due to stretching or compression? ••

8.7 Equation of motion

29. In an electric breakdown in air (a spark), an electron (inertia 9.11×10^{-31} kg) experiences an electric force of 4.83×10^{-13} N. What is the electron's acceleration? •

30. A 2.3-kg object experiences a single, time-dependent force $F_x(t) = (10 \text{ N/s})t - (20 \text{ N})$. What is the object's acceleration at $t = 0$, 2.0 s, and 4.0 s? •

31. Is it possible to figure out an object's past motion if you know its present position and velocity and the history of all the forces exerted on it? ••

32. The force exerted by expanding gases is what propels a shell out of a gun barrel. Suppose a force that has an average magnitude F propels the shell so that it has a speed v at the instant it reaches the end of the barrel. What magnitude force is required to double the speed of the shell in the same gun? ••

33. A 50-kg pneumatic hammer is fitted with a shock absorber at the top to lessen the impact on the worker. The vertical displacement of the hammer as a function of time is given by $x(t) = at^2 - bt^3$ (during the hammer's bounce), where $a = 15 \text{ m/s}^2$ and $b = 20 \text{ m/s}^3$. The positive x axis is directed upward. (a) Write an expression for the x component of the vector sum of forces exerted on the hammer as a function of time. (b) At which values of t is the x component of the vector sum of forces on the hammer positive, at which values of t is it negative, and at which values of t is it zero? ••

34. You push with a steady force of 19 N on a 48-kg desk fitted with casters (wheels that swivel) on its four feet. How long does it take you to move the desk 5.9 m across a warehouse floor? ••

35. In Figure P8.35, a 50-kg skier heads down a slope, reaching a speed of 35 km/h. She then slides across a horizontal snow field but hits a rough area. Assume the snow before the rough area is so slippery that you can ignore any friction between the skis and the snow. If the frictional force exerted by the snow in the rough area is 40 N, how far across the rough area does the skier travel before stopping? ••

Figure P8.35

36. A 2.34-kg cart on a long, level, low-friction track is heading for a small electric fan at 0.23 m/s. The fan, which was initially off, is turned on. As the fan speeds up, the magnitude of the force it exerts on the cart is given by at^2, where $a = 0.0200$ N/s². What is the speed of the cart 3.5 s after the fan is turned on? After how many seconds is the cart's velocity zero? •••

8.8 Force of gravity

37. What is the magnitude of the gravitational force exerted by Earth on a 1.0-kg brick when the brick is (a) in free fall and (b) resting on a table? •

38. Estimate the magnitude of the force of gravity exerted on (a) a packet of sugar, (b) a cup of water, (c) a packed carry-on suitcase, (d) an adult gorilla, and (e) a midsize car. •

39. The magnitude of the gravitational force exerted on the piano in Figure P8.39 is 1500 N. With how much force do you have to pull on the rope in order to hold up the piano (a) with the single pulley and (b) with the double pulley? •

Figure P8.39 (a)　　　(b)

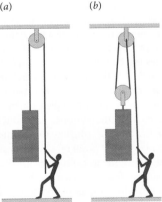

40. The piano in Figure P8.39 is being lowered with an acceleration of magnitude exactly $g/8$. To maintain this acceleration, with what force magnitude must the person pull on the rope (a) with the single pulley and (b) with the double pulley? ••

41. If the force of gravity exerted by Earth on a fully suited astronaut at Earth's surface is 1960 N, what is the magnitude of the gravitational force exerted on her (a) on Jupiter, where the acceleration due to gravity is 25.9 m/s², and (b) on the Moon, where the acceleration due to gravity is 1.6 m/s²? ••

42. You are climbing a rope straight up toward the ceiling. (a) What are the magnitude and direction of the force you must exert on the rope in order to accelerate upward at 1.5 m/s², assuming your inertia is 60 kg? (b) If the maximum tension the rope can support is 1225 N, what is the maximum inertia the rope can support at this acceleration if the inertia of the rope is so small that the gravitational force exerted on the rope can be ignored? ••

43. Two blocks are tied together by a cord draped over a pulley (Figure P8.43). (a) At first, the block on the table does not slip. How does the tension in the cord compare with the gravitational force exerted on the hanging block? (b) You tip the table a bit, and the block on the table begins to slip. Now how does the tension in the cord compare with the gravitational force exerted on the hanging block? Ignore any friction in the pulley. ••

Figure P8.43

44. Assume that the block on the table in Figure P8.43 has twice the inertia of the hanging block. (a) You give the block on the table a push to the right so that it starts to move. If the magnitude of the frictional force exerted by the table on the block is half the magnitude of the gravitational force exerted on the hanging block, what is the acceleration of the hanging block after you have stopped pushing the block on the table? (b) What would this hanging-block acceleration be if you had pushed the block on the table to the left instead? Consider only a short time interval after you stop pushing. ••

45. Assume that the block on the table in Figure P8.43 has half the inertia of the hanging block. (a) You push the block on the table to the right so that it starts to move. If the magnitude of the frictional force exerted by the table on the table block is half the magnitude of the gravitational force exerted on this block, what is the block's acceleration after you remove your hand? (b) What would this acceleration be if you had pushed the block to the left? Consider only a short time interval after you stop pushing. ••

46. A bag of concrete mix is tied to one end of a rope that is looped over a pulley attached to the ceiling, and the other end of the rope is tied to a pallet of lumber (Figure P8.46). The concrete mix has a greater inertia than the lumber. When both loads are released from rest so that the system is free to move, compare the tension in the rope with the magnitudes of the gravitational forces exerted by Earth on the lumber and on the concrete mix. Ignore any friction in the pulley. ••

Figure P8.46

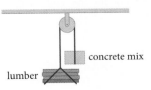

47. All blocks in Figure P8.47 are identical and you can ignore any friction in the pulleys. Rank the configurations in increasing order of tension in the rope. (Hint: Use free-body diagrams.) ••

Figure P8.47

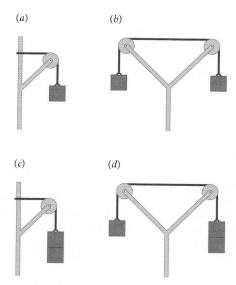

48. Picture an object in free fall. If the leading face of the object (which means the face closest to the ground) has a large surface area, air resistance becomes important. For low speeds, the force due to air resistance may be approximated as proportional to the object's velocity according to the expression $\vec{F}^a = -b\vec{v}$, where b is a number that depends on the shape and volume of the object and on the density of the air. (a) Show that, for an object of inertia m, there is a limiting speed (called the *terminal speed*) given by $v_t = mg/b$. (b) Show that, if the object is released from rest, its speed as a function of time is given by

$$v(t) = v_t \left[1 - e^{-(b/m)t} \right].$$

(c) What are the units of m/b? Sketch a graph of v as a function of time, with the horizontal axis in units of multiples of m/b. •••

49. At a munitions factory, chemicals M and P have to be added to a vat containing chemical C. Chemicals M and P have to be added simultaneously to prevent an accidental detonation. Your task, as the chemical engineer in charge, is to come up with a way of doing this, and you devise the pulley scheme shown in Figure P8.49, where block W is a counterweight whose inertia controls the speed at which the buckets containing chemicals M and P move. The idea is that when

Figure P8.49

a strap holding block W at rest is released, the block rises and the buckets containing M and P descend into the vat. In order to have the two buckets have an identical acceleration of magnitude $g/3$, what must be the ratios of the inertias m_W, m_M, and m_P? Ignore any friction in the pulleys. •••

8.9 Hooke's law

50. A 66-kg person experiences a gravitational force of about 660 N. Yet, if this person were to jump onto a spring scale, the scale would briefly read about 2400 N. Why does the scale register a force so much larger than the gravitational force exerted on the person? •

51. A 20-kg child stands in the center of a trampoline. (a) If the trampoline center is 0.11 m lower than before she got on, what is the spring constant of the trampoline? (b) Assuming the trampoline acts like a spring, how low would her 75-kg father sink if he stood alone in the center? •

52. Two springs have spring constants k_1 and $k_2 > k_1$. Connected as shown in Figure P8.52, they act like one spring. Compute the spring constant of the combination, and determine whether it is smaller than k_1, equal to k_1, between k_1 and k_2, equal to k_2, or greater than k_2. ••

Figure P8.52

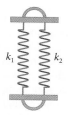

53. Two springs have spring constants k_1 and $k_2 > k_1$. Connected as shown in Figure P8.53, they act like one spring. Compute the spring constant of the combination. Is the combination smaller than k_1, equal to k_1, between k_1 and k_2, equal to k_2, or greater than k_2? ••

Figure P8.53

54. Two springs are hooked together end to end. When a 4.0-kg brick is suspended from one end of the combination, the combination stretches 0.15 m beyond its relaxed length. (a) What is the spring constant of the combination? (b) If the top spring stretches 0.10 m, what is the spring constant of each spring? ••

55. When a 5.0-kg box is hung from a spring, the spring stretches to 50 mm beyond its relaxed length. In an elevator accelerating upward at 2.0 m/s², how far does the spring stretch with the same box attached? ••

56. Suppose that when you lie on your bed, your body is supported by 30 of the springs in the mattress. Assuming that all the springs are identical and that your body presses on each spring equally, estimate their spring constant. (Experiment: Lie on your bed and see how far down you sink.) ●●

57. In the modified Atwood machine shown in Figure P8.57, each of the three blocks has the same inertia m. One end of the vertical spring, which has spring constant k, is attached to the single block, and the other end of the spring is fixed to the floor. The positions of the blocks are adjusted until the spring is at its relaxed length. The blocks are then released from rest. What is the acceleration of the two blocks on the right after they have dropped a distance d? Ignore any friction in the pulleys. ●●

Figure P8.57

58. Your boss hands you a bag containing four springs, A, B, C, and D, and tells you that they all have the same relaxed length. He wants you to rank them by the values of their spring constants. He locks you in a room with only the springs, a ruler, and a pad and pencil, and tells you to knock when you're done. Each spring has a hook on one end and a handgrip on the other. There is nothing in the room you can use to measure inertia.

After thinking for a few minutes, you hook springs A and B together, stand on one handgrip with your foot, and pull on the other handgrip with your hand until the combination spring has been stretched to double its original length (Figure P8.58). You notice that spring A makes up 65% of the stretched length and spring B makes up the other 35%.

Figure P8.58

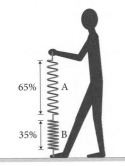

You work your way through the rest of the springs and tabulate the results:

Combination	Percent of stretched length
A/B	A 65, B 35
A/C	A 70, C 30
A/D	A 67, D 33

Based on these results, what are the numerical values in the ratio $k_A : k_B : k_C : k_D$? ●●●

59. A 5.0-kg red book is suspended from a spring attached to the ceiling of an elevator that is accelerating downward. Attaching the book causes the spring to stretch 71 mm beyond its relaxed length. You also have a yellow book of unknown inertia. When you attach the yellow book and the red one to the spring at the same time, the spring stretches 110 mm beyond its relaxed length. When the elevator is at rest, the combination of the two books stretches the spring 140 mm beyond its relaxed length. (a) What is the spring constant? (b) What is the inertia of the yellow book? (c) What is the elevator's acceleration? ●●●

8.10 Impulse

60. Why does a dropped dinner plate break when it lands on a tile floor but not when it lands on carpet? ●

61. If you try to pound a nail into a board with a 2.5-kg rubber mallet, you will have less luck than if you use a 0.8-kg steel hammer. Why? ●

62. In a baseball game, a batter hits the 0.150-kg ball straight back at the pitcher at 180 km/h. If the ball is traveling at 160 km/h just before it reaches the bat, what is the average force exerted by the bat on it if the collision lasts 5.0 ms? ●

63. In Figure P8.63, a heavy potted plant hangs by an enveloping hanger made of heavy yarn. (a) If you pull slowly on the tassel at the bottom, where is the hanger most likely to break? Why? (b) If you yank quickly on the tassel, where is the hanger most likely to break? ●●

Figure P8.63

64. Estimate the impulse delivered by a hammer to a nail being driven into a wood plank. ●●

65. A 1500-kg car goes from 0 to 100 km/h in 60 s. (a) What is the impulse delivered to the car? (b) What is the average vector sum of forces exerted on the car? ●●

66. A horizontal force F_{slide} is exerted on a 5.0-kg box sliding on a polished floor. As the box moves, the magnitude of F_{slide} increases smoothly from 0 to 5.0 N in 5.0 s. What is the box's speed at $t = 5.0$ s (a) if it starts from rest and (b) if at $t = 0$ it has a velocity of 3.0 m/s in the direction opposite the direction of F_{slide}? Ignore any friction between the box and the floor. ●●

67. Just before hitting the ground, a partially inflated 0.625-kg basketball has a speed of 3.30 m/s. Then it loses half of its kinetic energy as it bounces. (*a*) What is the ball's speed immediately after it bounces? (*b*) If the ball is in contact with the ground for 9.25 ms, what is the magnitude of the average force exerted by the ground on the ball? ••

68. A 1.2-kg ball dropped from a height of 3.0 m onto a steel plate rigidly attached to the ground bounces back up to a height of 2.5 m. (*a*) What is the impulse delivered to the ball by the plate? (*b*) What is the coefficient of restitution of the collision? ••

69. A 0.050-kg egg falls off a table that is 1.0 m high and lands on a tile floor. (*a*) What is the impulse delivered to the egg by the floor? (*b*) The egg lands on its side, and at the instant of impact the part of the shell farthest from the floor is 40 mm above the floor. The collision continues until the center of mass is on the floor. Estimate the time interval over which the collision lasts. (*c*) What is the average force exerted by the floor on the egg? (*d*) If the floor were carpeted, the collision time interval might be extended by as much as a factor of four as the carpet "gives" under the egg. What would the average force exerted on the egg be then? ••

70. When it crashes into a bridge support that does not move, a car goes from 80 km/h to 0 in 1.23 m. (*a*) What is the impulse delivered to the 70-kg driver by the seat belt, assuming the belt makes the driver's motion identical to the car's motion? (*b*) What is the average force exerted by the belt on the driver? (*c*) If the driver were not wearing the seat belt and flew forward until he hit the steering wheel, which stopped him in 0.0145 s, what would be the average force exerted by the steering wheel on him? (*d*) Would he be likely to survive the crash described in part *c*? Justify your answer by comparing the force exerted by the steering wheel on the driver to the gravitational force exerted on him. •••

71. A volleyball player serves the ball. The 0.27-kg ball was moving straight up and is at the high point of its trajectory when she hits it in a purely horizontal direction. The magnitude of the force exerted on the ball while her hand is in contact with it is given by

$$F(t)_{hb}^c = a\,t - b\,t^2,$$

where $a = 3.0 \times 10^5$ N/s and $b = 1.0 \times 10^8$ N/s². Her hand is in contact with the ball for 3.0 ms. What is the magnitude of (*a*) the impulse delivered to the ball, (*b*) the average force exerted by her hand on the ball, and (*c*) the maximum force exerted by her hand on the ball? (*d*) What is the velocity of the ball immediately after her hand loses contact with it? •••

8.11 Systems of two interacting objects

72. A 70-kg student is falling toward Earth. (*a*) Draw one free-body diagram for Earth and one for the student. Assume that air resistance can be ignored. (*b*) Calculate the student's acceleration and the vector sum of forces exerted on him. (*c*) Earth's inertia is 5.97×10^{24} kg. What is the magnitude of the planet's acceleration toward the student? •

73. Two 5.00-kg carts, one red and the other green, are 1.00 m apart on a surface where either cart experiences a constant frictional force of 5.00 N when it moves. Throughout the problem, the red cart is pushed with a constant force of 12.0 N toward the green one. What is the acceleration of (*a*) the center of mass before the two carts collide, (*b*) the red cart before they collide, and (*c*) the center of mass after they collide? •

74. Two 0.500-kg carts are 100 mm apart on a low-friction track. You push one of the carts with a constant force of 2.00 N directed so that the cart you push moves away from the other cart. Determine the acceleration of the center of mass of the two-cart system when (*a*) the carts do not interact and (*b*) the carts are connected by a spring of spring constant $k = 300$ N/m. ••

75. A force of magnitude *F* is exerted on the leftmost face of two blocks sitting next to each other on a slippery surface, with two of their faces touching. The inertia of the block on the right is *m*, and the inertia of the block on the left is *2m*. (*a*) What is the magnitude of the acceleration of each block? (*b*) What are the magnitude and direction of the contact force the left block exerts on the right block? (*c*) What are the magnitude and direction of the contact force the right block exerts on the left block? (*d*) How do your answers to parts *a*, *b* and *c* change if the positions of the two blocks are interchanged? ••

76. A 2.0-kg cart and an 8.0-kg cart are connected by a relaxed, horizontal spring of spring constant 300 N/m. You pull the 8.0-kg cart with some constant horizontal force. The separation between the carts increases for a short time interval, then remains constant as you continue to pull and the spring is stretched by 0.100 m. (*a*) What pulling force did you exert? (*b*) If you instead pull the 2.0-kg cart, what force must you exert to get the same stretch in the spring? ••

77. A 1500-kg truck and a 1000-kg car are parked with their rear bumpers nearly touching each other in a level parking lot. Both vehicles have their brakes off so that they are free to roll. A man sitting on the rear bumper of the truck exerts a constant horizontal force on the rear bumper of the car with his feet, and the car accelerates at 1.2 m/s². (*a*) What are the magnitude and direction of the acceleration of the center of mass of the car-truck system? (*b*) What vector sum of forces (magnitude and direction) is exerted on each vehicle? (*c*) What are the magnitude and direction of the acceleration of the truck? Ignore any friction between the tires and the parking lot surface. ••

78. A 70-kg skier is being towed on a rope behind a 450-kg snowmobile on a smooth, snow-covered surface at 10 m/s when the snowmobile hits a patch of muddy ground that brings it to a halt in 10 m. (*a*) What is the average acceleration of the snowmobile while it is slowing? (*b*) What is the average acceleration of the center of mass of the skier-snowmobile system? ••

79. A 4.0-kg block and a 2.0-kg block are connected to opposite ends of a relaxed spring of spring constant 300 N/m. The blocks are pushed toward each other for 0.50 s. The 4.0-kg block is pushed with a force of 50 N, and the 2.0-kg block is pushed with a force of 20 N. Ignore the inertia of the spring. (*a*) What is the acceleration of the center of mass of the two-block system while this pushing is going on? (*b*) When the blocks are released, what is the velocity of the 2.0-kg block if the 4.0-kg block is moving away from it at 5.0 m/s in the Earth reference frame? ••

8.12 Systems of many interacting objects

80. You push your 0.70-kg pillow across your bed with a constant force of 10 N. The bed provides a frictional force of 6.0 N. What is the acceleration of the center of mass of the pillow? •

81. You and a child are standing on a bathroom scale, which uses compression of a spring to measure the force exerted by your feet on the scale. What happens to the scale reading as you lift the child onto your shoulders? ••

82. You build a 2.0-kg mound of mashed potatoes on a kitchen scale (which is a spring scale). You then use a large spoon to press on the top of the mound at a constant force of 3.0 N. While you are doing this, the scale reading increases by 5.0%. (a) What is the acceleration of the center of mass of the potatoes? (b) Describe what happens to the shape of the mound and the scale reading as time goes by. ••

83. You use a rope of length ℓ to tie the front of a 1500-kg truck to the rear of a 1000-kg car and then use an identical rope of length ℓ to tie the front of the car to the rear of a 500-kg trailer. Each rope will break if its tension exceeds 2000 N. You attach a strong cable to the rear of the truck and use a winch to pull it with a force of magnitude F. (a) What is the acceleration of the three-vehicle system? (b) What is the maximum pulling force you can exert on the truck before a rope breaks? (c) If you now attach the cable to the front of the trailer and again pull the system, what is the maximum force with which you can pull it without snapping a rope? ••

84. A tugboat pulls two barges down the river. The barge connected to the tugboat, carrying coal, has an inertia of 2.0×10^5 kg, and the other barge, carrying pig iron, has an inertia of 3.0×10^5 kg. The resistive force between the coal barge and the water is 8.0×10^3 N, and the resistive force between the pig iron barge and the water is 10×10^3 N. The common acceleration of all three boats is 0.40 m/s². Even though the ropes are huge, the gravitational force exerted on them is much smaller than the pulling forces. (a) What is the tension in the rope that connects the tugboat to the coal barge? (b) What is the tension in the rope that connects the two barges? (c) Repeat parts a and b for the case in which the order of the barges is reversed. ••

85. A farmer is trying to haul three 500-kg trailers connected together with identical ropes that will break if their tension exceeds 2000 N. The first trailer is connected to his tractor with a rigid hitch. (a) If the tractor has to overcome a frictional force of 900 N per trailer to start them moving, what is the minimum force the tractor must exert on the first trailer to get them all rolling if the ropes are taut? (b) What is the maximum force the tractor can exert without breaking any rope? ••

86. A train engine is pulling four boxcars, each of inertia m. The engine can exert a force of magnitude F on what it is pulling. Assuming that friction can be ignored, what is the tension in each of the four couplers as the train starts off? ••

87. A red 10-kg cart is connected to a 20-kg cart by a relaxed spring of spring constant 60 N/m. The 20-kg cart is resting against another 10-kg cart, this one blue. All are on a low-friction track. You push the red 10-kg cart to the right, in the direction of the 20-kg cart, with a constant force of 10 N. (a) What is the acceleration of the center of mass of the three-cart system? (b) What is the acceleration of each cart the instant you begin to push? (c) What is the vector sum of the forces on each cart when the spring is compressed 100 mm? •••

Additional Problems

88. A toy wagon initially at rest is pulled by a child from one end of a driveway to the other end. The magnitude of the force the wagon exerts on the child is the same as the magnitude of the force the child exerts on the wagon. So, how does the child get the wagon moving? •

89. In years past, car frames were built as stiff as possible, but today's cars are built with "crumple zones" in the front and rear portions of the frame. What is the purpose of this design change? •

90. A 60-kg student is in an elevator moving downward with constant velocity. He uses a bathroom scale to measure the upward force exerted on his feet. What force magnitude does the scale read (a) when the elevator is traveling at constant velocity, (b) when the elevator slows to a stop with an acceleration of magnitude 2.0 m/s², and (c) when the elevator starts downward again with an acceleration of magnitude 2.0 m/s²? ••

91. You want to hang an object from the ceiling of an elevator that has a maximum acceleration of 4.0 m/s². (a) If you hang the object with fishing line that supports 45 N of force, what is the maximum inertia the object can have if the line is not to break? (b) What combination of slowing down, speeding up, going up, and going down of the elevator causes the greatest force to be exerted on the fishing line? ••

92. You need to lower a 45-kg safe from a window to the bed of a 1800-kg truck. You have a rope that is just long enough, but it will support no more than 42 kg. What must you be careful to do (or not to do) so that the rope does not break? ••

93. A 10-kg cart is connected to a 20-kg cart by a relaxed spring of spring constant 1000 N/m, and both carts are placed on a low-friction track. You push the 10-kg cart in the direction of the 20-kg cart with a constant force of 10 N. (a) What is the acceleration of the center of mass of the two-cart system at any instant? (b) What is the acceleration of each cart at the instant you begin to push? (c) What is the acceleration of each cart when the spring has its maximum compression? ••

94. A 5.0-kg block suspended from a spring scale is slowly lowered onto a vertical spring (Figure P8.94). (a) What does the scale read before the block touches the vertical spring? (b) If the scale reads 40 N when the bottom spring is compressed 30 mm, what is k for the bottom spring? (c) How far does the block compress the bottom spring when the scale reads 0? ••

Figure P8.94

5.0 kg

95. You need to lift a heavy load of inertia m_ℓ upward with an acceleration of magnitude $g/8$. You have two pulleys of inertia $m_p \ll m_\ell$, one attached to a beam above the loading dock and another that can be attached to the load. You consider a pulley system similar to that of Worked Problem 8.7, but you realize that these are double pulleys (each has two wheels side by side on a common axle) and you have a long piece of rope. ●●● CR

96. You and your uncle are fishing. You are using 22-N fishing line, which means the line can support that much tension without breaking. Your uncle challenges you to catch the biggest fish you can land in the boat with that line, with the provision that you have to hoist the fish out of the water with the line and not with a net. You finally hook a fish that looks to have the right inertia, and you start to reel it in slowly. Just as you get the fish up into the air, though, the line snaps and the fish gets away. You accuse your uncle of having behind your back switched your pole to a line weaker than the 22-N line you thought you were working with, but he denies it. He then points out that you were accelerating the fish upward at 1.0 m/s². ●●● CR

97. Your 70-kg friend always thinks he can dominate you in contests of strength because, even though you are a very athletic gymnast, your inertia is 63 kg. This time, while the gym class watches, he has attached a pulley to the ceiling of the gym, passed a rope over it, and tied one end of the rope around his waist. He hands you the other end of the rope, which is not quite long enough to reach the floor, and bets you that you cannot lift him off the ground by pulling on the rope. You realize that gravitational forces will favor your friend if the two of you are hanging on opposite sides of the pulley, but then you recall your amazing arm strength gained from years of work on gymnastic apparatus. ●●● CR

98. A burst of compressed air pushes a pellet out of a blowpipe. The force exerted by the air on the pellet is given by $F(t) = F_0 e^{-t/\tau}$, where τ is called a *time constant* because it has units of time. (*a*) What does F_0 represent? (*b*) What is the momentum of the pellet after an interval equal to one time constant has elapsed? (*c*) What is the momentum after an interval equal to five time constants has elapsed? (*d*) What is the final momentum of the pellet after a long time interval has elapsed? (*e*) If $\tau = 0.50$ ms, after how long a time interval has the momentum reached 95% of its final value? ●●●

99. *Thrust* is the force that pushes a rocket forward. It is a force exerted on the rocket as the engine expels hot gases from the rear of the rocket. For most rockets, thrust is variable rather than constant during the "burn time" (the time interval during which fuel is consumed and expelled). However, the thrust of a certain rocket is nearly constant during part of its burn time, beginning when the rocket has consumed 40% of its fuel and continuing until the rocket has consumed about 90% of its fuel. As the rocket consumes and expels its fuel, its inertia decreases. Describe the rocket's acceleration during this interval of constant thrust. ●●●

Answers to Review Questions

1. No. An opposing force of equal magnitude might also be exerted on the object. For example, a book at rest on a table experiences two forces, the downward gravitational force exerted by Earth and the upward contact force exerted by the table, and yet no change in motion occurs. Also, a force exerted on an object can decrease the object's speed, as when a crate sliding along the floor comes to rest as a result of the frictional force exerted on it.

2. The fact that the velocity is constant tells you that the vector sum of forces is zero. No direction need be specified for a zero vector.

3. The vector sum of all forces exerted on an object equals the time rate of change in the object's momentum.

4. No. Forces always form interaction pairs. Whenever object 1 exerts a force on object 2, object 2 exerts an equal and opposite force on object 1.

5. No. The forces in an interactive pair are exerted on the two objects of the pair. The forces asked about here are $F_{\text{by cup on table}}$ and $F_{\text{by coffee on cup}}$. Because these forces do not involve the same two objects, they do not form an interaction pair. (Each force is one member of an interaction pair, though, and you should be able to name the other member of each pair.)

6. A contact force is one that arises when two objects physically touch each other (which, as seen in Chapter 7, means the surfaces of the two objects must approach within a few atomic diameters of each other in order to create an interaction). A field force is one due to an interaction between two objects that occurs even when the objects are not touching each other. Contact forces require physical contact, while field forces do not.

7. (a) Contact (a frictional force); (b) field (the gravitational force exerted by Earth); (c) field (the magnetic force); (d) contact (air touching the sails); (e) contact (the pulling force exerted by the rubber band)

8. Yes. The only requirement for translational equilibrium is that the vector sum of the forces being exerted on the object is zero.

9. Watch for changes in its momentum (and use an inertial reference frame to make your observations). An object in translational equilibrium experiences no change in momentum.

10. First, a dot to represent the object that is the focus of the diagram. Second, a vector arrow, with its tail on the dot, for each force exerted on the object. Each arrow is labeled with an **F** carrying a superscript that represents the type of force and a subscript that represents first the object *by* which the force is exerted and then the object *on* which the force is exerted. And third, a separate vector arrow indicating the direction of the acceleration of the object that is the focus of the diagram.

11. Drawing a free-body diagram allows you to separate and focus on a particular object in a system of interacting objects and to account for and keep track of the forces exerted on that object. The goal is to ensure that in analyzing the problem you do not overlook any forces that *are* exerted on the object or include any forces that are *not* exerted on the object.

12. Because all the forces included in the vector sum should already be shown on the diagram. Thus adding an arrow representing the vector sum might lead to double counting of forces.

13. (a) As the magnitude of the force exerted on the spring increases, the spring length changes in proportion to the magnitude of the force. If the force is compressing the spring, the length decreases. If the force is stretching the spring, the length increases. (b) The magnitude of the force exerted by the spring on the object is identical to that exerted by the object on the spring, so the change in length is still proportional to the magnitude of the force exerted by the spring on the object.

14. An elastic force is a force exerted by an object that is reversibly deformed (either stretched or compressed).

15. The force is undiminished when the gravitational force on the rope, spring, or thread is much less than the pulling force being exerted and much less than the gravitational force on the block—in other words, when the rope, spring, or thread has negligible inertia.

16. Acceleration. There is in general no direct relationship between the vector sum of forces and any other aspect of motion, such as velocity or position.

17. The inertia of the object must be constant.

18. First law: In an inertial reference frame, an isolated object (one on which no forces are exerted) has a constant velocity. Second law: The time rate of change of an object's momentum equals the vector sum of the forces exerted on the object. Third law: When two objects A and B interact, the force exerted by A on B is equal in magnitude to the force exerted by B on A, but the two forces point in opposite directions.

19. (a) The force of gravity exerted on an object is proportional to the object's inertia. Because the sack of apples has larger inertia than a single apple, the force of gravity exerted on the sack is larger. (b) The free-fall acceleration (g) is the same for both the sack of apples and a single apple because the acceleration of each object is the vector sum of the forces exerted on the object divided by the object's inertia. Thus the effect of larger inertia cancels.

20. No. Equation 8.17 was derived for an object in free fall, but it holds for any object near Earth's surface, even one at rest. Objects subject to multiple forces may have accelerations that differ from g, but they still experience the same gravitational force mg.

21. For a given spring, the numerical value of k is the same for both stretching and compression.

22. The symbol represents the position of the free end when the spring is relaxed (neither stretched nor compressed).

23. The slope represents $1/k$, the inverse of the spring constant k.

24. The impulse delivered to an object equals the product of the vector sum of the forces exerted on the object and the time interval over which the forces are exerted.

25. It is necessary to take into account the time dependence of the variable force (or forces). This can be done by direct integration, as in *Principles* Eq. 8.26, or by determining the average value for each variable force and including that value in *Principles* Eq. 8.25.

26. An internal force is exerted between two objects that are part of the chosen system. An external force is exerted by an object external to the system on an object that is part of the system.

27. The center of mass accelerates as if both objects were located at the system's center of mass and the external force were applied at this location.

28. The vector sum is zero.

29. All parts of a rigid object experience the same acceleration, but for a deformable object different parts of the object have different velocities and different accelerations. For a rigid object, the acceleration of each part is the same as the acceleration of the center of mass, but the same is not true for a deformable object.

Answers to Guided Problems

Guided Problem 8.2 $5.2 \times 10^4\,\text{N/m}$

Guided Problem 8.4 37 mm

Guided Problem 8.6 (a) $a = g/6 = 1.6\,\text{m/s}^2$; (b) $\mathcal{T} = 57\,\text{N}$

Guided Problem 8.8 $\mathcal{T}_{\text{support}} = \frac{3}{2}(m_\text{p} + m_\text{b})(a + g) + m_\text{p}g$ reduces to $\mathcal{T}_{\text{support}} = \frac{3}{2}m_\text{b}(a + g)$ for $m_\text{p} = 0$

9

PRACTICE
Work

Chapter Summary   146

Review Questions   147

Developing a Feel   148

Worked and Guided Problems   149

Questions and Problems   157

Answers to Review Questions   163

Answers to Guided Problems   164

PRACTICE

Chapter Summary

Work done by a constant force (Sections 9.1, 9.2, 9.5, 9.6)

Concepts In order for a force to do work on an object, the point of application of the force must undergo a displacement.

The SI unit of work is the **joule** (J).

The work done by a force is positive when the force and the force displacement are in the same direction and negative when they are in opposite directions.

Quantitative Tools When one or more constant forces cause a particle or a rigid object to undergo a displacement Δx in one dimension, the work done by the force or forces on the particle or object is given by the **work equation:**

$$W = (\textstyle\sum F_x)\Delta x_F. \tag{9.9}$$

In one dimension, the work done by a set of constant nondissipative forces on a system of particles or on a deformable object is

$$W = \sum_n (F_{\text{ext}\,nx}\Delta x_{Fn}). \tag{9.18}$$

If an external force does work W on a system, the **energy law** says that the energy of the system changes by an amount

$$\Delta E = W. \tag{9.1}$$

For a closed system, $W = 0$ and so $\Delta E = 0$.

For a particle or rigid object, $\Delta E_{\text{int}} = 0$ and so

$$\Delta E = \Delta K. \tag{9.2}$$

For a system of particles or a deformable object,

$$\Delta K_{\text{cm}} = (\textstyle\sum F_{\text{ext}\,x})\Delta x_{\text{cm}}. \tag{9.14}$$

Energy diagrams (Sections 9.3, 9.4)

Concepts An **energy diagram** shows how the various types of energy in a system change because of work done on the system. (See the Procedure box on page 149.)

In choosing a system for an energy diagram, avoid systems for which friction occurs at the boundary because then you cannot tell how much of the thermal energy generated by friction goes into the system.

Variable and distributed forces (Section 9.7)

Concepts The force exerted by a spring is variable (its magnitude and/or direction changes) but nondissipative (no energy is converted to thermal energy).

The frictional force is dissipative and so causes a change in thermal energy. This force is also a distributed force because there is no single point of application.

Quantitative Tools The work done by a variable nondissipative force on a particle or object is

$$W = \int_{x_i}^{x_f} F_x(x)\,dx. \tag{9.22}$$

If the free end of a spring is displaced from its relaxed position x_0 to position x, the change in its potential energy is

$$\Delta U_{\text{spring}} = \tfrac{1}{2}k(x - x_0)^2. \tag{9.23}$$

If a block travels a distance d_{path} over a surface for which the magnitude of the force of friction is a constant F_{sb}^{f}, the energy dissipated by friction (the thermal energy) is

$$\Delta E_{\text{th}} = F_{\text{sb}}^{\text{f}}d_{\text{path}}. \tag{9.28}$$

Power (Section 9.8)

Concepts **Power** is the *rate* at which energy is either converted from one form to another or transferred from one object to another.

The SI unit of power is the **watt** (W), where $1 \text{ W} = 1 \text{ J/s}$.

Quantitative Tools The **instantaneous power** is

$$P = \frac{dE}{dt}. \tag{9.30}$$

If a constant external force $F_{\text{ext}\,x}$ is exerted on an object and the x component of the velocity at the point where the force is applied is v_x, the power this force delivers to the object is

$$P = F_{\text{ext}\,x}v_x. \tag{9.35}$$

Review Questions

Answers to these questions can be found at the end of this chapter.

9.1 Force displacement

1. What is the meaning of the term *work*?
2. Suppose you exert an external force on a system. Is work necessarily done on the system?

9.2 Positive and negative work

3. (*a*) A headwind pushes on a bicycle coasting eastward, slowing the bicycle down. What is the sign of the work done by the wind on the bicycle? (*b*) If the same wind pushes in the same direction on a bicycle initially at rest and moves the bicycle westward, what is the sign of the work done by the wind on the bicycle?
4. You throw a ball up into the air and then catch it. How much work is done by gravity on the ball while it is in the air?
5. Can the work done by a spring ever be negative?

9.3 Energy diagrams

6. In a system in which all the energy is mechanical, the initial and final energies are $K_i = 10.0 \text{ J}$, $K_f = 3.0 \text{ J}$, $U_i = +4.0 \text{ J}$, $U_f = +6.0 \text{ J}$. Sketch an energy diagram for the system.
7. A cart is at rest but free to move on a low-friction track. When you push and release the cart, it travels at constant speed to the right until it strikes a spring fixed to the end of the track. After a short time interval in contact with the spring, the cart moves back toward you at the same constant speed it had as it approached the spring. Sketch energy diagrams for the cart during these time intervals: (*a*) from the initial resting state until traveling at constant speed to the right, (*b*) from traveling at constant speed to the right until the spring is fully compressed, (*c*) from spring being fully compressed until traveling to the left at constant speed, (*d*) from the initial resting state until traveling to the left at constant speed.

9.4 Choice of system

8. Why should you avoid choosing a system that has a frictional force exerted across the system boundary?
9. When computing energies, should the gravitational interaction be associated with work, with potential energy, or with both?
10. What is the utility of analyzing an energy problem using more than one choice of system?

9.5 Work done on a single particle

11. State the relationship between work and energy in words.
12. Does the energy law (Eq. 9.1) apply to closed systems? To systems that are not closed?
13. Equations 9.8 and 9.9 are valid only for a particle. Why?
14. Discuss the similarities and differences in the momentum law (Eq. 4.18, $\Delta \vec{p} = \vec{J}$) and the energy law (Eq. 9.1, $\Delta E = W$).
15. In what sense is a crate a particle? Under what circumstances would it be inappropriate to treat a crate as a particle in working a physics problem?

9.6 Work done on a many-particle system

16. How does the energy law (Eq. 9.1) applied to a single particle differ from the energy law applied to a many-particle system? How does the work equation (Eq. 9.9) applied to a single particle differ from the work equation applied to a many-particle system?
17. Why is Eq. 9.18 restricted to nondissipative forces?
18. How does the change in kinetic energy of a many-particle system differ from the work done on the system by its environment?

9.7 Variable and distributed forces

19. When you plot the force exerted on a particle as a function of the particle's position, what feature of the graph represents the work done on the particle?
20. (*a*) In $W_{bs} = \frac{1}{2}k(x - x_0)^2$, the expression obtained in *Principles* Example 9.8 for the work done by a brick in compressing or stretching a spring, where does the factor $\frac{1}{2}$ come from? (*b*) If the force exerted by a load on a spring is $F_{\ell s\,x} = k(x - x_0)$ (Eq. 8.18) and the force displacement is $x - x_0$, why isn't W_{bs} just $F(x - x_0) = k(x - x_0)(x - x_0) = k(x - x_0)^2$?
21. You use Eq. 9.27 or Eq. 9.28 to calculate the thermal energy dissipated by friction as a crate slides across the floor and eventually comes to rest. Could this thermal energy be called the work done on the crate by the frictional force?

9.8 Power

22. How is instantaneous power related to average power?
23. When is the instantaneous power delivered to an object equal to the magnitude of the vector sum of the forces $\sum F_{\text{ext}}$ exerted on the object multiplied by the object's speed?

Developing a Feel

Make an order-of-magnitude estimate of each of the following quantities. Letters in parentheses refer to hints below. Use them as needed to guide your thinking.

1. The minimum work required to lift a large bag of groceries from floor to tabletop (D, J)
2. The energy dissipated when you stand on the top of an empty soda can and flatten the can (B, L, T)
3. The minimum work the engine must do to bring a fully loaded, 18-wheel truck from rest to freeway speed (A, K, Q)
4. The spring constant of a bungee cord (E, H, N, V)
5. The minimum power required to pitch a fastball like a professional baseball player (C, I, S)

6. The power provided to a car by the engine while cruising at freeway speed (F, M, U, Y)
7. The average electrical power used in your home during a typical day (R, G, W)
8. The electrical energy used in a typical U.S. home during a year (R, G, W, O, X)
9. The amount of chemical (source) energy needed to supply electrical power for residential consumption in the United States during one year (R, G, W, O, X, P, Z)

Hints

A. What is the inertia of a loaded 18-wheeler?
B. What is the gravitational force exerted by Earth on you?
C. What is the typical speed of a major league fastball?
D. What is the gravitational force exerted on a large bag of groceries?
E. What is the height of a typical bungee jump?
F. How many kilometers per liter of gasoline do you get cruising at freeway speed?
G. How much power is used during peak consumption hours?
H. What is the unstretched length of a bungee cord?
I. What is the inertia of a baseball?
J. What is the height of a typical table?
K. What is a typical freeway speed?
L. Through what displacement is the force exerted?
M. How much chemical (source) energy is available in 1 L of gasoline?
N. What is the relevant energy change?
O. How many seconds are in 1 year?
P. What is the number of residences in the United States?
Q. Does all the work done by the engine go into changing the truck's kinetic energy?
R. How much power is consumed by a typical appliance?
S. How long a time interval does a pitcher's forward throwing motion require?
T. How does the force you exert on the can relate to the gravitational force exerted by Earth on you?

U. What percentage of the energy in the gasoline is transferred to the drive wheels of the car?
V. What is a typical inertia of a bungee jumper?
W. During what percentage of a full day do you consume electricity at the peak consumption rate?
X. How is the amount of energy consumed related to the average power?
Y. How long a time interval is required to travel 100 km?
Z. What is the efficiency of source energy conversion, including transmission losses?

Key (all values approximate)

A. 3×10^4 kg; B. 7×10^2 N; C. 4×10^1 m/s; D. 1×10^2 N; E. 8×10^1 m; F. 1×10^1 km/L; G. 6×10^3 W; H. 3×10^1 m; I. 0.2 kg; J. 1 m; K. 3×10^1 m/s; L. 0.1 m; M. 4×10^7 J; N. gravitational potential energy becomes spring potential energy; O. 3×10^7 s; P. 90 million; Q. we assume it does in order to determine the minimum work; R. 100 W for lamps and most small appliances; 1000 W for refrigerators, microwave ovens, irons, toasters, bathroom heaters; 5000 W for central air conditioners; S. 0.5 s; T. about equal in magnitude; U. less than 20%; V. 7×10^1 kg; W. less than 30%; X. $\Delta E = P_{av}\Delta t$; Y. 1 h; Z. about 10%

Worked and Guided Problems

Procedure: Drawing energy diagrams

1. Specify the **system** under consideration by listing the components inside the system.
2. **Sketch** the system in its initial and final states. (The initial and final states may be defined for you by the problem, or you may have to choose the most helpful states to examine.) Include in your sketch any external forces exerted on the system that undergo a nonzero force displacement, and draw a dot at the point of application of each force.
3. Determine any nonzero **changes in energy** for each of the four categories of energy, taking into account the four basic energy-conversion processes illustrated in Figure 7.13:

   a. Did the speed of any components of the system change? If so, determine whether the system's kinetic energy increased or decreased and draw a bar representing ΔK for the system. For positive ΔK, the bar extends above the baseline; for negative ΔK, it extends below the baseline. (For some problems you may wish to draw separate ΔK bars for different objects in the system; if so, be sure to specify clearly the system component that corresponds to each bar and verify that the entire system is represented by the sum of the components.)

   b. Did the configuration of the system change in a reversible way? If so, draw a bar representing the change in potential energy ΔU for the system. If necessary, draw separate bars for the changes in different types of potential energy, such as changes in elastic and gravitational potential energy.

   c. Was any source energy consumed? If so, draw a bar showing ΔE_s. Source energy usually decreases, making ΔE_s negative, and so the bar extends below the baseline. Keep in mind that conversion of source energy is always accompanied by generation of thermal energy (Figure 7.13c and d).

   d. Does friction occur within the system, or is any source energy consumed? If so, draw a bar showing ΔE_{th}. In nearly all cases we consider, the amount of thermal energy increases, and so ΔE_{th} is positive.

4. Determine whether or not any **work** W is done by external forces on the system. Determine whether this work is negative or positive. Draw a bar representing this work, making the length of the bar equal to the sum of the lengths of the other bars in the diagram.

   If no work is done by external forces on the system, leave the bar for work blank, then go back and adjust the lengths of the other bars so that their sum is zero.

These examples involve material from this chapter but are not associated with any particular section. Some examples are worked out in detail; others you should work out by following the guidelines provided.

Worked Problem 9.1 Trash that

A janitor starts to push a 75-kg trash can that is initially at rest across a level surface by exerting a constant horizontal force of 50 N. After moving 10 m, the trash can has a speed of 2.0 m/s. (a) What are the magnitude and direction of the frictional force exerted by the surface on the can? (b) What power does the janitor provide at the 10-m mark?

❶ GETTING STARTED

(a) The work done by a force equals the magnitude of the force multiplied by the force displacement. The pushing force and the frictional force are both exerted over a known displacement to change the can's kinetic energy. The can accelerates, so the janitor must provide a force larger than the frictional force. The frictional force opposes the sliding motion, so it is directed in the opposite direction to the janitor's push. We could take the can as our system. However, that would mean the force of friction is exerted at the system boundary (and then we don't know how much of the dissipated energy ends up inside the system). So instead we choose the can and the floor as our system. Figure WG9.1 shows the can and floor system in its initial and final states, the force displacement Δx_F, and the single applied external force (friction is internal!).

We need to keep track of components, so we arbitrarily choose the positive x axis along the direction of motion of the can (to the right in Figure WG9.1)

Figure WG9.1

(b) Because we know the constant force exerted by the janitor and the can's speed once it has moved 10 m, we can use the product of force and speed to obtain the power delivered at the 10-m position.

PRACTICE

❷ **DEVISE PLAN** The surface does not move and so does not contribute to the system kinetic energy. Because the can is rigid, its contribution to the system kinetic energy is computed using the can's center-of-mass motion. The center of mass of the can experiences exactly the same displacement as any other point on the can, which is the force displacement. There is no potential energy to consider (no vertical motion to involve gravity, no springs) and no source energy in our system. The energy law allows us to account for the changes in kinetic energy, thermal energy, and work. We draw an energy diagram (Figure WG9.2) to summarize this information. We have an expression for each of these terms: The change in kinetic energy can be computed from given information, the frictional force is responsible for the change in thermal energy, and the janitor's force is responsible for the work done.

Figure WG9.2

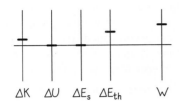

ΔK ΔU ΔE_s ΔE_{th} W

(a) We know the change in kinetic energy, the displacement, and the magnitude of the force exerted by the janitor. Thus the only unknown is the one we are asked to calculate: The magnitude of the frictional force. (b) Because the only external force that has a force displacement is exerted by the janitor, and that force is constant, the power at any instant is given by Eq. 9.35, $P = F_{ext\,x}v_x$. We know both the force and the speed at the requested location, so we can obtain the power.

❸ **EXECUTE PLAN**
(a) We begin with the energy law, Eq. 9.1:

$$\Delta E = \Delta K + \Delta U + \Delta E_s + \Delta E_{th} = W.$$

The thermal energy change ΔE_{th} is caused by the frictional force, so we can employ Eq. 9.27 to make the relationship explicit:

$$\Delta E_{th} = -F^f_{sc\,x}\Delta x_{cm}.$$

Similarly, we use Eq. 9.9 to express W in terms of the force exerted by the janitor and then substitute both expressions into the energy law:

$$\Delta K + 0 + 0 + (-F^f_{sc\,x}\Delta x_{cm}) = F^c_{jc\,x}\Delta x. \qquad (1)$$

Because the can is rigid, $\Delta x_{cm} = \Delta x$. Substituting this and our usual expression for kinetic energy into Eq. 1 and rearranging terms gives

$$\tfrac{1}{2}mv_f^2 - \tfrac{1}{2}mv_i^2 = F^c_{jc\,x}\Delta x + F^f_{sc\,x}\Delta x$$

$$\tfrac{1}{2}mv_f^2 - 0 = F^c_{jc\,x}\Delta x + F^f_{sc\,x}\Delta x$$

$$F^f_{sc\,x} = \frac{\tfrac{1}{2}mv_f^2}{\Delta x} - F^c_{jc\,x} = \frac{\tfrac{1}{2}(75\text{ kg})(2.0\text{ m/s})^2}{10\text{ m}} - (+50\text{ N}) = -35\text{ N}.$$

Because we chose to the right as the positive x direction in Figure WG9.1, the negative value for $F^f_{sc\,x}$ tells us that the direction of the frictional force is to the left. The magnitude of the frictional force is 35 N. ✔

(b) The power delivered by the janitor is

$$P = F_{ext\,x}v_x = (+50\text{ N})(+2.0\text{ m/s}) = 1.0 \times 10^2\text{ W}. ✔$$

❹ **EVALUATE RESULT** The frictional force is directed to the left and has a magnitude smaller than the force exerted by the janitor, as we expected.

We can also check our answer with kinematics because constant forces imply constant acceleration. The acceleration of the trash can is related to the initial and final speeds and the displacement by (see *Principles* Example 3.4)

$$a_x = \frac{v_{x,f}^2 - v_{x,i}^2}{2\Delta x} = \frac{(2.0\text{ m/s})^2 - 0^2}{2(10\text{ m})} = +0.20\text{ m/s}^2.$$

If we take just the can as our system, the vector sum of forces exerted on the can must therefore be

$$\sum F_{ext\,x} = ma_x = (75\text{ kg})(+0.20\text{ m/s}) = +15\text{ N}.$$

Knowing this value allows us to solve for the frictional force:

$$\sum F_{ext\,x} = F^c_{jc\,x} + F^f_{sc\,x} = +50\text{ N} + F^f_{sc\,x} = +15\text{ N}$$

$$F^f_{sc\,x} = +15\text{ N} - 50\text{ N} = -35\text{ N}$$

in agreement with our previous result.

The power we obtained is that of a 100-W light bulb. The amount of power required to lift a textbook at a constant 0.33 m/s is about 10 W, so this janitor is working about ten times as hard to accelerate the heavy trash can to 2.0 m/s by the end of the 10-m trip. Doing 1 chin-up per second, however, requires an average power of more than 200 W for a 60-kg individual, so the 100 W we obtained is not unreasonable.

Guided Problem 9.2 Delivering a piano

A moving company decides it's easier to hoist a 150-kg piano up to a second-floor apartment with a pulley rather than negotiate a narrow stairway. The piano is hoisted with a rope to the level of a window 5.3 m above the ground and temporarily brought to rest. How much work is done on the piano (a) by the rope and (b) by the force of gravity? (c) If it takes 1.0 min to hoist the piano, what average power must be supplied by the hoist?

❶ **GETTING STARTED**
1. What is a good choice of system? Should Earth be included in the system? Sketch the system in its initial and final states.
2. Can the piano be treated as a rigid object? How is the motion of its center of mass related to the motion of any other part of it?

3. Draw an energy diagram. Note that the piano is at rest both before and after it is hauled up.

4. Are there external forces exerted on the system? If so, what is the force displacement for each?

❷ DEVISE PLAN

5. How is the change of the system's energy related to the work done on the system by external forces? What formula expresses this relationship?

6. Is it reasonable to ignore any friction in the pulley so you can say that no energy is dissipated?

7. Is there enough information to solve for the unknowns?

❸ EXECUTE PLAN

8. Calculate the requested results.

9. Does the person running the hoist do work on the piano?

❹ EVALUATE RESULT

10. How does your calculated power compare with the power needed to walk up the stairs to the first floor in the same 1-min time interval?

11. Does the power calculated in part c seem reasonable for lifting a piano?

Worked Problem 9.3 Human cannonball

A circus performer of inertia m is launched into the air by a "cannon" that contains a spring platform for which the spring constant is k. The performer climbs into the cannon, compressing the spring. Then the spring is additionally compressed (by a winch) so that the initial position of the platform is a distance d below the position of the platform when the spring is relaxed (Figure WG9.3). (a) What maximum speed does the performer attain after launch? (b) How high above the relaxed position of the spring platform does the performer fly? Ignore any dissipative interactions, and analyze both parts of the problem using two systems: one that allows you to solve the problem using potential energy but not work and one that allows you to solve the problem using work but not potential energy.

Figure WG9.3

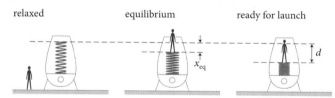

relaxed   equilibrium   ready for launch

❶ GETTING STARTED For a performer launched from rest by a compressed spring, we must determine the performer's maximum speed and maximum height. We must treat the performer as a particle or rigid object because there is no information that allows us to do otherwise. The release of the spring causes the performer to shoot straight up, and at the highest point of his trajectory he has zero instantaneous velocity as his direction of travel reverses. Because the force exerted by the spring is not constant, using a force approach to solve this problem would give us nonconstant acceleration, which means we cannot use constant-acceleration formulas to get the maximum height.

In choosing our systems, we must consider all the forces that might affect the performer's motion. We are told to ignore dissipative forces, and so it appears that the force exerted by the spring and the gravitational force are the crucial ones. This means we have two interaction pairs: spring-performer and Earth-performer. In order to use potential energy for the gravitational and spring forces, we must include both objects from each pair in our system. We can make this system closed and isolated by including the cannon and the ground underneath it. So system 1 is cannon, spring, performer, and Earth.

To solve the problem by analyzing the work done on the performer, we recall that only forces external to a system can do work on the system. Thus we define system 2 as being only the performer, so that the forces exerted on him are all external forces.

We know that the performer's speed is zero both initially and at his greatest height, and we know the formulas for the elastic and gravitational potential energies and for the work done by each force. The problem requires us to think about two different final instants during the motion: In part b we have $t_f =$ the instant the performer reaches his greatest height, but in part a we have $t_f =$ the instant he has his highest speed. We therefore analyze parts a and b separately, using the same initial position but a different final position in each case. But where, exactly, does the performer achieve the maximum speed? There are two forces that affect the speed of the performer: the gravitational force and the spring force. The gravitational force is constant and directed downward. The spring force is not constant and is directed upward. In order for the performer to accelerate upward, the spring force must exceed the gravitational force. As long as this is true, the performer gains speed by accelerating upward. The spring force gets smaller as the performer rises and the spring begins to relax. At some point the spring force must become smaller than the gravitational force so that the performer no longer accelerates upward but begins to accelerate downward, slowing his speed. Thus the maximum speed occurs when the spring force equals the gravitational force so that the performer is instantaneously in equilibrium. This is the position labeled "equilibrium" in Figure WG9.3 (though in that figure the performer is not moving).

Figure WG9.4 shows a time sequence of events, including the instant of launch, the equilibrium position of maximum speed, and the position of maximum height.

Figure WG9.4

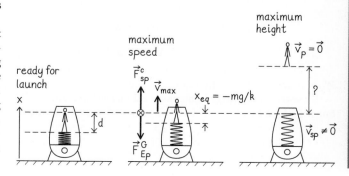

Now we produce an energy diagram for each system, beginning with the performer ready for launch and including the position of maximum speed as well as the position of maximum height (Figure WG9.5). Note that we separate the potential energy into two parts (elastic potential energy U_{sp} in the spring and the performer's gravitational potential energy U^G relative to Earth) and we do the same for work (work W_{sp} done by the spring on the performer and work W_g done by gravity on him).

Figure WG9.5

system 1: perfomer, spring, cannon, and Earth

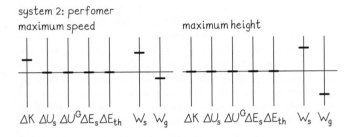

② **DEVISE PLAN** Using Figure WG9.5 as a guide, we write for the change in system 1 energy from launch to maximum speed

$$\Delta E_{sys\,1} = \Delta K + \Delta U + \Delta E_s + \Delta E_{th} = 0$$

$$\Delta K + \Delta U_{sp} + \Delta U^G + 0 + 0 = 0$$

$$\Delta K + \Delta U_{sp} + \Delta U^G = 0. \qquad (1)$$

For the maximum height computation, both the initial and final kinetic energies of the performer are zero. The spring and platform may continue to move up and down after the performer's feet leave the platform. However, any kinetic energy involved in the final motion of the spring and platform can be expressed as remaining spring potential energy because at the turning points of their motion the energy is entirely potential. This gives

$$\Delta E_{sys\,1} = \Delta K + \Delta U + \Delta E_s + \Delta E_{th} = 0$$

$$\Delta U_{sp} + \Delta U^G = 0. \qquad (2)$$

We have enough information to solve these equations.
For system 2, the energy accounting is still guided by Figure WG9.5, but it looks a bit different because now we are not looking at

any potential energy changes in the system. For maximum speed we have

$$\Delta E_{sys\,2} = \Delta K + \Delta U + \Delta E_s + \Delta E_{th} = W$$

$$\Delta K + 0 + 0 + 0 = W$$

$$\Delta K = W_g + W_{sp}. \qquad (3)$$

For maximum height, we do not have to concern ourselves with any energy remaining in the system due to the motion of the spring because the spring and platform are not part of system 2. This makes the accounting a bit easier:

$$\Delta E_{sys\,2} = \Delta K + \Delta U + \Delta E_s + \Delta E_{th} = W$$

$$0 + 0 + 0 + 0 = W$$

$$W_{sp} + W^G = 0 \qquad (4)$$

Because the spring force and the gravitational force are exerted over different force displacements, we track them separately. The gravitational force is exerted during the entire motion, but the spring force is exerted only during the time interval from the last instant at which the spring is in its launch (maximum compression) position to the instant when it reaches equilibrium for Eq. 3, and the position at which performer contact is lost for Eq. 4.

③ **EXECUTE PLAN**
(*a*) For either system 1 or system 2 maximum speed, the initial position is $x_i = -d$, and the final position is found by imposing the equilibrium condition (that the spring force and gravitational force cancel). The free-body diagram (see Figure WG9.4) can be used to solve for the final position. With the positive x axis directed upward and the origin at the relaxed position, both $\vec{x}_f$ and $\vec{g}$ have negative x components:

$$\sum \vec{F} = \vec{0}$$

$$\vec{F}^c_{sp} + \vec{F}^G_{Ep} = \vec{0}$$

$$-k(\vec{x}_f) + m(\vec{g}) = \vec{0}$$

$$-kx_f - mg = 0$$

$$x_f = x_{eq} = -\frac{mg}{k}.$$

The initial speed is zero, so for system 1 we can solve Eq. 1 for the final speed (maximum speed):

$$\Delta K + \Delta U_{sp} + \Delta U^G = 0$$

$$\tfrac{1}{2}mv_{max}^2 - \tfrac{1}{2}mv_i^2 + \tfrac{1}{2}kx_{eq}^2 - \tfrac{1}{2}kx_i^2 + mg(x_{eq}) - mg(x_i) = 0$$

$$\tfrac{1}{2}mv_{max}^2 - 0 + \tfrac{1}{2}k\left[\left(\frac{-mg}{k}\right)^2 - (-d)^2\right]$$

$$+ mg\left[\left(\frac{-mg}{k}\right) - (-d)\right] = 0$$

$$v_{max}^2 = \frac{k}{m}\left[d^2 - \left(\frac{m^2g^2}{k^2}\right)\right] - 2g\left(d - \frac{mg}{k}\right).$$

This expression can be simplified algebraically by combining the second and fourth terms on the right-hand side and then noting that the expression can be factored:

$$v_{max}^2 = \frac{k}{m}\left(d^2 - \frac{2mgd}{k} + \frac{m^2g^2}{k^2}\right)$$

$$v_{max}^2 = \frac{k}{m}\left(d - \frac{mg}{k}\right)^2$$

$$v_{max} = \sqrt{\frac{k}{m}}\left(d - \frac{mg}{k}\right). ✔$$

We now compute the maximum speed for system 2, using Eq. 3:

$$\Delta K = W_g + W_{sp}$$

$$\tfrac{1}{2}mv_{max}^2 - 0 = -mg\left(d - \frac{mg}{k}\right) + \tfrac{1}{2}k\left(d^2 - \frac{m^2g^2}{k^2}\right)$$

$$v_{max}^2 = -2g\left(d - \frac{mg}{k}\right) + \frac{k}{m}\left(d^2 - \frac{m^2g^2}{k^2}\right)$$

$$v_{max} = \sqrt{\frac{k}{m}}\left(d - \frac{mg}{k}\right). ✔$$

(b) For the maximum height calculation we must first determine the position at which performer contact is lost. The spring and platform exert a force on the performer given by Hooke's law—that is, a force whose magnitude is proportional to the stretch or compression of the spring, and whose direction is opposite to the displacement of the spring from its relaxed position. However, the performer is not attached to the platform, so the force applied to the performer can be in only the upward direction. That means the performer can experience a force from the spring only when the platform is below the position labeled "relaxed" in Figure WG9.3. We designate the final position of the spring and platform $x_{sp,f}$ and the final position of the performer $x_{p,f}$. The initial position of both objects is still the same, $x_i = -d$. Substituting Eq. 9.23 for the spring potential energy and Eq. 7.19, $\Delta U^G = mg\Delta x$, for the gravitational

potential energy in Eq. 2, we have

$$\Delta E_{sys\,1} = \Delta U_{sp} + \Delta U^G = 0$$

$$\tfrac{1}{2}kx_{sp,f}^2 - \tfrac{1}{2}kx_{sp,i}^2 + mgx_{p,f} - mgx_{p,i} = 0$$

$$0 - \tfrac{1}{2}k(-d)^2 + mgh - mg(-d) = 0$$

$$mgh = \tfrac{1}{2}kd^2 - mgd$$

$$h = \frac{kd^2}{2mg} - d. ✔$$

Now we repeat for system 2, noting that the spring force varies with position but the gravitational force does not. Equation 4 gives us

$$W_{sp} + W_g = 0$$

$$\int_{x_{sp,i}}^{x_{sp,f}} F_{sp\,x}^c dx + F_{Ep\,x}^G(x_{p,f} - x_{p,i}) = 0$$

$$+\tfrac{1}{2}k(-d)^2 - mg[+h - (-d)] = 0$$

$$\tfrac{1}{2}kd^2 - mg(h + d) = 0$$

$$h = \frac{kd^2}{2mg} - d. ✔$$

4 EVALUATE RESULT Having obtained the same answers with two systems gives us confidence in the results. As a further check, consider how the height depends on the spring constant: A larger k gives a greater height, as expected. We also expect the performer to rise higher if his inertia is smaller. The only place the performer's inertia appears in our expression for h is in the denominator, which is consistent with our expectation.

Our expression for the maximum speed tells us that d must exceed mg/k in order to get a positive result. This is reasonable because mg/k is the distance by which the spring would compress if the performer were simply standing on the platform; certainly that would not be enough for any launch! Additionally, our expression for h tells us that, to give a sensible result, d must exceed not mg/k but rather $2mg/k$. For small compressions beyond equilibrium, it is reasonable that the performer would not be launched at all—that is, would not lose contact with the platform, but would simply ride up and down with the spring and platform.

Guided Problem 9.4 Thrill seeking

A 50-kg woman goes bungee-jumping off a bridge that is 15 m above the ground. The bungee cords attached to her feet have a relaxed length of 5.5 m, and she is 1.8 m tall. What must the spring constant of the cords be if she is to just miss touching the ground with her head when she jumps?

1 GETTING STARTED

1. What system is appropriate for this problem? Sketch the initial and final situations, showing such details as the relaxed length of the cords, the woman's height, the bridge height, and an x axis.

2. As usual, several approaches are possible, some easier than others. Try energy methods first.

3. Make an energy diagram and a situation diagram for your system.

4. What assumptions do you make?

2 DEVISE PLAN

5. Write a general expression for the changes in energy during an appropriate time interval for the jump.

6. Check that the number of unknowns and the number of equations are the same.

3 EXECUTE PLAN

7. Do the algebra and isolate the spring constant.

8. Substitute your numerical values, including units.

4 EVALUATE RESULT

9. Examine your expression for the spring constant. What happens to k as the woman's inertia increases? As the bridge height increases?

10. Does the numerical answer make sense? Be sure it is the woman's head that just misses the ground and not her feet!

11. Are all assumptions you made reasonable?

Worked Problem 9.5 Working on the railroad

A train engine pulls a boxcar of inertia 12×10^3 kg, which in turn pulls a boxcar of inertia 17×10^3 kg. The train accelerates from rest to 11.1 m/s in 3.0 min. (*a*) How much work is done by the engine on the two-car combination? (*b*) How much work is done on each car individually? Ignore any friction effects.

❶ **GETTING STARTED** The work done by a force equals the magnitude of the force multiplied by the force displacement. We need to determine all forces exerted by any external objects on each car, and how far its point of application moves during the common 3.0-min time interval. The only displacement is horizontal, in the direction of motion of the train, which we will call the positive x direction. Thus only horizontal forces need be considered. Because the train cars are connected and move together, each point on each car has the same displacement as the engine. Therefore each applied force has the same force displacement, $\Delta \vec{x}_F$.

For part *a* we need to calculate the work done by the engine on the two-car combination, and so we take the two cars as our system. If we ignore any dissipative forces, there is one external horizontal force on the system, the pulling contact force $\vec{F}^c_{e1}$ exerted by the engine on the two-car combination. For part *b* we treat each car separately. Figure WG9.6 contains an initial and final sketch of each system, together with an energy diagram for each system.

Figure WG9.6

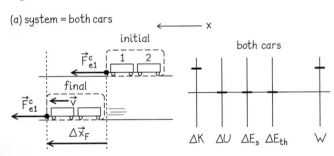

(a) system = both cars

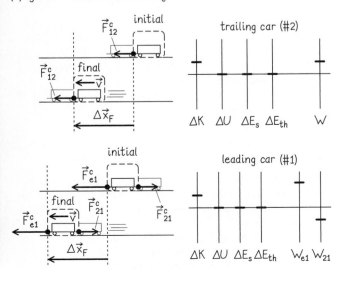

(b) system = each car individually

For the system consisting of the trailing car (car 2), there is also one external force that does work: the force exerted by the leading car (car 1). We call this force $\vec{F}^c_{12}$. For the system consisting of the leading car (car 1), there are two external forces that do work: the force exerted by the engine on car 1 ($\vec{F}^c_{e1}$) and the force exerted by car 2 on car 1 ($\vec{F}^c_{21}$). All three systems share some properties: No system has an internal source of energy, each system gains kinetic energy during the 3.0-min time interval of interest, and each experiences work done by external forces. Because we ignore any friction effects, we expect no thermal energy changes in any of the systems.

❷ **DEVISE PLAN** For each of the three systems we begin with the energy law, which in each case can be expressed as

$$\Delta E = \Delta K = W.$$

This expression tells us that the work done on each system depends on *only* the change in the velocity of that system. It does not depend on the time interval over which the force is exerted or on the magnitude of the acceleration. We can compute the change in kinetic energy directly in each case.

❸ **EXECUTE PLAN** Each system has zero initial kinetic energy, and each has a specific final kinetic energy. The difference of these values is the work done by external forces on each system.

(*a*) For the system of two cars, we obtain

$$W = \Delta K = K_f - K_i = \tfrac{1}{2}(m_1 + m_2)v^2 - 0$$
$$W = \tfrac{1}{2}(12 \times 10^3 \text{ kg} + 17 \times 10^3 \text{ kg})(11.1 \text{ m/s})^2$$
$$W = 1.8 \times 10^6 \text{ J.} ✔$$

(*b*) Here we must apply the energy law to each system (car) separately. For the leading car (car 1):

$$W = \Delta K = K_f - K_i = \tfrac{1}{2}(m_1)v^2 - 0$$
$$W = \tfrac{1}{2}(12 \times 10^3 \text{ kg})(11.1 \text{ m/s})^2$$
$$W_1 = 7.4 \times 10^5 \text{ J} ✔$$

For the trailing car (car 2):

$$W = \Delta K = K_f - K_i = \tfrac{1}{2}(m_2)v^2 - 0$$
$$W = \tfrac{1}{2}(17 \times 10^3 \text{ kg})(11.1 \text{ m/s})^2$$
$$W_2 = 1.0 \times 10^6 \text{ J.} ✔$$

❹ **EVALUATE RESULT** These two work values add up (to within rounding error) to the work done on the two-car system by the engine: $W_1 + W_2 = W$. This is reassuring. The only assumption we made is that the force exerted by the engine is the only external force exerted on the two-car system, and that assumption is valid because we were told to ignore friction effects. Based on our knowledge of the energy law, we should have recognized from the beginning that the time interval given was an extraneous piece of information.

Guided Problem 9.6 Fast as a speeding bullet

Because a fired bullet moves so quickly, determining its speed directly is difficult. Here is an indirect way. You fire the bullet (of inertia m_{bullet}) into a block of inertia m_{block} placed against a spring of spring constant k (Figure WG9.7), and the bullet becomes embedded in the block. If the spring compresses to a maximum distance d, what was the speed of the bullet? Ignore friction between the block and the surface on which it rests.

Figure WG9.7

❶ GETTING STARTED

1. How do you expect the compression distance to be related to the speed of the bullet? In other words, what should happen to the compression distance as the speed of the bullet is increased?

2. What can you say about the coefficient of restitution?
3. What kind of energy conversions occur after the collision?
4. Is it reasonable to consider the collision and the compression of the spring separately?

❷ DEVISE PLAN

5. How can you determine the speed of the block with the bullet embedded in it as the spring begins to compress?
6. For what system is the mechanical energy constant from the time immediately after the collision until the spring is fully compressed?

❸ EXECUTE PLAN

❹ EVALUATE RESULT

7. Does your expression for v_{bullet} show that the speed varies as you expect with changes in d, k, m_{bullet}, and m_{block}?

Worked Problem 9.7 Spring forward

One end of a spring of force constant k is attached to a wall, and the other end is attached to a block of inertia m sitting on a horizontal desk (Figure WG9.8). You pull on the block to stretch the spring a distance d past its relaxed position, and then hold the block at that position. A steady wind exerts on the block a constant force directed to the left, opposite the direction in which you moved the block. When you release the block, it moves to the left, experiencing a frictional force of magnitude F_{sb}^f until equilibrium is reached with the spring compressed a distance $2d$ from its relaxed position. What is the magnitude of the force exerted by the wind on the block?

Figure WG9.8

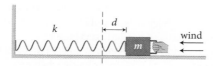

❶ GETTING STARTED Because three horizontal forces are exerted on the block (by the spring, friction, and the wind), we might be tempted to solve this as a force problem. We note, however, that the spring exerts a variable force, which means the acceleration is not constant and we cannot use constant-acceleration kinematics. Because we can calculate the potential energy of the spring and the amount of energy dissipated by the friction interaction, this problem might be solved by energy methods. The force of the wind is constant, and we can readily calculate any work done by this force. Choosing as our system the block, spring, and desk makes both the frictional and spring forces internal forces, so the only work done will be by the wind. Also, at maximum compression, the block's kinetic energy is zero, as it is the instant before you release it. Figure WG9.9 illustrates the initial and final situations and shows an energy diagram.

Figure WG9.9

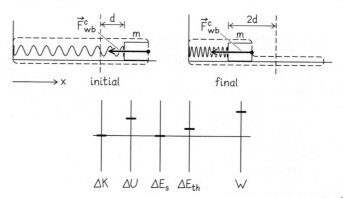

❷ DEVISE PLAN We begin with the energy law, $\Delta E = W$ (Eq. 9.1), with W being the work done on the system by the wind force:

$$\Delta E = W$$

$$\Delta K + \Delta U_{spring} + \Delta E_s + \Delta E_{th} = W. \qquad (1)$$

Our goal is to determine F_{wb}^c, the magnitude of the contact force exerted by the wind on the block, and we know from the definition of work that here $W = F_{wb}^c \Delta x_F$, where Δx_F is the displacement of the point of application of the force exerted by the wind. Because the wind exerts a force on the right face of the block, Δx_F is equal to the displacement Δx of the block (which, in turn, is equal to the displacement of the end of the spring). Thus, if we know each term on the left in Eq. 1, we can determine F_{wb}^c. There is no source energy for this system, and the initial and final speeds of the block are both zero, which makes $\Delta K = 0$. All we need to do is substitute into Eq. 1

the algebraic expressions for the spring's potential energy and the change in the amount of thermal energy dissipated by friction. We choose an x axis along the direction of motion of the block, with the positive direction pointing to the right in Figure WG9.9.

③ EXECUTE PLAN The potential energy of a spring displaced from its relaxed length is given by Eq. 9.23, $\Delta U_{spring}(x) = \frac{1}{2}k(x - x_0)^2$, and the energy dissipated by friction (Eq. 9.28) is the magnitude of the frictional force F_{db}^f exerted by the desk on the block multiplied by the distance traveled (which in this case is $|\Delta x|$). With x_0 representing the block's position with the spring relaxed, Eq. 1 becomes

$$0 + \tfrac{1}{2}k(x_f - x_0)^2 - \tfrac{1}{2}k(x_i - x_0)^2 + 0 + |F_{dbx}^f||\Delta x| = F_{wbx}^c\Delta x$$

$$\tfrac{1}{2}k(x_f - x_0)^2 - \tfrac{1}{2}k(x_i - x_0)^2 + |F_{dbx}^f||x_f - x_i| = F_{wbx}^c(x_f - x_i).$$

Now we substitute in values from Figure WG9.9:

$$\tfrac{1}{2}k(-2d - 0)^2 - \tfrac{1}{2}k(+d - 0)^2 + |F_{dbx}^f||-2d - (+d)|$$

$$= F_{wbx}^c[-2d - (+d)]$$

$$\tfrac{1}{2}k(4d^2) - \tfrac{1}{2}kd^2 + F_{db}^f(3d) = -F_{wb}^c(-3d)$$

$$\tfrac{3}{2}kd^2 + 3F_{db}^fd = 3F_{wb}^cd$$

$$F_{wb}^c = \tfrac{1}{2}kd + F_{sb}^f. \ ✔$$

④ EVALUATE RESULT We expect that as F_{sb}^f increases, the wind needs to blow harder in order to achieve the same final spring compression. Indeed, if F_{db}^f increases, F_{wb}^c increases, as predicted by our equation. As k decreases, less wind is needed to achieve the same compression, which is also reflected in our result.

The variation with d involves two opposing effects. As the distance d the spring is stretched increases, more initial elastic potential energy $(\frac{1}{2})(kd^2)$ is stored in the spring. However, an even greater amount $(\frac{1}{2})k(2d)^2 = 2kd^2$ must be stored in the maximally compressed spring. The larger d is, the greater the difference in spring potential energy that must be supplied by the external work done by the wind force. This is in agreement with our result.

Guided Problem 9.8 Darting about

A dart gun uses a compressed spring ($k = 2000$ N/m) to shoot a 0.035-kg dart horizontally at a target 1.0 m away. The spring is originally compressed 25 mm before the dart is released. (*a*) With what speed does the dart leave the gun? (*b*) If the dart penetrates the target to a distance of 10 mm, what average force is exerted on the dart to stop it?

① GETTING STARTED

1. How are the two questions related?
2. There are several approaches to solving this problem, some easier than others.
3. Choose an appropriate system and make an energy diagram.
4. Based on your chosen approach, which general equation is appropriate?

② DEVISE PLAN

5. Are there external forces exerted on your system? Should you worry about gravity in a horizontal flight over a short distance?

6. Check that you have enough equations for your unknowns.
7. How can you use the dart speed to determine the change in the dart's kinetic energy as it comes to a stop in the target?
8. What causes the dart to slow down? What formula relates the change in kinetic energy to the distance of penetration?

③ EXECUTE PLAN

④ EVALUATE RESULT

9. Examine the algebraic expression for the dart's speed. What happens as k gets bigger? Is this what you expect? What about when m (the inertia of the dart) gets bigger?
10. If the frictional force is the same at any depth, does the penetration depth you expect vary properly when making k or m bigger or smaller?
11. Is it reasonable to ignore the effect of gravity on the dart?

Questions and Problems

For instructor-assigned homework, go to MasteringPhysics®

Dots indicate difficulty level of problems: ● = *easy,* ●● = *intermediate,*
●●● = *hard;* CR = *context-rich problem.*

9.1 Force displacement

1. If an impulse is delivered to a system, is work necessarily done on the system? ●

2. Hitting a door with your bare fist hurts more than hitting a sofa cushion. In work terms, explain why this is so. ●

3. If you drop a brick from a height of 50 mm onto your toe, it probably won't hurt much, but if you drop the brick from a height of 0.5 m, it will hurt. The force of gravity exerted on the brick is the same in both cases. Explain why it hurts more from the increased height. ●●

4. An object moves with constant velocity. What can you say about the work done on a system that includes only this object? ●●

5. You hold a small steel ball in your hand above a laundry basket. Then you throw it straight up and watch as it rises, falls, and comes to rest in the pile of dirty laundry. Ignoring friction, discuss the forces exerted on the ball, the force displacements associated with those forces, and the work done by each force during this process. ●●

9.2 Positive and negative work

6. The *x* component of velocity of a particle as a function of time is shown in Figure P9.6. Over what intervals is the work done on the particle (*a*) positive, (*b*) negative, (*c*) zero? ●

Figure P9.6

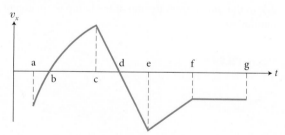

7. When you stand up from a seated position, you push down with your legs. Does this mean you do negative work when you stand up? ●

8. Mountain climbers know that it is harder to hike down a hill than to walk on level ground. Why is this so? ●●

9. As a car accelerates from rest to highway speed, a pedestrian watching the motion sees positive work done on the car because its kinetic energy increases. However, from the reference frame of someone driving in the same direction at highway speed, the car is initially moving (backward) and then becomes stationary, corresponding to negative work. Which observer is right? ●●

10. A system consists of two interacting particles 1 and 2, with particle 1 confined to an *x* axis and the system in equilibrium when particle 1 is at a given position on that axis. If displacing particle 1 in either direction from this position requires positive work to be done on the system by an external force, what is the shape of the potential energy curve (as a function of position) for the system in the region near equilibrium? ●●●

9.3 Energy diagrams

11. Does work done on a system necessarily change the system's kinetic energy? ●

12. Taking off on a runway, a small jet experiences a forward force of 90,000 N exerted by the engine and a backward force of 16,000 N exerted by the air. Choose a system and draw an energy diagram using the beginning and end of the runway as your initial and final positions. ●●

13. A block launched at speed *v* down an incline slides down the incline (there is friction between the block and incline) and into a spring (Figure P9.13), where it slows to zero speed as the spring compresses. (*a*) Draw an energy diagram for this process using a system consisting of the block, spring, Earth, and incline. If there is more than one object causing energy conversions, use a separate bar for each. (*b*) Repeat part *a* using a system of just the block and incline. (*c*) Why can't you use the block alone as your system? ●●

Figure P9.13

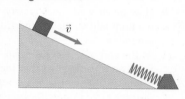

14. What, if anything, is wrong in the energy diagram in Figure P9.14 for a system comprising Earth, an inclined plane, and a block sliding along the incline? The block slides down the incline at constant speed. The only forces exerted on the block are the force of gravity and friction with the incline. ●●

Figure P9.14

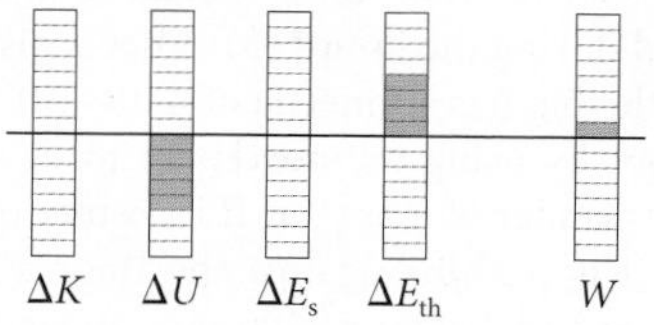

ΔK ΔU ΔE_s ΔE_{th} W

15. What possible scenario is described by the energy diagram in Figure P9.15? ●●●

Figure P9.15

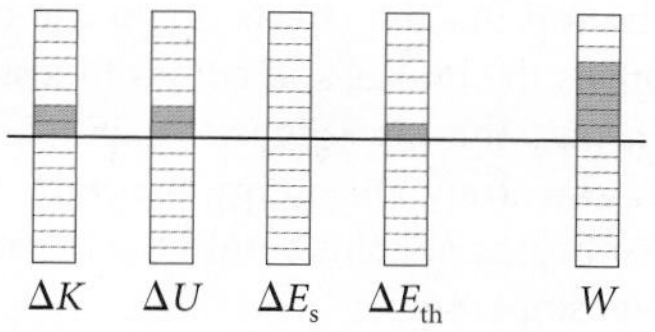

ΔK ΔU ΔE_s ΔE_{th} W

9.4 Choice of system

16. You are lifting a ball at constant velocity. (*a*) When the system is the ball, is work done on the system? If so, by what agent(s)? (*b*) Describe the potential energy of this system during the lift. (*c*) When the system comprises the ball and Earth, is work done on the system? If so, by what agent? (*d*) Describe how the potential energy of the ball-Earth system changes during the lift. ●

17. You throw a ball straight up into the air. If air resistance affects the motion of the ball, which takes longer: the upward trip or the downward trip? Analyze using three different systems. ●

PRACTICE

18. A tomato is launched vertically from a spring-loaded toy cannon placed on the pavement. The tomato rises to some maximum height and then falls back, but you have removed the cannon so that the tomato splats on the pavement. Discuss the energy conversions during this process using the following systems: (*a*) tomato, cannon, air, Earth, pavement; (*b*) tomato, cannon, air, pavement; (*c*) tomato, air, pavement; and (*d*) tomato. ••

19. Two identical balls collide elastically. If we treat each ball as a separate system, is the work done on each ball the same? ••

9.5 Work done on a single particle

20. How much work does gravity do on a 2.0-mg raindrop as the drop falls to the ground from a cloud 2000 m above the ground? •

21. At the end of a delivery ramp, a skid pad exerts a constant force on a package so that the package comes to rest in a distance *d*. When the pad is replaced by one that requires a distance of 2*d* to stop the same package, what happens to the length of time it takes for the package to stop? •

22. At the end of a delivery ramp, a skid pad exerts a constant force on a package so that the package comes to rest in a distance *d*. The ramp is changed so that the same package arrives at the skid pad at a higher speed and the stopping distance is 2*d*. What happens to the time interval required for the package to stop? •

23. A 55-kg acrobat must jump high and land on his brother's shoulders. To accomplish this, he leaps from a crouched position to a height where his center of mass is 1.20 m above the ground. His center of mass is 400 mm above the ground in the crouch and 900 mm above the ground when his feet leave the ground. (*a*) What is the average force exerted on him by the ground during the jump? (*b*) What is his maximum speed? ••

24. A beetle that has an inertia of 4.0×10^{-6} kg sits on the floor. It jumps by using its muscles to push against the floor and raise its center of mass. (*a*) If its center of mass rises 0.75 mm while it is pushing against the floor and then continues to travel up to a height of 300 mm above the floor, what is the magnitude of the force exerted by the floor on the beetle? (*b*) What is the beetle's acceleration during the motion? ••

25. A bicyclist pedals a distance *d* at constant speed on a level stretch of road. She then pedals up a hill and passes slowly over the top. As she coasts down the other side of the hill, she applies the brakes and comes to a stop when she is a distance *d* past the top (Figure P9.25). Choose an appropriate system and draw an energy diagram from the instant the bicyclist begins her climb until the instant she has stopped on the downslope. ••

Figure P9.25

26. In emergency braking, a certain car requires 7.0 m to come to a stop from an initial speed of 10 m/s. Use an argument based on work done on the car to determine the braking distance required if the car's initial speed is 30 m/s. ••

27. At instant t_0, you are sitting on a dock and you begin using a rope to lower a 4.0-kg lobster trap into the water, 1.4 m below. You lower the trap at a constant speed of 1.0 m/s. At instant t_1 it just reaches the water surface 1.4 m below the dock. You slacken the rope and the trap continues to move downward at a constant speed of 1.0 m/s until at instant t_2 the trap reaches the bottom of the 10-m-deep bay. Choose a system that includes the water, Earth, and the trap. (*a*) Do you do any work on the system in the interval from t_0 to t_2? (*b*) What forces are exerted on the trap during the interval from t_0 to t_2? (*c*) Draw an energy diagram for the interval from t_0 to t_2, showing a correct numerical scale. ••

28. A 1.1-kg lobster climbs into the trap of Problem 27, and at instant t_3 you begin to haul the trap up onto the dock. The trap moves at a constant 0.40 m/s through the water, and it breaks the surface of the water at instant t_4. You adjust your pull so that it continues moving at that speed until it reaches the height of the dock at instant t_5. Use a system of water, Earth, trap, and lobster. (*a*) Do you do any work on the system in the interval from t_3 to t_5? (*b*) What forces are exerted on the trap as it rises? (*c*) Draw an energy diagram for this system from instant t_3 to instant t_5. In your diagram, there are two kinds of energy you have to combine because you have no way to separate them from each other. Indicate this in your labels. ••

29. A 20-kg child runs up the amusement-park wave slide shown in Figure P9.29. When she is nearly at the top of the first hump, which is 0.90 m above the water, she flops down and slides over the top of the hump at 1.5 m/s. She slides down the other side of the hump and then up the slide, until she reverses motion at a height of 0.95 m above the water. What is the increase in thermal energy of the child and the slide due to the sliding from the top of the hump to the top of her motion? ••

Figure P9.29

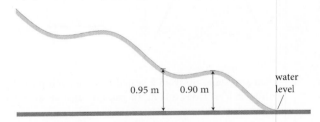

0.95 m 0.90 m

water level

30. In the block-and-tackle arrangement shown in Figure P9.30, three segments of the single rope pull on the block. (*a*) Show that the magnitude F_{pr}^c of the force exerted by a person on the rope to raise the block at constant speed is $mg/3$, where *m* is the inertia of the block. (*b*) One worker uses this arrangement to lift a heavy block, while another hauls up an identical block with a straight rope. After both blocks have been raised to the same second-floor window, has either worker done more work than the other? If so, which worker has done more work? ••

Figure P9.30

31. The pair of pulleys shown in Figure P9.31 is used to lift a 50-kg block 0.25 m. The constant acceleration of the block during the lift is +2.5 m/s², and the pulleys and rope have negligible inertia and negligible friction. What is the tension (*a*) in the pulling rope and (*b*) in the rope from which the block hangs? (*c*) What force is exerted by the upper mount on the top pulley? (*d*) How much work is done by the person pulling the rope in lifting the block 0.25 m? (*e*) By how much does the kinetic energy of the block change in this process? (*f*) How much work is done on the block by the rope attached to it? (*g*) How much work does the ceiling do in this process? (*h*) How much work does Earth do on the block? (*i*) Draw energy diagrams for three systems during the lifting: block; block and Earth; block, Earth, pulleys, rope, and person. ●●●

Figure P9.31

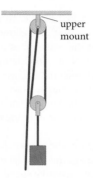

upper mount

9.6 Work done on a many-particle system

32. You push a blob of gelatin with a constant force of 3.0 N across a wet table on which it slides easily. Because the blob shape distorts, its center of mass moves only 30 mm during the time interval in which the point of application of your force moves 50 mm. (*a*) What is the work done by you on the blob? (*b*) By how much does your force change the kinetic energy of the center of mass? ●

33. A 1000-kg car traveling at 5.0 m/s runs into the side of an underpass. The concrete wall is not affected, but the car crumples such that in the time interval from the instant of impact to the instant at which the car comes to rest, its center of mass travels forward 0.50 m. What are (*a*) the magnitude of the average force exerted on the car, (*b*) the work done by the wall on the car, and (*c*) the change in kinetic energy of the car's center of mass? ●

34. A 60-kg ice skater stands facing a wall with his arms bent and then pushes away from the wall by straightening his arms. At the instant at which his fingers lose contact with the wall, his center of mass has moved 0.50 m, and at this instant he is traveling at 3.0 m/s. What are (*a*) the average force exerted by the wall on him, (*b*) the work done by the wall on him, and (*c*) the change in the kinetic energy of his center of mass? ●

35. Two 0.50-kg carts, one red and one green, sit about half a meter apart on a low-friction track. You push on the red one with a constant force of 2.0 N for 0.15 m and then remove your hand. The cart moves 0.35 m on the track and then strikes the green cart. (*a*) What is the work done by you on the two-cart system? (*b*) How far does the system's center of mass move while you are pushing the red cart? (*c*) By what amount does your force change the kinetic energy of the center of mass of the system? ●●

36. Two 0.50-kg carts are pushed toward each other from starting positions at either end of a 6.0-m low-friction track. Each cart is pushed with a force of 3.0 N, and that force is exerted for a distance of 1.0 m. (*a*) What is the work done on the two-cart system? (*b*) What is the change in kinetic energy of the system? (*c*) What is the kinetic energy of the center of mass of the system? ●●

37. You have a 2.0-m chain lying on the floor alongside ten 0.10-kg cubical blocks, each 0.20 m on a side. Each block is resting on the floor, and the inertia of the chain is 1.0 kg. Which process requires less work: lifting one end of the chain so that the chain hangs vertically with the bottom link just touching the floor, or stacking the blocks in a column ten blocks tall? ●●

38. A 1.0-kg cart and a 0.50-kg cart sit at different positions on a low-friction track. You push on the 1.0-kg cart with a constant 2.0-N force for 0.15 m. You then remove your hand, and the cart slides 0.35 m and strikes the 0.50-kg cart. (*a*) What is the work done by you on the two-cart system? (*b*) How far does the system's center of mass move while you are pushing the 1.0-kg cart? (*c*) By what amount does your force change the kinetic energy of the system's center of mass? ●●

39. Two 1.0-kg blocks, one gray and one tan, are lined up along a horizontal *x* axis. The gray block is at $x = -4.0$ m, and the tan one is at $x = +4.0$ m. A constant force of 1.0 N is exerted on the gray block in the positive *x* direction through a distance of 2.0 m. Taking the two blocks as the system and ignoring friction, calculate the following over the interval required for the gray block to move from $x = -4.0$ m to $x = -2.0$ m: (*a*) the work done on the system, (*b*) the change in the energy of the system, (*c*) the change in the kinetic energy of the system's center of mass, and (*d*) the value of $F\Delta x_{cm}$, where $F = 1.0$ N and Δx_{cm} is the displacement of the system's center of mass. (*e*) If the blocks collide totally inelastically, what is the energy of the system after the collision? ●●

40. Two identical 0.50-kg carts, each 0.10 m long, are at rest on a low-friction track and are connected by a spring that is initially at its relaxed length of 0.50 m and is of negligible inertia. You give the cart on the left a push to the right (that is, toward the other cart), exerting a constant 5.0-N force. You stop pushing at the instant when the cart has moved 0.40 m. At this instant, the relative velocity of the two carts is zero and the spring is compressed to a length of 0.30 m. A locking mechanism keeps the spring compressed, and the two carts continue moving to the right. (*a*) What is the work done by you on the two-cart system? (*b*) How far does the system's center of mass move while you are pushing the left cart? (*c*) By what amount do you change the system's kinetic energy? ●●

41. Because the soles of your shoes have cleats, you can exert a forward force of 100 N even on slippery ice. A 10-kg picnic cooler is at rest on a frozen pond, and you want to get it to shore, up the bank, and stuck in the snow. From past experience, you know that, in order to move up the bank and stick, the cooler needs to be moving at 3.0 m/s at the instant it starts up the bank. Standing some distance from the cooler and not wanting to walk to it, you decide to throw 1.0-kg snowballs at it to get it moving. If each snowball smashes into the cooler in a totally inelastic collision, what minimum number of snowballs must you throw? ●●●

PRACTICE

42. A bullet of inertia m traveling at speed v is fired into a wooden block that has inertia $4m$ and rests on a level surface. The bullet passes through the block and emerges with speed $v/3$, taking a negligible amount of the wood with it. The block moves to the right but comes to rest after traveling a distance d. (a) What is the magnitude of the frictional force between the block and the surface while the block is moving? (b) What is the ratio of the energy dissipated as the bullet passes through the block to the energy dissipated by friction between the surface and the bottom face of the block? •••

43. You fire a bullet of inertia m into the block shown in Figure P9.43. The bullet is initially traveling at speed v, and the inertia of the block is $4m$. The surface on which the block sits is rough, and the spring has spring constant k. The bullet becomes embedded in the block. If the spring compresses a maximum distance d, what is the magnitude of the frictional force exerted by the surface on the block while the block is moving? •••

Figure P9.43

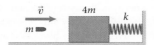

9.7 Variable and distributed forces

44. Spring B is stiffer than spring A. Which one has more energy stored in it if you (a) compress both springs with the same force and (b) compress both springs the same displacement from their relaxed lengths? •

45. Which arrangement in Figure P9.45 requires more work to give the block a rightward displacement $\vec{d}$? The springs, blocks, and surfaces are identical in the two arrangements and the springs are initially at their relaxed length. •

Figure P9.45

46. Estimate the work done by the force plotted in Figure P9.46 for an object displaced from $x = 1.0$ m to $x = 3.0$ m. •

Figure P9.46

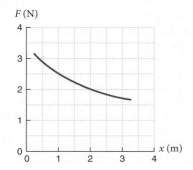

47. Stretching a certain spring 0.10 m from its relaxed length requires 18 J of work. How much more work does it take to stretch this spring an additional 0.10 m? •

48. A force varies with time according to the expression $F = a\Delta t$, where $a = 2.0$ N/s. From this information, can you determine the work done on a particle that experienced this force over a displacement of 0.50 m? ••

49. In pushing a 0.024-kg dart into a toy dart gun, you have to exert an increasing force that tops out at 6.0 N when the spring is compressed to a maximum value of 0.12 m. (a) What is the launch speed of the dart when fired horizontally? (b) Does your answer change if the dart is fired vertically? ••

50. The work done by a certain force is given by $W(\Delta x) = a\Delta x + b(\Delta x)^3$. Write an expression for the force as a function of Δx. ••

51. A 6.0-kg bowling ball is held 10 mm above a mattress and then released from rest and allowed to fall, sinking into the mattress. Model the mattress as a single spring with spring constant $k = 500$ N/m. (a) Calculate the ball's kinetic energy, the gravitational potential energy of the system (ball, spring, and Earth), and the elastic potential energy at the following compressions: 0, 0.050 m, 0.10 m, 0.15 m, 0.20 m. (b) What is the maximum compression of the mattress? Ignore the inertia of the spring. ••

52. In a carnival game, the player throws a ball at a haystack. For a typical throw, the ball leaves the hay with a speed exactly one-half of the entry speed. (a) If the frictional force exerted by the hay is a constant 6.0 N and the haystack is 1.2 m thick, derive an expression for the typical entry speed as a function of the inertia of the ball. Assume horizontal motion only, and ignore any effects due to gravity. (b) What is the typical entry speed if the ball has an inertia of a 0.50 kg? ••

53. The force exerted on a certain object varies with the object's position according to the function $F_x(x) = ax^2 + bx^3$ where $a = 3.0$ N/m² and $b = -0.50$ N/m³. What is the work done on the object by this force as the object moves from $x = -0.40$ m to $x = 2.0$ m? ••

54. You devise a wind-up car powered by a spring for trips to the grocery store. The car has an inertia of 500 kg and is 4.2 m long. It should be able to accelerate from rest to 20 m/s at least 50 times before the spring needs winding. The spring runs the length of the car, and a full winding compresses it to half this length. In order to meet the acceleration requirement, what must the value of the spring constant be? ••

55. A 0.15-kg cart on wheels is at rest at the origin of an x axis between (but not connected to) two relaxed springs aligned along that axis. When the cart is pushed to the left of the origin, the left spring ($k_l = 4.0$ N/m) is compressed and exerts a force on the cart. When the cart is pushed to the right of the origin, the right spring ($k_r = 6.0$ N/m) is compressed and exerts a force on the cart. (a) Plot the force exerted on the cart as a function of its displacement. (b) If the cart is given an initial speed of 4.0 m/s at the origin $x = 0$, it might be headed to the left or to the right. How far does it go in each case? (c) What is the work done on the cart as it moves from $x = 0$ to its extreme right position and from $x = 0$ to its extreme left position? Ignore any friction effects. •••

56. To complement the baseball-pitching machines at the sports complex where you work, you have designed a catching machine using a mitt, a spring, and a latch that locks the spring in place when the ball has stopped moving. Your boss wants to know the speed of the baseballs coming out of a pitching machine that's been breaking down lately, and you figure you can use the compression of the spring in your catching machine to determine this speed. The mitt that catches the balls has three times the inertia of one ball. Derive an expression for the speed of a thrown ball in terms of the ball's inertia, the distance the spring is compressed, and the spring constant. Take the catch to be a totally inelastic collision between mitt and ball. ●●●

9.8 Power

57. Suppose a jogger has to exert a force of 25 N against air resistance to maintain a velocity of +5.0 m/s. At what rate is the jogger expending energy? ●

58. A 35-kg girl climbs a 10-m rope in 25 s. What is her average power? ●

59. Hiking trails on steep slopes often zigzag back and forth rather than running in a straight path up the slope. What does having a zigzag path accomplish, given that getting to the top of the slope requires the same amount of energy regardless of the path? ●

60. Your 1000-kg car, moving at 7.0 m/s, approaches the bottom of a hill that is 20 m high (Figure P9.60). To save gas, you use on average only 3.3 kW of engine power, realizing that half of the energy delivered by the engine and half of the initial kinetic energy will be dissipated. If you calculated correctly, your car will just barely make it over the hill. What time interval is required for your car to travel from the bottom to the top of the hill? ●●

Figure P9.60

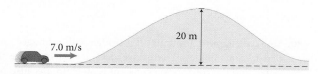

61. A constant, nondissipative external force is exerted on a particle. Is the same work done on the particle each second? ●●

62. A team of dogs accelerates a 200-kg dogsled from 0 to 5.0 m/s in 3.0 s. (a) What is the magnitude of the force exerted by the dogs on the sled? (b) What is the work done by the dogs on the sled in the 3.0 s? (c) What is the instantaneous power of the dogs at the end of the 3.0 s? (d) What is their instantaneous power at 1.5 s? Assume that the acceleration is constant. ●●

63. A 20-kg child wants to slide along a horizontal yard toy that consists of a slick plastic sheet covered with a thin layer of water to reduce friction to almost zero. To cause him to slide, you squirt him with a stream of water from a hose. (a) If the water from the hose exerts a constant force of 5.0 N on the child, what is the work done by the water in the interval from $t = 0$ to $t = 1.0$ s? (b) What is the instantaneous power delivered by the water at $t = 1.0$ s? ●●

64. A cog system on the beginning segment of a roller coaster needs to get 25 occupied cars up a 100-m vertical rise over a time interval of 60 s. Each car experiences a gravitational force of 5670 N. The cars start at rest and end up moving at 0.50 m/s. (a) How much work is done by the cog system on the cars? (b) What average power must the cog system supply? (c) What would happen to the amount of work done by the cog system if the cars had to be lifted the same distance in 30 s? ●●

65. A box slides across a frozen pond toward the left shore as two children standing on opposite shores pull on the box with ropes. The child on the left shore pulls with a force of 3.0 N, and the child on the right shore pulls with a force of 2.0 N. At a certain instant, the speed of the box is 3.0 m/s. (a) At that instant, what is the power resulting from each force? (b) What is the power resulting from both forces? (c) Do your answers to parts a and b change as the box slides? ●●

66. A motor must lift a 1000-kg elevator cab. The cab's maximum occupant capacity is 400 kg, and its constant "cruising" speed is 1.5 m/s. The design criterion is that the cab must achieve this speed within 2.0 s at constant acceleration, beginning from rest. (a) When the cab is carrying its maximum capacity, at what rate must the motor deliver energy to get the cab up to cruising speed? (b) At what constant rate must the motor supply energy as the fully loaded cab rises after attaining cruising speed? ●●●

Additional Problems

67. An elevator operated by an electric motor rises at a constant speed. What is the work done on the elevator as it rises a distance h? ●

68. Is it possible for a force to be exerted on an object *along the direction of the object's motion* and still produce no change in the object's kinetic energy? If yes, give an example. If no, explain why not. ●

69. A shopping cart that has an inertia of 12 kg when empty is loaded with 38 kg of groceries. A child pushing the loaded cart loses control, and the cart rolls into a concrete lamppost, which sustains no damage. The cart is moving at 2.0 m/s when it strikes the lamppost, and the groceries all crash forward on impact, moving the center of mass forward by 0.10 m before everything comes to a halt. What are (a) the average force exerted by the post on the cart-groceries system, (b) the work done by the post on the system, and (c) the change in kinetic energy of the system's center of mass? ●

70. Is it possible for the kinetic energy of an object to remain constant even when the vector sum of the forces exerted on the object is not zero? If yes, give an example. If no, explain why not. ●●

71. To be rescued from the roof of a burning building, a 70-kg man must be able to jump straight up and catch hold of a rope ladder hanging from a helicopter. He has to raise his center of mass 2.00 m above the roof to do so. (a) If he crouches down such that his center of mass is 0.50 m above the roof and his feet leave the roof when his center of mass is 1.05 m above the roof, what minimum force must he exert on the roof to reach the ladder? (b) A typical candy bar contains about 850 kJ of food energy. How many candy bars does this jump require? ●●

72. About how much work does a trash collector do in the course of an 8-hour shift? ••

73. A constant, nondissipative external force is applied to a solid block, and it is found that the rate at which energy is delivered by the force is constant as the block moves horizontally. What does this imply for the motion of the block? ••

74. An object is suspended from the ceiling by a spring aligned along a vertical x axis, with upward defined as the positive x direction. When the spring is at its relaxed length, the object is at the origin of the axis. The object is pulled down to position $-x_1$, held at rest, and released. Consider its motion from the instant of release until the instant it returns to $-x_1$. Ignore friction, air drag, and the inertia of the spring. (a) What force(s) is exerted on the object during the round trip, and in what direction(s)? (b) How much work is done on the object by the spring as the object travels away from $-x_1$ and back to $-x_1$? ••

75. You are midway up hill A in Figure P9.75. Across the valley is hill B, the summit of which is slightly higher than your present spot. You want to roll a ball so that it crests hill B. To get the ball moving, you can roll it either up or down hill A, giving the ball the same initial speed in both cases. (a) Which roll direction is more likely to get the ball to crest B, or does it matter? Ignore any energy dissipation. (b) In what way would you change your answer if you had to take energy dissipation into account? ••

Figure P9.75

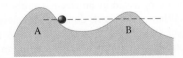

76. A 10-kg dog jumps up in the air to catch a ball. The dog's center of mass is normally 0.20 m above the ground, and he is 0.50 m long. The lowest he can get his center of mass is 0.10 m above the ground, and the highest he can get it before he can no longer push against the ground is 0.30 m. If the maximum force the dog can exert on the ground in pushing off is 2.5 times the gravitational force Earth exerts on him, how high can he jump? ••

77. Assume that when you stretch your torso vertically as much as you can, your center of mass is 1.0 m above the floor. The maximum force you can exert on the floor in pushing off is 2.3 times the gravitational force Earth exerts on you. How low do you have to crouch in order to jump straight up and have your center of mass be 2.0 m above the floor? Is this crouch practical? ••

78. The 2000-kg cab of a freight elevator is rated to carry a maximum load of 1200 kg, but a careless worker fills the cab with freight that has a combined inertia of 1400 kg. The worker then raises the cab 3.1 m at a steady 2.0 m/s. You notice an unfamiliar whine and come over to investigate. As supervisor, you know that the motor has a 10% tolerance in its rated power output, and you must decide whether there's going to be trouble. ••• CR

79. Taking a sip of coffee one day, you start thinking about the effort it takes to do so. Your forearm is a type of lever. As shown in Figure P9.79, the biceps muscle is attached to the forearm bone about 50 mm in front of the elbow joint. The cup of coffee in your hand is 350 mm from the elbow, and you wonder about the work done by your biceps on your forearm as you raise the cup very slowly. You realize that you can use energy principles to estimate the forces involved. ••• CR

Figure P9.79

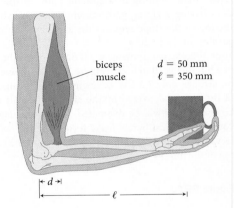

biceps muscle

$d = 50$ mm
$\ell = 350$ mm

80. You are planning to jump off a bridge with a bungee cord tied to your ankles. The bridge deck is 150 m above the water, and the cord spring constant is 40 N/m. Because you must fall the length of the unstretched cord before it begins to stretch, you realize that the unstretched length of the bungee cord, which is adjustable, has to be adjusted based on your inertia. You enjoy thrills and want to reverse your motion just above the water surface. ••• CR

Answers to Review Questions

1. Work is the change in the energy of a system due to external forces.

2. No. In order for work to be done on a system, an external force must be exerted on the system and *the point of application of that force must move.* If your force is exerted on a concrete wall, for example, there may be no observable displacement, which means you do no work on the wall.

3. (*a*) Negative, because the wind force points in one direction (toward the west) and the force displacement (picture it as the point where the wind "touches" the front of the bicycle) is in the opposite direction. (*b*) Positive, because the direction of the force displacement is the same as the direction of the force.

4. None, because the negative work done by gravity on the ball as it rises is canceled by the positive work done by gravity on the ball as it falls back into your hand.

5. Yes, as when you step on a spring scale. As you compress the spring, it exerts an upward force on you. You are moving downward, though, and so the point of application of the force exerted by the spring on you is displaced downward.

6. Mechanical energy is kinetic and potential energy; there must be no changes in thermal or source energy for this system. $\Delta K = K_f - K_i = +3.0\ \text{J} - (+10.0\ \text{J}) = -7.0\ \text{J}.\ \Delta U = U_f - U_i = +6.0\ \text{J} - (+4.0\ \text{J}) = +2.0\ \text{J}.$ The sum of these changes must equal the work done, which allows us to draw the work bar on the system's energy diagram:

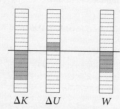

$\Delta K \quad \Delta U \quad W$

7. With the cart as the system, both your hand and the spring exert external forces during the motion, and both involve nonzero force displacements. Because the cart returns at the same speed, no thermal energy is dissipated along the track or at the spring. Because your hand is outside the system, there is no source energy to consider, and because the spring is outside the system, there is no potential energy to consider. Thus only kinetic energy and work energy bars must be drawn. No numbers are given, so the resulting diagrams cannot be precisely scaled, but they must look like this:

(*a*)   (*b*)

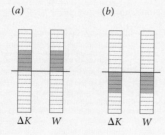

$\Delta K \quad W \qquad \Delta K \quad W$

(*c*)   (*d*)

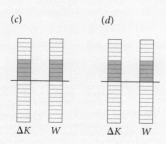

$\Delta K \quad W \qquad \Delta K \quad W$

8. First, the point of application of a frictional force is not well defined but involves contact across many points on two surfaces. This makes it impossible to determine the force displacement. Second, the work done by friction changes the thermal energy of not just the system but also the other object in the interaction pair that is outside the system; you do not generally know how to compute the exact split in this thermal energy, so it is not possible to attribute a specific thermal energy change to the system.

9. Never with both. Which form you use depends on your choice of system. If both parts of the interaction pair (Earth and an object) are included in your system, you must use potential energy rather than work. If only half of the interaction pair is in your system, then you must use work rather than potential energy.

10. Using more than one viewpoint gives us a better understanding of what energy transfers and conversions take place. Also, comparing the different results provides a consistency check.

11. The work done on a system, which is the product of the external force exerted on the system and the displacement of its point of application, is equal to the change in the system's energy.

12. The energy law applies to any system. In the case of a closed system, there are no external forces to do work on the system, so the energy of the system remains constant ($\Delta E = 0$). For a system that is not closed, external forces can do work on the system, so the energy of the system may change by the amount of this work ($\Delta E = W$). In either case the energy law ($\Delta E = W$) applies.

13. In deriving this equation, it was necessary to assume that the only possible change in energy was a change in kinetic energy: $\Delta E = \Delta K$. This is not true in general, but it is true for a particle (which has no internal structure and hence cannot have internal energy changes).

14. Both laws express conservation principles. Neither energy nor momentum can be created or destroyed, but either can be transferred into or out of a system. The momentum law states that any change in the momentum of a system must be due to such a transfer, called impulse. The energy law states that any change in the energy of a system must be due to such a transfer, called work. For an isolated system, no momentum transfer is allowed, while for a closed system, no energy transfer is allowed. The laws are thus similar. The momentum law, though, involves vector values. The equation requires careful attention to vector components of momentum and impulse. The energy law involves scalars; there are no components of energy or work to account for, even though computation of work or energy might require attention to components of vectors during some of the intermediate steps.

15. As long as the internal energy changes are negligible, we can think of the crate as a particle. If there is substantial internal energy change (in a collision, for example), the crate cannot be treated as a particle.

16. The energy law has the same form, applied in the same way in both cases: $\Delta E = W$. The work equation has different forms in the two cases. For the single particle, there is only one displacement possible, so all external forces exerted on the particle have the same force displacement, and the work equation is the vector sum of forces multiplied by this common displacement: $W = (\sum F_x)\Delta x$. For a many-particle system, different particles may experience different external forces and different force displacements, so the work equation must keep track of each separate force and displacement:

$$W = \sum_{n} (F_{\text{ext }nx}\Delta x_{Fn}).$$

17. Because dissipative forces generally dissipate energy on both sides of the system boundary in such a way that the precise amount of energy transferred into or out of the system is not calculable.

18. Because energy may appear in forms other than kinetic, the work done on a many-particle system is not necessarily equal to the change in the system's kinetic energy. The work done is the sum of the external forces times the displacement of the point of application of each force (Eq. 9.18), and the change in kinetic energy is the sum of the external forces times the displacement of the center of mass of the system (Eq. 9.14).

19. The area under the curve is the magnitude of the work done on the particle.

20. (a) The factor $\frac{1}{2}$ comes from the integration formula $\int x^m \, dx = x^{m+1}/(m + 1)$ with $m = 1$. (b) Because the force exerted by the spring is a function of how much the spring is stretched or compressed, the same must be true of the force exerted on the spring. Because $F_{\text{on spring}}$ varies, the simple formula $W = F(x - x_0)$ cannot be used. However, the factor $\frac{1}{2}$ in the expression for $W_{\text{on spring}}$ is reasonable in the sense that we expect the average force to be $(\frac{1}{2})k(x - x_0)$, and so $W_{\text{on spring}}$ is $(\frac{1}{2})k(x - x_0)(x - x_0)$.

21. No. The frictional force dissipates energy in both the crate and the floor. Equation 9.28 was derived for a block-surface system, meaning that it contains both the energy change experienced by the block and the energy change experienced by the surface over which the block moves. We do not know how much of this energy to attribute to the block alone, so we cannot in good conscience call this the "work done by the frictional force on the block." We cannot even call it the "work done by the frictional force on the block-surface system" because the frictional force is not external to this system.

22. Instantaneous power is power delivered at any given instant, and average power is power delivered over a time interval Δt. Instantaneous power is average power in the limit that the energy change occurs over a time interval (Δt) that approaches zero. Average power is a ratio of energy change to time interval, while instantaneous power is a derivative of energy with respect to time.

23. When $\sum F_{\text{ext}}$ is constant.

Answers to Guided Problems

Guided Problem 9.2 (a) 7.8×10^3 J; (b) -7.8×10^3 J; (c) 1.3×10^2 W

Guided Problem 9.4 2.5×10^2 N/m

Guided Problem 9.6 $v_{\text{bullet}} = \dfrac{\sqrt{k(m_{\text{block}} + m_{\text{bullet}})}}{m_{\text{bullet}}} d$

Guided Problem 9.8 (a) 6.0 m/s; (b) 63 N

10

PRACTICE
Motion in a Plane

Chapter Summary   166

Review Questions   168

Developing a Feel   169

Worked and Guided Problems   170

Questions and Problems   179

Answers to Review Questions   187

Answers to Guided Problems   188

PRACTICE

Chapter Summary

Vectors in two dimensions (Sections 10.1, 10.2, 10.6, 10.9)

Concepts To add two vectors $\vec{A}$ and $\vec{B}$, place the tail of $\vec{B}$ at the head of $\vec{A}$. Their sum is the vector drawn from the tail of $\vec{A}$ to the head of $\vec{B}$.

To subtract vector $\vec{B}$ from vector $\vec{A}$, reverse the direction of $\vec{B}$ and then add the reversed $\vec{B}$ to $\vec{A}$.

Any vector $\vec{A}$ can be written as

$$\vec{A} = A_x\hat{\imath} + A_y\hat{\jmath}, \qquad (10.5)$$

where A_x and A_y are the *components* of $\vec{A}$ along the x and y axes. The x component of $\vec{A}$ is the projection of $\vec{A}$ on the x axis, and the y component is the projection of $\vec{A}$ on the y axis. The components A_x and A_y are signed numbers and can be negative.

Quantitative Tools The magnitude of any vector $\vec{A}$ is

$$A \equiv |\vec{A}| = +\sqrt{A_x^2 + A_y^2} \qquad (10.6)$$

and the angle θ that $\vec{A}$ makes with the positive x axis is given by

$$\tan\theta = \frac{A_y}{A_x}. \qquad (10.7)$$

If $\vec{R} = \vec{A} + \vec{B}$ is the vector sum of $\vec{A}$ and $\vec{B}$, the components of $\vec{R}$ are

$$R_x = A_x + B_x$$
$$R_y = A_y + B_y. \qquad (10.9)$$

The **scalar product** of two vectors $\vec{A}$ and $\vec{B}$ that make an angle ϕ when placed tail to tail is

$$\vec{A} \cdot \vec{B} \equiv AB\cos\phi. \qquad (10.33)$$

Projectile motion in two dimensions (Section 10.7)

Concepts For a projectile that moves near Earth's surface and experiences only the force of gravity, the acceleration has magnitude g and is directed downward. The projectile's horizontal acceleration is zero, and the horizontal component of its velocity remains constant.

At the highest point in the trajectory of a projectile, the vertical velocity component v_y is zero, but the vertical component of the acceleration is g and is directed downward.

Quantitative Tools If a projectile is at a position (x, y), its **position vector** $\vec{r}$ is

$$\vec{r} = x\hat{\imath} + y\hat{\jmath}. \qquad (10.10)$$

If a projectile has position components x_i and y_i and velocity components $v_{x,i}$ and $v_{y,i}$ at one instant, then its acceleration, velocity, and position components a time interval Δt later are

$$a_x = 0$$
$$a_y = -g$$
$$v_{x,f} = v_{x,i} \qquad (10.17)$$
$$v_{y,f} = v_{y,i} - g\Delta t \qquad (10.18)$$
$$x_f = x_i + v_{x,i}\Delta t \qquad (10.19)$$
$$y_f = y_i + v_{y,i}\Delta t - \tfrac{1}{2}g(\Delta t)^2 \qquad (10.20)$$

Collisions and momentum in two dimensions (Section 10.8)

Concepts Momentum is a vector, so in two dimensions changes in momentum must be accounted for by components. This means two equations, one for x and one for y components of momentum. The coefficient of restitution is a scalar and is accounted for by a single equation.

Quantitative Tools

$$\Delta p_x = \Delta p_{1x} + \Delta p_{2x} = m_1(v_{1x,f} - v_{1x,i}) + m_2(v_{2x,f} - v_{2x,i}) = 0 \qquad (10.21)$$
$$\Delta p_y = \Delta p_{1y} + \Delta p_{2y} = m_1(v_{1y,f} - v_{1y,i}) + m_2(v_{2y,f} - v_{2y,i}) = 0. \qquad (10.22)$$

Forces in two dimensions (Sections 10.2, 10.3)

Concepts In two-dimensional motion, the component of the acceleration parallel to the instantaneous velocity changes the speed and the component of the acceleration perpendicular to the instantaneous velocity changes the direction of the instantaneous velocity.

When choosing a coordinate system for a problem dealing with an accelerating object, if possible make one of the axes lie along the direction of the acceleration.

Friction (Sections 10.4, 10.10)

Concepts When two surfaces touch each other, the component of the contact force perpendicular (normal) to the surfaces is called the **normal force** and the component parallel (tangential) to the surfaces is called the **force of friction.**

The direction of the force of friction is such that the force opposes *relative* motion between the two surfaces.

When the surfaces are not moving relative to each other, we have **static friction.** When they are moving relative to each other, we have **kinetic friction.**

The magnitude of the force of kinetic friction is independent of the contact area and independent of the relative speeds of the two surfaces.

Quantitative Tools The maximum magnitude of the force of static friction between any two surfaces 1 and 2 is proportional to the normal force:

$$(F^s_{12})_{\max} = \mu_s F^n_{12}, \tag{10.46}$$

where μ_s is the unitless **coefficient of static friction.** This upper limit means that the magnitude of the frictional force must obey the condition

$$F^s_{12} \leq \mu_s F^n_{12}. \tag{10.54}$$

The magnitude of the force of kinetic friction is also proportional to the normal force:

$$F^k_{12} = \mu_k F^n_{12}, \tag{10.55}$$

where $\mu_k \leq \mu_s$ is the unitless **coefficient of kinetic friction.**

Work (Sections 10.5, 10.9)

Concepts For a sliding object, the normal force does no work because this force is perpendicular to the direction of the object's displacement.

The force of kinetic friction is a nonelastic force and thus causes energy dissipation.

The force of static friction is an elastic force and so causes no energy dissipation.

The work done by gravity is independent of path.

Quantitative Tools The work done by a constant nondissipative force when the point of application of the force undergoes a displacement $\Delta \vec{r}_F$ is

$$W = \vec{F} \cdot \Delta \vec{r}_F. \tag{10.35}$$

The work done by a variable nondissipative force when the point of application of the force undergoes a displacement is

$$W = \int_{\vec{r}_i}^{\vec{r}_f} \vec{F}(\vec{r}) \cdot d\vec{r}. \tag{10.44}$$

This is the line integral of the force over the path traced out by the point of application of the force.

For a variable dissipative force, the change in the thermal energy is

$$\Delta E_{\text{th}} = -\int_{\vec{r}_i}^{\vec{r}_f} \vec{F}(\vec{r}_{\text{cm}}) \cdot d\vec{r}_{\text{cm}}. \tag{10.45}$$

When an object descends a vertical distance h, no matter what path it follows, the work gravity does on it is

$$W = mgh. \tag{10.40}$$

PRACTICE

Review Questions

Answers to these questions can be found at the end of this chapter.

10.1 *Straight* is a relative term

1. A plane flying east drops a package (without a parachute) to workers at a logging camp. What does the path of the package look like to a logger standing on the ground? What does the path look like to the pilot? Where is the plane relative to the logger when the package lands at his feet: east of where he stands, west of where he stands, or directly overhead?

2. A passenger in a speeding train drops a peanut. Which is greater: the magnitude of the acceleration of the peanut as measured by the passenger, or the magnitude of the acceleration of the peanut as measured by a person standing next to the track?

10.2 Vectors in a plane

3. Why does describing motion that is not along a straight line require two reference axes?

4. How does adding two vectors that lie in a plane and are neither parallel nor antiparallel to each other differ from adding two vectors that lie along a straight line?

5. Describe how to subtract a vector 1 from a vector 2 when the two do not lie along a straight line.

6. Is vector addition commutative? Is vector subtraction commutative?

7. Describe the procedures for adding and subtracting two vectors.

8. What does it mean when the instantaneous velocity of an object at instant t_1 is not parallel to the average velocity during a short time interval $\Delta t = t_2 - t_1$?

9. Describe the effect of the parallel and perpendicular components of the acceleration of an object on the object's velocity.

10.3 Decomposition of forces

10. What is the meaning of "normal" and "tangential" components of a contact force?

11. Sketch a free-body diagram for a block resting on an inclined surface using (a) vector forces and (b) normal (y) and tangential (x) force components.

12. When choosing a rectangular coordinate system for an object moving in two dimensions and being accelerated during that motion, why should you choose a system in which one axis lies along the direction of the acceleration?

10.4 Friction

13. Compare and contrast the normal force with the force of static friction and the force of kinetic friction.

14. You push horizontally on a crate at rest on the floor, gently at first and then with increasing force until you cannot push harder. The crate does not move. What happens to the force of static friction between crate and floor during this process?

15. You are trying to push a heavy desk, and it's not budging. A friend remarks, "You're not pushing it hard enough to overcome its inertia." What's wrong with that statement, and what's really going on?

10.5 Work and friction

16. Why is it all right to choose a system for which static friction occurs at the system boundary but a bad idea to do so for kinetic friction?

17. Describe one case in which a force of static friction does no work and one case in which a force of static friction does work. Make your examples different from the ones given in the *Principles* volume.

18. Explain what is wrong with this statement: The force of friction always opposes motion.

10.6 Vector algebra

19. In polar coordinates, can r be negative? If not, why not?

20. Suppose the angle between a vector $\vec{B}$ and the y axis of a rectangular coordinate system is θ. Write an expression for B_x in terms of the magnitudes of B and θ.

21. How are the component vectors $\vec{A}_x$ and $\vec{A}_y$ of vector $\vec{A}$ and the x and y components A_x and A_y of $\vec{A}$ related?

10.7 Projectile motion in two dimensions

22. A baseball player hits a fly ball that has an initial velocity for which the horizontal component is 30 m/s and the vertical component is 40 m/s. What is the speed of the ball at the highest point of its flight?

23. What is the shape of the path of an object launched at an angle to the vertical, assuming that only the force of gravity is exerted on the object?

24. Does the maximum height achieved by a projectile depend on both the x and y components of its launch velocity? Does the horizontal range depend on both components?

10.8 Collisions and momentum in two dimensions

25. How do the principles and equations describing collisions in two dimensions differ from the principles and equations describing collisions in one dimension?

26. Why is it possible to completely determine the outcome of a one-dimensional collision of two objects given only the inertias and initial velocities of each object and the coefficient of restitution, but it is not possible to do this in two dimensions?

10.9 Work as the product of two vectors

27. Under what circumstances could the scalar product of two vectors be zero?

28. Describe how to calculate the work done by a constant nondissipative force as an object moves along a path from position 1 to position 2, and identify which quantities in your statement are scalars and which are vectors.

29. Describe how to compute the work done by a variable nondissipative force as an object moves from position 1 to position 2.

10.10 Coefficients of friction

30. Discuss the difference between contact area and effective contact area, and explain why this distinction is relevant for frictional forces exerted by two surfaces on each other.

31. A fellow student tells you that, according to a measurement he did, the concrete surface of the road has a coefficient of kinetic friction of 0.77. What's wrong with this statement?

32. Often a wedge-shaped doorstop won't hold a door open unless you kick it until it sits tightly under the door. What does forcing the wedge into a tight fit accomplish?

33. In a panic situation, many drivers make the mistake of locking their brakes and skidding to a stop rather than applying the brakes gently. A skidding car often takes longer to stop. Why?

Developing a Feel

Make an order-of-magnitude estimate of each of the following quantities. Letters in parentheses refer to hints below. Use them as needed to guide your thinking.

1. The sum of the displacement vector from New York to Cheyenne, WY and the displacement vector from Cheyenne, WY to Los Angeles (O, D, I)
2. The work done in one day by a moving sidewalk conveyor in a major airport (AA, H, T, W)
3. The maximum tension you can generate in a rope tied between two trees by pushing steadily against the rope midpoint (CC, Z, E, T, P)
4. The stopping distance for a police car in hot pursuit on the freeway under dry conditions (E, J, L, S)
5. The maximum range of a rifle bullet fired horizontally over level ground from a standing position (G, M, Q)
6. The maximum horizontal range of a thrown baseball (C, R)
7. The average force of static friction exerted on a person by a moving sidewalk conveyor in a major airport (AA, H, T, BB, A)
8. The maximum angle of inclination for which your physics book will not slip down a tilted wooden table (U, Y)
9. The maximum distance a hockey puck can slide on smooth horizontal ice (N, B, F, K, V, X)

Hints

A. How is the work done by constant external forces related to the force displacement?
B. What is the inertia of a hockey puck?
C. What is the average speed of a major league fastball?
D. What does the sum of these displacements represent?
E. What is a reasonable estimate of the coefficient of static friction?
F. What is the inertia of a hockey stick plus the inertia of a player's upper body?
G. What is the initial height of the rifle bullet above the ground?
H. What are the speeds of a person walking and a person walking on a moving sidewalk?
I. What is the displacement vector from Cheyenne to Los Angeles?
J. What is a hot-pursuit freeway speed?
K. At what speed is the hockey stick being swung just before it hits the puck?
L. What is the magnitude of the acceleration produced by the vector sum of the forces exerted on the car when the brakes are applied?
M. For how long is the bullet in the air?
N. How could you determine the launch speed of the puck?
O. How do you apply the graphical method to obtain the sum of two vectors?
P. About how far can you displace the center of the rope by pushing?
Q. What is the bullet's speed as it exits the rifle barrel?
R. How should the x and y components of the launch velocity compare with each other?
S. Given the average (constant) acceleration and the car's initial speed, how can you get the car's displacement from the instant the brakes are applied to the instant the car stops?
T. What is the inertia of a person?
U. What is the coefficient of static friction for a book cover on wood?
V. What is the coefficient of kinetic friction between puck and ice?

W. How many people ride a moving sidewalk in a major airport in one day?
X. How does the puck's displacement relate to the other information you know?
Y. How is the magnitude of the force of static friction related to the magnitude of the normal force?
Z. What is the most useful free-body diagram?
AA. Ignoring dissipation, how is work related to kinetic energy change for an object subject to an external force?
BB. What is the length of a moving sidewalk in an airport?
CC. What is a reasonable length for a rope tied between two trees?

Key (all values approximate)

A. Eq. 9.9, $W = \Sigma F_x \Delta x_F$; B. 0.2 kg; C. 4×10^1 m/s; D. the displacement from New York to Los Angeles; E. less than 1.0 for most situations, but use 1.0 as a reasonable estimate; F. 2×10^1 kg; G. less than 2 m; H. walking, 1 m/s, walking on moving sidewalk, 2 m/s; I. 2×10^3 km west southwest; J. 5×10^1 m/s; K. 2×10^1 m/s; L. use a free-body diagram to show that $a \approx g$; M. 0.6 s, the time interval it takes to fall from the initial height; N. treat stick-puck hit as an elastic collision; O. place the relevant vectors head to tail on a scale diagram; the vector sum is a vector from the tail of the first vector to the head of the last; P. 1 m; Q. 5×10^2 m/s; R. they should be equal for maximum range; S. Eq. 1 in *Principles* Example 3.4: $\Delta x = (v_{x,f}^2 - v_{x,i}^2)/2a_x$; T. 7×10^1 kg; U. similar to wood on wood, about 0.5; V. 0.1; W. 7×10^3; X. consider the relationship between the initial kinetic energy of the puck and the energy dissipated by friction; Y. Eq. 10.54, $F_{12}^s \leq \mu_s F_{12}^n$; Z. free-body diagram of a small piece of rope at its midpoint; AA. $W = \Delta K$; BB. 4×10^1 m; CC. 3×10^1 m because longer ropes cannot be pulled tight enough that it doesn't sag much under its own weight

Worked and Guided Problems

Procedure: Working with frictional forces

1. Draw a free-body diagram for the object of interest. Choose your x axis parallel to the surface and the y axis perpendicular to it, then decompose your forces along these axes. Indicate the acceleration of the object.
2. The equation of motion in the y direction allows you to determine the sum of the y components of the forces in that direction:

$$\Sigma F_y = ma_y.$$

Unless the object is accelerating in the normal direction, $a_y = 0$. Substitute the y components of the forces

from your free-body diagram. The resulting equation allows you to determine the normal force.

3. The equation of motion in the x direction is

$$\Sigma F_x = ma_x.$$

If the object is not accelerating along the surface, $a_x = 0$. Substitute the x components of the forces from your free-body diagram. The resulting equation allows you to determine the frictional force.

4. If the object is not slipping, the normal force and the force of static friction should obey Inequality 10.54.

These examples involve material from this chapter but are not associated with any particular section. Some examples are worked out in detail; others you should work out by following the guidelines provided.

Worked Problem 10.1 Mosquito trail

The position of a mosquito along a curved path is given by

$$\vec{r} = (at^3 - bt)\hat{\imath} + (c - dt^4)\hat{\jmath}, \tag{1}$$

where $a = 20.0$ mm/s^3, $b = 50.0$ mm/s, $c = 60.0$ mm, and $d = 70.0$ mm/s^4. (a) Plot the mosquito's trajectory in the time interval from $t = -2.00$ s to $t = +2.00$ s, and interpret your graph. (b) Calculate the mosquito's position, velocity, and acceleration at $t = +2.00$ s. (c) What is the orientation of the line tangent to the curved path at this instant?

❶ **GETTING STARTED** We are given position as a function of time in vector form (Eq. 1). The trajectory we are asked for in part a is the curve describing the mosquito's path in two-dimensional space. To obtain the trajectory we must extract the x and y position values from Eq. 1 and plot y versus x. Position, velocity, and acceleration are related by derivatives. We can obtain the position needed for part b directly from Eq. 1, and we can take time derivatives of Eq. 1 to obtain the velocity and acceleration at $t = +2.00$ s. We know that the velocity of any object is always tangent to its trajectory, so for part c, the orientation of the tangent line should match the direction of the velocity we computed in part b.

❷ **DEVISE PLAN** The trajectory is a plot of the y component of the position versus the x component of the position, and we have the information about each component at any instant in Eq. 1. (a) To draw this curve, we need a table of the position components for several values of t. We can choose specific values of t and substitute those values into the expressions for the x and y components of position:

$$x(t) = at^3 - bt \tag{2}$$

$$y(t) = c - dt^4. \tag{3}$$

(b) We know that velocity and position are related by $\vec{v} = d\vec{r}/dt$, and acceleration and velocity by $\vec{a} = d\vec{v}/dt$. We need to evaluate these time derivatives at $t = +2.00$ s. (c) The tangent to the trajectory at $t = +2.00$ s is in the same direction as the velocity at that instant. We can therefore specify the orientation of the tangent line by computing the angle relative to the x axis that satisfies $\tan \theta = v_y/v_x$.

❸ **EXECUTE PLAN**
(a) We construct a table for the x and y components of the mosquito's position at instants that are close enough so that we can see how to interpolate. To get a rough idea of the shape, we begin with a few equally spaced instants of time: $t = -2.00$ s, -1.00 s, 0, $+1.00$ s, and $+2.00$ s. The results are

t (s)	x (mm)	y (mm)
−2.00	−60.0	−1060
−1.00	+30.0	−10.0
0	0	+60.0
+1.00	−30.0	−10.0
+2.00	+60.0	−1060

This choice of times provides a uniform spread of x values, but the y values are rather widely spread to be useful for interpolation. Now that we know the range of values, we can set the scales of the axes on our graph (which will be quite different for x and y). We also see the need to fill in a few more time values to specify the variation of y, perhaps one extra time between the central table rows and at least two extra values between the outer rows (where y grows dramatically). Inserting various values for t into Eqs. 2 and 3, we end up with

t (s)	x (mm)	y (mm)
-2.00	-60.0	-1060
-1.50	$+7.50$	-294
-1.00	$+30.0$	-10.0
-0.500	$+22.5$	$+55.6$
0	0	$+60.0$
$+0.50$	-22.5	$+55.6$
$+1.00$	-30.0	-10.0
$+1.50$	-7.50	-294
$+2.00$	$+60.0$	-1060

These data points yield the curve shown in Figure WG10.1.

Figure WG10.1

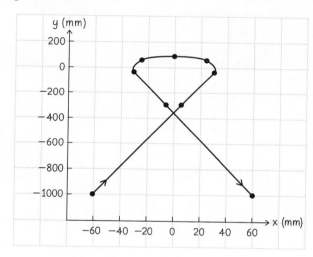

The physical interpretation of Figure WG10.1 is that there are three basic motions. First, the mosquito flies in from the left at high speed, indicated by the large spacing between $t = -2.00$ s and $t = -1.00$ s. That these dots get closer and closer together around $t = -1.00$ s tells us that the mosquito slows down. The more closely spaced horizontal portion indicates a slow speed as the mosquito backtracks around the origin. Then it flies off to the right at increasing speed. ✔

(b) Equation 1 gives us the mosquito's position at $t = +2.00$ s. The velocity at that instant is

$$\vec{v} = \frac{d}{dt}\left[(at^3 - bt)\hat{i} + (c - dt^4)\hat{j} \right]$$

$$= (3at^2 - b)\hat{i} - 4dt^3\hat{j}$$

and the acceleration is

$$\vec{a} = \frac{d}{dt}\left[(3at^2 - b)\hat{i} - 4dt^3\hat{j} \right]$$

$$= 6at\hat{i} - 12dt^2\hat{j}.$$

At $t = +2.00$ s, these equations give

$$\vec{r}(+2.00 \text{ s}) = \left[(20.0 \text{ mm/s}^3)(2.00 \text{ s})^3 - (50.0 \text{ m/s})(2.00 \text{ s}) \right]\hat{i}$$

$$+ \left[60.0 \text{ m} - (70.0 \text{ m/s}^4)(2.00 \text{ s})^4 \right]\hat{j}$$

$$= +(60.0 \text{ mm})\hat{i} - (1.06 \times 10^3 \text{ mm})\hat{j} \checkmark$$

$$\vec{v}(+2.00 \text{ s}) = +(190 \text{ mm/s})\hat{i} - (2.24 \times 10^3 \text{ mm/s})\hat{j} \checkmark$$

$$\vec{a}(+2.00 \text{ s}) = (240 \text{ mm/s}^2)\hat{i} - (3.36 \times 10^3 \text{ mm/s}^2)\hat{j}. \checkmark$$

(c) The orientation of the line tangent to the trajectory at $t = +2.00$ s is the same as the orientation of the velocity vector at that instant because the velocity always points along the direction of motion. To determine the orientation of a vector in a plane, we use $\theta = \tan^{-1}(v_y/v_x)$. Taking our v_x and v_y values from the velocity expression we derived in part b, we have

$$\theta = \tan^{-1}\left(\frac{-2240 \text{ mm/s}}{190 \text{ mm/s}} \right) = -85.2°.$$

This is the angle between the positive x axis and the line tangent to the trajectory. ✔

❹ EVALUATE RESULT Judging from Figure WG10.1, the answers are not unreasonable. The mosquito's speed is fast, mostly in the $\pm y$ direction, and it is accelerating in this general direction. We could have constructed a larger table with more values of t, but the curve we obtained seems to be smooth enough not to hide any surprises.

We can also compare the calculated angle of the tangent line at $t = +2.00$ s with the information in our xy table. At $t = +2.00$ s, the mosquito's horizontal position is $x = 60.0$ mm. We see from Figure WG10.1 that the tangent to the trajectory at that x position does have a negative slope, consistent with our negative angle of $-85.2°$. However, is the size of the angle consistent with the graph? The angle between the tangent to the curve of Figure WG10.1 at $x = 60.0$ mm and the x axis looks to be approximately $-45°$ rather than $-85°$. We have to be careful, though! Notice that the scales for the x and y axes are different. If we had used the same scale on both axes, the angle would look like $-85°$. On the other hand, the resulting graph would contain mostly blank space, with the entire trajectory squeezed into a narrow vertical band along the central portion of the x axis. Having so much blank space in a graph would be a clear sign that some re-scaling is needed.

Guided Problem 10.2 Cannonball trajectory

A cannonball fired from the top of a cliff at instant $t = 0$ has an initial velocity of magnitude v_i and is aimed at an angle θ above the horizontal (Figure WG10.2). (a) Someone who wants to know the path of the ball after it leaves the cannon asks you for an equation describing the motion. Using the parameters given in Figure WG10.2, derive the equation

$$y(x) = y_i + (\tan\theta)(x - x_i) - \frac{g}{2(v_i\cos\theta)^2}(x - x_i)^2. \qquad (1)$$

(b) If the ball lands a distance d from the cliff base, what is the height h from which the cannon ball is fired? Assume the distance between the cliff edge and the cannon's muzzle tip is negligibly small.

Figure WG10.2

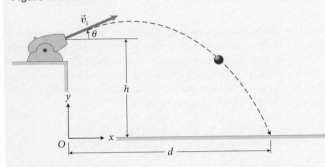

① GETTING STARTED

1. What type of motion is this?
2. What is the acceleration of the ball after it leaves the cannon? What are the sign and magnitude of a_x and of a_y? If the acceleration is constant, which kinematics equations can you use to derive Eq. 1?

② DEVISE PLAN

3. Write separate expressions for $x(t)$ and $y(t)$, the ball's position coordinates as a function of time.
4. For Eq. 1, you want the y coordinate as a function of x rather than of t. Can you solve for t in terms of $(x - x_i)$? Make the appropriate replacement of variables in the equation for y.
5. Can the x and y components of the initial velocity be expressed as functions of v_i and θ? Substitute this information as needed.

③ EXECUTE PLAN

6. Do the algebra to finish deriving the trajectory equation.
7. For part b, choose a suitable origin for your coordinate system. It should be different from the random origin shown in Figure WG10.2.
8. For your choice of origin, replace the generic symbols x_i, y_i, x, and y with symbols specific to part b, such as d and h.

④ EVALUATE RESULT

9. Does Eq. 1 have the expected dependence of y on x?
10. Does the value of h behave as you would expect as v_i, θ, and d change?

Worked Problem 10.3 Moving day

A mover pushes a bench of inertia m up a ramp that extends from the ground to the deck of a moving van and makes an angle θ with the ground (Figure WG10.3). None too smartly, he pushes horizontally with a force that has a magnitude equal to half the magnitude of the gravitational force exerted on the bench. The coefficient of kinetic friction between the ramp and the bench is μ_k. (a) Derive an expression for the acceleration of the bench in terms of θ and μ_k. (b) What must the coefficient of kinetic friction be (expressed in terms of θ) if the speed of the bench is constant? (c) Write an expression for the work done by the mover on the bench as the bench moves a distance d up the ramp.

Figure WG10.3

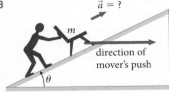

direction of mover's push

Figure WG10.4

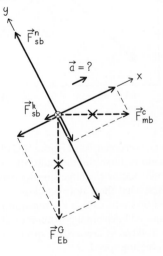

① **GETTING STARTED** The bench is the object we must focus on, and several forces are exerted on it: the mover's pushing force, the gravitational force, the frictional force, and the normal force exerted by the ramp surface. We must remember that the pushing force is directed horizontally, not up the ramp. The bench moves only parallel to the ramp, so the component of acceleration perpendicular to the ramp is zero. This information is summarized in the free-body diagram of the bench (Figure WG10.4).

Because the direction of the pushing force (horizontal) is not the same as the direction of the motion, we must take care to get the force components right. We choose a coordinate system tilted like the ramp surface and show components for any vector that is not parallel to an axis. We construct our components by drawing arrows for them along the x and y directions and then crossing out the original vectors. Because the speed is constant up the ramp in part b, we know the acceleration is zero in that part, but it is probably not zero in part a. However, the acceleration might be directed either up or down the ramp. The question mark at the acceleration in Figure WG10.4 reinforces our uncertainty on this point.

❷ DEVISE PLAN

(*a*) We have no numerical values, but using the free-body diagram to construct the vector sum of forces should give us an expression for the acceleration.

Because they each have a component that opposes the *x* component F_{mbx}^c of the force exerted by the mover, both the frictional force $\vec{F}_{sb}^k$ and the gravitational force $\vec{F}_{Eb}^G$ affect the acceleration. Applying some geometry and the information given about some of these forces, we should be able to use $\sum F_{by}$ to express the remaining forces in terms of θ and μ_k and then use $\sum F_{bx}$ to obtain an expression for the acceleration. We can eliminate one unknown because the magnitude of the frictional force depends on the magnitude of the normal force according to the relationship $F_{sb}^k = \mu_k F_{sb}^n$.

Because the normal force is in the direction of the *y* axis, we will employ the vector sum of forces in the *y* direction to calculate it. There is no acceleration in the *y* direction, and so $\sum F_{by} = ma_y = m(0) = 0$. In addition to the normal force, the gravitational force and the force exerted by the mover have components in the *y* direction. The only remaining variable is the acceleration, which we symbolize as a_x. We can isolate this variable to obtain the expression we need, and the expression should reveal the correct direction for the acceleration.

(*b*) Constant velocity requires zero acceleration. If we set this expression for acceleration equal to zero, it should be possible to obtain an expression for the coefficient of friction.

(*c*) The work done by a constant external force can be computed as the scalar product of the force and the force displacement. These quantities are not parallel to each other, so we must carefully keep track of components. Because we have information about both the magnitude and direction of each of these quantities, the work done can be computed from Eq. 10.35 regardless of the direction of the acceleration.

❸ EXECUTE PLAN (*a*) With our choice of axes orientation, the acceleration is along the *x* axis, and we start by analyzing the vector sum of the forces in the *x* direction. Because four forces are exerted on the bench, there should be four terms initially in the sum. We are given that $F_{mb}^c = \frac{1}{2}F_{Eb}^G = \frac{1}{2}mg$, and so

$$\sum F_{bx} = ma_x$$

$$F_{sbx}^k + F_{sbx}^n + F_{mbx}^c + F_{Ebx}^G = ma_x$$

$$-F_{sb}^k + 0 + (+F_{mb}^c \cos\theta) + (-F_{Eb}^G \sin\theta) = ma_x$$

$$-\mu_k F_{sb}^n + \tfrac{1}{2}mg\cos\theta - mg\sin\theta = ma_x. \tag{1}$$

The vector sum of the forces in the *y* direction gives us

$$\sum F_{by} = ma_y$$

$$F_{sby}^k + F_{sby}^n + F_{mby}^c + F_{Eby}^G = m(0)$$

$$0 + (+F_{sb}^n) + (-F_{mb}^c \sin\theta) + (-F_{Eb}^G \cos\theta) = 0$$

$$F_{sb}^n = \tfrac{1}{2}mg\sin\theta + mg\cos\theta.$$

Substituting this value for F_{sb}^n in Eq. 1 yields

$$-\mu_k\left(\tfrac{1}{2}mg\sin\theta + mg\cos\theta\right) + \tfrac{1}{2}mg\cos\theta - mg\sin\theta = ma_x$$

$$a_x = \tfrac{1}{2}g\cos\theta - \mu_k g\cos\theta - \tfrac{1}{2}\mu_k g\sin\theta - g\sin\theta$$

$$a_x = g\cos\theta\left(\tfrac{1}{2} - \mu_k\right) - g\sin\theta\left(1 + \tfrac{1}{2}\mu_k\right). \tag{2}$$

Thus the acceleration is in either the positive or negative *x* direction, depending on the relative size of the terms in Eq. 2. Expressing the result in vector form, we have

$$\vec{a} = \left[g\cos\theta\left(\tfrac{1}{2} - \mu_k\right) - g\sin\theta\left(1 + \tfrac{1}{2}\mu_k\right)\right]\hat{\imath}. ✔$$

(*b*) If the bench's speed is constant as it goes up the ramp, a_x must be zero. To determine the value of μ_k for which this is true, we set a_x equal to zero in Eq. 2 and solve for μ_k:

$$0 = g\cos\theta\left(\tfrac{1}{2} - \mu_k\right) - g\sin\theta\left(1 + \tfrac{1}{2}\mu_k\right)$$

$$= \tfrac{1}{2}g\cos\theta - \mu_k g\cos\theta - \tfrac{1}{2}\mu_k g\sin\theta - g\sin\theta$$

$$\mu_k\cos\theta + \tfrac{1}{2}\mu_k\sin\theta = \tfrac{1}{2}\cos\theta - \sin\theta$$

$$\mu_k(2\cos\theta + \sin\theta) = \cos\theta - 2\sin\theta$$

$$\mu_k = \frac{\cos\theta - 2\sin\theta}{2\cos\theta + \sin\theta}. ✔$$

(*c*) Because the bench is rigid and moves along a straight line, the displacement of the location where the mover exerts his force, the force displacement, is the same as the center-of-mass displacement— namely, a distance *d* up the ramp. The work done by the mover's force is

$$W = \vec{F}\cdot\Delta\vec{r}_F = \vec{F}_{mb}^c\cdot\vec{d} = F_{mb}^c d\cos\theta = \tfrac{1}{2}mgd\cos\theta. ✔$$

❹ EVALUATE RESULT Because of the plus and minus signs in Eq. 2, a_x can be positive or negative depending on the relative sizes of μ_k and θ. Because μ_k is always associated with a minus sign in our expression, it tends to make the acceleration negative (down the ramp), as we expect. For small angles, $\cos\theta$ is positive and much larger than $\sin\theta$, making the acceleration positive for small values of μ_k. (If $\mu_k > \frac{1}{2}$, the acceleration must be negative.) As θ increases, $\cos\theta$ decreases, which causes a decrease in $g\cos\theta(\frac{1}{2} - \mu_k)$, the only positive term in 2, relative to the increasing $\sin\theta$ term. Thus increasing θ drives a_x to negative values, which is what we expect: As the slope gets steeper, we expect the bench to slow down rather than speed up. The value of μ_k in part *b* balances these effects to achieve zero acceleration. Note that the work done by the mover is positive because $\cos\theta$ is positive for any inclination between 0 and 90°. This is true regardless of the magnitude or direction of the bench's acceleration, as long as it moves up the ramp.

PRACTICE

Guided Problem 10.4 Push that shed

Three winter anglers on an ice-fishing trip push a 120-kg shed out onto the ice. They are somewhat careless and push in different directions. As seen from above, these forces are $\vec{F}^c_{1s} = 32$ N directed 30° east of north, $\vec{F}^c_{2s} = 55$ N directed due north, and $\vec{F}^c_{3s} = 41$ N directed 60° west of north. What are the magnitude and direction of the shed's acceleration?

❶ GETTING STARTED

1. Because this is a force problem involving three vectors, consider which is more appropriate: graphical addition or addition by components.
2. Because forces are involved, a free-body diagram of the shed is useful. Is it possible to draw only the forces in the plane of the ice? If so, why is it a good idea to set things up this way?

❷ DEVISE PLAN

3. Choose two coordinate axes. Does it matter which directions you choose for your axes, given that the direction of the acceleration is unknown? Is one orientation of your coordinate system more appropriate than other orientations, given the information supplied?

4. Write the two components of Newton's second law for forces parallel to the ice.
5. Do you have enough information to compute the value of each component of the three forces in your coordinate system?

❸ EXECUTE PLAN

6. Add components to obtain an x component and y component for the vector sum of the forces.
7. How are the components of the acceleration related to the force components?
8. Can you compute the magnitude and direction of the acceleration (relative to north) given the x and y components of the acceleration?

❹ EVALUATE RESULT

9. Did you make any assumptions? If so, decide whether they are reasonable or should be reconsidered.
10. Is the acceleration direction consistent with what your intuition tells you about the overall effect of the three forces?
11. Is the magnitude of the acceleration reasonable?

Worked Problem 10.5 Spring up that incline

The ramp in Figure WG10.5 has a rough surface and makes an angle θ with the horizontal. The spring attached to the support at the bottom of the ramp has a force constant k. You place a block of inertia m against the free end of the spring and then push the block against the spring until the block has moved a distance d from its position when the spring is at its relaxed length. When you release the block from rest, it slides up the ramp and eventually leaves the spring behind, stopping finally a distance ℓ from its position just before release. Derive an expression for ℓ in terms of k, d, m, θ, and the coefficient of kinetic friction μ_k between block and ramp.

Figure WG10.6

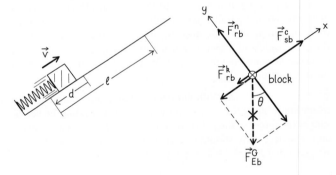

Figure WG10.5

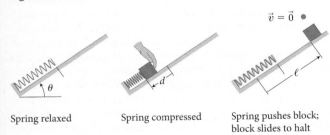

Spring relaxed   Spring compressed   Spring pushes block; block slides to halt

❶ GETTING STARTED Four forces are exerted on the block: the constant force of gravity, the variable spring force, the constant normal force, and the dissipative frictional force. (The phrase *rough surface* in the problem statement is what tells us we must consider friction.) A good first step is to draw a free-body diagram that shows all four forces (Figure WG10.6). To get an

expression for the magnitude of the force of kinetic friction, we need to know the magnitude of the normal force. We include in our diagram a tilted coordinate system to help us calculate this force and the components of any vector not parallel to one of our coordinate axes.

Because the spring force varies with position, we should investigate whether we can use work-energy methods, accounting for the action of the spring through the equation for potential energy. Thus a situation diagram is called for, showing the initial compression distance d and the final position of the block a distance ℓ up the ramp (Figure WG10.7). With this approach, the initial elastic potential energy of the compressed spring is converted to gravitational potential energy as the block rises and to thermal energy dissipated through friction. Although the kinetic energy changes during the process, it is the same at start and finish: zero.

Figure WG10.7

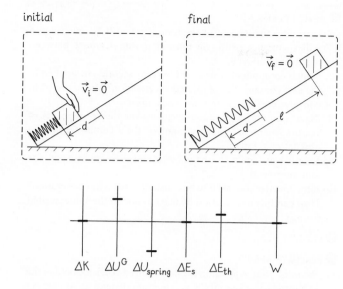

initial   final

ΔK ΔU^G ΔU_{spring} ΔE_s ΔE_{th} W

② **DEVISE PLAN** If we choose the block, spring, ramp, and Earth as our system, there are no external forces, and therefore no work is done on the system:

$$\Delta E = W = 0$$

$$\Delta K + \Delta U_{\text{spring}} + \Delta U^G + \Delta E_{\text{th}} = 0. \qquad (1)$$

This expression will contain all our variables, and so with a little algebra we should be able to isolate ℓ, the variable we are after.

The thermal energy dissipated is the product of the magnitude of the force of kinetic friction and the distance ℓ over which the block moves:

$$\Delta E_{\text{th}} = F^k_{\text{rb}}\ell = (\mu_k F^n_{\text{rb}})\ell. \qquad (2)$$

To calculate the magnitude F^n_{rb}, we note that the normal force is aligned along the y axis. We employ the vector sum of force components along this axis and the fact that along this axis $a_y = 0$:

$$\sum F_{by} = ma_y = m(0) = 0. \qquad (3)$$

Equations 1–3 should give us all the information we need to derive an expression for ℓ.

③ **EXECUTE PLAN** Substituting an expression that contains the quantities given in the problem statement for each term in Eq. 1 gives us

$$\left(\tfrac{1}{2}mv_f^2 - \tfrac{1}{2}mv_i^2\right) + \left[\tfrac{1}{2}k(x_f - x_0)^2 - \tfrac{1}{2}k(x_i - x_0)^2\right]$$

$$+ (mgh_f - mgh_i) + F^k_{\text{rb}}\ell = 0$$

$$(0 - 0) + \left[\tfrac{1}{2}k(0)^2 - \tfrac{1}{2}k(-d)^2\right] + mg(h_f - h_i) + \mu_k F^n_{\text{rb}}\ell = 0$$

$$-\tfrac{1}{2}kd^2 + mg(+\ell \sin \theta) + \mu_k F^n_{\text{rb}}\ell = 0$$

$$\ell(mg \sin \theta + \mu_k F^n_{\text{rb}}) = \tfrac{1}{2}kd^2, \qquad (4)$$

where x_0 is the block's position when the spring is at its relaxed length and we have used h to designate vertical height coordinates so as not to mistakenly use the tilted y coordinate. Equation 4 looks pretty close to what we are after (an expression for ℓ in terms of $k, d, m, \theta,$ and μ_k). All we need to derive is an expression for F^n_{rb}, which we obtain from Eq. 3, the vector sum of the forces along the y axis. Four forces are exerted on the block, so

$$\sum F_{by} = F^k_{\text{rb}y} + F^n_{\text{rb}y} + F^c_{\text{sb}y} + F^G_{\text{Eb}y} = 0$$

$$0 + (+F^n_{\text{rb}}) + 0 + (-F^G_{\text{Eb}} \cos \theta) = 0$$

$$F^n_{\text{rb}} = F^G_{\text{Eb}} \cos \theta = mg \cos \theta.$$

Substituting this expression for F^n_{rb} into Eq. 4 gives us what we're after:

$$\ell(mg \sin \theta + \mu_k mg \cos \theta) = \tfrac{1}{2}kd^2$$

$$\ell = \frac{kd^2}{2mg(\sin \theta + \mu_k \cos \theta)}. \checkmark$$

④ **EVALUATE RESULT** We expect that the greater the force of friction (in other words, the higher the value of μ_k), the shorter the distance ℓ the block travels, as this expression predicts. We also expect that for any fixed value of μ_k, the distance ℓ increases as θ decreases. When $\theta = 0$ (the block sits on a horizontal surface), $\sin \theta = 0$ and ℓ has a large value. Because $\sin \theta$ increases faster than $\cos \theta$ decreases, the $\theta = 0$ value of ℓ is in fact its maximum value (assuming that $\mu_k < 1$).

The larger the spring constant k or the greater the compression distance d, the farther the block should go, which is what our expression predicts. Our result behaves the way we would expect.

Note that the x components of the four forces in this problem contribute to the work done on the system. The two y components, $F^n_{\text{rb}y}$ and $F^G_{\text{Eb}y}$, do not. Their lack of contribution to the system energy is not just because they cancel each other, however. Let us consider the effect of the normal force, which is perpendicular to the direction of motion. The scalar product of two vectors is nonzero only when the vectors have a component in the same direction, so $\vec{F}^n_{\text{rb}} \cdot \Delta \vec{r}_F = 0$ for the normal force as well as for any other force that is always perpendicular to the direction of motion. This means that such forces cannot rearrange energy within the system.

Guided Problem 10.6 Moving a load

A variation of the Atwood machine (see Figure WG8.6) can be used to haul a load up an incline. Using this variation, you must design a coal cart of inertia m_c that moves up an incline set at an angle θ with the horizontal. A rope attached to the cart is thrown over a pulley at the top of the incline, and a metal block of inertia m_b is hung vertically at the other end of the rope as a counterweight. (*a*) In terms of θ and the two inertias, what is the magnitude of the acceleration of the cart if the block alone is used to draw the cart up the incline? (*b*) In terms of θ and m_c, what must the inertia of the block be if the cart is to move at constant speed?

❶ GETTING STARTED

1. Sketch the physical situation.
2. What variable must you determine in part *a*? Which approach is more relevant for determining it: force analysis or energy analysis? Can the result of part *a* be adapted to solve part *b*?
3. How many objects are involved in the motion? How are their motions related?
4. When you analyze forces for two or more objects, each must have its own free-body diagram and coordinate system, leading to a set of force-acceleration equations for each object.
5. Put the acceleration of each object in your diagrams, and recall that it is useful to have one coordinate axis parallel to an object's acceleration, with the axes oriented such that the motion of one object along a positive axis in its coordinate system corresponds to the motion of the other object along its corresponding positive axis.

❷ DEVISE PLAN

6. How does the tension in the rope segment between cart and pulley compare with the tension in the segment between block and pulley?
7. Which component, *x* or *y*, of Newton's second law applied to the cart contains the requested acceleration? Do you need to consider the other component equation?
8. Make sign and magnitude decisions for the components of vectors where possible (for example, the *x* component of the gravitational force).
9. Repeat your analysis for the block and then count equations and unknowns.
10. If the speed of the cart is to be constant, what is its acceleration? How can this information be used to obtain the value needed for m_b?

❸ EXECUTE PLAN

❹ EVALUATE RESULT

11. Examine your expression for the required acceleration. Do the magnitude and sign of the acceleration depend as you expect on m_c and m_b? On θ?
12. Do your answers to questions 11 make sense in the limit $\theta = 0$? In the limit $\theta = 90°$?

Worked Problem 10.7 Accident investigation

While on a test flight at an altitude of 2000 m, a 1500-kg unmanned drone moving eastward at 60 m/s exploded into three pieces. Satellite data show that the explosion occurred exactly 50 km east of Reno, Nevada. Beginning directly beneath the explosion site, searchers find a 310-kg piece 1200 m to the south and 600 m to the east. An 830-kg piece is found 400 m to the north and 1500 m to the east. Assume that air resistance is negligible and that all three pieces hit the ground simultaneously. What were the velocities of the three pieces immediately after the explosion, and where should investigators look for the third piece?

❶ **GETTING STARTED** We are seeking the final velocities of three pieces and the final position of one piece. The initial velocity and position of the drone are given, as well as the final positions of two of the three pieces. That still leaves a lot of unknowns! Moreover, this appears to be a three-dimensional problem, something we have not yet encountered. However, each piece is a projectile that moves in a plane (though each moves in a different plane). But it gets better because we have some very useful clues. For example, the assumption that all three pieces hit the ground at the same time will allow us to immediately solve for motion in the vertical direction. Then it will be necessary only to analyze two horizontal components of the motion by using the techniques we already have at our disposal.

Let's organize our thoughts by drawing before and after satellite views (looking down from high above) of the system (Figure WG10.8). The figure shows the drone immediately before the explosion and the known final positions of two of the pieces.

Figure WG10.8

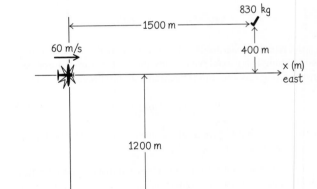

The initial velocity is also indicated, and we set up coordinates with the origin directly under the explosion site, with the positive *x* direction to the east, the positive *y* direction upward (not shown), and the positive *z* direction to the south. We can treat the explosion like an explosive separation, and so the drone is an isolated system. Therefore the momentum of the pieces must be the same before and after the explosion.

② DEVISE PLAN Because we assume the three pieces hit the ground simultaneously, in the absence of air resistance each must have the same final vertical component of velocity. This is easily seen from the vertical position equation for a projectile, Eq. 10.20:

$$y_f = y_a + v_{y,a}\Delta t - \tfrac{1}{2}g(\Delta t)^2.$$

The subscripts are "a" for immediately after the explosion and "f" for final (just before hitting the ground). Each piece started at the same height y_a, each reached the same final height ($y_f = 0$), and each experienced the same values of g and Δt, so the only unspecified variable is the velocity just after the explosion $v_{y,a}$. The equation thus demands the same value of vertical velocity immediately after the explosion for each piece. Moreover, this vertical velocity must be zero! Prior to the explosion, the system had zero velocity, and thus zero momentum, in the vertical direction. After the explosion, the vertical momentum must still be zero. This might be arranged by having some of the pieces ejected upward and some downward, but then they would not hit the ground at the same time. All three pieces therefore fall from rest as far as the vertical motion is concerned:

$$v_{1y,a} = v_{2y,a} = v_{3y,a} = 0.$$

This allows us to compute the time of fall, which together with the given position information and simple kinematics should give all three x and z components of velocity just after the explosion.

The location of the missing piece can be found using the momentum equation. Alternatively, we might compute it by considering that the center of mass of a system moves with the system momentum.

③ EXECUTE PLAN We carry one extra digit in all of our calculations so as not to accumulate rounding error, but we report the results to two significant digits to match the precision of the given information.

The time of fall from 2000 m for each piece is identical:

$$y_f = y_a + v_{y,a}\Delta t - \tfrac{1}{2}g(\Delta t)^2 = y_a + (0)\Delta t - \tfrac{1}{2}g(\Delta t)^2$$

$$= y_a - \tfrac{1}{2}g(\Delta t)^2$$

$$(\Delta t)^2 = \frac{y_a - y_f}{\tfrac{1}{2}g} = \frac{+2000 \text{ m}}{\tfrac{1}{2}(9.8 \text{ m/s}^2)}$$

$$\Delta t = 20.2 \text{ s}.$$

This leads immediately to the x and z components of velocity after the explosion of the two pieces whose final positions we know:

$$x_{1f} = x_{1a} + v_{1x,a}\Delta t$$

$$v_{1x,a} = \frac{x_{1f} - x_{1a}}{\Delta t} = \frac{+600 \text{ m} - 0}{20.2 \text{ s}} = +29.7 \text{ m/s} = +30 \text{ m/s} ✔$$

$$x_{2f} = x_{2a} + v_{2x,a}\Delta t$$

$$v_{2x,a} = \frac{x_{2f} - x_{2a}}{\Delta t} = \frac{+1500 \text{ m} - 0}{20.2 \text{ s}} = +74.2 \text{ m/s} = +74 \text{ m/s} ✔$$

$$z_{1f} = z_{1a} + v_{1z,a}\Delta t$$

$$v_{1z,a} = \frac{z_{1f} - z_{1a}}{\Delta t} = \frac{+1200 \text{ m} - 0}{20.2 \text{ s}} = +59.4 \text{ m/s} = +59 \text{ m/s} ✔$$

$$z_{2f} = z_{2a} + v_{2z,a}\Delta t$$

$$v_{2z,a} = \frac{z_{2f} - z_{2a}}{\Delta t} = \frac{-400 \text{ m} - 0}{20.2 \text{ s}} = -19.8 \text{ m/s} = -20 \text{ m/s}. ✔$$

The velocity of the missing piece can now be obtained from the momentum equation, and its location from the same kinematics employed above. The inertia of the missing piece is $1500 \text{ kg} - 310 \text{ kg} - 830 \text{ kg} = 360 \text{ kg}$. We have

$$\vec{p}_i = \vec{p}_a$$

$$m_{\text{plane}}\vec{v}_i = m_1\vec{v}_{1a} + m_2\vec{v}_{2a} + m_3\vec{v}_{3a}.$$

We already know that the y components of velocities are zero immediately after the explosion, so we use the x and z component equations:

$$m_{\text{plane}}v_{x,i} = m_1v_{1x,a} + m_2v_{2x,a} + m_3v_{3x,a}$$

$$v_{3x,a} = \frac{m_{\text{plane}}v_{x,i} - m_1v_{1x,a} - m_2v_{2x,a}}{m_3}$$

$$= \frac{(1500 \text{ kg})(+60 \text{ m/s}) - (310 \text{ kg})(29.7 \text{ m/s})}{360 \text{ kg}}$$

$$- \frac{(830 \text{ kg})(74.2 \text{ m/s})}{360 \text{ kg}}$$

$$= +53 \text{ m/s} ✔$$

$$m_{\text{plane}}v_{z,i} = m_1v_{1z,a} + m_2v_{2z,a} + m_3v_{3z,a}$$

$$v_{3z,a} = \frac{m_{\text{plane}}v_{z,i} - m_1v_{1z,a} - m_2v_{2z,a}}{m_3}$$

$$= \frac{(1500 \text{ kg})(0) - (310 \text{ kg})(59.4 \text{ m/s})}{360 \text{ kg}}$$

$$- \frac{(830 \text{ kg})(-19.8 \text{ m/s})}{360 \text{ kg}}$$

$$= -5.50 \text{ m/s}. ✔$$

The signs in our values for v_3 indicate that piece 3 is moving in the positive x (eastward) and negative z (northward) directions after the explosion. To finish our task—finding the location of piece 3—we apply kinematics. In the 20 s before it struck the ground, piece 3 traveled to this final resting place:

$$x_{3f} = x_{3a} + v_{3x,a}\Delta t$$

$$= 0 + (+53.3 \text{ m/s})(20.2 \text{ s})$$

$$= +1077 \text{ m} = 1.1 \times 10^3 \text{ m} ✔$$

$$z_{3f} = z_{3a} + v_{3z,a}\Delta t$$

$$= 0 + (-5.50 \text{ m/s})(20.2 \text{ s})$$

$$= -111 \text{ m} = -1.1 \times 10^2 \text{ m}. ✔$$

PRACTICE

④ **EVALUATE RESULT** The calculated velocity components are all comparable in magnitude to the original velocity of the drone. Piece 3 should be found at approximately 1100 m east and 110 m north of the origin (or the point of explosion). The magnitude of the displacement of piece 3 is neither huge nor tiny compared with the displacements of the other two pieces. We can check our result by recalling that the motion of the center of mass of a system is affected only by forces acting from outside the system. Thus the center of mass of the drone follows the parabolic path of a projectile launched at altitude 2000 m with eastward velocity 60 m/s:

$$x_{cm,f} = x_{cm,i} + v_{i,x}\Delta t = 0 + (60 \text{ m/s})(20.2 \text{ s})$$

$$= 1212 \text{ m} = 1.2 \times 10^3 \text{ m}.$$

We can compute the location of the center of mass of the pieces using the results obtained above:

$$x_{cm} = \frac{m_1 x_{1f} + m_2 x_{2f} + m_3 x_{3f}}{m_{plane}}$$

$$= \frac{(310 \text{ kg})(+600 \text{ m}) + (830 \text{ kg})(+1500 \text{ m})}{1500 \text{ kg}}$$

$$+ \frac{(360 \text{ kg})(+1100 \text{ m})}{1500 \text{ kg}}$$

$$= +1218 \text{ m} = 1.2 \times 10^3 \text{ m}.$$

This method can also verify that the z component of the center-of-mass position is indeed zero, if we recall that small differences in large numbers cause reduced significant digits:

$$z_{cm} = \frac{m_1 z_{1f} + m_2 z_{2f} + m_3 z_{3f}}{m_{plane}}$$

$$= \frac{(310 \text{ kg})(+1200 \text{ m}) + (830 \text{ kg})(-400 \text{ m})}{1500 \text{ kg}}$$

$$+ \frac{(360 \text{ kg})(-111 \text{ m})}{1500 \text{ kg}}$$

$$= +0.0267 \text{ m} = 0.$$

Under the given assumptions about direction of motion and constancy of velocities, the answers we obtain are not unreasonable.

Guided Problem 10.8 Hockey with a bang

Someone puts a firecracker into a hockey puck and slides it onto frictionless ice while you watch. The explosion breaks the puck into exactly two pieces, and you see the two pieces slide across the ice without rising into the air. Piece P (inertia 60 g) moves with speed 4.5 m/s at an angle of 22° south of east. Piece Q (inertia 110 g) moves with speed 1.9 m/s at an angle of 52° north of west. Compute the initial velocity of the hockey puck before the explosion. Ignore the inertia of the firecracker.

❶ **GETTING STARTED**
1. What are you asked to determine?
2. What is an appropriate choice of system?
3. Which fundamental principle(s) would be most useful for this analysis?
4. Make a sketch showing the appropriate initial and final conditions.

❷ **DEVISE PLAN**
5. Write the appropriate equation(s) that express the principle you chose to make the analysis.
6. Are there sufficient equations to compute the unknowns?

❸ **EXECUTE PLAN**
7. Insert known values and evaluate your expression(s).

❹ **EVALUATE RESULT**
8. Is it possible that the hockey puck was at rest before the explosion?
9. Is more than one answer possible? Are your values reasonable?

Questions and Problems

For instructor-assigned homework, go to MasteringPhysics®

Dots indicate difficulty level of problems: • = *easy,* •• = *intermediate,* ••• = *hard;* CR = *context-rich problem.*

10.1 *Straight* is a relative term

1. Suppose a missile moving along a trajectory has a constant vertical speed and a constant horizontal acceleration. What is the shape of this trajectory as seen by a stationary observer? •

2. Riding in a train on horizontal tracks, you notice a boy drop a baseball from the roof of his house nearby. (*a*) What is the shape of the trajectory you see for the ball if the train is moving at constant speed? (*b*) Would you see the trajectory as a straight line if the train were accelerating? ••

3. You are performing an experiment similar to the one shown in *Principles* Figure 10.1. The ball is dropped from a height of $y_i = 0.200$ m on the pole above the cart just as the pole and cart pass your origin of coordinates. The ball lands inside the cart at the instant the cart passes the location $x = +0.0600$ m. The lab exercise includes, among other things, that you determine the distance traveled by the ball from when it is released until it lands in the cart. Your lab partner immediately says "0.200 m," but the girl at the next table says "0.0600 m." The lab partners just behind you are arguing about whether it is 0.209 m or 0.214 m. You have not done the calculation, but of these four numbers, which could be correct? ••

10.2 Vectors in a plane

4. Is it possible for the acceleration and velocity vectors of an object to be *always* perpendicular to each other? If no, explain why not. If yes, describe the motion. •

5. A baseball player hits a long fly ball to center field. What is the direction of the ball's acceleration at the highest point of its flight? (Take air resistance into consideration.) •

6. Can an object be accelerated without changing its kinetic energy? •

7. (*a*) Can two vectors of unequal lengths add up to zero? (*b*) What about three vectors of unequal lengths? ••

8. Three swimmers who all swim at the same speed discuss how to cross a river in the shortest amount of time. Swimmer A will swim straight across the river at a right angle to the current. Swimmer B reasons that the current will carry A downstream, meaning that A will cover a greater distance to get across and therefore will take a longer time interval. B says he will aim at an upstream angle such that, allowing for the current, he will reach the other side directly across from where he starts, thus covering the shortest distance and arriving first. Swimmer C, reasoning that the time interval needed for B to cross will be longer than B expects because some of B's effort will be spent battling the current, plans to aim at a downstream angle, so that the current assists rather than opposes him. This way he will be traveling at the highest speed and get across first. Which swimmer gets across first? ••

9. Figure P10.9 shows the position of a regularly flashing light on top of a car (bird's-eye view) as the car moves from (a) to (f). Describe the acceleration of the car at each labeled position based on the information in the figure. ••

Figure P10.9

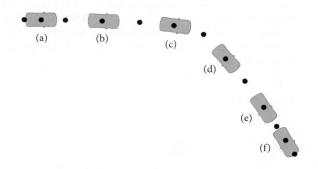

10.3 Decomposition of forces

10. Which is harder to slow down: a car going up a hill or a car going down a hill? Why? •

11. A sweater hanging on a clothesline makes the line sag in the center. Could you tighten the clothesline enough (assuming whatever material for the line you want) to eliminate the sag completely? ••

12. A sign is suspended by two wires as shown in Figure P10.12. Is the tension in each wire *larger than, equal to,* or *smaller than* the gravitational force exerted on the sign by Earth? ••

Figure P10.12

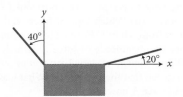

13. A man is crossing a roof that is inclined at an angle of 30° above the horizontal. His path takes him parallel to the roof ridge but 4.0 m (along the roof surface) below the peak. He is supported by two ropes, one attached to an object above each end of the roof. The left rope is attached to a chimney at 1.0 m above the peak of the roof; the right rope is attached to a flagpole at 3.0 m above the peak of the roof. The chimney rope runs parallel to the roof surface, but the flagpole rope does not. Both ropes reach the man at waist level. When the man is halfway across the 10-m-wide roof, describe each force that acts on the man, and identify which have components in each of three perpendicular directions: parallel to the roof ridge, normal to the roof surface, and tangential to the roof surface. •••

10.4 Friction

14. A common parlor trick is to yank a tablecloth hard so that any dishes and glasses on the cloth stay put instead of falling onto the floor. (*a*) Explain how this happens. (*b*) Why, if you pull on the cloth too slowly, do the dishes and glasses get dragged off the table? •

15. When a car accelerates gradually (no squealing tires), is the friction between tires and road kinetic or static? •

16. Figure P10.16 represents two stroboscopic pictures (taken at the same flash rate) of a block moving along a ramp that exerts a frictional force on the block. In one sketch, the block is released from rest at the top of the ramp and moves downward. In the other sketch, the block is placed at the bottom of the ramp and shoved up the ramp. Which sketch is for the downward motion? Which is for the upward motion? •

Figure P10.16

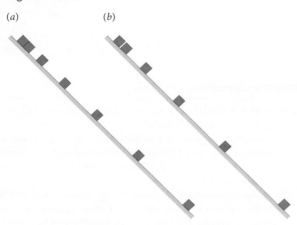

(a)　　　　(b)

17. You keep a chalkboard eraser pressed against the chalkboard by using your finger to exert a horizontal force on the back of the eraser. (a) Which type of force (call it force A) keeps the eraser from falling? (b) What is the magnitude of force A? (c) Does the magnitude of force A (which keeps the eraser from falling) depend on how hard you press? (d) Now release your finger so that the eraser falls. What happens to the magnitude of force A now? ••

18. In Figure P10.18, all surfaces experience friction. Is the magnitude of the horizontal component of the force exerted by A on B greater than, equal to, or less than the magnitude of the horizontal force exerted by A on C if the system of blocks is (a) moving at a constant velocity to the right and (b) accelerating to the right due to an external force exerted on B? ••

Figure P10.18

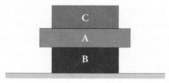

19. A factory worker pulls on a rope attached to a pallet loaded with a crate, dragging the combination across a rough factory floor. Draw a free-body diagram for the pallet and one for the crate (a) when the pallet is on level ground and (b) when the pallet is being pulled up an incline. The velocity is constant in both situations. ••

10.5 Work and friction

20. What is the direction of the frictional force exerted on a coffee cup (a) sitting on a stationary table, (b) being pulled to the right across a table, and (c) sitting on a table while the table is dragged to the right? •

21. A worker places a circular saw on top of a long board that is lying on the floor. Then she and a coworker lift the ends of the board and carry it across the construction site, placing it across two sawhorses at its destination. Discuss the work done by the normal force and the force of static friction by the board on the saw during this process. •

22. A resort uses a rope to pull a 55-kg skier up a 15° slope at constant speed for 100 m. (a) Determine the tension in the rope if the snow is slick enough to allow you to ignore any frictional effects. (b) How much work does the rope do on the skier? ••

23. A friend claims that her car can accelerate from a stop to 60 mi/h (26.8 m/s) in 5.1 s, but the speedometer is broken. You decide to test her claim by riding with her, and you bring along a small metal washer, a short length of string, a protractor, and a pen that can write on glass. Sitting in the passenger seat, you tie the washer to one end of the string and then let the string-washer combination hang straight down by rolling up the passenger-door window to pinch the free end of the string. While the car is stationary, you draw a line on the window alongside the vertical string. (a) How large is the angle the string makes with the line you drew if the car accelerates at the rate your friend claims? (b) If an identical washer does not slide on the horizontal dashboard while this is happening, what is the ratio of the frictional force to the gravitational force exerted on this washer? ••

24. You drive a car onto a 100-vehicle ferry, park, and have a short nap. When the ferry arrives at its destination, you start the car and drive off to begin your vacation. Estimate the magnitude of work done by the forces of static friction between the car tires and either the roadway or the ferry surface during each portion of this journey. ••

25. Two children slide down playground slides of identical heights but different angles, one shallow and one steep. (a) Suppose the slides are so slippery that the force of friction is negligibly small. Which child, if either, moves faster at the bottom of the slide? Which one, if either, arrives at the bottom first? (b) Now take into account the kinetic friction between the child and the slide. Which child, if either, moves faster at the bottom of the slide? ••

10.6 Vector algebra

26. For the vectors $\vec{A} = 2.0\hat{\imath} + 3.0\hat{\jmath}$ and $\vec{B} = -4.0\hat{\imath} + 5.0\hat{\jmath}$, calculate (a) $\vec{A} + \vec{B}$ and (b) $\vec{A} - \vec{B}$. (c) If the direction (but not the magnitude) of $\vec{B}$ is allowed to change by any amount you like, how large can $|\vec{A} + \vec{B}|$ be? How small? •

27. For the vectors $\vec{A} = 3.0\hat{\imath} + 2.0\hat{\jmath}$ and $\vec{B} = -2.0\hat{\imath} + 2.0\hat{\jmath}$, determine (a) $\vec{A} + \vec{B}$ and (b) the magnitude of $|\vec{A} + \vec{B}|$. •

28. Determine the polar coordinates of vectors (a) $\vec{A} = 3.0\hat{\imath} + 2.0\hat{\jmath}$ and (b) $\vec{B} = -2.0\hat{\imath} + 2.0\hat{\jmath}$. •

29. Your directions on a scavenger hunt map say to walk 36 m east, then 42 m south, then 25 m northwest. With the positive x direction being east, what is your displacement (a) in polar coordinates and (b) in Cartesian coordinates? ••

30. In Figure P10.30, if the tension in the cable attaching the platform to the building on the right is 800 N, what are (a) the tension in the cable attaching the platform to the building on the left and (b) the inertia of the platform? ••

Figure P10.30

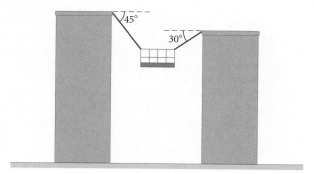

31. You leave your house and walk east for 1.0 h, northeast for 1.5 h, south for 1.0 h, and southwest for 2.5 h, always moving at the same speed. Realizing it is going to get dark soon, you head directly home. How long does it take to walk directly home if your speed stays the same as it was on every leg of the walk? ••

32. An object's displacement is given by $\vec{r}(t) = (At + Bt^3)\hat{\imath} + (C - Dt^2)\hat{\jmath}$. (a) Write the expression for $\vec{v}(t)$ for this motion and the expression for $\vec{a}(t)$. (b) Using the values $A = 2.0$ m/s, $B = 0.10$ m/s^3, $C = 5.0$ m, and $D = 0.20$ m/s^2, plot the object's position on a graph that has x as the horizontal axis and y as the vertical axis for instants $t = 0$, 1.0 s, 2.0 s, and 3.0 s. At these positions, draw the velocity and acceleration vectors. ••

33. Translate the position function $\vec{r}(t) = A \cos(\omega t)\hat{\imath} + A \sin(\omega t)\hat{\jmath}$ into polar coordinates, where A and ω are constants. ••

34. A plane travels in a straight line from position A to position B in 65 min, moving at an average speed of 400 km/h. In a car traveling from A to B, the driver finds that the trip is 600 km long by the route he is forced to take, which goes straight south for a certain distance and then straight west for a longer distance. At what angle south of west does the plane travel? ••

35. A child rides her bike five blocks east and then three blocks north. It takes her 15 min, and each block is 160 m long. What are (a) the magnitude of her displacement, (b) her average velocity and (c) her average speed? ••

36. (a) In a Cartesian coordinate system that has axes x and y, the coordinates of a certain point P are (x, y). What are the coordinates of P in a Cartesian coordinate system that has axes x' and y' if the origins coincide but the x' axis is tipped above the x axis by an angle φ? (b) If the x and y coordinates of P are (5.0 m, 2.0 m) and $\varphi = 30°$, determine (x', y'). •••

37. (a) Suppose you have two arrows of equal length on a tabletop. If you can move them to point in any direction but they must remain on the tabletop, how many distinct patterns are possible such that the arrows, treated as vectors, sum to zero? [Note: If a pattern cannot be rotated on the tabletop to match another pattern in your list, it is a distinct pattern.] (b) Repeat part a for three arrows, assuming the angle between any adjacent pair must be equal. (c) Repeat for four arrows, keeping the angle between neighboring arrows the same throughout the pattern. (d) Do you recognize a relationship between the number of arrows (vectors) and the number of distinct patterns? •••

10.7 Projectile motion in two dimensions

38. A ball is hurled horizontally out of a window 10 m off the ground with an initial speed of 15 m/s. How far from the building does the ball hit the ground? •

39. Which is the best representation in Figure P10.39 of the trajectory of a cantaloupe thrown horizontally off a bridge? What is wrong with the other paths? •

Figure P10.39

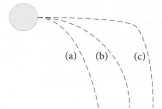

40. A bullet is fired horizontally from a high-powered rifle. At the same instant, a bullet that was resting on top of the rifle falls off. Which bullet hits the ground first? •

41. A rifle is aimed horizontally at a target 100 m away, and the bullet leaves the rifle barrel at 650 m/s. If the gun is aimed right at the bull's-eye, by how much does the bullet miss the center? •

42. On a rifle that has a telescopic sight, the telescope is usually not parallel to the barrel of the rifle. The angle the telescope makes with the barrel has to be adjusted for the distance to the target. Explain why. ••

43. A ball rolls off a table and lands on the floor. The horizontal distance between the position at which the ball lands and the edge of the table is 0.50 m, and the tabletop is 0.80 m above the floor. (a) How long does it take the ball to reach the floor? (b) What is the ball's speed when it leaves the table? ••

44. A package is dropped from a helicopter traveling at 15 m/s (horizontally) at an altitude of 200 m, but the parachute attached to the package fails to open. (a) How long does it take for the package to hit ground? (b) How far does the package travel horizontally before it lands? (c) What is the speed of the package just before it lands? ••

45. The velocity of an object is given in SI units by $\vec{v} = (at - bt^2)\hat{\imath} + c\hat{\jmath}$, with $a = 14$ m/s^2, $b = 10$ m/s^3, and $c = 22$ m/s. (a) If the initial position of the object at $t = 0$ is at the origin ($x_i = y_i = 0$), when, if ever, does the object return to the origin? (b) When, if ever, is the velocity zero? (c) When, if ever, is the acceleration zero? ••

46. A cannon launches two shells at the same speed, one at 55° above horizontal and one at 35° above horizontal. Which shell, if either, has the longer range? Which shell, if either, is in the air longer? Assume level ground and ignore air resistance. ••

47. A golf ball is hit from the tee and travels above level ground. Accounting for air resistance, where is the horizontal location of the peak of the flight: at a position that is less than half the range, at a position that is half the range, or at a position that is more than half the range? ••

48. A cannon fires a shell at an angle such that the shell's initial velocity has a horizontal component of 20 m/s and a vertical component of 35 m/s. Draw the approximate location of the shell at the instants 0, 1.0 s, 2.0 s, 3.0 s, and 4.0 s. At each location, draw the horizontal and vertical components of the velocity. ••

49. Taking air resistance into account, draw the acceleration vector of a golf ball at five positions along its trajectory from tee to green. Assume for simplicity that the force of air resistance has a constant magnitude. ••

50. How far does a fastball released from the pitcher's hand at a speed of 42 m/s fall as it travels from pitcher to batter, a distance of 18.4 m? (Your answer gives you a good idea of why the pitcher stands on a mound of dirt.) ••

51. The fish in Figure P10.51, peering from just below the water surface, spits a drop of water at the grasshopper and knocks it into the water. The grasshopper's initial position is 0.45 m above the water surface and 0.25 m horizontally away from the fish's mouth. If the launch angle of the drop is 63° with respect to the water surface, how fast is the drop moving when it leaves the fish's mouth? ••

Figure P10.51

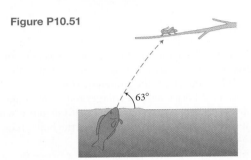

52. A burglar evading the police is running as fast as he can on a building's flat rooftop. At the edge, he sees another flat rooftop across an alley. The horizontal distance across the alley is 8.0 m, the other rooftop is 3.0 m lower than the one on which the burglar is standing, and he can run the 50-m dash in 5.6 s. Can he make the leap when moving at full speed? ••

53. You throw a ball at an angle of 30° above the horizontal at a wall 20 m away. The ball's initial speed is 15 m/s, and it leaves your hand at a height of 1.5 m above the ground. (a) How long does the ball take to get to the wall? (b) How high up on the wall does it hit? (c) What are the horizontal and vertical components of its velocity as it hits the wall? (d) When it hits, has it already passed the highest point in its trajectory? ••

54. Figure P10.54 shows a friend standing on the roof of a building that is 51.8 m tall. The roof is square and measures 20 m on a side. You want to shoot a paintball so that it lands on the roof and startles your friend, using a gun that shoots

Figure P10.54

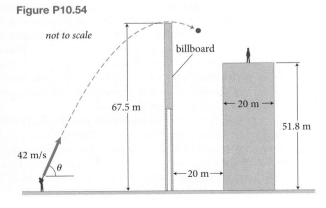

paintballs at a muzzle speed of 42 m/s. The only problem is a slim billboard 67.5 m high between you and the roof, 20 m in front of the building. You position yourself in front of the billboard such that when you hold the gun 1.5 m above the ground and fire, the paintball just barely gets over the billboard at the highest point in its trajectory. (a) At what angle θ above the horizontal do you need to shoot the ball to clear the billboard? (b) What is your horizontal distance from the billboard? (c) How long does the paintball take to move from the highest point in its trajectory to the height of the roof? (d) Does the ball strike the roof? (e) What is the speed of the ball when it strikes? •••

55. A brush fire is burning on a rock ledge on one side of a ravine that is 30 m wide. A fire truck sits on the opposite side of the ravine at an elevation 8.5 m above the burning brush. The firehose nozzle is aimed 35° above horizontal, and the firefighters control the water velocity by adjusting the water pressure. Because the water supply at a wilderness fire is limited, the firefighters want to use as little as possible. At what speed should the stream of water leave the hose so that the water hits the fire on the first shot? •••

56. You are standing on a slope of 20° to the horizontal and are going to throw a ball at 15 m/s up the incline (Figure P10.56). If you throw the ball at 35° with respect to the horizontal, at what distance up the incline from your feet does it land? Assume the ball leaves your hand directly overhead, 2.0 m above the slope. •••

Figure P10.56

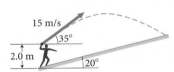

57. A block of inertia m sits at the origin of an xy coordinate system aligned along an incline that makes an angle ϕ with respect to the horizontal (Figure P10.57). You launch the block up the incline, with initial speed v_i directed at an angle θ with respect to the x axis in the plane of the incline. (a) What are the magnitude and direction of the block's acceleration expressed in terms of the indicated coordinate system? (b) Derive an expression for the block's velocity as a function of time, expressed in unit vector notation for the indicated coordinate system. (c) What is the maximum value of the block's displacement Δy along the y axis? (d) What is the block's maximum displacement Δx along the x axis? Ignore friction. •••

Figure P10.57

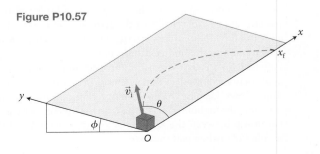

10.8 Collisions and momentum in two dimensions

58. A shell explodes prematurely in midair. What happens to the center of mass of the shell? •

59. A cue ball of inertia m is given a speed v before it collides elastically with a full rack of 15 stationary balls in a game of pool. If each of the 16 balls has an inertia m, what are (a) the *average* speed of each ball after they fly apart and (b) the *average* momentum of each ball after they fly apart?" ••

60. Object A, at rest on a low-friction table, is struck a glancing blow by object B. Show that, if the collision is elastic and the two objects have equal inertias, the angle between the final directions of motion of the two objects is 90°. ••

61. Disk 1 (of inertia m) slides with speed 1.0 m/s across a low-friction surface and collides with disk 2 (of inertia $2m$) originally at rest. Disk 1 is observed to bounce off at an angle of 15° to its original line of motion, while disk 2 moves away from the impact at an angle of 55°. (a) Calculate the unknown final speed of each disk. (b) Is this collision elastic? ••

62. A system of inertia 0.50 kg consists of a spring gun attached to a cart. The system is at rest on a horizontal low-friction track. A 0.050-kg projectile is loaded into the gun, then launched at an angle of 40° with respect to the horizontal plane. With what speed does the cart recoil if the projectile rises 2.0 m at its maximum height? ••

63. A spring ($k = 3800$ N/m) is compressed between two blocks: block 1 of inertia 1.40 kg and block 2 of inertia 2.00 kg. The combination is held together by a string (not shown in Figure P10.63). The combination slides *without spinning* across low-friction ice at 2.90 m/s when suddenly the string breaks, allowing the spring to expand and the blocks to separate. Afterward, the 2.00-kg block is observed to move at a 34.0° angle to its initial line of motion at a speed of 3.50 m/s, while the smaller block moves off at an unknown speed and angle. Neither block is rotating after the separation, and you can ignore the inertias of the spring and the string relative to those of the blocks. (a) Determine the velocity of block 1 after the separation. (b) Determine the original compression of the spring, $x - x_0$, from its relaxed length. ••

Figure P10.63

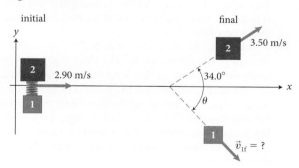

64. Disk P (inertia 0.40 kg) moves at an unknown velocity across a low-friction horizontal surface and collides with disk Q (inertia 0.70 kg), which is initially at rest. After the collision, the two (now slightly dented) disks move apart without spinning. Velocity information is provided in the initial and final top-view diagrams (Figure 10.64). (a) What was the initial velocity of disk P? (b) What fraction of the initial kinetic energy is converted during the collision? •••

Figure P10.64

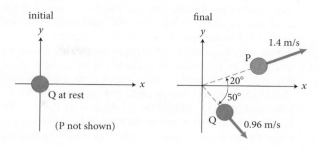

10.9 Work as the product of two vectors

65. If the angle between vectors $\vec{A}$ and $\vec{B}$ is 165° and if $A = 3.0$ m and $B = 2.5$ m, what is the value of $\vec{A} \cdot \vec{B}$? •

66. Vector $\vec{A}$ has a magnitude of 3.5 units, and vector $\vec{B}$ has a magnitude of 11 units. If the value of $\vec{A} \cdot \vec{B}$ is 22.4 units squared, what is the angle between $\vec{A}$ and $\vec{B}$? •

67. Vector $\vec{A}$ has a magnitude of 5.0 units in the negative y direction. Vector $\vec{B}$ has a positive x component of 3.0 units and a negative y component of 7.0 units. (a) What is the angle between the vectors? (b) Determine $\vec{A} \cdot \vec{B}$. •

68. A force $\vec{F} = F_x \hat{\imath} + F_y \hat{\jmath}$ with $F_x = 50$ N and $F_y = 12$ N is exerted on a particle as the particle moves along the x axis from $x = 1.0$ m to $x = -5.0$ m. (a) Determine the work done by the force on the particle. (b) What is the angle between the force and the particle's displacement? ••

69. A force $\vec{F} = F_x \hat{\imath} + F_y \hat{\jmath}$ with $F_x = 3.0$ N and $F_y = 2.0$ N is exerted on a particle that undergoes a displacement $\vec{D} = D_x \hat{\imath} + D_y \hat{\jmath}$ with $D_x = 2.0$ m and $D_y = -2.0$ m. (a) Determine the work done by the force on the particle. (b) What is the angle between $\vec{F}$ and $\vec{D}$? ••

70. Calculate $\vec{C} \cdot (\vec{B} - \vec{A})$ if $\vec{A} = 3.0 \hat{\imath} + 2.0 \hat{\jmath}$, $\vec{B} = 1.0 \hat{\imath} - 1.0 \hat{\jmath}$, and $\vec{C} = 2.0 \hat{\imath} + 2.0 \hat{\jmath}$. ••

71. You are playing with a toy in which a marble launched by a spring rolls along a track full of loop-the-loops and curves with almost no energy dissipated. You want the marble to have a speed of 0.70 m/s when it arrives at the top of a loop that is 0.30 m higher than the marble's position when you launch it. (a) What initial speed must the marble have? (b) If the marble has an inertia of 5.0 g and the spring has a spring constant of 13 N/m, how far back do you need to pull the spring on the level launching chute to get the marble to have this initial speed? ••

72. You are unloading a refrigerator from a delivery van. The ramp on the van is 5.0 m long, and its top end is 1.4 m above the ground. As the refrigerator moves down the ramp, you are on the down side of the ramp trying to slow the motion by pushing horizontally against the refrigerator with a force of 300 N. How much work do you do on the refrigerator during its trip down the ramp? ••

73. A 20-kg boy slides down a smooth, snow-covered hill on a plastic disk. The hill is at a 10° angle to the horizontal, and the slope is 50 m long. (a) If the boy starts from rest, what is his speed at the bottom of the hill? (b) His sister then hauls him and the disk from the bottom to the top by means of a rope attached to the disk. What average force must she exert on the rope if the boy is going 0.50 m/s when he reaches the top? ••

74. Two children pull a third child on a tricycle by means of two ropes tied to the handlebars. The combined inertia of the child and tricycle is 35 kg. One child pulls with a constant force of 100 N directed to the right of the straight-ahead direction at an angle of 45°. (*a*) If the second child pulls with a force of 80 N, at what angle to the left of the straight-ahead direction must he pull to make the tricycle move straight ahead? (*b*) With the second child pulling at the angle you calculated in part *a*, what is the acceleration of the tricycle? (*c*) What is the work done on the tricycle by the two ropes if the children pull constantly like this while the tricycle moves 2.0 m? ••

75. If $\vec{A} = A_x\hat{\imath} + A_y\hat{\jmath}$ and $\vec{B} = B_x\hat{\imath} + B_y\hat{\jmath}$, show that $\vec{A} \cdot \vec{B} = A_xB_x + A_yB_y$. (You may wish to use the fact that the scalar product satisfies the distributive property: $\vec{a} \cdot (\vec{b} + \vec{c}) = \vec{a} \cdot \vec{b} + \vec{a} \cdot \vec{c}$.) •••

76. Starting from rest, an intern pushes a 45-kg gurney 40 m down the hall with a constant force of 80 N directed downward at an angle of 35° with respect to the horizontal. (*a*) What is the work done by the intern on the gurney during the 15-m trip? (*b*) How fast is the gurney going when it has moved 15 m? (*c*) How much time elapses during the 15-m journey? Ignore friction. •••

77. You know you can provide 500 W of power to move large objects. You need to move a 50-kg safe up to a storage loft, 10 m above the floor. (*a*) With what average speed can you pull the safe straight up? (*b*) How much work do you do on the safe in doing this? (*c*) With what average speed can you pull the safe up a 30° incline? (*d*) How much work do you do on the safe moving if you slide it up the incline to the loft? •••

10.10 Coefficients of friction

78. A bartender gives a full mug of beer an initial speed v_{full}, and the mug stops in front of a patron sitting at the end of the bar. This patron later asks for a mug only half full. The bartender stands at the same position as when she sent the full mug to the patron and gives the mug an initial speed v_{half}. If the mug is to again stop in front of the patron, which is true: $v_{half} < v_{full}$, $v_{half} = v_{full}$, or $v_{half} > v_{full}$? •

79. The heavy crate in Figure P10.79 has plastic skid plates on its bottom surface and a tilted handle attached to one side. Which is easier: pushing the crate or pulling it? Assume your force is exerted along the incline of the handle in either case. •

Figure P10.79

80. You want to walk down your icy driveway without sliding. If the incline of the driveway is 15° from the horizontal, what must the coefficient of static friction be between your shoes and the ice? •

81. Moving a 51-kg box across a floor, you discover that it takes 200 N of force to get the box moving, and then 100 N keeps it moving at constant speed. What are the coefficients of static and kinetic friction between box and floor? •

82. Starting from rest, you push your physics book horizontally along a table. Plot the magnitude of the force of friction exerted on the book as a function of the magnitude of the force you exert. Include both static and kinetic cases in the range of force magnitudes you consider. ••

83. A resort uses a rope to pull a 55-kg skier up a 40° slope at constant speed for 100 m. (*a*) Calculate the tension in the rope if the coefficient of kinetic friction between snow and skis is $\mu_k = 0.20$. (*b*) How much work does the rope do on the skier? (If you did Problem 22, you might be interested in comparing your answers here with those in Problem 22.) ••

84. When moving on level ground, cross-country skiers slide their skis along the snow surface to stay moving. The coefficients of friction for a given set of skis and given snow conditions can be modified by various types of waxes. In order to move across the snow as fast as possible, (*a*) should you choose a wax that makes the coefficient of static friction between skis and snow as high as possible or as low as possible? (*b*) Should you choose a wax that makes the coefficient of kinetic friction between these two surfaces as high as possible or as low as possible? ••

85. A janitor is pushing an 11-kg trashcan across a level floor at constant speed. The coefficient of friction between can and floor is 0.10. (*a*) If the janitor is pushing horizontally, what is the magnitude of the force he exerts on the can? (*b*) If he pushes at an angle of 30° down from the horizontal, what must the magnitude of his pushing force be to keep the can moving at constant speed? ••

86. A hockey puck on the ice starts out moving at 10.50 m/s but after 40.00 m has slowed to 10.39 m/s. (*a*) What is the coefficient of kinetic friction between ice and puck? (*b*) On this same ice, what is the speed after 40.00 m when an identical puck is glued on top of the first one and the combination again starts at 10.50 m/s? ••

87. The coefficient of kinetic friction between tires and dry pavement is about 0.80. Assume that while traveling at 27 m/s you lock your brakes and as a result the only horizontal force on the car is the frictional one. (*a*) How many seconds does it take you to bring your car to a stop? (*b*) If the road is wet and the coefficient of kinetic friction between tires and pavement is only 0.25, how long does it take? (*c*) How far do you travel in this time interval? ••

88. A block of inertia m is placed on an inclined plane that makes an angle θ with the horizontal. The block is given a shove directly up the plane so that it has initial speed v, and the coefficient of kinetic friction between the block and the plane surface is μ_k. (*a*) How far up the plane does the block travel before it stops? (*b*) If the coefficient of static friction between block and surface is μ_s, what maximum value of θ allows the block to come to a halt somewhere on the plane and not slide back down? ••

89. A 1.0-kg block on a horizontal tabletop is pushed against the free end of a spring (the other end is attached to a wall) until the spring is compressed 0.20 m from its relaxed length. The spring constant is $k = 100$ N/m, and the coefficient of kinetic friction between block and tabletop is 0.20. When the block is released from being held against the compressed spring, how far does the block travel before coming to rest? ••

PRACTICE

90. A man exerts a constant force to pull a 50-kg box across a floor at constant speed. He exerts this force by attaching a rope to the box and pulling so that the rope makes a constant angle of 36.9° above the horizontal. The coefficient of kinetic friction for the box-floor interface is $\mu_k = 0.10$. What is the work done by the man as he moves the box 10 m? ••

91. An inclined plane that makes an angle of 30° with the horizontal has a spring of spring constant 4500 N/m at the bottom (Figure P10.91). A 2.2-kg block released near the top of the plane moves down the plane and compresses the spring a maximum of 0.0240 m from its relaxed length. (a) Ignoring any frictional effects, calculate the distance the leading edge of the block travels from the instant the block is released to the instant it briefly stops against the compressed spring. (b) Now suppose there is friction between the block and the surface of the plane, with $\mu_k = 0.10$. If the block is again placed somewhere on the plane and allowed to slide down and compress the spring by the same 0.0240 m, how far does the leading edge of the block travel now? ••

Figure P10.91

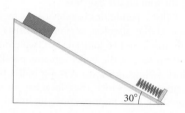

92. You push down on a book of inertia m resting on a table with a force directed at an angle θ away from vertical. The coefficient of static friction between book and table is μ_s. If θ is not larger than some critical value, you cannot get the book to slide no matter how hard you push. What is that critical value? •••

93. You push a 2.0-kg block up a ramp by exerting a 100-N force directed parallel to the ramp, which is at a 30° angle to the horizontal. (a) Ignoring any effects due to friction, calculate the block's speed when you have pushed it 2.0 m if its upward speed when you began pushing was 2.0 m/s. (b) Now consider friction. If the coefficient of kinetic friction between block and ramp is 0.25, what is the block's speed after you have pushed it 2.0 m? (c) With this value of μ_k, how much farther up the ramp does the block move once you stop pushing? •••

94. A platform that rolls on wheels and has inertia m_p is attached to a wall by a horizontal spring of spring constant k. A load of inertia m_ℓ sits on the platform, and the coefficient of static friction between platform and load is μ_s. If you pull the platform away from the wall so that the spring is stretched a distance x from its relaxed length and then let go, the platform-spring combination bounces back and forth. (a) At what position(s) in the motion is the load most likely to fall off? (b) If you want the load to stay in place held only by friction, what is the maximum distance x_{max} you can stretch the spring from its relaxed length and have the load stay on when you release the platform-spring combination? •••

95. You have to specify the power output of a motor for a ski tow rope that will carry 20 skiers at a time. The grade of the ski slope is 32° above horizontal, and the average coefficient of kinetic friction between skis and snow is 0.12. If the tow speed is to be 3.0 m/s and each skier has an inertia of 60 kg (a reasonable average value), what must the minimum power output of the motor be? •••

96. A spring of spring constant k is attached to a support at the bottom of a ramp that makes an angle θ with the horizontal. A block of inertia m is pressed against the free end of the spring until the spring is compressed a distance d from its relaxed length. Call this position A. The block is then released and moves up the ramp until coming to rest at position B. The surface is rough from position A for a distance $2d$ up the ramp, and over this distance the coefficient of kinetic friction for the two surfaces is μ. Friction is negligible elsewhere. What is the distance from A to B? •••

Additional Problems

97. A sports car skids to a stop, leaving skid marks 290 m long. If the coefficient of kinetic friction between tires and pavement is 0.50, how fast was the car going before the skid? •

98. If you were to throw a penny horizontally as hard as you could from the top floor of the Empire State Building, about how far from the base would it land? •

99. Plot the range of a projectile as a function of the launch angle above horizontal. •

100. Watering the garden, you need to have the water reach farther and instinctively raise the hose nozzle to increase the range of the water stream. You realize that this works because increasing the angle the nozzle makes with the ground increases the time interval during which the water is in flight, thereby letting the horizontal motion continue longer. This maneuver works only up to a point, however. Why? ••

101. A particle initially at the origin of an xy coordinate system and having an initial velocity of $(40 \text{ m/s})\hat{\imath}$ experiences a constant acceleration of $\vec{a} = a_x\hat{\imath} + a_y\hat{\jmath}$, with $a_x = -1.0 \text{ m/s}^2$ and $a_y = -0.50 \text{ m/s}^2$. (a) What is the maximum x coordinate of the particle's motion? (b) What is its velocity when it reaches this position? (c) What is its y coordinate at this instant? ••

102. Describe the properties of two vectors for which (a) $\vec{A} + \vec{B} = \vec{A} - \vec{B}$ and (b) $|\vec{A}| + |\vec{B}| = |\vec{A} + \vec{B}|$. ••

103. Three forces are exerted on a 2.00-kg block initially at rest on a slippery surface: a 100-N force directed along the positive x axis, a 50.0-N force that makes an angle of 30.0° with the positive x axis, and a 144-N force that makes an angle of 190° with the positive x axis. (a) What is the vector sum of the forces exerted on the block? (b) What is the work done on the block in 10.0 s? ••

104. An airline pilot begins a trip to Duluth from an airport located 1500 km south of Duluth. Her air speed is 260 m/s, but a wind blows from west to east at 40 m/s that takes her off course if she flies directly north. She has the choice of heading slightly westward so that the wind causes her to end up at Duluth or heading due north until she reaches a position directly east of Duluth and then heading west into the wind to Duluth. Which route takes a shorter time interval? ••

PRACTICE

105. The coefficient of static friction between layers of snow is 3.7 under ideal conditions. What is the steepest angle a snow field can form (θ in Figure P10.105) before there is an avalanche? ••

Figure P10.105

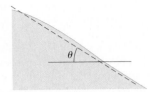

106. Having seen in *Principles* Example 10.6 that there is a simple expression for the horizontal range of a projectile across level ground, you begin to wonder whether there is an expression for the horizontal range of an object that is thrown with initial speed v_i at an angle θ above the horizontal from the edge of a cliff of height h. ••• CR

107. You have inherited property in Vermont that would make an excellent ski resort. One of the slopes has a cliff on the other side of the hill, and this gives you a money-saving idea. Instead of a chair lift or motorized tow rope, you decide to attach a pulley to the top of the cliff and then drape the tow rope over the pulley, with one end of the rope temporarily secured at the base of the ski slope and a counterweight attached to the end that hangs over the pulley at the cliff edge. The plan is to release the rope and pull two skiers (with the inertia for the pair kept between 100 kg and 200 kg) up the 400-m slope, which has an incline angle of 35° and a coefficient of kinetic friction between skis and snow no larger than 0.10. You guess that customers get nervous if they move faster than 5.0 m/s, and you wonder if a single counterweight can satisfy these constraints. ••• CR

108. You are investigating a rural accident. The evidence at the scene suggests that a car was traveling west when struck by a van traveling south. The crushed vehicles stuck together, then came to rest at the end of skid marks of length 14 m along a 76° line as shown in Figure P10.108. (The figure is a top view.) The ground is flat with no curbs, and you estimate a coefficient of kinetic friction $\mu_k = 0.70$ for the crushed vehicles along the path of the skid. You look up the inertias of these model vehicles and find that $m_c = 1400$ kg and $m_v = 6500$ kg. You know that the speed limit is 35 mi/h on both roads, which allows you to decide about driver citations. ••• CR

Figure P10.108

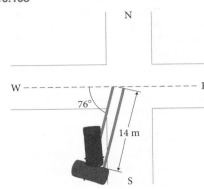

109. A 0.45-kg soccer ball is kicked at an angle θ above the ground with initial speed v_i. (a) Determine the height of the ball as a function of the horizontal distance it travels down the field. (b) For $\theta = 30°$, what initial speed must the ball have in order to land 48 m from where it is kicked? (c) If the foot is in contact with the ball for 0.15 s, what is the average force exerted by the foot on the ball? •••

110. A ballistic pendulum is a device for measuring bullet speeds. One of the simplest versions consists of a block of wood hanging from two long cords. (Two cords are used so that the bottom face of the block remains parallel to the floor as the block swings upward.) A 9.5-g bullet is fired into a ballistic pendulum in which the block has an inertia of 5.0 kg, and the block rises 60 mm above its initial position. (a) What is the speed of the bullet just before it hits the block? (b) How much energy is dissipated in the collision? What forms does this dissipated energy take? •••

111. A 0.010-kg bullet traveling at 300 m/s hits a 2.0-kg ballistic pendulum like the one described in Problem 110. However, the block is not thick enough for this bullet, and the bullet passes through the block, exiting with one-third of its original speed. How high above its original position does the block rise? •••

Answers to Review Questions

1. The logger sees the package fall in a curved trajectory. The pilot sees it fall in a straight line directly under the plane. The plane is directly above the logger when the package lands at his feet.
2. The accelerations are the same to both observers—namely, g.
3. This type of motion has two vector components, one vertical and one horizontal, and we need a reference axis for each.
4. In both cases, you place the tail of the second vector at the head of the first vector, and their sum is the vector extending from the tail of the first vector to the head of the second one. The difference is that in more than one dimension the magnitude of the sum depends on the angle between the vectors.
5. Subtraction is the same as if the vectors did lie along a straight line: Reverse the direction of vector 1 and then add that reversed vector to vector 2.
6. Yes, $\vec{A} + \vec{B} = \vec{B} + \vec{A}$. No, $\vec{A} - \vec{B} \neq \vec{B} - \vec{A}$.
7. To add two vectors $\vec{A}$ and $\vec{B}$, draw a scale model (magnitude and direction) of vector $\vec{A}$, then draw a scale model of vector $\vec{B}$ such that its tail is placed at the head of vector $\vec{A}$. The sum is a vector drawn from the tail of $\vec{A}$ to the head of $\vec{B}$. To subtract vector $\vec{B}$ from vector $\vec{A}$, reverse the direction of vector $\vec{B}$ and then add it to vector $\vec{A}$ using the addition process. The difference is a vector drawn from the tail of $\vec{A}$ to the head of the reversed $\vec{B}$.
8. The object is not moving in a straight line during the time interval Δt, but it is moving on a curved path.
9. The component of acceleration parallel to an object's velocity changes the magnitude of the velocity (speed) but does not change its direction. The component of acceleration perpendicular to the velocity changes the direction of the velocity but does not change its magnitude.
10. The normal component of this force is perpendicular to the surface of contact, and the tangential component is parallel to this surface.
11. (a)   (b)

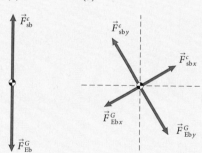

12. With one axis parallel to the acceleration, the components of force along the other axis add to zero, simplifying the algebra.
13. All three forces are components of the contact force experienced by two surfaces in contact with each other. The normal force is the component of the contact force perpendicular to the two surfaces. The force of friction, static or kinetic, is the component of the contact force parallel to the surfaces. The direction of any normal force always opposes interpenetration of the two surfaces; the direction of any frictional force always opposes slipping between the two surfaces. The magnitudes of the normal force and the force of static friction adjust within a range of possible values from zero to some maximum allowed value. Exceeding the maximum allowable normal force results in the failure (breaking) of one or both materials. When this happens, the surface can no longer exert a normal force and so its magnitude becomes zero. Exceeding the maximum allowable magnitude of the force of static friction results in slipping, in which case the force of static friction is replaced by a force of kinetic friction.

14. The force of static friction increases in magnitude as you push harder, so that its magnitude remains equal to the horizontal force you exert on the crate. The direction of the force of static friction opposes the direction of your push at all times.
15. Static friction rather than inertia is what needs to be overcome. The magnitude of your pushing force is smaller than the maximum magnitude of the frictional force the floor can exert on the desk.
16. The force of static friction is an elastic force and hence causes no dissipation of energy. The force of kinetic friction is not elastic and does cause energy dissipation. If this dissipation occurs at the system boundary, we have no means of knowing what fraction of the dissipated energy stays within the system and what fraction is transferred out of the system, so we cannot balance the energy equation.
17. Many answers are possible. The force of static friction between tires and road is responsible for accelerating a car, but it does no work on the car because there is zero force displacement. (Unless there is slipping, which would no longer involve static friction, the point of application of the force does not move.) When you carry a stack of books, the static friction between the top book and the one underneath it is responsible for accelerating the top book whenever you speed up or slow down. Work is done on the top book by the force of static friction because there is a nonzero force displacement.
18. Friction need not oppose motion, as in the case of the books in Review Question 17. What friction does always oppose is slipping, or *relative* motion between the surfaces in contact.
19. No, because in polar coordinates r represents a distance, not a position vector, and distances are always positive.
20. $B \sin \theta$. The sine function is the ratio of the opposite side to the hypotenuse, and the cosine is the ratio of the adjacent side to the hypotenuse. Because angles are generally given between a vector (hypotenuse) and the x axis, the cosine function is usually associated with an x component. Not in this case! (Something helpful for remembering which component is associated with $\sin \theta$ and which is associated with $\cos \theta$: Starting at the vector arrow, draw an arc across θ to the component that forms the other side of the angle. This component is the one associated with $\cos \theta$.)
21. A component vector is one of a set of vectors into which a given vector may be decomposed. Thus the sum of the component vectors equals the given vector. A vector component is the signed scalar that, together with an appropriate unit vector, describes the magnitude and direction of a component vector. The component vectors are related to the x and y components of the same vector by *Principles* Equation 10.5:

$$\vec{A} = \vec{A}_x + \vec{A}_y = A_x \hat{\imath} + A_y \hat{\jmath}.$$

This relationship reflects the option of considering vector addition as (scalar) addition of vector components in a chosen rectangular coordinate system, or (vector) addition of component vectors that lie along the chosen coordinate axes.
22. At the highest point, the vertical component of the velocity is 0, and so the ball's speed is given by the unchanged horizontal component: 30 m/s.
23. The path is parabolic. More specifically, $y(x)$ is a quadratic function.
24. The maximum height depends on only the vertical component of the launch velocity. However, a problem might involve a fixed launch speed with a variable angle of launch, in which case the horizontal and vertical components of the launch velocity are interdependent. The horizontal range of a projectile does depend on both components of launch velocity: the x component in order to move quickly horizontally, and the y component in order to remain in the air long enough to travel far.

25. The principles are the same: The momentum of the system is unchanged by the collision, and the coefficient of restitution determines the relative velocities of the colliding objects after the collision in terms of the relative velocities before the collision. The difference is that, in two dimensions, each momentum vector has two components, giving two equations that must be satisfied rather than one, as in one dimension. The relative velocities also have two components in two dimensions, but their relationship described by the coefficient of restitution still involves a single equation.

26. In one dimension there are two unknown final velocity components and two equations relating initial and final values. Momentum remains unchanged, and the initial and final relative velocities are related by the coefficient of restitution. In two dimensions these same principles provide three equations: two component equations to ensure that the momentum is unchanged, plus the single coefficient of restitution equation relating the initial and final relative velocities. However, in two dimensions there are in general four unknown final velocity components, two for each object. Thus in two dimensions at least one more piece of information is required for a complete solution.

27. The magnitude of one or both of the vectors might be zero, or the vectors might be perpendicular to each other, in which case the factor $\cos \phi$ in Eq. 10.33 is zero because $\phi = 90°$.

28. The work (a scalar) is the scalar product of the constant force (a vector) and the force displacement (a vector) between the two positions. It is equal to the product of the magnitudes (scalars) of these two vectors and the cosine of the angle between them.

29. The force displacement $\Delta \vec{r}_F$ cannot be used. Instead, the path along which the object moves from position 1 to position 2 must be broken up into infinitesimally small force displacements $d\vec{r}$. The work done is computed as a line integral of the scalar product of force and force displacement over the path from position 1 to position 2: $W = \int_1^2 \vec{F} \cdot d\vec{r}$.

30. The contact area is the macroscopic area over which the surfaces touch each other. The effective contact area is the actual area of microscopic contact between the two surfaces. The effective contact area is many thousands of times smaller than the contact area in everyday experience, and it is this effective contact area that increases as the normal force increases. The contact area does not increase with increasing normal force; hence, the force of friction is independent of the contact area. The available frictional force is dependent, as you would expect, on the effective contact area and hence on the normal force exerted by the surfaces on each other.

31. He has to tell you what material he had in contact with the concrete in order for the information to have any value. The coefficients of friction depend on the material of *both* surfaces in contact, not just one.

32. Kicking in the wedge increases the normal force exerted by the floor on the wedge and hence also increases the magnitude of the force of static friction exerted by the floor on the wedge.

33. Skidding involves kinetic friction because the tires and the road are sliding past each other. Rolling involves static friction because the tire and the road do not slip. The magnitude of the force of kinetic friction between any two surfaces is typically smaller than the maximum value of the magnitude of the force of static friction between the surfaces, and so it is the latter you want working for you when you stop a car.

Answers to Guided Problems

Guided Problem 10.2 (*b*) $h = \dfrac{gd^2}{2v_i^2 \cos^2 \theta} - d \tan \theta$

Guided Problem 10.4 0.88 m/s^2 at $11°$ W of N

Guided Problem 10.6 (*a*) $a = \dfrac{m_b - m_c \sin \theta}{m_b + m_c} g$; (*b*) $m_b = m_c \sin \theta$

Guided Problem 10.8 $\vec{v}_{p,i} = (0.85 \text{ m/s})\hat{\imath} + (0.37 \text{ m/s})\hat{\jmath}$

11

PRACTICE
Motion in a Circle

Chapter Summary   190

Review Questions   191

Developing a Feel   192

Worked and Guided Problems   193

Questions and Problems   200

Answers to Review Questions   208

Answers to Guided Problems   208

Chapter Summary

Rotational kinematics (Sections 11.1, 11.2, 11.4)

Concepts During **rotational motion**, all the particles in an object follow circular paths around the *axis of rotation*.

The **rotational velocity** ω_ϑ of an object is the rate at which the object's rotational coordinate ϑ changes.

The **rotational acceleration** α_ϑ is the rate at which an object's rotational velocity changes.

Quantitative Tools When an object travels a distance s along the circumference of a circle of radius r, the object's **rotational coordinate** ϑ is a unitless quantity defined as s divided by the circle's radius:

$$\vartheta \equiv \frac{s}{r}. \tag{11.1}$$

The arc distance s is measured from the positive x axis. To measure ϑ we need to choose a direction of increasing ϑ and a zero, just as we need to specify a direction of increasing x and an origin to measure position along an axis.

For any rotating object, the rotational velocity and rotational acceleration are

$$\omega_\vartheta = \frac{d\vartheta}{dt} \tag{11.6}$$

$$\alpha_\vartheta = \frac{d\omega_\vartheta}{dt} = \frac{d^2\vartheta}{dt^2}. \tag{11.12}$$

Translational variables for rotating objects (Sections 11.1, 11.4)

Concepts The velocity $\vec{v}$ of an object moving along a circle is always perpendicular to the object's position vector $\vec{r}$ measured from the axis of rotation.

The tangential component v_t of the velocity is tangent to the circle. The radial component v_r of the velocity is zero.

An object moving in a circle has a nonzero acceleration (even if its speed is constant) because the direction of the velocity changes.

An inward force is required to make an object move in a circle, even at constant speed.

Quantitative Tools The tangential and radial components of the velocity of an object moving along a circular path are

$$v_t = r\omega_\vartheta \tag{11.10}$$

$$v_r = 0. \tag{11.18}$$

The radial component of the acceleration is

$$a_r = -\frac{v^2}{r}. \tag{11.16}$$

This radial component is called the **centripetal acceleration** and is directed toward the center of the circle. It can also be written as

$$a_r = -r\omega^2.$$

The tangential component of the acceleration is

$$a_t = r\alpha_\vartheta. \tag{11.23}$$

The magnitude of the acceleration is

$$a = \sqrt{a_r^2 + a_t^2}. \tag{11.21}$$

Constant rotational acceleration (Section 11.4)

Concepts If the tangential acceleration a_t of a rotating object is constant, its rotational acceleration α_ϑ is also constant. In only that case, the rotational kinematics relationships for *constant rotational acceleration* apply.

Quantitative Tools If an object with constant rotational acceleration α_ϑ initially has a rotational coordinate ϑ_i and a rotational velocity $\omega_{\vartheta,i}$, then after a time interval Δt its rotational coordinate and rotational velocity are

$$\vartheta_f = \vartheta_i + \omega_{\vartheta,i}\Delta t + \tfrac{1}{2}\alpha_\vartheta(\Delta t)^2 \tag{11.26}$$

$$\omega_{\vartheta,f} = \omega_{\vartheta,i} + \alpha_\vartheta\Delta t. \tag{11.27}$$

Rotational inertia (Sections 11.3, 11.5, 11.6)

Concepts **Rotational inertia** is a measure of an object's tendency to resist any change in its rotational velocity. The rotational inertia depends on the inertia of the object and on how that inertia is distributed. The SI units of rotational inertia are kilograms-meters-squared ($\text{kg} \cdot \text{m}^2$).

Quantitative Tools The rotational inertia I of a rotating particle of inertia m is

$$I = mr^2, \tag{11.30}$$

where r is the distance from the particle to the rotation axis. For an extended object, the rotational inertia is

$$I = \int r^2 dm. \tag{11.43}$$

The **parallel-axis theorem:** If I_{cm} is the rotational inertia of an object of inertia m about an axis A through the object's center of mass, the rotational inertia I of the object about an axis parallel to A and a distance d away from A is

$$I = I_{cm} + md^2. \qquad (11.53)$$

Rotational kinetic energy and angular momentum (Section 11.5)

Concepts Rotational kinetic energy is the kinetic energy of an object due to its rotational motion.

Angular momentum L_ϑ is the capacity of an object to make other objects rotate.

A particle can have angular momentum even if it is not rotating.

The law of **conservation of angular momentum** says that angular momentum can be transferred from one object to another but cannot be created or destroyed. The angular momentum of an object or system is constant when no tangential forces are exerted on it.

Quantitative Tools The rotational kinetic energy of an object that has rotational inertia I and rotational speed ω is

$$K_{rot} = \tfrac{1}{2}I\omega^2. \qquad (11.31)$$

The angular momentum of an object that has rotational inertia I and rotational velocity ω_ϑ is

$$L_\vartheta = I\omega_\vartheta. \qquad (11.34)$$

The angular momentum of a particle of inertia m and speed v about an axis of rotation is

$$L = r_\perp mv, \qquad (11.36)$$

where $r_\perp$ is the perpendicular distance from the axis to the line of action of the particle's momentum. The distance $r_\perp$ is called the **lever arm distance** of the momentum relative to the axis.

Review Questions

Answers to these questions can be found at the end of this chapter.

11.1 Circular motion at constant speed

1. Give an example of (*a*) an object that has translational motion but no rotational motion, (*b*) an object that has rotational motion but no translational motion, and (*c*) an object that has both rotational motion and translational motion.
2. Is it possible for an object to have a nonzero acceleration if the object is traveling (*a*) at constant velocity and (*b*) at constant speed?
3. For an object moving along a circular path, what is the difference between the object's rotational coordinate ϑ and its polar angle θ?
4. For an object in circular motion at constant speed, describe the directions of the object's position vector (relative to the center of the circular trajectory), velocity vector, and acceleration vector at a given instant.

11.2 Forces and circular motion

5. If you are sitting in the passenger seat of a car that makes a quick left turn, your shoulders seem to lean to the right. What causes this apparent rightward motion?
6. Why shouldn't you use a rotating reference frame when analyzing forces?
7. What can you say about the vector sum of the forces exerted on an object that moves in a circle at constant speed? How does this vector sum of forces vary with speed and radius?

11.3 Rotational inertia

8. Is rotational inertia an intrinsic property of an object? Explain your answer.
9. Why does a tightrope walker carry a long pole?

11.4 Rotational kinematics

10. Does an object moving in a circle always have centripetal acceleration? Does it always have rotational acceleration? Does it always have tangential acceleration?
11. What is the mathematical expression that gives the relationship between rotational coordinate ϑ and polar angle θ?

12. Explain the distinctions among rotational acceleration, tangential acceleration, and centripetal acceleration.
13. For an object moving in circular motion, write an expression showing the relationship between (*a*) its arc length along the circle s and its rotational coordinate ϑ, (*b*) its velocity v_t and rotational velocity ω_ϑ, and (*c*) its tangential acceleration a_t and its rotational acceleration α_ϑ.
14. You are whirling a ball on the end of a string in a horizontal circle above your head. How does the ball's centripetal acceleration change if you increase the speed v so that the time interval needed to complete a revolution is halved?

11.5 Angular momentum

15. Define *rotational inertia* for a particle moving in a circle. Define *lever arm distance* for a particle moving near an axis of rotation.
16. Describe the relationship between the rotational kinetic energy of an object moving in circular motion and the object's rotational inertia.
17. How is angular momentum L related to momentum mv for an object of inertia m moving with constant velocity?
18. If both the rotational inertia I and the rotational speed ω of an object are doubled, what happens to the object's rotational kinetic energy?
19. What is the meaning of the statement: Angular momentum is conserved?

11.6 Rotational inertia of extended objects

20. Your physics book can be rotated around three mutually perpendicular axes passing symmetrically through its center. About which axis is the rotational inertia smallest? About which axis is it largest?
21. A diatomic molecule, which has a dumbbell shape, can rotate about three symmetrical axes that pass through the center of the molecule. Describe them. Rotation at a specific rotational speed about which axis gives the molecule the lowest kinetic energy?
22. State the parallel-axis theorem in words.

Developing a Feel

Make an order-of-magnitude estimate of each of the following quantities. Letters in parentheses refer to hints below. Use them as needed to guide your thinking.

1. The speed v of a point on the equator as Earth rotates (D, P)
2. The rotational inertia of a bowling ball about an axis tangent to its surface (A, R, X)
3. Your rotational inertia as you turn over in your sleep (V, C)
4. The angular momentum around the axle of a wheel/tire combination on your car as you cruise on the freeway (E, I, O, AA, S)
5. The angular momentum of a spinning ice skater with each arm held out to the side and parallel to the ice (G, X, N, U)
6. The speed you would need to orbit Earth in a low orbit (F, P)
7. The magnitude of the force exerted by the Sun on Earth to hold Earth in orbit (B, L, T, Z)
8. The kinetic energy associated with Earth's rotation (Z, P, D)
9. The angular momentum, about a vertical axis through your house, of a large car driving down your street (H, Y, M)
10. The kinetic energy of a spinning yo-yo (K, W, J, Q)

Hints

A. What is the inertia of a bowling ball?
B. How long a time interval is needed for Earth to make one revolution around the Sun?
C. What simple geometric shape is an appropriate model for a sleeping person?
D. What is Earth's rotational speed?
E. What is the combined inertia of the wheel and tire?
F. What is the relationship between force and acceleration for this orbit?
G. How can you model the skater's shape during her spin?
H. What is the inertia of a midsize car?
I. What is the radius of the tire?
J. How many turns are needed to rewind the yo-yo?
K. What is the yo-yo's rotational inertia?
L. What is the radius of Earth's orbit?
M. What is the perpendicular distance from the house to the car's line of motion?
N. What is the skater's rotational inertia with arms held out?
O. How can you model the combined rotational inertia of the wheel and tire?
P. What is Earth's radius?
Q. What is the final rotational speed?
R. What is the radius of a bowling ball?
S. What is the rotational speed of the tire?
T. What is the required centripetal acceleration?

U. What is the skater's initial rotational speed?
V. What is your inertia?
W. When thrown, how long a time interval does the yo-yo take to reach the end of the string?
X. What is needed in addition to the formulas in *Principles* Table 11.3 in order to determine this quantity?
Y. What is a typical speed for a car moving on a city street?
Z. What is Earth's inertia?
AA. What is a typical freeway cruising speed?

Key (all values approximate)

A. 7 kg; B. 1 y = 3×10^7 s; C. solid cylinder of radius 0.2 m; D. period = 24 h, so $\omega = 7 \times 10^{-5}$ s^{-1}; E. 10^1 kg; F. from Eqs. 8.6, 8.17, and 11.16, $\sum \vec{F} = m\vec{a}$, so $mg = mv^2/r$; G. a solid cylinder with two thin-rod arms of inertia 4 kg held out perpendicularly; H. 2×10^3 kg; I. 0.3 m; J. 2×10^1 turns; K. 6×10^{-5} kg · m^2 (with yo-yo modeled as solid cylinder); L. 2×10^{11} m; M. 2×10^1 m; N. 4 kg · m^2; O. between MR^2 (cylindrical shell representing tire) and $MR^2/2$ (solid cylinder representing wheel)—say, $3MR^2/4$; P. 6×10^6 m; Q. about twice the average rotational speed, or $\omega = 5 \times 10^2$ s^{-1}; R. 0.1 m; S. no slipping, so $\omega = v/r \approx 10^2$ s^{-1}; T. 8×10^{-3} m/s^2; U. $\omega \approx 10$ s^{-1}; V. 7×10^1 kg; W. 0.5 s; X. the parallel-axis theorem; Y. 3×10^1 mi/h; Z. 6×10^{24} kg; AA. 3×10^1 m/s

Worked and Guided Problems

Table 11.3 Rotational inertia of uniform objects of inertia M about axes through their center of mass

Rotation axes oriented so that object could roll on surface: For these axes, rotational inertia has the form cMR^2, where $c = I/MR^2$ is called the *shape factor*. The farther the object's material from the rotation axis, the larger the shape factor and hence the rotational inertia.

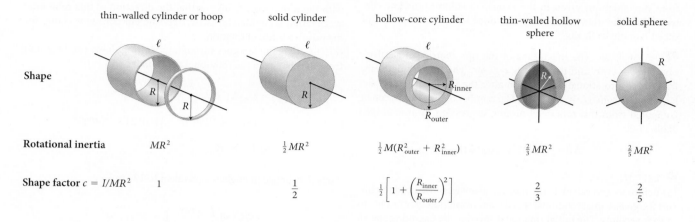

	thin-walled cylinder or hoop	solid cylinder	hollow-core cylinder	thin-walled hollow sphere	solid sphere
Rotational inertia	MR^2	$\frac{1}{2}MR^2$	$\frac{1}{2}M(R_{outer}^2 + R_{inner}^2)$	$\frac{2}{3}MR^2$	$\frac{2}{5}MR^2$
Shape factor $c = I/MR^2$	1	$\dfrac{1}{2}$	$\dfrac{1}{2}\left[1 + \left(\dfrac{R_{inner}}{R_{outer}}\right)^2\right]$	$\dfrac{2}{3}$	$\dfrac{2}{5}$

Other axis orientations

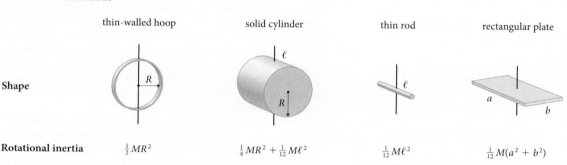

	thin-walled hoop	solid cylinder	thin rod	rectangular plate
Rotational inertia	$\frac{1}{2}MR^2$	$\frac{1}{4}MR^2 + \frac{1}{12}M\ell^2$	$\frac{1}{12}M\ell^2$	$\frac{1}{12}M(a^2 + b^2)$

These examples involve material from this chapter but are not associated with any particular section. Some examples are worked out in detail; others you should work out by following the guidelines provided.

Worked Problem 11.1 The fan

The ceiling fan in Figure WG11.1 has blades that are 0.60 m long and is spinning at 80 rev/min. It is turned off and comes to rest in 40 s. (*a*) What are the speed v and the magnitude a_c of the centripetal acceleration of the tip of any blade just before the fan is switched off? (*b*) After it is switched off, what is the fan's average rotational acceleration? (*c*) How many revolutions does the fan complete before coming to rest?

❶ **GETTING STARTED** We are given information about initial and final rotational speeds, $\omega_i = 80$ rev/min and $\omega_f = 0$. We must obtain kinematic quantities, and this sounds like a straightforward rotational kinematics problem. We approach this problem as we would a kinematics problem involving translational motion. We begin with a motion diagram for the tip of one of the blades (Figure WG11.2). We note that the fan is slowing down, which means that the points

Figure WG11.1

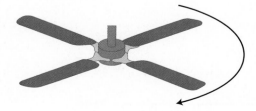

Figure WG11.2

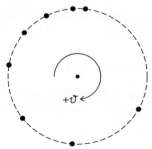

on the circle representing the blade tip at different instants get closer together as time goes on. We have to pick a direction for the positive sense of the rotation, which we arbitrarily choose to be clockwise as we look up at the fan. This is the equivalent of choosing the direction of the x axis in a translational motion problem. With this choice for direction, the initial rotational velocity is positive. The rotational variables we need should be directly available from the equations given in the *Principles* volume, and the centripetal acceleration can be obtained once we have computed the speed v of the blade tip.

❷ **DEVISE PLAN** The magnitude of the tip's centripetal acceleration is $a_c = v^2/r$, and its speed is $v = \omega r$. The magnitude of the fan's average rotational acceleration after the fan is turned off can be obtained from $\alpha_\vartheta = \Delta\omega_\vartheta / \Delta t$. The number of revolutions can be computed from this rotational motion expression from *Principles* Table 11.2:

$$\Delta\vartheta = \omega_{\vartheta,i}\Delta t + \tfrac{1}{2}\alpha_\vartheta(\Delta t)^2.$$

❸ **EXECUTE PLAN**

(*a*) Before we can calculate the numerical answers, we need to convert the given initial rotational speed from revolutions per minute to inverse seconds. The initial rotational speed is the operating speed, which is constant until the fan turns off. Therefore its initial instantaneous value is equal to the average (operating) value:

$$\omega_i = |\omega_{\vartheta,i}| = \left|\frac{\Delta\vartheta}{\Delta t}\right| = \left(\frac{1}{1\,\text{rad}}\right)\frac{\Delta\theta(\text{rad})}{\Delta t}$$

$$= 80\,\frac{\text{rev}}{\text{min}}\left(\frac{2\pi\,\text{rad}}{\text{rev}}\right)\left(\frac{1}{1\,\text{rad}}\right)\left(\frac{1\,\text{min}}{60\,\text{s}}\right) = 8.4\,\text{s}^{-1}.$$

Before the fan starts to slow down, the speed of the blade tip is

$$v_i = \omega_i r = (8.4\,\text{s}^{-1})(0.60\,\text{m}) = 5.0\,\text{m/s}. \checkmark$$

The magnitude of the initial centripetal acceleration of the tip is

$$a_c = \frac{v_i^2}{r} = \frac{(5.0\,\text{m/s})^2}{0.60\,\text{m}} = 42\,\text{m/s}^2. \checkmark$$

(*b*) From our choice of clockwise as the positive rotation sense, the initial rotational velocity is positive: $\omega_{\vartheta,i} = +\omega_i$. The average rotational acceleration during spin-down is

$$\alpha_\vartheta = \frac{\Delta\omega_\vartheta}{\Delta t} = \frac{\omega_{\vartheta,f} - \omega_{\vartheta,i}}{\Delta t} = \frac{0 - (+8.4\,\text{s}^{-1})}{40\,\text{s}} = -0.21\,\text{s}^{-2}, \checkmark$$

with the minus sign telling us that the direction of this rotational acceleration is counterclockwise, opposite our clockwise choice for the positive sense of rotation.

(*c*) The angle the fan goes through in slowing down can be found from the change in the rotational coordinate:

$$\Delta\vartheta = \omega_{\vartheta,i}\Delta t + \tfrac{1}{2}\alpha_\vartheta(\Delta t)^2$$

$$= (+8.4\,\text{s}^{-1})(40\,\text{s}) + \tfrac{1}{2}(-0.21\,\text{s}^{-2})(40\,\text{s})^2$$

$$= 1.7 \times 10^2.$$

Thus the change in angle is equivalent to 170 rad, or

$$(170\,\text{rad})\left(\frac{1\,\text{rev}}{2\pi\,\text{rad}}\right) = 27\,\text{rev}. \checkmark$$

❹ **EVALUATE RESULT** The negative sign in our result for part *b*, $-0.21\,\text{s}^{-2}$, indicates that the rotational acceleration is causing the rotational velocity to change in the direction opposite our chosen positive direction. Because the initial rotational velocity is positive, the negative value for rotational acceleration means the rotational velocity is becoming less positive and its magnitude is decreasing; in other words, the fan is slowing down. This is also consistent with a negative tangential acceleration of the tip.

The numerical values do not seem unusually large or small for a ceiling fan. It was necessary to treat the average rotational acceleration as if it were a constant rotational acceleration in order to use the kinematic equation for the angle the blade tip turns through as the fan slows down. This is permissible because, just as with translational motion, an average acceleration produces the same effect as a constant acceleration of the same magnitude.

Guided Problem 11.2 Swift shuttle

A space shuttle follows a circular orbit at an altitude of 300 km above the ground and has a period of 90.5 min. What are the shuttle's (*a*) rotational speed ω, (*b*) speed v, and (*c*) centripetal acceleration magnitude a_c?

❶ **GETTING STARTED**
1. How is rotational speed related to the given information?
2. Are kinematic equations sufficient to provide the relationships you need?
3. Sketch a motion diagram for the shuttle. How should the dots representing its position at various instants be spaced?

❷ **DEVISE PLAN**
4. Write an expression for the speed in terms of the rotational speed.
5. What is the numerical value of the orbit radius?
6. Write an expression for the magnitude of the centripetal acceleration.

❸ **EXECUTE PLAN**
7. Substitute numerical values for the known quantities. Watch out for units!

❹ **EVALUATE RESULT**
8. Is the speed reasonable for a spacecraft? How does the acceleration compare to g?

Worked Problem 11.3 Careful on that curve!

A highway has a flat section with a tight level curve of radius 40 m (Figure WG11.3). If the coefficient of static friction between tires and pavement surface is 0.45, what is the maximum speed v at which a car of inertia m can negotiate the curve safely—in other words, without the tires slipping?

Figure WG11.3

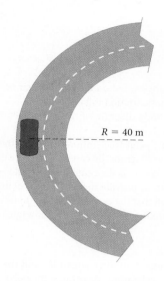

$R = 40$ m

❶ **GETTING STARTED** We know from experience that if it goes too fast in the curve, the car goes off the outer edge of the road. This tells us there must be an inward force exerted on the car to provide centripetal acceleration and thus keep it moving in circular motion. In this case, that force is the frictional force. (Note that even though the tires are moving, the force is the force of *static* friction because the tires are not sliding on the road surface.) Let us use Newton's second law, keeping in mind that there is an acceleration toward the center of the curve. The first step is to produce a free-body diagram for the car and choose our coordinate directions (Figure WG11.4).

Figure WG11.4

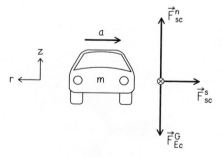

The frictional force is always directed opposite the direction the car would slide if there were no friction, which would be outward in this case. Thus the frictional force is aimed inward, supplying the necessary centripetal component of force directed toward the center of the curve (to the right in Figure WG11.4).

❷ **DEVISE PLAN** Because the acceleration and the frictional force are along our chosen radial r axis, we begin with Newton's second law along that axis:

$$\sum F_r = ma_r$$
$$F_{sc,r}^n + F_{sc,r}^s + F_{Ec,r}^G = ma_r$$
$$0 + (-F_{sc}^s) + 0 = m(-a_c)$$
$$F_{sc}^s = ma_c \qquad (1)$$

The centripetal acceleration magnitude is related to the car's speed through $a_c = v^2/R$, where R is the (fixed) radius of the curve. The frictional force is related to the normal force through Eq. 10.54, $F_{sc}^s \le \mu_s F_{sc}^n$. At the maximum speed, just before slipping, $\le$ is replaced by $=$ and we have

$$F_{sc,max}^s = \mu_s F_{sc}^n. \qquad (2)$$

This means we have enough equations to solve for the maximum speed v if we know the normal force. Because the normal force is along the z axis, we use Newton's second law along that axis to determine its magnitude.

❸ **EXECUTE PLAN** Starting with Eq. 1 in the form

$$F_{sc}^s = ma_c = m\frac{v^2}{R},$$

we have, at maximum speed,

$$F_{sc,max}^s = m\frac{v_{max}^2}{R}.$$

From Eq. 2 we make the substitution

$$\mu_s F_{sc}^n = m\frac{v_{max}^2}{R}. \qquad (3)$$

To get the magnitude of the normal force, we use Newton's second law in the z direction:

$$\sum F_z = ma_z$$
$$F_{sc,z}^n + F_{sc,z}^s + F_{Ec,z}^G = ma_z$$
$$(+F_{sc}^n) + 0 + (-mg) = m(0)$$
$$F_{sc}^n = mg.$$

We substitute this result into Eq. 3 to get

$$\mu_s mg = m\frac{v_{max}^2}{R}$$

$$v_{max} = \sqrt{\mu_s gR}$$

$$= \sqrt{(0.45)(9.8 \, \text{m/s}^2)(40 \, \text{m})}$$

$$= 13 \, \text{m/s} = 30 \, \text{mi/h}. \checkmark$$

PRACTICE

❹ **EVALUATE RESULT** The most remarkable feature of the result we obtained for v_{max} is that v_{max} does not depend on the car's inertia. This means that the maximum safe speed is 30 mi/h for both a compact car and a cement truck. Does this make sense? It seems that a truck should go more slowly because more force is required to keep it moving in a circle with the same acceleration as that exerted on a compact car. Although that is true, the frictional force is proportional to the normal force, which in turn is proportional to the inertia. Therefore the inertia cancels out.

A larger coefficient of friction μ_s means you can go faster without slipping, as is also the case for a bigger (straighter) curve with larger radius R. This is exactly what $v_{max} = \sqrt{\mu_s gR}$ says.

No assumptions were made, but we should review our decision to use the maximum available force of static friction. In order for the tires to slip, we must require an inward force greater than friction can provide, and so our use of maximum values is correct. The numerical value for the maximum speed is plausible and raises no red flag to cause us to question our numerical result.

Guided Problem 11.4 It's in the bank

Highway curves are often banked to reduce a vehicle's reliance on friction when negotiating the turn: On a banked curve, there is a centripetal component of normal force acting on the vehicle. When the angle and speed are such that friction plays no role in a vehicle's motion in the curve, the nature of the road surface is immaterial and thus the posted speed limit applies in both wet and dry weather. Suppose the posted speed limit is 100 km/h on a curve of radius 180 m. At what angle θ to the horizontal should the curve be banked so that reliance on friction is not necessary?

❶ **GETTING STARTED**

1. Draw a cutaway view that shows the bank angle θ (Figure WG11.5).

Figure WG11.5

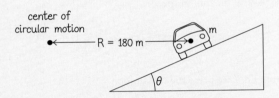

cutaway view

center of
circular motion
← R = 180 m ——— m
θ

2. Do you notice the similarity to Worked Problem 11.3 and to the roller skater leaning into a curve in *Principles* Example 11.4?

❷ **DEVISE PLAN**

3. What forces are exerted on the car? Remember that you want the force of static friction to be zero.

4. Draw a free-body diagram of the car as seen in the cutaway view.

5. Show the centripetal acceleration that keeps the car moving in a circle in your diagram. Be careful about the direction of this acceleration. Where is the center of the circle?

6. Which directions provide the best choice of an xy coordinate system? Recall that the acceleration should lie along one axis.

7. Count the unknowns. Does the bank angle appear in your equations?

❸ **EXECUTE PLAN**

8. From the component force equations and the relationship of centripetal acceleration magnitude to speed, perform some algebra to solve for the bank angle.

9. Get the numerical value for the angle. Don't forget units.

❹ **EVALUATE RESULT**

10. Examine your algebraic expression for the bank angle. How does the required angle change as the speed increases? As the radius of the curve increases? Are these variations what you expect?

11. Does the angle depend on a vehicle's inertia? Is this reasonable?

12. Consider the numerical value. Is this a reasonable angle? How steep are the angles you usually see on highways?

Worked Problem 11.5 Wind power

The commercial wind turbine in Figure WG11.6 consists of three 6400-kg blades, each 35.0 m long. It generates 1.0 MW of electrical power when it rotates with very little friction at a constant speed in the usually steady wind. However, when the wind suddenly stops, the turbine spins to a stop in 100 s. From this information, estimate the following for the situation when the wind is blowing steadily: (a) the rotational kinetic energy of the turbine, (b) its rotational speed, and (c) the magnitude of its angular momentum.

❶ **GETTING STARTED** When the rotational speed ω is constant, the turbine converts wind energy to electrical energy without changing its own rotational kinetic energy. When the wind stops, the turbine's rotational kinetic energy is converted to electrical energy for as long as that rotational kinetic energy lasts. It is this conversion of energy, not friction, that slows the turbine down when the wind stops. We can gain information about the constant-speed state by backtracking from the spin-down data.

Figure WG11.6

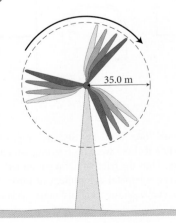

35.0 m

② **DEVISE PLAN** When the wind is steady, the turbine rotates with a constant rotational speed and has a certain amount of rotational kinetic energy. Once the wind dies, that kinetic energy is converted to electrical energy as the turbine spins down. During the spinning down, the work done by its environment on the turbine is equal to the change in the rotational kinetic energy:

$$\Delta K_{\text{rot}} = W_{\text{ct}}.$$

In Chapter 9 we saw that the work done by the environment on a system has the same magnitude as the work done by the system on its environment, but the two carry opposite signs. Therefore the work done by the environment on the turbine during the spin-down interval is the negative of the work W_{tc} done by the turbine on the environment during that interval. This work is the average power generated by the turbine multiplied by the time interval the turbine takes to come to rest:

$$K_{\text{rot,f}} - K_{\text{rot,i}} = W_{\text{ct}}$$
$$= -W_{\text{tc}}$$
$$= -P_{\text{av}}\Delta t. \qquad (1)$$

Knowing that the final rotational kinetic energy is zero, we can determine the initial rotational kinetic energy if we can reasonably estimate the average power during spin-down. Once we have $K_{\text{rot,i}}$, we can get the initial rotational speed ω_i and the magnitude of the angular momentum from

$$K_{\text{rot,i}} = \tfrac{1}{2} I_t \omega_i^2 \qquad (2)$$
$$L_\vartheta = I_t \omega_{\vartheta i}. \qquad (3)$$

③ **EXECUTE PLAN** (*a*) We suspect that the rate of power generation decreases as the blades slow down, but we are not given sufficient information to make this statement quantitative. So we need to make a reasonable assumption. Let us say that the average power generated during spin-down is half the constant-speed power of 1.0 MW: $P_{\text{av}} = 500$ kW. Then we can say from Eq. 1 that

$$K_{\text{rot,f}} - K_{\text{rot,i}} = -P_{\text{av}}\Delta t$$
$$0 - K_{\text{rot,i}} = -P_{\text{av}}\Delta t$$
$$K_{\text{rot,i}} = P_{\text{av}}\Delta t$$
$$= (5.0 \times 10^5 \text{ J/s})(100 \text{ s})$$
$$= 5.0 \times 10^7 \text{ J}$$
$$= 50 \text{ MJ}. ✔$$

(*b*) To obtain the rotational speed ω_i when the wind is blowing at a constant speed, we solve Eq. 2 for this variable:

$$\omega_i^2 = \frac{2K_{\text{rot,i}}}{I_t}.$$

First we must obtain the rotational inertia for the turbine, which can be modeled as three rods whose ends abut the turbine axle. For a rod of length ℓ and inertia m, the rotational inertia about an axis through one end of the rod is $m\ell^2/3$ (see *Principles* Example 11.11). Thus the rotational inertia for the three-blade turbine is $I_t = 3(m\ell^2/3) = m\ell^2 = (6400 \text{ kg})(35 \text{ m})^2 = 7.84 \times 10^6 \text{ kg} \cdot \text{m}^2$. The steady-state rotational speed of the turbine is then

$$\omega_i = \sqrt{\frac{2K_{\text{rot, i}}}{I_t}}$$
$$= \sqrt{\frac{2(5.0 \times 10^7 \text{ J})}{7.84 \times 10^6 \text{ kg} \cdot \text{m}^2}}$$
$$= 3.6 \text{ s}^{-1}. ✔$$

(*c*) For the angular momentum, Eq. 3 gives

$$|L_{\vartheta,i}| = |I_t \omega_{\vartheta,i}|$$
$$= I_t \omega_i$$
$$= (7.84 \times 10^6 \text{ kg} \cdot \text{m}^2)(3.6 \text{ s}^{-1})$$
$$= 2.8 \times 10^7 \text{ kg} \cdot \text{m}^2/\text{s}. ✔$$

④ **EVALUATE RESULT** By being careful in distinguishing between work done *by* the turbine and work done *on* it, we have gotten the signs right: We did not wind up with ω_i^2 being negative. The initial rotational speed $\omega_i = 3.6 \text{ s}^{-1}$ corresponds to a rotational period $T_i = 2\pi/\omega_i = 1.7$ s. This seems reasonable for a large commercial wind turbine and is consistent with observations if you have driven by a wind farm: The blades take a couple of seconds to complete each revolution.

We made several approximations: that the turbine blades are rod-shaped, that the rotational inertia of the turbine hub can be ignored, and that the average power during spin-down is half the constant-speed power. This last approximation is the most uncertain, but the average power should be some fraction of the constant-speed power because it must be between the initial and final values. Instead of saying the average power is half the constant-speed value, we might have said one-third or two-thirds. Those choices are both reasonable and would affect the numerical answers but not the spirit of our solution.

The hub's rotational inertia should not be significant because the hub radius is much smaller than the length of the blades. Modeling the blades as rods introduces some error, but the alternative is a very complicated integration procedure requiring specific information about the blades' shape. The error is not greater than the uncertainty introduced by the assumption that $P_{\text{av}} = P_{\text{constant speed}}/2$.

Guided Problem 11.6 Hit that door

You stand in front of a door of inertia m_d and width ℓ_d as it swings toward you with rotational speed ω_i. You throw a ball of inertia m_b and speed v_b toward the door so that it strikes the door perpendicularly at a distance d from the hinged edge, as shown in the overhead view (Figure WG11.7). The ball bounces directly back toward you with one-fourth its original speed. (a) What is the final rotational speed ω_f of the door after the collision? (b) What must the speed v_b of the ball be at impact if you want the door to swing away from you after the ball hits?

Figure WG11.7

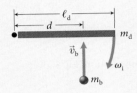

① GETTING STARTED

1. What is the best choice of system? Is this system isolated? Closed?
2. Modify Figure WG11.7 to obtain a two-part sketch of the system before and after the collision. Because rotational quantities

have signs, include a curved arrow about the rotation axis to indicate the direction you choose for the positive sense of rotation.

② DEVISE PLAN

3. How many unknown quantities do you seek in part a? Are there additional unknowns for part b?
4. What quantity is constant in this collision? Write the algebraic expression stating the equality of the initial and final values. Make sure to account for all the objects in your system on both sides of the equation.
5. What is the most convenient choice of location for the reference axis? Select the appropriate signs for each term in your equation. Base your decisions about signs on the chosen direction of positive rotation, even for the ball moving in a straight line.

③ EXECUTE PLAN

6. Solve for the door's final rotational speed ω_f.
7. How is ω_f different if the door swings away from you instead of toward you after the ball hits?

④ EVALUATE RESULT

8. Does the equation you get scale as you expect with changes in m_b, m_d, v_i, ℓ_d, and d? For instance, how does ω_f behave as the ball's inertia gets very large or very small?

Worked Problem 11.7 Just unwind

A block of inertia m_b is attached to a very light string wound about a uniform disk of inertia m_d and radius R_d, which rotates on a horizontal axle (Figure WG11.8). If the block is released from rest, what are its speed $v_{b,f}$ and acceleration a_b after it has fallen a distance d? Ignore any friction force exerted on or by the axle.

Figure WG11.8

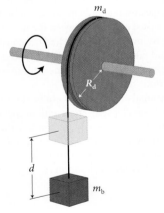

① GETTING STARTED Here we have two kinds of motion: the rotational motion of the disk about the axle and the translational motion of the block falling under the influence of gravity. We might try applying conservation of energy or angular momentum to determine the block's speed, but we will likely need Newton's second law or kinematic equations to get the block's acceleration. If we choose the disk, block, and string as our system, then the system's

angular momentum about the axle does not remain constant—the block and disk both contribute increasing angular momentum in the same direction. We have no experience in dealing with possible changes in Earth's angular momentum, and so perhaps we should try conservation of energy. If we choose the disk, block, string, and Earth as our system, our system is closed.

② DEVISE PLAN We can determine the speed of the block from its kinetic energy $\frac{1}{2}mv^2$. It gains kinetic energy through the conversion of gravitational potential energy as it falls. Some of the initial potential energy must also go into the kinetic energy of the disk, however, because the rotational and translational motions are connected by the string. Fortunately, we know the relationship: Because of the connecting string, a point on the outer rim of the disk must always have the same speed v as the block. Thus from Eq. 11.10 $v_b = \omega_d R_d$. Because there is no dissipation of energy and no source energy is involved, we know that the mechanical energy $K + U$ of the system is constant:

$$K_f + U_f = K_i + U_i.$$

(Remember that the absence of any subscripts denoting disk or block tells you that these variables are for the system as a whole.)

We can use energy accounting to determine the block's final speed $v_{b,f}$ after it has fallen a distance d. We can then use kinematic equations to get the block's acceleration a_b. The disk can be thought of as a very short solid cylinder, and so we can use the solid cylinder rotational inertia from Table 11.3: $I = mR^2/2$.

Because we want to use the formula for gravitational potential energy (mgy), we choose the positive y axis upward because that direction was assumed in the derivation of this equation (Eq. 7.21).

We also make a diagram to represent all the information given in a convenient way (Figure WG11.9).

Figure WG11.9

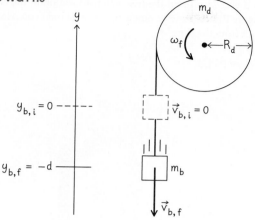

For convenience, we place the origin of our y axis at the initial level of the block, making $y_{b,i} = 0$ (and thus $U_{b,i} = m_b g y_{b,i} = 0$). Because the system starts at rest, $K_i = 0$. After the block has fallen a distance d, it is at $y_{b,f} = -d$, information that we put in our diagram. We can get the block's final speed $v_{b,f}$ by using this final position $-d$ and the relationship between the block's speed and the disk's rotational speed: $v_{b,f} = v_{d,f} = R_d \omega_{d,f}$.

❸ EXECUTE PLAN For our system (disk, block, string, Earth), energy conservation tells us that

$$K_f + U_f = K_i + U_i$$

$$\tfrac{1}{2} m_b v_{b,f}^2 + \tfrac{1}{2} I_d \omega_{d,f}^2 + m_b g y_{b,f} + m_d g y_{d,f}$$
$$= 0 + 0 + m_b g y_{b,i} + m_d g y_{d,i}. \qquad (1)$$

Because the vertical position of the disk does not change, $m_d g y_{d,f} = m_d g y_{d,i}$, and so we drop these two terms from the equation. Next, we eliminate the factor containing ω by modifying Eq. 11.10 to $\omega_{d,f} = v_{d,f}/R_d$ and then replacing $v_{d,f}$ with $v_{b,f}$ because any point on the disk rim has the same speed as the block. Then, with I_d replaced by $m_d R_d^2 / 2$, Eq. 1 becomes

$$\tfrac{1}{2} m_b v_{b,f}^2 + \tfrac{1}{2}\left(\tfrac{1}{2} m_d R_d^2\right)\left(\frac{v_{b,f}}{R_d}\right)^2 + m_b g(-d) = m_b g(0)$$

$$\tfrac{1}{2} m_b v_{b,f}^2 + \tfrac{1}{4} m_d v_{b,f}^2 = m_b g d$$

$$v_{b,f}^2 = \frac{2m_b}{m_b + \tfrac{1}{2} m_d} g d$$

$$v_{b,f} = \sqrt{\frac{2m_b}{m_b + \tfrac{1}{2} m_d} g d}. \checkmark \qquad (2)$$

Because we do not know the time interval needed for the block to travel the distance d but we do know the initial and final speeds, we can use *Principles* Eq. 3.13, $v_f^2 = v_i^2 + 2a(x_f - x_i)$, for the block's motion along the y axis to get its acceleration:

$$v_{by,f}^2 = v_{by,i}^2 + 2a_{by}(y_f - y_i)$$

$$\frac{2m_b}{m_b + \tfrac{1}{2} m_d} g d = (0)^2 + 2a_{by}(-d - 0) = -2a_{by}d$$

$$a_{by} = -\frac{m_b}{m_b + \tfrac{1}{2} m_d} g. \checkmark$$

❹ EVALUATE RESULT If it were not connected to the disk by the string, the block would be in free fall and its speed after it had fallen a distance d would be the familiar free-fall result $v = \sqrt{2gd}$. In our case, the speed is lower because of the presence of the term $\tfrac{1}{2} m_d$ in the denominator of Eq. 2. We expect this lower speed because some of the gravitational potential energy the system loses as the block falls must go into increasing the rotational kinetic energy of the disk rather than all of it going into increasing the kinetic energy of the block. We also expect that if $m_d \ll m_b$, the disk should have little effect. If we set $m_d = 0$ to get the limiting case, we indeed get the expected free-fall speed.

The negative sign in our acceleration expression indicates that the acceleration is downward because the positive y axis points upward, as we expect for a falling block that starts from rest. Because not all of the system's potential energy goes into translational kinetic energy, the block cannot fall as fast as it would if there were no rotational diversion. Therefore an acceleration magnitude less than g makes sense.

Guided Problem 11.8 Globe pulling

You have a very large globe of Earth that has an inertia of 20 kg and a radius of 0.50 m. It can rotate freely about its polar axis, and you want to start it spinning by wrapping a string around its equator and then pulling on the string. You plan to attach one end of the string to the equator with a bit of museum wax, so that it will not slip until all the string is unwound. If you are restricted to exerting a continuous pulling force of magnitude 2.0 N, how much string do you need to wrap tightly around the globe to make it spin at 0.50 rev/s?

❶ GETTING STARTED

1. Draw a sketch of the globe and string showing how the pulling force is exerted.
2. Is your force tangential to the globe surface? Which quantity should you focus on: angular momentum or energy? Can you compute the effect of your force on the quantity you choose?

3. What must you assume about whether the string slips as it is pulled?

❷ DEVISE PLAN

4. Write an equation for the quantity you named in step 2. Does the rotational speed appear in this expression?
5. What is the globe's rotational inertia? (Is the globe more like a thin spherical shell or a solid sphere?) How does this variable come into play in this problem?
6. How does the string length come into your calculation? Would a tangential force applied continuously at a single point on the equator (such as a force produced by a small rocket motor attached to the globe) produce the same effect?

❸ EXECUTE PLAN

❹ EVALUATE RESULT

Questions and Problems

For instructor-assigned homework, go to MasteringPhysics® (MP)

Dots indicate difficulty level of problems: • *= easy,* •• *= intermediate,* ••• *= hard;* CR *= context-rich problem.*

11.1 Circular motion at constant speed

1. For a given level of technology, there is a maximum density of "bits" of information that can be stored on a computer's hard-drive disk. The data are stored in a tight spiral, so that "tracks" are essentially circles of increasing radius from inner to outer parts of the disk. As the disk spins at a constant rate, data are read at a constant rate regardless of the radius of the track being read. Which determines the disk capacity: the innermost track or the outermost track? •

2. What is the ratio of the rotational speed of Earth's daily rotation to that of its annual revolution? •

3. When one race car passes another during a race, the passing car usually tries to pass on the inner lane of a curve in the track. Why? •

4. The toy race track in Figure P11.4 has three lanes, and cars cannot change lanes. The centerlines of any two adjacent lanes are 100 mm apart, and the innermost lane has a radius of 1.00 m. (*a*) If each car has the same average speed, compare the time intervals needed for the cars to complete 20 laps. (*b*) If all three cars start beside each other and finish the race at the same instant, how do their average speeds compare? ••

Figure P11.4

5. Plot, as a function of time, the rotational velocity of your car's wheels as you (*a*) back out of your driveway, (*b*) drive down a residential street, (*c*) turn onto a highway, (*d*) drive on the highway for a while, (*e*) exit the highway and stop at a traffic light, (*f*) drive on a city street, and (*g*) park at your favorite Chinese restaurant. ••

6. Hold a quarter down flat on a table with a finger. Then place a second quarter flat on the table, touching one point on the edge of the first quarter. If you roll the second quarter around the first so that the edges do not slip, describe the revolution and rotation as the second quarter returns to its original location on the table. •••

11.2 Forces and circular motion

7. Suppose the expression for centripetal acceleration is of the form $a_c = v^p r^q$, where p and q are constants. By analyzing the units in this expression, determine the numerical values of p and q. •

8. Standing in front of a tree, you whirl your sling with a rock in it. If you wish to hit the tree, at what location in the circle in Figure P11.8 (nearest point a, b, c, or d) should you release the rock from the sling? •

Figure P11.8

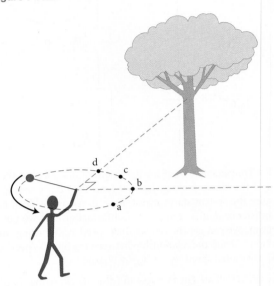

9. The Ferris wheel in Figure P11.9 is turning at constant speed. Draw and label free-body diagrams showing the forces on passengers at the numbered positions. Be careful to show correct relative lengths on your vector arrows. •

Figure P11.9

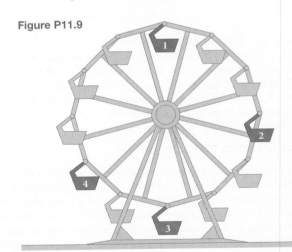

10. To see some of the trouble you can get into with noninertial reference frames, try this experiment. You and a friend stand opposite each other on the edge of a playground merry-go-round and get it rotating. While it is rotating, you attempt to toss a ball to your friend. As seen from above, what does the path of the ball look like (*a*) to an observer standing at the top of a nearby slide, (*b*) to you, and (*c*) to your friend? (*d*) If you did not know you were in a rotating reference frame, how would you explain the path of the ball in part *b*? ••

11. A salad spinner (Figure P11.11) is used to dry lettuce that has been washed. What physics is going on as the spinner does its job? ••

Figure P11.11

12. A race car is negotiating a curve on a banked track. There is a certain speed $v_{critical}$ at which friction is not needed to keep the car on the track in the curve. (This is the speed the car should travel if there is an oil slick on the track, for example.) Draw a free-body diagram showing the forces exerted on the car when it is moving at speed (a) $v = v_{critical}$, (b) $v > v_{critical}$, and (c) $v < v_{critical}$. ••

13. In a certain elaborate roller coaster, the cars are not on tracks but inside a U-shaped track much like a bobsled run. The cars are thus free to swing left and right across the U, climbing the walls as they do so. A student on the ride holds a string with a chunk of metal tied to the free end, with the chunk dangling between his knees as the ride begins. When the car enters a tight left turn, does the chunk swing toward the student's left knee, toward his right knee, or in some other direction? Ignore any friction between cars and track. ••

14. The ball in Figure P11.14 is attached to a vertical rod by two strings of equal strength (able to support much more than one of these balls) and equal length. The strings are light and do not stretch. The rod begins to rotate with its rotational speed increasing slowly, causing the ball to revolve with slowly increasing speed. (a) Which string breaks first? (b) To justify your answer to part a, draw a free-body diagram for the ball, indicating all forces exerted on it and their relative magnitudes. ••

15. Tie a heavy metal washer to one end of a string, hold the string by the other end, and whirl it in a horizontal circle above your head. Pay close attention to where your hand is and to where the washer is as you speed up the rotation. Ignore air resistance. (a) Draw the motion of your hand and the motion of the washer. At a given instant, draw the location of the washer and your hand and the string between them. (b) As you speed up the rotation, what force is responsible for the acceleration of the washer? Is there a tangential component to this acceleration? A radial component? (c) When the washer achieves some given rotational speed, you cease increasing that speed. Describe the motion of your hand now. •••

11.3 Rotational inertia

16. Floor lamps usually have a base with large inertia, while the long body and top have much less inertia. If you want to show off by spinning a floor lamp like a baton, where should you grab the lamp? •

17. A gymnast's backflip is considered more difficult to do in the layout (straight body) position than in the tucked position. Why? •

18. An ice cube and a rubber ball are both placed at one end of a warm cookie sheet, and the sheet is then tipped up. The ice cube slides down with virtually no friction, and the ball rolls down without slipping. If the ball and the ice cube have the same inertia, which one reaches the bottom first? •

19. (a) Is there a choice of axes of rotation for a rotating bowling ball and a rotating baseball such that the baseball has a greater rotational inertia? If so, describe the axes. If not, explain why not. (b) Repeat for a revolving bowling ball and a revolving baseball. ••

20. You are designing a uniform thin rectangular door to fill a 1 m wide × 2 m tall opening, and you want it to be easy to open. (a) Considering only vertical and horizontal axes, what placement of the axis of rotation would result in the smallest rotational inertia? (b) Is such an axis practical? ••

21. You stand on the edge of a playground carousel while your friends help you get it going. After they stop pushing, you walk toward the center. (a) What happens to the rotational inertia of the carousel-you system as you move inward from the edge? (b) How does this change in rotational inertia affect the motion of the carousel? ••

11.4 Rotational kinematics

22. A car's engine is turning the crankshaft at 6000 rev/min. What is the rotational speed ω? •

23. What is the rotational speed of a watch's second hand? Its hour hand? •

24. If your tires are worn, does your odometer (which is keyed to the turns of the axle) underestimate or overestimate the number of miles you drive? •

25. Moving at its maximum safe speed, an amusement park carousel takes 12 s to complete a revolution. At the end of the ride, it slows down smoothly, taking 2.5 rev to come to a stop. What is the magnitude of the rotational acceleration of the carousel while it is slowing down? ••

Figure P11.14

at rest   rotating

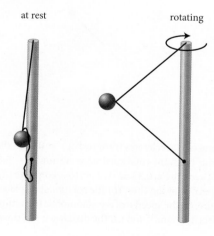

26. You have a weekend job selecting speed limit signs to put at road curves. The speed limit is determined by the radius of the curve and the bank angle. For a turn of radius 400 m and a 7.0° bank angle, what speed limit should you post so that a car traveling at that speed negotiates the turn successfully even when the road is wet and slick? ••

27. As you drive clockwise around a turn (as viewed from overhead), you see backed-up traffic ahead and so you slow down. Which force diagram in Figure P11.27 best illustrates your acceleration? ••

Figure P11.27

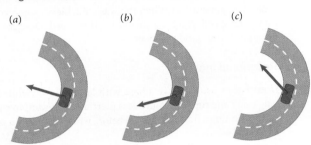

   (a)   (b)   (c)

28. At the end of a ride, a carnival carousel decelerates smoothly from some initial rotational speed ω to a stop. (a) Plot, as a function of time, the speed v of a carousel horse at radius R and of another horse at radius $2R$. (b) Plot the magnitude of the centripetal acceleration of the two horses as a function of time. ••

29. You attach one end of a string of length ℓ to a small ball of inertia m. You attach the string's other end to a pivot that allows free revolution. You hold the ball out to the side with the string taut along a horizontal line, as in Figure P11.29. (a) If you release the ball from rest, what is the tension T in the string as a function of the angle θ swept through? (b) What maximum tension should the string be able to sustain if you want it not to break through the ball's entire motion? ••

Figure P11.29

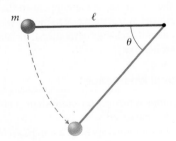

30. A pendulum is made of a bob dangling from a lightweight string of length ℓ. The bob is pulled sideways so that the string makes an angle θ_i with respect to the vertical, then released. As it swings down, what is the rotational speed of the bob as a function of the changing angle θ? ••

31. Assume that Earth's orbit around the Sun is a perfect circle. Earth's inertia is 5.97×10^{24} kg, the radius of its orbit is

1.50 × 10^{11} m, and its orbital period is 365.26 days. (a) What is the magnitude of its centripetal acceleration around the Sun? (b) What are the magnitude and direction of the force necessary to cause this acceleration? ••

32. A ball is suspended by a string from the top of a vertical rod as shown in Figure P11.32. The rod starts at rest and then begins to rotate with a small constant rotational acceleration. Qualitatively plot as a function of time the magnitude of (a) the vertical component and (b) the horizontal component of the tensile force exerted on the ball by the string. ••

Figure P11.32

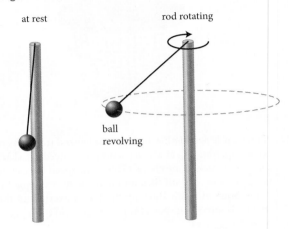

at rest   rod rotating

ball revolving

33. You hold a small ice cube near the top edge of a hemispherical bowl of radius 100 mm. When you release the cube from rest, what is the magnitude of its acceleration at the instant it reaches the bottom of the bowl? Ignore friction. ••

34. A 0.100-kg ball attached to a 0.20-m rod of very small inertia moves in a vertical circle in such a way that its speed remains constant. Let T_1, T_2, T_3, and T_4 be the tensions in the rod at positions 1, 2, 3, and 4 in Figure P11.34. (a) Rank the tensions in order of magnitude, greatest first. (b) What rotational speed is required at position 3 to keep T_3 barely greater than zero? (c) At that speed, what is the tension at position 1? ••

Figure P11.34

35. An automobile accelerates from rest at $t = 0$ such that its tires undergo a constant rotational acceleration $\alpha = 5.8 \text{ s}^{-2}$. The radius of each tire is 0.33 m. At $t = 10$ s, compute (a) the rotational speed ω of the tires, (b) the rotational displacement $\Delta\vartheta$ of each tire, (c) the speed v of the automobile (assuming the tires stay perfectly round), and (d) the distance it has traveled. ••

36. For the tread on your car tires, which is greater: the tangential acceleration when going from rest to highway speed as quickly as possible or the centripetal acceleration at highway speed? ●●

37. A roller-coaster car initially at a position on the track a height h above the ground begins a downward run on a long, steeply sloping track and then goes into a circular loop-the-loop of radius R whose bottom is a distance d above the ground (Figure P11.37). Ignore friction. (*a*) What is the car's speed when it reaches the bottom of the loop? (*b*) What is the magnitude of the normal force exerted on the car at that instant? (*c*) What is its speed when its position is one-quarter of the way around the loop? (*d*) What is the magnitude of the normal force exerted on it at the one-quarter position? (*e*) What is the car's acceleration at the one-quarter position? ●●●

Figure P11.37

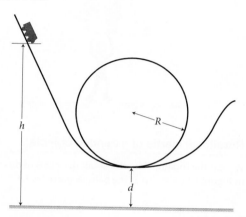

38. (*a*) Consider a circular vertical loop-the-loop on a roller coaster. Show that if a car coasts without power around the loop, the difference between the normal force exerted by the car on a passenger at the top of the loop and the normal force exerted by the car on her at the bottom of the loop is six times the gravitational force exerted on her. (*b*) This difference of $6mg$, referred to as "six gees," is quite a difference for the body to tolerate. To avoid this stress, vertical loops are teardrop-shaped rather than circular, designed so that the centripetal acceleration is constant all around the loop. How must the radius of curvature R change as the car's height h above the ground increases in order to have this constant centripetal acceleration? Express your answer as a function: $R = R(h)$. ●●●

39. A ball is put into a cone and made to move at a constant speed of 3.00 m/s in a horizontal circle of radius 0.500 m (Figure P11.39). (*a*) What is the centripetal component of the ball's acceleration? (*b*) What is the tangential component of its

Figure P11.39

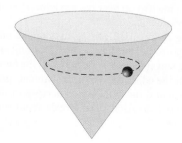

acceleration? (*c*) What force counteracts the force of gravity to keep the ball moving in a horizontal circle? (*d*) Use these insights to determine the height h at which the ball is circling above the bottom of the cone. [Hint: This is equivalent to determining the angle the cone makes with its vertical axis.] ●●●

40. A race car driving on a banked track that makes an angle θ with the horizontal rounds a curve for which the radius of curvature is R. (*a*) As described in Problem 12, there is one speed $v_{critical}$ at which friction is not needed to keep the car on the track. What is that speed in terms of θ and R? (*b*) If the coefficient of friction between tires and road is μ, what maximum speed can the car have without going into a skid when taking the curve? ●●●

11.5 Angular momentum

41. A 0.25-kg ball attached to a string is being spun around on a horizontal surface at a rotational speed of 2.5 rev/s. What is the ball's kinetic energy if the string is 1.0 m long? ●

42. Three small, identical 1.0-kg pucks are attached to identical 0.50-m strings, tied together at a common center as shown in Figure P11.42. When whirled in circular motion at angular speed 3.0 s^{-1}, what are (*a*) the rotational kinetic energy and (*b*) the magnitude of the angular momentum about the common center? ●

Figure P11.42

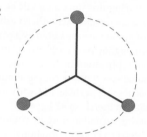

43. As you stand on the sidewalk outside an apartment building, a 2.0-kg bird flies toward you at 2.0 m/s along a straight line that passes 3.0 m above your head and runs parallel to the street. A friend is reclining with his head 4.0 m directly above your head, watching from a balcony. (*a*) What angular momentum do you observe the bird to have, using your head as a reference point? (*b*) What angular momentum does your friend observe the bird to have, using his own head as a reference point? ●

44. Puck A, of inertia m, is attached to one end of a string of length ℓ, and the other end of the string is attached to a pivot so that the puck is free to revolve on a smooth horizontal surface. Puck B, of inertia $8m$, is attached to one end of a string of length $\ell/2$, and the other end of the string is attached to a second pivot so that B is also free to revolve. In each case, the puck is held as far as possible from the pivot so that the string is taut and then given an initial velocity $\vec{v}$ perpendicular to the string. (*a*) How does the magnitude of the angular momentum of puck A about its pivot compare with that of puck B about its pivot? (*b*) How does the rotational kinetic energy of A compare with that of B? ●●

45. You have a pail of water with a rope tied to the handle. If you whirl the pail in a vertical circle fast enough, none of the water spills out, even when the pail is upside down. Explain why. ●●

46. Consider the roller-coaster car in Problem 37. At what minimum value of the starting height h above the ground must the car begin its journey if it is to remain on the track at the top of the loop? ●●

47. In Figure P11.47, two identical pucks B and C, each of inertia m, are connected by a rod of length ℓ and negligible inertia that is free to rotate about its center. Then puck A, of inertia $m/2$, comes in and hits B. After the collision, in which no energy is dissipated, what are (a) the rotational speed of the dumbbell and (b) the velocity $\vec{v}$ of A? ••

Figure P11.47

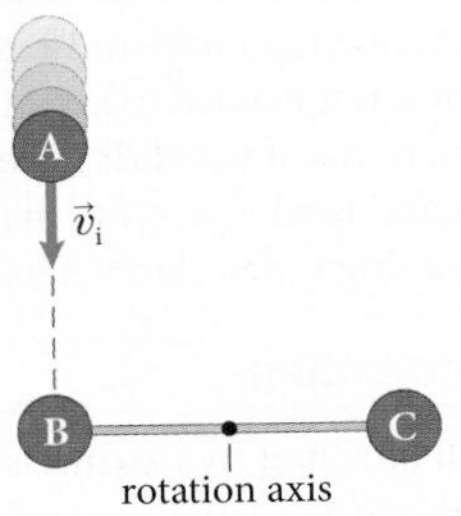

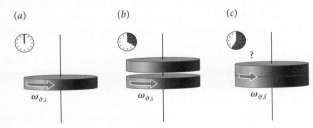

rotation axis

48. A small puck on an air table revolves in a circle with rotational speed ω, held at radius r by a weighted string that passes through a hole in the table. You slowly pull down on the weighted end of the string, decreasing the radius r. Assume that the angular momentum of the puck about the hole remains constant. (a) What is the rotational speed when half the string has been taken up? (b) What has happened to the speed v during this time interval? ••

49. Two skaters skate toward each other, each moving at 3.3 m/s. Their lines of motion are separated by a perpendicular distance of 2.0 m. Just as they pass each other (still 2.0 m apart), they link hands and spin about their common center of mass. What is the rotational speed of the couple about the center of mass? Treat each skater as a particle, one with an inertia of 75 kg and the other with an inertia of 48 kg. ••

50. A record player turntable initially rotating at $33\frac{1}{3}$ rev/min is braked to a stop at a constant rotational acceleration. The turntable has a rotational inertia of 0.020 kg · m². When it is switched off, it slows down to 75% of its initial rotational speed in 5.0 s. (a) How long does it take to come to rest? (b) How much work has to be done on the turntable to bring it to rest? ••

51. A disk of rotational inertia I about the central axis shown in Figure P11.51 is rotating about this axis with initial rotational velocity $\omega_{\vartheta,i}$ on low-friction bearings. A second identical disk is held at rest a few millimeters directly above the first disk and suddenly dropped. After some slipping, the two discs are observed to have a common rotational speed about the original axis of rotation. What is the magnitude of the final rotational velocity of the combined disks? ••

Figure P11.51

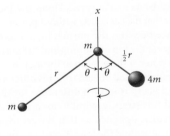

(a)   (b)   (c)

$\omega_{\vartheta,i}$ $\omega_{\vartheta,i}$ $\omega_{\vartheta,f}$

52. A popular ride at some playgrounds consists of two circular disks joined by rungs (Figure P11.52). The center of the unit is attached to a horizontal axle around which the unit rotates.

The rotational inertia of the unit is 200 kg · m², and its radius is 1.00 m. A 30.0-kg child runs and grabs the bottom rung of the ride when it is initially at rest. Because the child can pull herself up to meet the rung at waist height, treat her as a particle located on the rung. (a) If the friction between unit and axle is minimal and can be ignored, what is the minimum speed v the child must have when she grabs the bottom rung in order to make the unit rotate about its axle and carry her over the top of the unit? (b) Is this a reasonable speed for a child? •••

Figure P11.52

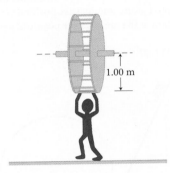

1.00 m

11.6 Rotational inertia of extended objects

53. What is the rotational inertia about the x axis of the rigid object in Figure P11.53? (Treat the balls as particles.) •

Figure P11.53

x

m

$\frac{1}{2}r$

r θ θ

$4m$

m

54. A cubical wire framework in which each edge is 0.25 m long has a 0.20-kg steel ball at each of its eight corners. What is the rotational inertia of this framework around an axis along one edge? Ignore the inertias of the wires. •

55. What is the formula for the rotational inertia of a thin spherical shell of inertia m and radius R about an axis tangent to any point on its surface? •

56. A deer's leg is very slender near the hoof, with most of the muscle bunched up near the hip joint. How does this work to the deer's advantage? •

57. A 20-kg child running at 1.4 m/s jumps onto a playground merry-go-round that has inertia 180 kg and radius 1.6 m. She is moving tangent to the platform when she jumps, and she lands right on the edge. What is the rotational speed of the merry-go-round and the child if the merry-go-round started from rest? Ignore any friction in the axle about which the platform rotates. ••

PRACTICE

58. Earth has an inertia of 5.97×10^{24} kg and a radius of 6.37×10^6 m. (*a*) Calculate its rotational inertia about the rotation axis, assuming uniform density. (*b*) The rotational inertia of Earth about its rotation axis has been measured to be 8.01×10^{37} kg·m^2. What does the difference between this value and the value you calculated in part *a* tell you about the distribution of inertia throughout the planet? ●●

59. The rotational inertia of any flat object about an axis perpendicular to the object is equal to the sum of the rotational inertias about two mutually perpendicular axes in the plane of the object, if all three axes pass through a common point. Using this "perpendicular-axis theorem," determine (*a*) the rotational inertia of a uniform hoop of inertia *m* and radius *R* about a diameter of the hoop, and (*b*) the rotational inertia of a uniform square sheet of inertia *m* and side length *a* about a line drawn from the midpoint of one edge to the midpoint of the opposite edge. ●●

60. The structure in Figure P11.60 is made of three identical rods of uniform density. Rank the rotational inertias of the structure, from smallest to largest, about the four rotation axes represented by the dashed lines. ●●

Figure P11.60

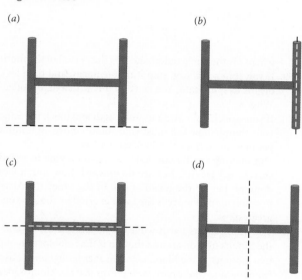

61. Global warming might result in the melting of the polar ice caps. What effect would this melting have on the length of the day? ●●

62. A hollow cylinder, a solid cylinder, and a billiard ball are all released at the top of a ramp and roll to the bottom without slipping. (*a*) Rank them, from least to greatest, according to the fraction of the kinetic energy that is rotational as they roll. (*b*) What are the ratios of their speeds when they reach the bottom of the ramp? ●●

63. Estimate the rotational inertia of a 70-kg athlete about a horizontal axis passing through her waist from left to right. ●●

64. A 0.20-kg block and a 0.25-kg block are connected to each other by a string draped over a pulley that is a solid disk of inertia 0.50 kg and radius 0.10 m. When released, the 0.25-kg block is 0.30 m off the ground. What speed does this block have when it hits the ground? ●●

65. You accidentally knock a full bucket of water off the side of the well shown in Figure P11.65. The bucket plunges 15 m to the bottom of the well. Attached to the bucket is a light rope that is wrapped around the crank cylinder. How fast is the handle turning when the bucket hits bottom? The inertia of the bucket plus water is 12 kg. The cylinder has a radius of 0.080 m and inertia of 4.0 kg. ●●

Figure P11.65

66. A block of inertia *m* is attached to a block of inertia 2*m* by a light string draped over a uniform disk of inertia 3*m* and radius *R* that can rotate on a horizontal axle (Figure P11.66). Initially, the block of inertia *m* is held in place so that the string is taut. When this block is released and so is free to move, what is its speed after it has risen a height *h*? Ignore friction between disk and axle. ●●

Figure P11.66

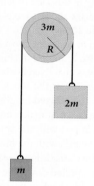

67. A thick-walled cylinder of inertia *m* has an outer radius R_{outer}, an inner radius $R_{\text{inner}} = R_{\text{outer}}/2$, and a length ℓ (Figure P11.67). It rotates around an axis attached to its outside along its length. What is its rotational inertia about this axis? ●●

Figure P11.67

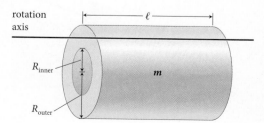

68. A thin 0.15-kg rectangular sheet has 40-mm short sides and 80-mm long sides. Determine the rotational inertia of the sheet about an axis perpendicular to it and located (*a*) at the center of a long side, (*b*) at the center of a short side, and (*c*) at a corner. ••

69. One type of wagon wheel consists of a 2.0-kg hoop fitted with four 0.80-kg thin rods placed along diameters of the hoop so as to make eight evenly spaced spokes. For a hoop of radius 0.30 m, what is the rotational inertia of the wheel about an axis perpendicular to the plane of the wheel and through the center? ••

70. A thin rod, 0.83 m long, is pivoted such that it hangs vertically from one end. You want to hit the free end of the rod just hard enough to get the rod to swing all the way up and over the pivot. How fast do you have to make the end go? ••

71. Big Ben in London is the most accurate mechanical clock of its size. The 300-kg hour hand is 2.7 m long, and the 100-kg minute hand is 4.2 m long. Calculate the rotational kinetic energy of the two-hand system, treating each hand as if it were a thin rod. ••

72. Consider an object that has rotational inertia I about some arbitrary axis of rotation. The *radius of gyration* of the object is the distance from this axis where all the object's inertia could be concentrated and still produce the same rotational inertia I about the axis. What is the radius of gyration of (*a*) a solid disk of radius R about an axis perpendicular to the plane of the disk and passing through its center, (*b*) a solid sphere of radius R about an axis through its center, and (*c*) a solid sphere of radius R about an axis tangent to the sphere? •••

73. A thin piece of metal in the shape of a right triangle has inertia m. Its perpendicular sides are of lengths ℓ and 2ℓ. Determine the rotational inertia about an axis that runs (*a*) along side ℓ and (*b*) along side 2ℓ. •••

74. A block of inertia m is attached to a light string wound around a uniform disk also of inertia m (Figure P11.74). The disk has radius R and rotates on a fixed horizontal axle through its center. The opposite side of the block is attached to a spring of force constant k and relaxed length ℓ attached at the bottom of an inclined surface that makes an angle θ with the horizontal. The block is at rest with the spring stretched a distance d from its relaxed length, with the block held in this position by a clamp. When the clamp is loosened so that the block is free to move, the block is pulled down the plane by the spring. What is the speed of the block when the spring is at half of its relaxed length? Ignore friction. •••

Figure P11.74

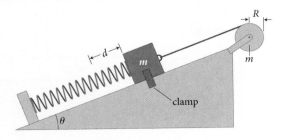

75. A certain baseball bat is a tapered cylinder that increases in diameter from 0.040 m at one end to 0.070 m at the other end and is 0.84 m long. The bat has uniform density and inertia 0.85 kg. What is the rotational inertia of this bat around an axis that passes perpendicularly through the skinny end? •••

Additional Problems

76. In the laboratory, you are trying to put pennies on a rotating platform so that they don't slide off. Your partner tells you, "Pennies on the outside stay on better because $a_c = v^2/r$. For larger r, the centripetal acceleration is less, and so the frictional force has an easier time holding a penny in place out there." Is his reasoning correct? •

77. Solar and lunar eclipses are caused by a fortuitous relationship between the sizes of the Moon and Sun and their distances from Earth. Look up data for the Earth-Moon and Earth-Sun distances and the diameters of the Moon and the Sun. (*a*) Calculate the *angular size* of the Sun's disk as seen by an observer on Earth, where *angular size* is the angle *subtended* by the Sun's disk. (*b*) Calculate the angular size of the Moon's disk. •

78. Show that the centripetal acceleration of an object undergoing circular motion at constant speed can be expressed as

$$a_c = \frac{4\pi^2 r}{T^2},$$

where r is the circle radius and T is the period of the motion. •

79. If you spin a raw egg, stop it momentarily, and then let go, it begins to spin again. You can't do this with a hard-boiled egg. Why? •

80. If you stretch your arm out full length and block the Sun with your thumb, how fast must your thumb move, in millimeters per minute, to track the Sun's motion? ••

81. Out looking for Tarzan, Jane swings on a vine to speed her search. (*a*) During her arc downward, how much work is done on her by the tensile force in the vine? (*b*) At which position along the arc is the force exerted by the vine on Jane greatest? ••

82. The Moon orbits Earth while Earth orbits the Sun, and thus the Moon's motion around the Sun is the combination of those two motions. Is the Moon's motion ever *retrograde* if observed from the Sun, where *retrograde* means that the Moon appears to move in the direction opposite the direction of Earth's motion relative to the distant stars? (You need to look up values for the Earth-Sun distance and the Earth-Moon distance to answer this question.) ••

83. Suppose several particles are in concentric circular orbits under the influence of some centripetal component of force, the magnitude of which depends on distance from the center of the motion. Each particle has a different orbital radius r and a corresponding period T. How does the force depend on r (*a*) if T is proportional to $\sqrt{r}$, (*b*) if T is proportional to $r^{3/2}$, and (*c*) if T is independent of r? ••

84. Approximate the Sun as a uniform sphere of radius 6.96×10^8 m, rotating about its central axis with a period of 25.4 days. If, at the end of its life, the Sun collapses inward to form a uniform dwarf star that is approximately the same size as Earth, what will the period of the dwarf's rotation be? ••

85. An open door of inertia m_d and width ℓ_d is at rest when it is struck by a thrown ball of clay of inertia m_b that is moving at speed v_b at the instant it strikes the door (Figure P11.85). The impact location is a distance $d = \frac{2}{3}\ell_d$ from the rotation axis through the hinges. The clay ball strikes perpendicular to the door face and sticks after it hits. What is the rotational speed ω of the door-ball system? Do not ignore the inertia m_b. ••

Figure P11.85

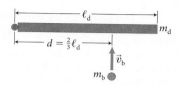

86. Your aunt owns an amusement park, and she wants you to add a circular loop to an existing roller-coaster ride. The first hill for the existing roller coaster is 55 m tall, and your aunt wants you to build, right after this hill, the tallest loop possible without having the cars fall out of the track or the passengers fall out of the cars. You think for a minute and realize what the minimum speed at the top of the loop has to be, and this gives you what you need to design the loop. ••• CR

87. The spacecraft in the movie *2001: A Space Odyssey* has a rotating cylinder to create the illusion of gravity, inside of which the crew walks and exercises. While watching this old movie, you realize that the radius of the cylinder is roughly three times a crew member's height. You imagine this cylinder rotating sufficiently rapidly to replicate Earth's gravity, but you worry about the difference between the acceleration at the top of the astronaut's head and the acceleration at her feet. Not wanting lightheaded astronauts, you phone NASA with your new design. ••• CR

88. A turntable of radius 0.320 m is fitted with an electromagnet that causes steel objects to stay in place on the table. A triangular mark on the rim shows the rotational position of the table relative to the surroundings (Figure P11.88). The turntable completes one rotation in 0.360 s, and the rotation is counterclockwise in the view shown in the figure. A 0.125-kg steel rod of length 0.160 m is placed on the table so that one end is at the center and the rod is oriented radially outward, as shown in the figure. The electromagnet is switched on, and the turntable is started. After the table reaches its steady-state rotational speed, the electromagnet is switched off at an instant when the rod lies along the positive x axis. Describe the ensuing motion of the rod. •••

Figure P11.88

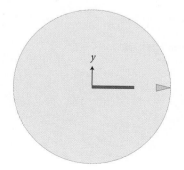

89. You have a toy top that plays music if the rotational speed is at least 1800 rev/min. The way to get this top spinning is to pull on a string wrapped around a spindle of diameter of 6.0 mm. The top has a rotational inertial of about 5.0×10^{-3} kg·m². The string is 1.2 m long. You want to hear the top sing. ••• CR

90. A dart of inertia m_d is fired such that it strikes with speed v_d, embedding its tip in the rim of a target that is a uniform disk of inertia m_t and radius R_t. The target is initially rotating clockwise in the view shown in Figure P11.90, with rotational speed ω about an axis that runs through its center and is perpendicular to its plane. Assume that the dart's inertia is concentrated at its tip. What is the final rotational speed of the target if the dart strikes (*a*) tangent to the target rim as in Figure P11.90*a* and (*b*) normal to the rim as in Figure P11.90*b*? •••

Figure P11.90

(*a*)   (*b*)

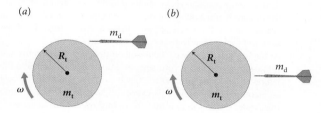

91. A smooth-faced cube of inertia m and side length d sliding along a highly polished floor at speed v strikes a door threshold, and the collision makes the cube tilt as shown in Figure P11.91. As the cube tips over, its center of mass is located at a variable horizontal distance x. What is the rotational speed of the cube as a function of x? The rotational inertia of a cube about one edge is $\frac{2}{3}md^2$. [Hint: Call the vertical distance traveled by the center of mass h.] •••

Figure P11.91

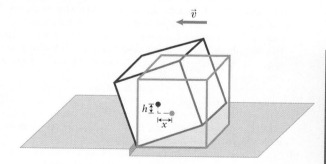

Answers to Review Questions

1. Many answers possible. Some are: (*a*) falling brick, racehorse on home stretch, cruising airplane; (*b*) spinning compact disc, Ferris wheel, roulette wheel; (*c*) thrown Frisbee, propeller on outboard motor; bowling ball headed toward pins.

2. (*a*) No, because constant velocity means zero acceleration. (*b*) Yes, because a change in direction means nonzero acceleration even if there is no change in speed.

3. The rotational coordinate specifies the location of the object on the circle in terms of a unitless number that increases by 2π for each revolution of the object. The polar angle specifies the location of the object on the circle in units of radians, degrees, or revolutions, typically measured from the positive x axis to the radius vector that locates the object.

4. The position vector points from the center of the circular trajectory to the object at the chosen instant. The velocity vector is perpendicular to the position vector, tangent to the circular trajectory in the direction of the object's motion (along the tangential axis). The acceleration vector points opposite the position vector, toward the center of the circular trajectory.

5. At the beginning of the car's curving motion, your shoulders continue going straight but your legs and lower body are pulled to the left by the frictional force exerted by the car seat. Because the car is accelerating into the curve (leftward) in the Earth reference frame, the car door comes in to meet your right shoulder, an occurrence you mistake for your shoulders traveling out of the curve (rightward) in the car's reference frame.

6. Because a rotating reference frame is accelerating and therefore noninertial. From *Principles* Chapter 6 we know that the methods we use in this text for monitoring energy and momentum changes are not valid in noninertial reference frames.

7. A vector sum of the forces must be directed toward the center of the circle to make the object move at a constant speed in circular motion. The vector sum of the forces provides the required centripetal acceleration, and it increases with increasing speed and decreases with increasing radius.

8. No. An intrinsic property (see *Principles* Section 4.6) is one that we cannot change without changing the object. The value we calculate for a given object's rotational inertia depends on what we choose as the axis of rotation. The rotational inertia of a thin hoop, for instance, is mr^2 for rotation about an axis perpendicular to the hoop face but only $mr^2/2$ for rotation about an axis parallel to the hoop face. Same hoop, different values of I, meaning I is not an intrinsic property.

9. The rotation axis the walker is concerned about is a horizontal axis passing through her feet. Without the pole, all the material in the system (just her) is near the axis. With the pole, some of the material of the walker-pole system is far from the axis, making it more difficult for her to rotate about that axis and fall.

10. Yes. Circular motion always involves centripetal acceleration. The other two forms of acceleration may or may not be involved in the motion. If speeding up or slowing down, the object has both rotational acceleration and tangential acceleration; these are related by a factor of the radius, so they always coexist.

11. The rotational coordinate is the polar angle (expressed in units of radians) divided by one radian:

$$\vartheta = \frac{\theta\,(\text{rad})}{1\,\text{rad}}.$$

12. Rotational acceleration has units of inverse seconds squared and measures the rate at which the rotational speed ω changes with time. Tangential acceleration has units of acceleration (m/s^2) and measures the rate at which the speed v changes with time. They are related in that the magnitude of the tangential acceleration equals the product of the radius and rotational acceleration. Centripetal acceleration has units of acceleration (m/s^2) and measures the rate at which the *direction* of velocity $\vec{v}$ changes with time. Centripetal acceleration is the radial component of acceleration (toward the center of the circle), while tangential acceleration is the component of acceleration tangent to the circle.

13. These relationships are found in *Principles* Table 11.1: (a) $s = r\vartheta$, (b) $v_t = r\omega_\vartheta$, (c) $a_t = r\alpha_\vartheta$.

14. If the time interval is halved, the speed v is doubled. Equation 11.15, $a_c = v^2/r$, tells you that doubling the speed quadruples the centripetal acceleration.

15. The rotational inertia of a particle is defined as the product of the particle's inertia and the square of its radius of circular motion. The lever arm distance (or lever arm) is the perpendicular distance between the line of action of the momentum of the particle and the rotation axis.

16. The rotational kinetic energy of an object is related to its rotational inertia in the same way that the kinetic energy of a particle is related to its inertia, with rotational inertia substituted for inertia and rotational speed substituted for speed:

$$\tfrac{1}{2}mv^2 \leftrightarrow \tfrac{1}{2}I\omega^2.$$

17. The angular momentum of this object is the product of its momentum and the perpendicular distance from the line of action of its momentum to the axis of rotation: $L = I\omega = mvr_\perp$ (Eqs. 11.34 and 11.36).

18. It increases by a factor of 8. The doubling of I in $K_{\text{rot}} = \tfrac{1}{2}I\omega^2$ causes K_{rot} to increase by a factor of 2, and the doubling of ω causes K_{rot} to increase by a factor of 4.

19. Angular momentum can be neither created nor destroyed (but it can be transferred from one object to another).

20. Smallest rotational inertia: axis running through the center of the book and parallel to its largest dimension. Largest rotational inertia: axis passing through the center of the book and perpendicular to both covers.

21. Two axes pass through the molecule's center perpendicular to the bond connecting the two atoms; the third axis runs parallel to the bond. Rotation about either perpendicular axis involves a sizable rotational inertia. Rotation around the parallel axis involves a smaller rotational inertia and therefore a smaller kinetic energy.

22. The rotational inertia of an object about any axis A is equal to the rotational inertia of the same object about an axis that runs through its center of mass and is parallel to axis A plus a term equal to the product of the object's inertia and the square of the perpendicular distance between the two axes.

Answers to Guided Problems

Guided Problem 11.2 (*a*) $\omega = 1.16 \times 10^{-3}\,\text{s}^{-1}$,
(*b*) $v = 7.73 \times 10^3\,\text{m/s}$, (*c*) $a_c = 8.94\,\text{m/s}^2$

Guided Problem 11.4 $\theta_{\text{bank}} = 23.6°$

Guided Problem 11.6 (*a*) $\omega_f = \dfrac{15m_b v_b d}{4m_d \ell_d^2} - \omega_i$, (*b*) $v_b > \dfrac{4m_d \ell_d^2 \omega_i}{15m_b d}$

Guided Problem 11.8 $\ell = 8.2\,\text{m}$

12

PRACTICE
Torque

Chapter Summary 210

Review Questions 212

Developing a Feel 213

Worked and Guided Problems 214

Questions and Problems 221

Answers to Review Questions 230

Answers to Guided Problems 231

Chapter Summary

Torque (Sections 12.1, 12.5)

Concepts The **torque** due to a force exerted on an object is the tendency of the force to give rotational acceleration to the object. The torque about some pivot is the product of the magnitude of the force and the lever arm distance. The SI units for torque are $N \cdot m = kg \cdot m^2/s^2$.

An object is in *translational equilibrium* if the vector sum of all the external forces exerted on it is zero. An object is in **rotational equilibrium** if the vector sum of the external torques is zero. If the vector sums of both the forces and the torques are zero, the object is in **mechanical equilibrium.**

Quantitative Tools If $\vec{r}$ is the position vector from a pivot to the location at which a force $\vec{F}$ is exerted on an object and θ is the angle between $\vec{r}$ and $\vec{F}$, the torque τ produced by the force about the pivot is

$$\tau \equiv rF \sin \theta = r_\perp F = rF_\perp, \tag{12.1}$$

where $r_\perp$ is the component of $\vec{r}$ perpendicular to $\vec{F}$ and $F_\perp$ is the component of $\vec{F}$ perpendicular to $\vec{r}$.

Translational equilibrium: $\sum \vec{F}_{ext} = \vec{0}$.

Rotational equilibrium: $\sum \tau_{ext\,\vartheta} = 0$.

Mechanical equilibrium:

$$\sum \tau_{ext\,\vartheta} = 0 \quad \text{and} \quad \sum \vec{F}_{ext} = \vec{0}. \tag{12.14}$$

Rotation of a rigid object (Sections 12.1, 12.2, 12.5–12.7)

Concepts In **free rotation**, an object rotates about its center of mass. The translational and rotational motions are analyzed separately, and the torques and rotational inertia are computed about the center of mass. In *rotation about a fixed axis*, an object is constrained to rotate about a physical axis. Torques and rotational inertia are computed about that axis.

When an object is **rolling without slipping** across a surface, there is no relative motion at the location where the object touches the surface. The object rotates about its geometric center as it translates across the surface. Torques and rotational inertia are computed about the object's center of mass, which for a symmetrical object is the geometric center.

The external force exerted on a rolling object changes the object's center-of-mass kinetic energy. The external torque on a rolling object changes the object's rotational kinetic energy.

Quantitative Tools For both particles and extended bodies, the vector sum of the torques when rotation is about a fixed axis is

$$\sum \tau_{ext\,\vartheta} = I\alpha_\vartheta. \tag{12.10}$$

For an object of radius R that is rolling without slipping, the motion of the center of mass is described by

$$v_{cmx} = R\omega_\vartheta \tag{12.19}$$

$$a_{cmx} = R\alpha_\vartheta \tag{12.23}$$

$$\sum F_{extx} = ma_{cmx}$$

$$\sum \tau_{ext\,\vartheta} = I\alpha_\vartheta. \tag{12.10}$$

The change in an object's rotational kinetic energy resulting from torques is

$$\Delta K_{rot} = \left(\sum \tau_{ext\,\vartheta} \right) \Delta \vartheta \quad \text{(constant torques, rigid object).} \tag{12.31}$$

The kinetic energy of a rolling object is

$$K = K_{cm} + K_{rot} = \tfrac{1}{2}mv_{cm}^2 + \tfrac{1}{2}I\omega^2 \tag{12.33}$$

and the change in the kinetic energy of a rolling object is

$$\Delta K = \Delta K_{cm} + \Delta K_{rot}. \tag{12.34}$$

Angular momentum (Sections 12.1, 12.5)

Concepts A **rotational impulse** J_ϑ is the amount of angular momentum transferred to a system from the environment by external torques.

If the sum of the external torques due to forces exerted on a system is zero, the angular momentum of the system remains constant.

Quantitative Tools External torque caused by forces exerted on an object causes the object's angular momentum L_ϑ to change:

$$\sum \tau_{ext\,\vartheta} = \frac{dL_\vartheta}{dt}. \tag{12.12}$$

The **angular momentum law** says that the change in the angular momentum of an object is equal to the **rotational impulse** given to the system:

$$\Delta L_{\vartheta} = J_{\vartheta}. \tag{12.15}$$

If the constant external torques on a system last for a time interval Δt, the **rotational impulse equation** says that the rotational impulse is

$$J_{\vartheta} = \left(\sum \tau_{\text{ext}\,\vartheta} \right) \Delta t. \tag{12.17}$$

The law of conservation of angular momentum states that if $\sum \tau_{\text{ext}\,\vartheta} = 0$, then $dL_{\vartheta}/dt = 0$. This means that

$$\sum \tau_{\text{ext}\,\vartheta} = \frac{dL_{\vartheta}}{dt} = 0 \Rightarrow \Delta L_{\vartheta} = 0. \tag{12.13}$$

Rotational quantities as vectors (Sections 12.4, 12.8)

Concepts A **polar vector** is a vector associated with a displacement. An **axial vector** is a vector associated with a rotation direction. This vector points along the rotation axis, and its direction is given by the **right-hand rule** for axial vectors: When you curl the fingers of your right hand along the direction of rotation, your outstretched thumb points in the direction of the vector that specifies that rotation.

Right-hand rule for vector products: When you align the fingers of your right hand along the first vector in a vector product and curl them from that vector to the second vector in the product through the smaller angle between the vectors, your outstretched thumb points in the direction of the vector product.

The magnitude of the vector product of two vectors is equal to the area of the parallelogram defined by them.

Quantitative Tools The magnitude of the **vector product** of vectors $\vec{A}$ and $\vec{B}$ that make an angle $\theta \leq 180°$ between them when they are tail to tail is

$$|\vec{A} \times \vec{B}| = AB \sin \theta. \tag{12.35}$$

If $\vec{r}$ is the vector from the origin of a coordinate system to the location where a force $\vec{F}$ is exerted, the torque about the origin due to $\vec{F}$ is

$$\vec{\tau} = \vec{r} \times \vec{F}. \tag{12.38}$$

If $\vec{r}$ is the vector from the origin of a coordinate system to a particle that has momentum $\vec{p}$, the angular momentum of the particle about the origin is

$$\vec{L} = \vec{r} \times \vec{p}. \tag{12.43}$$

Review Questions

Answers to these questions can be found at the end of this chapter.

12.1 Torque and angular momentum

1. Can a small force produce a greater torque than a large force?
2. A friend pushes a doorknob to open the door for you, and the door swings all the way open in 2 s. On the way back, you return the favor but push not on the knob but close to the hinged edge. Although the door again opens all the way in 2 s, your friend notices your exertion and comments, "You had to exert a lot more torque doing it that way." Is he right?
3. What reference point should you choose for computing torques for a stationary object that is not mounted on a pivot, axle, or hinge?
4. Describe a case where the line of action of a force exerted on an object that can rotate either freely or about a fixed axis passes through the pivot even though the point of application of the force is not at the pivot. What is the torque in this case?

12.2 Free rotation

5. Can an object in free rotation rotate or revolve about any point you might choose?
6. Why is the concept of center of mass so useful in analyzing rotational motion?

12.3 Extended free-body diagrams

7. When you hold your left forearm parallel to the floor, what is the direction of the force exerted by the humerus bone on the elbow joint?
8. What is the main difference between a standard free-body diagram and an extended free-body diagram? Why is it necessary to use extended free-body diagrams when we study rotational motion?
9. If what you need in a problem is an extended free-body diagram, why draw a standard free-body diagram first? Isn't it possible to use an extended free-body diagram to compute vector sums of forces as well as vector sums of torques?

12.4 The vectorial nature of rotation

10. Describe the right-hand rule for determining a vector direction for the rotation of an object.
11. What is the direction of the rotational velocity vector for the second hand of a clock hanging on the wall?
12. Why are the words *clockwise* and *counterclockwise* insufficient to describe rotation in three dimensions?

12.5 Conservation of angular momentum

13. Which statement is correct: "In mechanical equilibrium, the vector sum of all the torques and forces is zero" or "In mechanical equilibrium, the vector sum of the torques is zero and the vector sum of the forces is zero."
14. Suppose a single force (one that is not canceled by any other forces) is exerted on an object. Is it possible for this single force to change both the object's momentum and its angular momentum?
15. A baton-twirler tosses her spinning baton up into the air. While it is in the air, is the baton's momentum constant? Is its angular momentum constant?
16. What is rotational impulse, and how is it related to the angular momentum law?
17. Under what circumstances is angular momentum conserved? Under what circumstances does the angular momentum of a system remain constant?

12.6 Rolling motion

18. For a wheel that is rolling without slipping, what is the relationship between the wheel's rotational speed and the speed v of its center of mass?
19. If two gears are in contact, with their teeth meshing, what do they have in common: rotational speed, tangential speed v_t of the teeth, or some other quantity?
20. What is the shape factor for a rolling object, and what does it measure?
21. An object rolls without slipping down a hill. How are the magnitudes of its rotational acceleration α_ϑ and the acceleration of its center of mass a_{cm} related?

12.7 Torque and energy

22. Explain how to compute the work done on a rotating object when you know both the sum of the torques caused by the external forces exerted on the object and the number of rotations the object makes in the time interval of interest.
23. If an object's motion is both translational and rotational, how should its kinetic energy be determined?
24. For an object that rolls without slipping, describe what effect static friction has on the work done on the object and on the object's energy.

12.8 The vector product

25. Draw a vector arrow on a sheet of paper, draw another vector arrow on a second sheet, and then lay the sheets on your desk in any arbitrarily chosen orientations. What angle does the vector representing the vector product of these two vectors make with the desktop? Does your answer depend on the orientation of the two vectors?
26. How is the magnitude of the vector product of two vectors related to the vectors?
27. What is $\vec{A} \cdot (\vec{A} \times \vec{B})$ equal to?
28. If you accidentally compute a torque using $\vec{F} \times \vec{r}$ rather than $\vec{r} \times \vec{F}$, what is wrong with your result?
29. What is the vector relationship between the angular momentum and the momentum for an isolated particle?

Developing a Feel

Make an order-of-magnitude estimate of each of the following quantities. Letters in parentheses refer to hints below. Use them as needed to guide your thinking.

1. The magnitude of the torque caused by the force you exert to open your bedroom door (C, O)
2. The magnitude of the torque caused by the force required to accelerate a sports car from 0 to 60 mi/h in 4 s (E, R, K)
3. The maximum torque caused by the gravitational force exerted on your bicycle when you lean into a tight turn (A, V)
4. The length of a lever you need to raise the front end of a car in which the engine is in the front (B, H, N, W)
5. The gear ratio (ratio of front gear radius to rear gear radius) for the highest gear of a racing bicycle (D, I, S, Q)
6. The largest feasible gear ratio for a hand-held electric drill (G, L, U)
7. The maximum useful torque that can be delivered to each of a car's two drive wheels (Y, P, K)
8. The work required to remove and replace five lug nuts as you change a flat tire on your car (F, X)
9. The maximum torque caused by the force you exert on a diving board when you dive into a pool (Z, T, J, M)

Hints

A. What is the distance from the pivot to the center of mass?
B. What is the distance from the axis of rotation to the car's center of mass?
C. What is the lever arm distance?
D. What is the diameter of a bicycle wheel?
E. What acceleration is needed?
F. What torque through what angle is needed to "break" each nut loose?
G. What is the minimum number of teeth on a gear to allow smooth turning?
H. What is the distance from the car's axis of rotation to the point where the lever touches the car?
I. At what speed can a strong cyclist move in the highest gear?
J. What is the average force you exert on the diving board?
K. What is the radius of each wheel?
L. If each tooth is 1 mm thick, what minimum radius is needed for the smaller (driveshaft) gear?
M. What distance does the diving board extend beyond any support?
N. What is the ratio of the required lifting force to the gravitational force exerted on you by Earth?
O. What is the force magnitude?
P. What normal force (needed to support the car) is exerted on each drive wheel?
Q. What is the minimum comfortable time interval needed for the cyclist's foot to complete one cycle?
R. What is the vector sum of the forces required?
S. What is the rotational speed of the wheel?
T. What is the time interval between the instant you hit the board on your last bounce and the instant the board is at maximum depression?

U. What maximum gear radius could fit comfortably in the drill casing?
V. What is your maximum angle of lean?
W. What is the minimum distance from the point where you pivot the lever to the location where the lever touches the car?
X. What torque through what angle is needed to unscrew each loosened lug nut?
Y. What coefficient of friction can be used to move the car forward most effectively?
Z. If you jump high on your last bounce, with what speed do you hit the end of the diving board?

Key (all values approximate)

A. 1 m; B. axis at rear wheels, center of mass just behind front wheels, so 2 m; C. 0.7 m; D. 0.7 m; E. 7 m/s^2; F. $1 \times 10^2\,\text{N} \cdot \text{m}$ through one-quarter of a turn; G. 1×10^1; H. about 3 m; I. 2×10^1 m/s; J. consider both gravitational force and momentum change divided by the time interval, about 2×10^3 N for an 80-kg person; K. 0.3 m; L. about 3 mm, allowing for spacing between teeth; M. less than 3 m; N. 14 to 1; O. 1×10^1 N; P. 4×10^3 N, or one-third of the gravitational force for a front-wheel-drive car; Q. 0.4 s; R. 0.7 times the gravitational force exerted on the car—say, 7 kN; S. $6 \times 10^1\,\text{s}^{-1}$; T. 0.3 s; U. 3×10^1 mm; V. 40° away from vertical; W. 0.3 m; X. 8 N · m through 5 revolutions; Y. to avoid wasteful slipping, use the coefficient of static friction, which may be close to 1; Z. 3 m/s

Worked and Guided Problems

Procedure: Extended free-body diagrams

1. Begin by making a standard free-body diagram for the object of interest (the *system*) to determine what forces are exerted on it. Determine the direction of the acceleration of the center of mass of the object, and draw an arrow to represent this acceleration.
2. Draw a cross section of the object in the plane of rotation (that is, a plane perpendicular to the rotation axis) or, if the object is stationary, in the plane in which the forces of interest lie.
3. Choose a reference point. If the object is rotating about a hinge, pivot, or axle, choose that point. If the object is rotating freely, choose the center of mass. If the object is stationary, you can choose any reference point. Because forces exerted at the reference point cause no torque, it is most convenient to choose the point where the largest number of forces is exerted or where an unknown force is exerted. Mark the location of your reference point and choose a positive direction

of rotation. Indicate the reference point in your diagram by the symbol ⊙.
4. Draw vectors to represent the forces that are exerted on the object and that lie in the plane of the drawing. Place the tail of each force vector at the point where the force is exerted on the object. Place the tail of the gravitational force exerted by Earth on the object at the object's center of mass.* Label each force.
5. Indicate the object's rotational acceleration in the diagram (for example, if the object accelerates in the positive ϑ direction, write $\alpha_\vartheta > 0$ near the rotation axis). If the rotational acceleration is zero, write $\alpha_\vartheta = 0$.

---

*Is the gravitational force really exerted at the center of mass? Suppose the gravitational force exerted by Earth were exerted at some other point. The force would then cause a permanent torque about the center of mass, and any object dropped from rest would begin spinning, which is not true.

These examples involve material from this chapter but are not associated with any particular section. Some examples are worked out in detail; others you should work out by following the guidelines provided.

Worked Problem 12.1 Using a wrench

Loosening the rusted lug nuts on a car tire requires 220 N·m of torque. (*a*) A 60-kg mechanic holds a wrench horizontally on one nut. The wrench is 400 mm long, and she holds it at the free end and exerts a force equal to the magnitude of the gravitational force exerted on her. If the direction of the force she exerts is straight downward, how much torque does she cause on the nut? (*b*) Suppose one nut is oriented such that the wrench makes a 37° angle with the horizontal, as in Figure WG12.1. By exerting the identical force she exerted in part *a*, can the mechanic loosen this nut with the 400-mm wrench? If not, how long a wrench must she use to loosen the nut?

Figure WG12.1

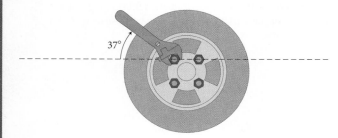

❶ **GETTING STARTED** Given that this problem involves torques, we follow the procedure for drawing extended free-body diagrams at the top of this page. We begin by making a free-body diagram for the wrench (Figure WG12.2*a*). The wrench is subject to three

forces: a downward force exerted by the mechanic on the wrench, an upward contact force exerted by the nut on the wrench, and a downward force of gravitation exerted by Earth on the wrench. Because the gravitational force on the wrench is negligible compared to the other two forces, we ignore this force in this problem. Next we draw extended free-body diagrams for the wrench for both parts of this problem (Figure WG12.2*b* and *c*). In both cases we choose the nut as the reference point. With that choice of reference point, only the force $\vec{F}^c_{mw}$ causes a torque.

Figure WG12.2

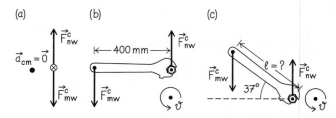

❷ **DEVISE PLAN** The magnitude of the torque caused by a force exerted on an object is

$$\tau = |\vec{r} \times \vec{F}| = rF \sin \theta, \tag{1}$$

where θ is the angle between the line of action of the force and the position vector that points from the rotation axis to the point of application of the force. The magnitude of the contact force exerted by the mechanic on the wrench is the magnitude of the gravitational force exerted on her: $F^c_{mw} = F^G_{Em} = m_m g$. The lever arm distance $r_\perp$ is the wrench length ℓ because the point of application of her force is the very end of the wrench. In part a the angle between $\vec{r}$ and $\vec{F}^c_{mw}$ is 90°.

The force magnitude and direction remain the same in part b, but the angle θ changes. We note that the 37° angle the wrench makes with the horizontal is *not* the angle we use to calculate the torque. We want the angle between the downward force and the wrench, which is $\theta = 90° - 37°$. We can use Eq. 1 for part b, with $r = \ell$ in this case, if we solve for the wrench length: $\ell = \tau/(F \sin \theta) = \tau/(F^G_{Em} \sin \theta)$.

❸ **EXECUTE PLAN**

(*a*) The magnitude of the gravitational force exerted on the mechanic is $m_m g = (60 \text{ kg})(9.8 \text{ m/s}^2) = 588 \text{ N}$. The torque caused by the force exerted by the mechanic is then

$$\tau = (0.400 \text{ m})(588 \text{ N})(\sin 90°) = 235 \text{ N} \cdot \text{m}$$

$$= 2.4 \times 10^2 \text{ N} \cdot \text{m}. ✔$$

(*b*) With the wrench making a 37° angle with the horizontal but the force exerted vertically downward, the angle between the line of action of the force and the wrench is $\theta = 90° - 37° = 53°$. The torque her force causes now is

$$\tau = (0.400 \text{ m})(588 \text{ N})(\sin 53°) = 188 \text{ N} \cdot \text{m}.$$

This is insufficient to loosen the rusted nut. She can loosen it with a longer wrench, though. The minimum wrench length ℓ to achieve the needed 220-N · m torque is

$$\ell = \frac{\tau}{F^G_{Em} \sin \theta} = \frac{220 \text{ N} \cdot \text{m}}{(588 \text{ N})(\sin 53°)} = 0.47 \text{ m}. ✔$$

A slightly longer wrench will loosen the nut, or the mechanic can just put a piece of hollow pipe over the end of her 400-mm wrench to increase the lever arm distance a bit.

❹ **EVALUATE RESULT** Our answers are close to the needed torque given in the problem statement, 220 N · m, which gives us confidence in our results, but what other bounds can we establish to test their reasonableness? Consider the force $\vec{F}^c_{wm}$ exerted by the wrench on the mechanic in part a. By Newton's third law, this force is equal in magnitude to the force exerted by the mechanic on the wrench, but in the opposite direction. Making full use of the gravitational force exerted on the mechanic in our calculation means we assume that no support force other than $\vec{F}^c_{wm}$ is necessary to keep the mechanic from falling—the wrench provides all the required support. In fact, this is the greatest force she can exert on the wrench without invoking some creative arrangement. In order to exert a downward force greater than the gravitational force, the mechanic would need to be able to push upward against some nearby fixed object, perhaps with her feet. With no mention of such an effort in the problem statement, we assume she did no such contortions.

Exerting a horizontal force requires some frictional force to compensate, or a nearby object on which the mechanic can push horizontally with her feet in order to maintain a zero sum of horizontal forces exerted on herself. Coefficients of friction rarely exceed 1, so again the gravitational force is the upper limit for a typical person to exert on a wrench. In practice, the maximum force possible with a person's hands is likely to be a bit smaller than the gravitation force. Most people must stand on the wrench in order to supply a force as great as the gravitational force exerted on them by Earth.

Guided Problem 12.2 At the grindstone

A circular grindstone, radius 0.17 m, is used to sharpen a knife. The knife is pressed against the grindstone with a normal force of 20 N, and the coefficient of kinetic friction between knife and stone is 0.54. What torque must the motor driving the grindstone maintain in order to keep the stone rotating at constant speed?

❶ **GETTING STARTED**

1. Consider the force exerted by the knife on the grindstone. Does any component of this force cause a torque about the grindstone's axle?

❷ **DEVISE PLAN**

2. To maintain a constant rotational speed, what must be the vector sum of the torques caused by the forces exerted on the grindstone? What two torques are involved?

3. Does the normal force exerted by the knife on the grindstone contribute to the torque equation?

❸ **EXECUTE PLAN**

4. Compute the magnitude of the torque that must be canceled by the motor.

❹ **EVALUATE RESULT**

5. Examine the expression for torque used in step 4. What happens to the torque as the normal force changes? As the coefficient of friction changes? Are these two trends reasonable?

6. Consider your numerical result. Could a motor supply this amount of torque? Could a person supply this amount of torque with a foot pedal? (If you have trouble answering these two questions, review the Developing a Feel questions in this chapter.)

Worked Problem 12.3 Accelerating bicycle

A bicyclist initially traveling at a steady 3.0 m/s accelerates at a constant 1.1 m/s². The time interval over which the acceleration takes place is long enough for the wheels to make 20 rotations. If the wheels have a radius of 0.30 m, what is their rotational speed after the acceleration stops?

❶ GETTING STARTED The appropriate diagram for kinematics problems is a motion diagram (Figure WG12.3). The bicyclist moves horizontally while the wheel both rotates and moves horizontally, with increasing rotational speed as the bicyclist accelerates. This is a simple kinematics problem, but the key is to establish a relationship between the given translational quantities and the desired rotational quantities. We use the fact that for rolling without slipping, there is a relationship between the distance the center of each wheel travels and the number of rotations the wheel makes.

Figure WG12.3

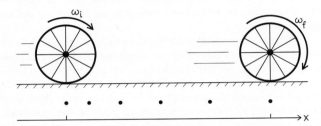

❷ DEVISE PLAN The acceleration of the bicycle is the acceleration of the center of each wheel, which is equal to the tangential acceleration of a point on the rim. The relationships between translational and rotational quantities for rolling without slipping are

$$x = \vartheta R$$
$$v_{cmx} = \omega_\vartheta R$$
$$a_{cmx} = \alpha_\vartheta R.$$

Because the acceleration of the bicycle is constant, the rotational acceleration is also constant. The rotational speed of the wheels is described by the equation

$$\omega_{\vartheta,f}^2 = \omega_{\vartheta,i}^2 + 2\alpha_\vartheta(\vartheta_f - \vartheta_i). \tag{1}$$

Thus we have enough information to compute $\omega_{\vartheta,i}$ and α_ϑ, and $\vartheta_f - \vartheta_i$ is the 20 rotations expressed as a change in the rotational coordinate.

❸ EXECUTE PLAN We use Eq. 1 to determine the final rotational speed, substituting translational variables for $\omega_{\vartheta,i}$ and α_ϑ:

$$\omega_f^2 = \left(\frac{v_{cmx,i}}{R}\right)^2 + 2\left(\frac{a_{cmx}}{R}\right)(\vartheta_f - \vartheta_i).$$

The final rotational speed is therefore

$$\omega_f = \left[\left(\frac{3.0 \text{ m/s}}{0.30 \text{ m}}\right)^2 + 2\left(\frac{1.1 \text{ m/s}^2}{0.30 \text{ m}}\right)(20 \text{ rev})(2\pi/\text{rev})\right]^{1/2}$$

$$= 32 \text{ s}^{-1}. ✔$$

❹ EVALUATE RESULT The algebraic equation for ω_f shows that it is greater for larger accelerations and higher initial speeds. Both of these trends are reasonable. The reasonableness of the numerical result can be checked by noting that the final speed is $v = \omega R = (32 \text{ s}^{-1})(0.30 \text{ m}) = 9.6 \text{ m/s} = 34 \text{ km/h} = 21 \text{ mi/h}$, which is a reasonable speed for a cyclist.

Guided Problem 12.4 Inclined race

A solid disk and a hoop, which have different radii and different inertias, are set near the top of a ramp and then released (Figure WG12.4). The ramp is inclined at an angle of 4.0° and is 1.8 m long. (a) Which object reaches the bottom first? (b) After the first object reaches the bottom, how many seconds pass before the other object reaches the bottom?

Figure WG12.4

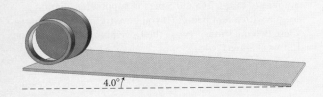

❶ GETTING STARTED
1. What types of motion are involved here?
2. Consider possible approaches to part a: conservation laws, other fundamental principles, or kinematics alone. Repeat

for part b. If more than one approach is feasible, you may be able to use one method to check an answer you obtained using another method.

❷ DEVISE PLAN
3. Draw a diagram showing the physical quantities needed to describe the approach you have decided to use. Write a general equation based on your diagram for the acceleration and rotational acceleration of each object. Consider whether you need separate diagrams for parts a and b.
4. How does the shape factor for a hoop compare with that for a cylinder (remember that the disk is a very short cylinder)?
5. Which object experiences the greater rotational acceleration: the one with the larger shape factor or the one with the smaller shape factor?
6. For each object, write an expression that relates the acceleration to the time interval the object takes to reach the bottom. Repeat for rotational acceleration. Do you now have enough information to solve the problem?

③ EXECUTE PLAN

7. Solve the equations you wrote in step 6 for the time interval each object takes to roll down the ramp.

8. Subtract your expression for the time interval for the object that reaches the bottom first from the time interval for the object that reaches the bottom second. How do you know from this difference whether you are wrong about which object reaches the bottom first?

④ EVALUATE RESULT

9. Examine your expression for how the time interval needed to reach the bottom changes as you change the physical conditions. What happens as the length of the ramp increases? As the angle of inclination increases? Are these trends reasonable?

10. Consider your numerical result. Is the order of magnitude about right?

Worked Problem 12.5 Falling yo-yo

A yo-yo is basically a spool consisting of an inner small axle and two outer disks, with a string wrapped around the axle. Calculate the acceleration of a 0.130-kg yo-yo if it's allowed to drop while the free end of the string is held at a fixed distance above the floor. Each disk has a radius of 17.0 mm, which means the yo-yo radius is the same, $R_Y = 17.0$ mm, and the axle radius is $R_a = 5.00$ mm. Ignore the inertia of the axle and assume that the entire 0.130-kg inertia is for the two disks.

① GETTING STARTED We know that the yo-yo accelerates as it falls because of the gravitational force. At the same time, however, the string tends to pull the yo-yo upward. We also know that the yo-yo increases its spin rate as it falls, an increase that must be due to a vector sum of torques caused by these two opposing forces. This information implies that we have to use force and torque laws to get our acceleration.

② DEVISE PLAN The torque caused by the force exerted by the string converts gravitational potential energy to kinetic energy. However, energy methods give us the yo-yo's speed after it has dropped a given distance rather than the instantaneous acceleration requested. We can get this acceleration from Newton's second law, except we do not know the value of the tension in the string. Another equation is needed.

If we assume that the string does not slip, we can relate the rotational acceleration to the desired acceleration via $a_{cm\,x} = R_a\alpha_\vartheta$ and then get the rotational acceleration from the vector sum of the external torques, $\sum \tau_{ext\,\vartheta} = I\alpha_\vartheta$ (Eq. 12.10).

Although the acceleration of the yo-yo's center of mass does not depend on the locations of the points of application of the forces, its rotational acceleration does. Thus we need both a free-body diagram and an extended free-body diagram for the yo-yo (Figure WG12.5). The only forces exerted are the downward gravitational force exerted by Earth $\vec{F}^G_{EY}$, the point of application of which is at the yo-yo's center of mass; and the upward force exerted by the string $\vec{F}^c_{sY}$, which is directed tangent to the axle.

Figure WG12.5

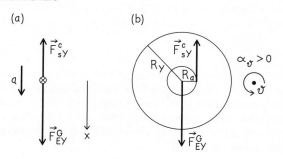

(a)   (b)

We have to be careful about the signs of our accelerations. The easiest way to do this is to choose directions for our axes such that positive translation corresponds to positive rotation. Because we know that the yo-yo accelerates downward, we choose the positive x axis pointing downward. A positive downward translation corresponds in Figure WG12.5 to a counterclockwise rotation as the string unwinds, so we put a counterclockwise curved arrow labeled ϑ in our diagram. We note that the only force that produces a nonzero torque about the center of mass is the force exerted by the string, because the force of gravity has a zero lever arm distance about this axis.

③ EXECUTE PLAN We begin with the equation of motion based on the vector sum of forces:

$$F^c_{sY\,x} + F^G_{EY\,x} = m_Y a_{cm\,x}.$$

Substituting in the sign and magnitude for each component gives us

$$-F^c_{sY} + F^G_{EY} = m_Y a_{cm\,x}. \tag{1}$$

We suspect that $a_{cm\,x}$ must be positive, but let us see if the final result bears this out.

The rotational equation of motion for the components along the axis specified by the curved arrow for our rotational coordinate system is

$$\sum \tau_{ext\,\vartheta} = \tau_{sY\,\vartheta} + \tau_{EY\,\vartheta} = I\alpha_\vartheta$$

$$+R_a F^c_{sY} + 0 = \left(\tfrac{1}{2} m_Y R_Y^2\right)\alpha_\vartheta, \tag{2}$$

where we have used the fact that the yo-yo is approximately the shape of a solid cylinder, for which the expression for rotational inertia is $I = mR^2/2$.

The relationship between the rotational acceleration and the acceleration of the center of mass is

$$a_{cm\,x} = \alpha_\vartheta R_a.$$

Note that the axle radius is the appropriate radius to use here. Solving this equation for α_ϑ, substituting the result into Eq. 2, and dividing through by R_a, we get

$$+R_a F^c_{sY} + 0 = \left(\tfrac{1}{2} m_Y R_Y^2\right)\left(\frac{a_{cm\,x}}{R_a}\right)$$

$$+F^c_{sY} = \left(\tfrac{1}{2} m_Y R_Y^2\right)\left(\frac{a_{cm\,x}}{R_a^2}\right). \tag{3}$$

Next we substitute Eq. 3 into Eq. 1, solve the result for a_{cmx}, and insert numerical values:

$$-\left(\tfrac{1}{2}m_Y R_Y^2\right)\left(\frac{a_{cmx}}{R_a^2}\right) + F_{EY}^G = m_Y a_{cmx}$$

$$F_{EY}^G = m_Y a_{cmx} + \left(\tfrac{1}{2}m_Y R_Y^2\right)\left(\frac{a_{cmx}}{R_a^2}\right)$$

$$m_Y g = m_Y\left(1 + \tfrac{1}{2}\frac{R_Y^2}{R_a^2}\right)a_{cmx}$$

$$a_{cmx} = \frac{g}{1 + \tfrac{1}{2}\dfrac{R_Y^2}{R_a^2}}$$

$$a_{cmx} = \frac{9.80 \text{ m/s}^2}{1 + \tfrac{1}{2}(0.0170 \text{ m})^2/(0.00500 \text{ m})^2} = 1.45 \text{ m/s}^2. \checkmark$$

❹ **EVALUATE RESULT** The expression we obtained for a_{cmx} tells us that the acceleration of the yo-yo depends on the radius R_a of the axle and on the shape factor $c = \tfrac{1}{2}$ (from Table 11.3). If R_a decreases so that more of the yo-yo's inertia is concentrated at the end pieces (or if R_Y increases), the yo-yo accelerates more slowly. As we expected, our expression shows that a_{cmx} can only be positive. The numerical value of the acceleration is reasonable: A value quite a bit smaller than the free-fall value makes sense because the force exerted by the string impedes the falling motion.

Guided Problem 12.6 Fickle yo-yo

The yo-yo in Figure WG12.6, which has inertia m_Y, outer radius R_Y, and axle radius R_a, sits on a rough horizontal surface. Depending on the angle ϕ at which the string is pulled with tension T, the yo-yo rolls either to the left or to the right in the drawing. (*a*) For a given angle ϕ, determine expressions for the magnitude and direction of the acceleration. (*b*) What is the critical angle at which the yo-yo changes the direction in which it rolls?

Figure WG12.6

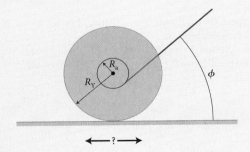

❶ **GETTING STARTED**
1. What sort of motion results when the string is pulled at the critical angle? (Hint: What motion could be intermediate between rolling right and rolling left?)
2. What sort of diagram is appropriate for analyzing torques? Analyzing forces? How many cases must you draw separately?

3. Identify all the forces exerted on the yo-yo. What is the direction of the frictional force? (Hint: In which direction would the yo-yo slide if the surface were ice?)

❷ **DEVISE PLAN**
4. What is a good choice for the axis about which to calculate the torques?
5. Draw both a horizontal x axis and a curved arrow for your selection of a positive sense of rotation about your chosen axis.
6. Write the second-law force and torque equations. Account for all forces and torques. Are any zero? How many unknowns do you have?
7. What is a reasonable shape factor to use for the yo-yo?
8. What condition must be met at the critical angle if the yo-yo rolls neither left nor right?
9. Derive an expression for the critical angle in terms of known quantities.

❸ **EXECUTE PLAN**
10. Make sign and magnitude decisions for each force component and each torque component. Attempt a solution.

❹ **EVALUATE RESULT**
11. Does the coefficient of friction or the magnitude of the frictional force play a role? Is this surprising? Consider possible hidden assumptions.
12. What do your equations tell you about the motion as ϕ approaches zero? As ϕ approaches 90°?
13. Obtain a spool or yo-yo and give this a try. Can you reproduce the behavior exhibited by the equations?

Worked Problem 12.7 Leaning ladder of physics

A ladder of length $\ell = 13.0$ m leans against a smooth vertical wall. The base rests on rough ground at a distance $d_w = 5.00$ m from the wall, and the top is at height $h = 12.0$ m above the ground (Figure WG12.7). The magnitude of the force of gravity exerted on the ladder is 324 N. A rope is attached to the ladder at a distance $d_r = 2.00$ m up from the base, and a man pulls horizontally on this rope with a force of magnitude 390 N. In order for the ladder not to slip as he pulls, what are (*a*) the vertical component and (*b*) the horizontal component of the force exerted by the ground on the ladder?

Figure WG12.7

1 **GETTING STARTED** A sketch has been provided with the problem, so we can focus on which physical principles elucidate this situation. Because nothing is changing the ladder's state of motion, this is a case of mechanical equilibrium. Therefore the sum of the forces exerted on the ladder must be zero, and the sum of the torques caused by these forces must be zero. Three force components of unknown magnitude are exerted on the ladder, so we should get three component equations from our force and torque analysis.

2 **DEVISE PLAN** As usual, we draw a free-body diagram and an extended free-body diagram (Figure WG12.8). Because we need to deal with components of forces and torques, we have to specify axes and a rotational direction. Let us choose the positive x direction to be to the right, the positive y direction to be up, and the pivot about which to calculate the torques to be the base of the ladder, with positive rotation counterclockwise. We put these two sign notations in our diagram.

Figure WG12.8

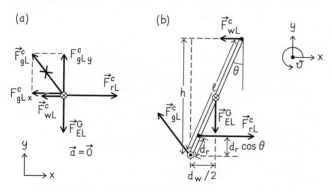

We know the magnitudes and directions of the gravitational force and the force exerted by the man, so we draw those to scale first in our free-body diagram. The man exerts a force to the right, and its magnitude exceeds that of the gravitational force. That means we suspect that in the absence of friction the bottom of the ladder would slide to the right. Thus we draw the contact force

exerted by the ground as pointing upward and to the left. If this turns out to be wrong, our equations will provide a correction by virtue of a negative answer.

It is useful to identify a relevant angle for our analysis, and let us select the one between the top of the ladder and the wall, calling it θ. The ladder forms a 5–12–13 right triangle with the wall, and this angle satisfies the relationships $\sin \theta = \frac{5.00}{13.0}$ and $\cos \theta = \frac{12.0}{13.0}$ (making $\theta = 22.6°$).

We want to know the force exerted by the ground on the ladder. We know that because the wall is smooth we can ignore any frictional force exerted by the wall. This means the wall exerts no vertical force, only a horizontal (normal) contact force. We can therefore look at the sum of the forces exerted on the ladder in the vertical direction and not worry about any force in this direction exerted by the wall. The sum of the torques about the base of the ladder must be zero because there is no rotation. We know that the force exerted by the wall is directed to the left in the orientation shown in Figure WG12.7, and we know that the horizontal forces must balance because the ladder isn't accelerating in the x direction. This gives us enough equations to solve for all unknowns.

3 **EXECUTE PLAN**
(*a*) Because the sum of the vertical components of the four forces must be zero, we write

$$\Sigma F_y = F^c_{\text{wL}y} + F^G_{\text{EL}y} + F^c_{\text{rL}y} + F^c_{\text{gL}y} = 0.$$

The vertical component of the contact force exerted on the ladder by the ground is the normal force, so we call its magnitude $F^n_{\text{gL}} = |F^c_{\text{gL}y}|$. (The horizontal component is the friction force.) Thus we get

$$\Sigma F_y = 0 + (-F^G_{\text{EL}}) + 0 + F^n_{\text{gL}} = 0$$

$$F^n_{\text{gL}} = |F^c_{\text{gL}y}| = F^G_{\text{EL}} = 324 \text{ N.} ✔$$

Note that the direction of $F^c_{\text{gL}y}$ must be upward because the only other vertical force (the gravitational force) is downward.

(*b*) To determine the horizontal component $F^c_{\text{gL}x}$, we need two equations because there are two unknowns in the equation for the horizontal sum of the forces: $F^c_{\text{gL}x}$ (the one we have to determine) and $F^c_{\text{wL}x}$. Thus we begin with torques. The sum of the ϑ components of the torques caused by the four forces about the rotation axis must be zero because there is no rotation. Using our generic symbol $r_\perp$ for lever arm distances, we write our first torque equation in the general form

$$\Sigma \tau_\vartheta = \pm r_{\text{wL}\perp} F^c_{\text{wL}} \pm r_{\text{EL}\perp} F^G_{\text{EL}} \pm r_{\text{rL}\perp} F^c_{\text{rL}} \pm r_{\text{gL}\perp} F^c_{\text{gL}} = 0.$$

We note that the term $r_{\text{gL}\perp} F^c_{\text{gL}}$, the torque caused by the force exerted by the ground on the ladder, must be zero because the lever arm distance for this torque is zero for a pivot at the base of the ladder.

Our chosen direction for positive rotation determines the sign of each torque. We determine these signs by looking at Figure WG12.8*b* to see which way each force tends to rotate the ladder about the pivot. Remember that counterclockwise in the figure is positive rotation:

$$\sum \tau_\vartheta = +hF_{wL}^c + \left(-\tfrac{1}{2}d_w F_{EL}^G\right) + \left[-(d_r \cos\theta)F_{rL}^c\right] + 0 = 0$$

$$(12.0 \text{ m})F_{wL}^c - \tfrac{1}{2}(5.00 \text{ m})(324 \text{ N})$$

$$- (2.00 \text{ m})\left(\tfrac{12.0}{13.0}\right)(390 \text{ N}) = 0.$$

This gives us the magnitude of the force exerted by the wall:

$$F_{wL}^c = \frac{\tfrac{1}{2}(5.00 \text{ m})(324 \text{ N}) + (2.00 \text{ m})\left(\tfrac{12.0}{13.0}\right)(390 \text{ N})}{12.0 \text{ m}} = 128 \text{ N}.$$

The sum of the horizontal forces is

$$\sum F_x = F_{wLx}^c + F_{ELx}^G + F_{rLx}^c + F_{gLx}^c = 0$$

$$(-F_{wL}^c) + 0 + (+F_{rL}^c) + F_{gLx}^c = 0,$$

where we have used the known information about the directions of the forces exerted by the wall and the rope. We now solve for the horizontal component of the force exerted on the ladder by the ground:

$$F_{gLx}^c = +F_{wL}^c - F_{rL}^c = 128 \text{ N} - 390 \text{ N} = -262 \text{ N}. \checkmark$$

④ EVALUATE RESULT As we suspected, the man pulls on the rope hard enough that the ladder would slide toward him if the ground weren't holding it back. So the ground must supply a horizontal force in the direction opposite the direction of the force exerted by the rope, which means in the negative x direction. Thus, the negative value for F_{gLx}^c is reasonable and agrees with our free-body diagram. If the man had pulled with less force, we might have obtained a positive value. We also know that the ground needs to be pushing up on the ladder (if it were not, the ladder would fall through the ground), so our $+324$ N result in part a is reasonable. However, let's double-check our answer by calculating the torque about the ladder's center of mass, which should also be zero:

$$\sum \tau_\vartheta = \left[+\left(\tfrac{1}{2}\ell - d_r\right)\cos\theta\right]F_{rL}^c + \left[+\tfrac{1}{2}\ell \cos\theta\right]F_{wL}^c$$

$$+ \left[+\tfrac{1}{2}\ell \cos\theta\right]F_{gLx}^c + \left[-\tfrac{1}{2}\ell \sin\theta\right]F_{gLy}^c.$$

Note the absence of any contribution by the gravitational force here. Why is this so?

When we substitute numerical values, we get zero, verifying our earlier torque result. This is one of the nice things about mechanical equilibrium problems. You can pick whichever pivot is easiest to balance around to get an initial answer and then pick a different pivot to check that answer.

Guided Problem 12.8 Moving a refrigerator

Your new refrigerator, of inertia m, has been delivered and left in your garage. As shown in Figure WG12.9, it has length ℓ in vertical dimension and each side of its square base is of length d. You need to slide it across the rough garage surface to get it into your house. The coefficients of static and kinetic friction between base and garage surface are almost equal, so you approximate $\mu = \mu_s = \mu_k$. You push horizontally at a height h above the surface, exerting a force just big enough to keep the refrigerator moving. You dislike bending over, so you push at the highest possible point that will not cause the refrigerator to tip as it slides. Thus the refrigerator is always on the verge of tipping. (a) Where along the base of the refrigerator is the effective point of application of the normal force exerted by the garage surface on the refrigerator; that is, at what location can you picture the normal force as being concentrated? (b) If the refrigerator is not to tip, and if its center of mass is at its center, what is the maximum value h_{max} at which you can push?

Figure WG12.9

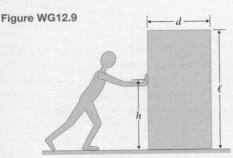

① GETTING STARTED

1. What condition must be met in order for the refrigerator not to rotate?
2. How can you determine the force you push with to just keep the refrigerator moving?
3. Which force(s) tend to tip the refrigerator, and which tend to prevent it from tipping?

② DEVISE PLAN

4. Draw a free-body diagram and an extended free-body diagram for the refrigerator. Indicate a sign for each coordinate axis (x, y, and ϑ) so that you can correctly determine the signs of the components.
5. What is the lever arm distance of the normal force $\vec{F}_{fr}^n$ exerted by the floor?
6. How does the height at which you push affect the point of application of $\vec{F}_{fr}^n$?
7. Is there enough information to solve for the value of the lever arm distance of $\vec{F}_{fr}^n$ at which the refrigerator begins to tip?
8. What condition exists just before tipping begins?

③ EXECUTE PLAN

④ EVALUATE RESULT

9. In your expression for the lever arm distance, does each term have a sign that is physically plausible?
10. Does your answer make sense if μ is reduced to zero or increased to 1.0? What if d becomes very large or very small?

Questions and Problems

Dots indicate difficulty level of problems: • = easy, •• = intermediate, ••• = hard; CR = context-rich problem.

12.1 Torque and angular momentum

1. To open a stuck jar lid, wearing rubber gloves sometimes works. Why does that help? •

2. You are using a steel crowbar to lift a big rock, with a smaller stone as a fulcrum at the center of the crowbar, but you are not quite able to lift the rock. What can you change about this setup to lift the rock? •

3. The specifications for tightening the bolts on a car engine's valve cover are given in terms of torque rather than in terms of how much force should be exerted on the tightening wrench. Why is torque a better specification here? •

4. When the wrench you are working with does not loosen a nut, you can sometimes succeed by slipping a length of pipe over the end of the wrench and pushing at the end of the pipe. Why does this work? ••

5. In which cases illustrated in Figure P12.5 does pushing in the way and at the location shown cause a torque about a vertical axis passing through the hinges? The dot shows the point of application of the force in each case, and the dashed lines help indicate the line of action of each force. Forces (*a*), (*b*), and (*f*) are exerted parallel to the gate; forces (*c*), (*d*), and (*e*) are exerted perpendicular to the gate. ••

Figure P12.5

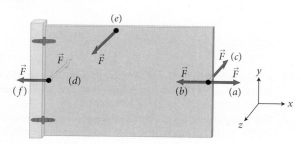

6. Rank, from greatest to smallest, the magnitudes of the torques about the wrench head in Figure P12.6 caused by exerting the same force at the different positions shown. ••

Figure P12.6

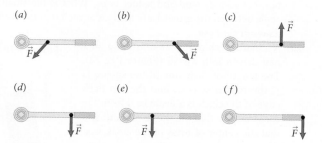

7. Which way is easier to balance the asymmetrical baton of Figure P12.7 on your finger? Explain your choice. ••

Figure P12.7

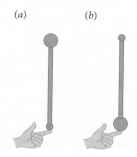

8. Explain why, when a car makes a sharp turn on an unbanked road, the wheels on the inside of the turn tend to come off the ground. ••

9. Figure P12.9 is a top view of a hand-driven carousel in a park. Four parents of the children riding on the device each push on the carousel in different ways. The forces exerted by each of the four parents and the positions at which the forces are exerted are labeled A–D. Rank the parents' pushes in order of increasing torque they cause on the carousel. ••

Figure P12.9

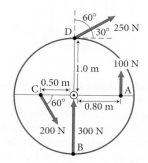

10. The uniform rod shown in Figure P12.10 is fixed to an axis going through its center, so that it is free to rotate clockwise or counterclockwise. It has four hooks: one at each end of the rod and one halfway between the center (axis) and each end. You need to hang four objects from this rod, with one and only one object on each hook, after which the rod must be balanced. The objects have masses M, $3M$, $5M$, and $9M$, respectively. How can this be accomplished? •••

Figure P12.10

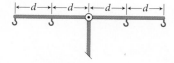

12.2 Free rotation

11. A very thin bar of length ℓ and negligible inertia connects two blocks A and B that have inertias $4m$ and m, respectively (Figure P12.11). When this rod-block system is made to spin with no translational motion and no mechanical pivot, each block moves through a circular path in space, with circumferences C_A and C_B. Determine the ratio C_A/C_B. ••

Figure P12.11

12. The system shown in Figure P12.12 consists of two balls A and B connected by a thin rod of negligible mass. Ball A has three times the inertia of ball B and the distance between the two balls is ℓ. If the system has a translational velocity of v in the x direction and is spinning counterclockwise at an angular speed of $\omega = 2v/\ell$, determine the ratio of the instantaneous speeds of the two balls v_A/v_B at the moment shown. ••

Figure P12.12

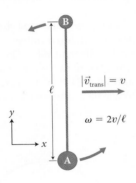

13. Put your index fingers about 1 m apart and rest a meterstick or similar object on top of them. Then, while holding the meterstick parallel to the floor, slowly draw your fingers toward each other. You will find that first one finger slips easily under the stick, then the other finger, then the first one again, then the other one again. No matter how much alternating takes place, however, your fingers *always* meet at the center of the stick. Explain what is happening. •••

12.3 Extended free-body diagrams

14. To change a light bulb, you climb halfway up a stepladder. Draw an extended free-body diagram for the ladder. •

15. Which refrigerator in Figure P12.15 is in the greatest danger of tipping over? Note positions of center of mass. •

Figure P12.15

(a)   (b)   (c)

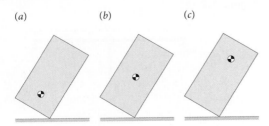

16. Which configuration of the four uniform boxes in Figure P12.16 makes the hand-truck easier to handle? Explain your answer. ••

Figure P12.16

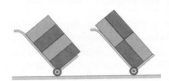

17. A father and son are carrying a sofa down the stairs, with the son leading. Consider the sofa's center of mass to be at the geometric center. (a) Draw an extended free-body diagram for the sofa. (b) Who carries the greater share of the load? (c) Would your answer to part b be the same if, instead of the sofa, the men were carrying a sheet of plywood? ••

18. In the mobile in Figure P12.18, what are the inertias of the giraffe and of the elephant in terms of the inertia m_m of the monkey? ••

Figure P12.18

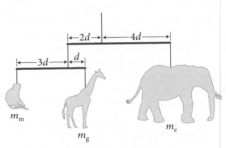

19. A painter leans a ladder at some angle against a smooth wall, with the foot of the ladder resting on wall-to-wall carpeting. Why might the ladder slip as he climbs it? ••

20. You are pushing a cart across the room, and the cart has wheels at the front and the back. Your hands are placed on top of the cart at the center (left to right) of the top edge, pushing horizontally. There is friction between the wheels and the floor. Is the normal force between the floor and the front wheels greater than, smaller than, or equal to the normal force between the floor and the rear wheels? Explain your answer using an extended free-body diagram. ••

21. The top of a ladder of inertia m rests against a smooth wall, and the foot rests on the ground. The coefficient of static friction between ground and ladder is μ_s. What is the smallest angle between the ground and the ladder such that the ladder does not slip? •••

22. A square clock of inertia m is hung on a nail driven into a wall (Figure P12.22). The length of each side of the square is ℓ, the thickness is w, and the top back edge of the clock is a distance d from the wall. Assume that the wall is smooth and that the center of mass of the clock is at the geometric center. Obtain an expression for the magnitude of the normal force exerted by the wall on the clock. •••

Figure P12.22

12.4 The vectorial nature of rotation

23. Beginning mechanics sometimes are taught a rule for tightening and untightening bolts: Righty-tighty, lefty-loosey. Explain why this memory device works. •

24. What is the direction of the rotational velocity vector of the jar lid when you open a jar of pickles? •

25. Gear A in Figure P12.25 is spinning such that its rotational velocity vector points in the negative *x* direction. It is lowered so that its edge comes in contact with the edge of gear B. What is the direction of the rotational velocity caused by the force that A exerts on B? •

Figure P12.25

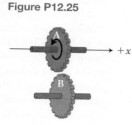

26. An experienced pool player can control the placement of a ball after it hits an edge bumper by putting spin, or "English," on the ball. This is accomplished by aiming the cue stick so that it hits the ball either left or right of the ball's center. Suppose a player gives a ball English by hitting it to the left of center (left from his perspective). Which path in Figure P12.26 does the ball take after it collides with the bumper? Account for this behavior. ••

Figure P12.26

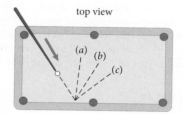

top view

27. The beach ball shown in Figure P12.27 undergoes a rotation of 90° around the *x* axis and then a rotation of 90° around the *y* axis. A second, identically colored ball undergoes first the rotation of 90° around the *y* axis and then the 90° rotation around the *x* axis. What additional rotations are necessary to put the second ball in exactly the same orientation as the first? ••

Figure P12.27

12.5 Conservation of angular momentum

28. Why do quarterbacks throw the football with significant spin about its long axis? •

29. A 1.5-m uniform rod is being used to balance two buckets of paint, each of inertia *m*, one at each end of the rod. (*a*) Where is the pivot located? (*b*) If paint is removed from one bucket until its inertia is *m*/4, where must the pivot now be placed in order to keep the rod balanced? Ignore the inertia of the rod. •

30. A 0.20-kg turntable of radius 0.20 m spins about a vertical axis through its center. A constant rotational acceleration causes the turntable to accelerate from 0 to 28 revolutions per second in 8.0 s. Calculate (*a*) the rotational acceleration and (*b*) the torque required to cause this acceleration. •

31. A 44-kg child runs at 3.0 m/s tangent to a stationary 180-kg playground merry-go-round that has a radius of 1.2 m. The child then jumps on and grabs hold, causing the merry-go-round to rotate. Determine the resulting rotational velocity of the merry-go-round. ••

32. Standing on the edge of a rotating playground carousel, you take sight on the center of the carousel and walk directly toward that position. (*a*) What happens to the rotational speed of the carousel as you walk? (*b*) In the system consisting of only the carousel (a nonisolated system), any change in angular momentum is due to a torque. Because the hub of the carousel is well lubricated, the only candidate for what causes this torque is the force of friction exerted by your feet on the carousel. "But," you say, "I walked radially inward. How could I possibly cause a torque with my feet?" What's going on? ••

33. Standing with your toes touching the base of a wall, press your nose to the wall and try to stand on tiptoe. What happens and why? ••

34. Draw the angular momentum vector for a tetherball, taking as your origin the center of the pole. Is the angular momentum constant? If not, what force causes the torque that changes the angular momentum? ••

35. Imagine that an asteroid 1 km in diameter collides with Earth. Estimate the maximum fractional change in the length of the day due to this collision. ••

36. You want to hang a 10-kg sign that advertises your new business. To do this, you use a pivot to attach the base of a 5.0-kg beam to a wall (Figure P12.36). You then attach a cable to the beam and to the wall in such a way that the cable and beam are perpendicular to each other. The beam is 2.0 m long and makes an angle of 37° with the vertical. You hang the sign from the end of the beam to which the cable is attached. (*a*) What must be the minimum tensile strength of the cable (the minimum amount of tension it can sustain) if it is not to snap? (*b*) Determine the horizontal and vertical components of the force exerted by the pivot on the beam. ••

Figure P12.36

37°

pivot

37. The top end of a two-by-four piece of lumber that is 3.0 m long is leaned at a height of 1.8 m against a smooth wall so that the bottom end makes an angle of 37° with the floor. If this board has an inertia of 4.6 kg, what are (*a*) the normal force exerted by the floor on it and (*b*) the normal force exerted by the wall on it? ••

38. Your physics instructor calls you up to the front of the class to be a demonstration assistant. She has you stand on a turntable that is free to rotate and hands you a spinning bicycle wheel, as shown in Figure P12.38. Letting go of the top of the wheel's axle, you then stop its spinning with your free hand. What happens when you stop the wheel? ••

Figure P12.38

39. Helicopters have a small tail rotor as well as the large main rotor. Why is the tail rotor needed? ••

40. A 35-kg child stands on the edge of a 400-kg playground merry-go-round that is turning at the rate of 1 rev every 2.2 s. He then walks to the center of the platform. If the radius of the platform is 1.5 m, what is its rotational speed once he arrives at the center? ••

41. An ice figure skater starts out spinning at 0.85 revolutions per second with her arms outstretched, as shown in Figure P12.41. She wears lightly weighted bracelets to enhance the spin-up effect when she pulls her arms in. (a) Calculate her final rotational speed if her rotational inertia is 3.6 kg·m² with her arms outstretched and 1.1 kg·m² with her arms pulled close to her body. (b) Determine the increase in her rotational kinetic energy. (c) Where does this added energy come from? ••

Figure P12.41

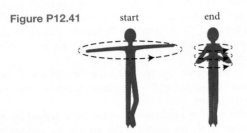

42. A 25-kg child starts at the center of a playground merry-go-round that has a radius of 2.0 m and rotational inertia of 500 kg·m² and walks out to the edge. If the merry-go-round has a rotational speed of 0.20 s⁻¹ when she is at the center, what is its rotational speed when she gets to the edge? ••

43. In the setup in Figure P12.43, does pulling the string downward while the ball is revolving make the following properties of the ball increase, decrease, or stay the same: (a) angular momentum, (b) rotational speed, (c) rotational kinetic energy? ••

Figure P12.43

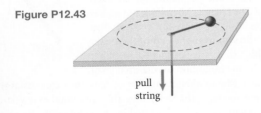

pull
string

44. Your physics instructor has you sit on a chair that is free to rotate and hands you a spinning bicycle wheel (Figure P12.44). What happens when you flip the wheel over so that the end of the axle that initially pointed up now points down? ••

Figure P12.44

45. A putty ball of inertia m moving at speed v slides on a slick surface along a path that is perpendicular to the long axis of a rod of length ℓ and inertia 2m that is at rest on the surface. The ball collides with one end of the rod and sticks to it. (a) Just as the ball sticks, where is the center of mass of the ball-rod system? After the collision, (b) what are the magnitude and direction of the velocity of the center of mass and (c) what is the rotational speed of the system? Ignore any effects due to friction. •••

46. A 4.5-kg bowling ball is perched on a concrete ledge directly below your dorm room window, with the side of the ball opposite the holes touching the wall. Wanting to hold the ball in place so that it doesn't roll off and land on somebody, you manage to hook one of the holes with a wire and exert a purely tangential (and vertical) force on the ball. If the coefficient of static friction between ball and ledge is the same as that between ball and wall, $\mu_s = 0.50$, what is the maximum upward force you can exert so that the ball does not rotate and you lose your hold? Even though the ball has holes drilled in it, assume a uniform distribution of inertia. •••

47. Disk A in Figure P12.47 has radius R_A and thickness h and is initially rotating clockwise, as viewed from above, at $\omega_i/2$. Disk B, made of the same material as A, has radius $R_B = R_A/2$ and thickness h and is initially rotating counterclockwise at ω_i. The two disks are constrained to rotate about the same axis, which runs through their centers. There is friction between the disks, and so once B slides down and touches A, they spin at the same rotational speed. (a) What is their common final rotational velocity? (b) If the frictional force is the only force that causes the system to lose energy, what is the fractional change in the kinetic energy? •••

Figure P12.47

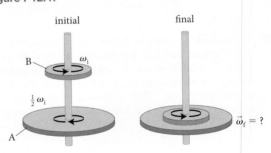

48. A 25-kg ladder of length 5.0 m leans against a smooth wall and makes an angle of 50° with the ground. A 75-kg man starts to climb the ladder. If the coefficient of static friction between ground and ladder is 0.50, what distance along the ladder can the man climb before the ladder starts to slip? •••

12.6 Rolling motion

49. An object rolls without slipping onto a surface where the coefficient of friction between object and surface is twice as great as that needed to prevent slipping. Describe the subsequent motion. •

50. To get better acceleration, you want to reduce the inertia of your bicycle. You have your choice of spending $45 on only one of three light-alloy items, each reducing the inertia of your bicycle by the same amount: a new seat post, a new set of pedals, or a new set of wheel rims. Which is the best buy? •

51. A 3.0-kg solid ball rolls without slipping down a ramp inclined to the horizontal at an angle of 30°. What are (a) the acceleration of the ball's center of mass and (b) the magnitude of the frictional force exerted on the ball? ••

52. A 5.0-kg solid cylinder of radius 0.25 m is free to rotate about an axle that runs along the cylinder length and passes through its center. A thread wrapped around the cylinder is pulled straight from the cylinder so as to unwrap with a steady tensile force of 20 N. As the thread unwinds, the cylinder rotates and there is no slippage between thread and cylinder. If the cylinder starts from rest at $t_i = 0$, calculate (a) its rotational velocity at $t_f = 5.0$ s and (b) the angle through which it has turned at $t_f = 5.0$ s. Ignore any friction in the axle. ••

53. A disk and a hoop roll down an inclined plane. If the plane is inclined at an angle of 30° from the horizontal, what is the minimum coefficient of friction required so that neither object slips? ••

54. A 2.0-kg solid cylinder of radius 0.45 m rolls without slipping down a ramp inclined at an angle of 60° to the vertical. Calculate (a) the acceleration of the cylinder's center of mass, (b) the cylinder's rotational acceleration, and (c) the time interval required for it to travel 35 m down the incline if it starts from rest. (d) What is the rotational velocity of the cylinder at the end of the time interval found in part c? ••

55. Two cans of pumpkin roll down a loading dock ramp. They have identical inertias, but one of them has a larger radius (and a shorter length) than the other. Which one, if either, makes it down the ramp first? ••

56. You deliver a horizontal impulse to a cue ball with a cue stick while playing pool. If the ball is to roll without slipping, at what height h above its center (in terms of its radius R) must you strike it? •••

57. Two balls of the same radius and same inertia roll down an inclined plane, starting from rest. One ball is hollow, and the other is solid. What is the ratio of the time intervals the two balls require to reach the bottom? ••

58. In the Atwood machine shown in Figure P12.58, the pulley radius is 0.10 m, and the rotational inertia of the pulley is 0.15 kg·m². Calculate (a) the acceleration of the blocks, (b) the tension in the cord on the left, and (c) the tension in the cord on the right. Ignore the cord's inertia. •••

Figure P12.58

2.0 kg

5.0 kg

59. A block of inertia m sits on a smooth surface (Figure P12.59). A light string is attached to it and placed over a solid pulley of inertia $3m$ and radius R, and a ball of inertia m is attached to the free end of the string. (a) If there is no slippage between string and pulley, determine the magnitude of the block's acceleration. (b) If you detach the ball and pull down on the hanging string with a force of magnitude mg, what is the magnitude of the block's acceleration? •••

Figure P12.59

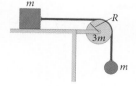

60. A 320-g can of soup is 10.8 cm tall and has a radius of 3.19 cm. (a) Calculate the can's rotational inertia about its axis of symmetry, assuming the can to be a solid cylinder. (b) When released from rest at the top of a ramp that is 3.00 m long and makes an angle of 25° with the horizontal, the can reaches the bottom in 1.40 s. What is the value of rotational inertia obtained from this experiment? (c) Compare the calculated and experimentally determined rotational inertias and suggest possible reasons for the difference. •••

12.7 Torque and energy

61. A 0.20-kg solid cylinder is released from rest at the top of a ramp 1.0 m long. The cylinder has a radius of 0.15 m, and the ramp is at an angle of 15° with the horizontal. What is the rotational kinetic energy of the cylinder when it reaches the bottom of the ramp? •

62. A solid 50-kg cylinder has a radius of 0.10 m. What minimum work is required to get the cylinder rolling without slipping at a rotational speed of 20 s⁻¹? •

63. A 1.0-kg ring with an inner radius of 0.060 m and an outer radius of 0.080 m is sent rolling without slipping up a ramp that makes an angle of 30° with the horizontal. If the initial speed of the ring is 2.8 m/s, how far along the ramp does it travel before it comes to a stop? ••

64. A certain car engine delivers enough force to create 380 N·m of torque when the engine is operating at 3200 revolutions per minute. Calculate the average power delivered by the engine at this rotation rate. ••

65. A 680-kg disk of radius 1.2 m is mounted on a fixed axle. A force exerted on the disk causes a constant torque that gives the disk a rotational acceleration magnitude of 0.30 s⁻² for 5.0 s. If the disk had an initial rotational speed of 4.5 s⁻¹, how much work was done to provide the torque, assuming no energy is lost to friction? ••

66. A large steel bar of length ℓ is hinged at one end to a wall. A mechanic holds the other end so that the bar is parallel to the ground and places a penny on the bar right at the end he is holding. (a) What is the rotational acceleration of the bar when he lets go? (b) Does the penny remain in contact with the bar after it is released? ••

67. A light, unstretchable string is wound around the perimeter of a 4.0-kg disk that has a radius of 0.50 m and is free to rotate about an axle that runs perpendicular to the disk face and through the disk center. A 2.0-kg block is connected to the free end of the string and allowed to slide down a ramp that makes an angle of $\theta = 37°$ with the vertical (Figure P12.67). Calculate the magnitude of the disk's rotational acceleration. Ignore friction between block and ramp. ••

Figure P12.67

θ

68. A 5.0-kg hollow cylinder of radius 0.25 m rotates freely about an axle that runs through its center and along its long axis. A cord is wrapped around the cylinder and is pulled straight from the cylinder with a steady tensile force of 50 N. As the cord unwinds, the cylinder rotates, with no slippage between cord and cylinder. (a) Calculate the work done by the tensile force as the cylinder rotates through 1000 rad. (b) If the cylinder starts from rest, calculate its rotational speed after it has rotated through 1000 rad. ••

PRACTICE

69. A marble of inertia m is held against the side of a hemi-spherical bowl as shown in Figure P12.69 and then released. It rolls without slipping. The initial position of the marble is such that an imaginary line drawn from it to the center of curvature of the bowl makes an angle of 30° with the vertical. The marble radius is $R_m = 10$ mm, and the radius of the bowl is $R_b = 100$ mm. Determine the rotational speed of the marble about its center of mass when it reaches the bottom. ••

Figure P12.69

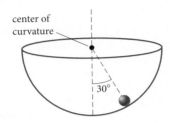

center of curvature

30°

70. A 3.0-kg disk of radius 50 mm rolls down a ramp inclined at an angle of 28° with the vertical. If the disk starts out at rest and the coefficients of static and kinetic friction between the ramp and the disk are both 0.50, what is the rotational speed of the disk after it has traveled 1.5 m? ••

71. An almost-conical toy top (Figure P12.71) that has radius $R = 20$ mm and inertia 0.125 kg is spun up using a force of 5.0 N and a string that is 1.0 m long. What are (a) the work done on the top, (b) its kinetic energy, and (c) its final rotational speed? The shape factor for this top about its axis of symmetry is 0.35. ••

Figure P12.71

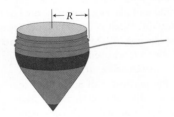

← R →

72. Archimedes' screw, one of the first mechanical devices invented for lifting water, is a very large screw fitted tightly inside a shaft (Figure P12.72). The bottom of the device is placed in a pool of water. As someone turns the handle to make the screw turn, water is carried up along the ridges of the screw and comes out the top of the shaft and into a storage tank. As the handle turns, the work done by the force exerted on the handle is converted to gravitational potential energy of the water-Earth system. Let's say you want to take a shower using this device. You figure that your shower will use 44 L of water, and so you have to raise this amount to the storage tank 2.5 m above the pool. Every time you turn the handle, you exert a force that causes a torque of 12 N · m. How many times must you turn the handle? ••

Figure P12.72

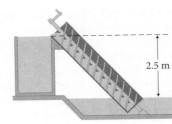

2.5 m

73. You attach a 0.50-m length of string to a 50-g puck and pass the other end of the string through a hole in the center of a table. Grasping the string under the table, you pull just enough string through the hole so that, when your friend gives the puck a sideways push, it moves in a circle of radius $r_i = 0.45$ m at 1.5 m/s. You then slowly pull on the string, decreasing the circle's radius to 0.20 m. (a) What is the puck's speed when $r = 0.20$ m? (b) What is the tension in the string as a function of r? What is its value when $r = 0.20$ m? (c) How much work is done in moving the puck from the 0.45-m circle to the 0.20-m circle? ••

74. A 30-kg solid sphere of radius 0.12 m is rolling without slipping on a horizontal surface at 2.0 m/s. (a) What average torque is required to stop the sphere in 5.0 rev without inducing skidding? (b) If this torque is caused by a soft braking bumper that is lowered down until it just makes contact with the top of the sphere, what is the magnitude of the frictional force between bumper and sphere? •••

75. A marble that has a radius of 10 mm is placed at the top of a globe of radius 0.80 m. When released, the marble rolls without slipping down the globe. Determine the angle from the top of the globe to the location where the marble flies off the globe. •••

76. A 4.0-kg bowling ball is thrown down the alley with a speed of 10.0 m/s. At first the ball slides with no rotation. The coefficient of friction between ball and alley surface is 0.20. (a) How long a time interval does it take for the ball to achieve pure rolling motion? (b) What is the ball's translational speed at the end of this time interval? •••

77. You shove a cube of inertia m and side length d so that it slides along a smooth table with speed v_i (Figure P12.77a). The cube then hits a raised lip at the end of the table. After it hits the lip, the cube begins to rotate about it (Figure 12.77b). (a) Show that the magnitude of the cube's angular momentum about the lip before the collision is $L = mdv_i/2$. (b) Explain why the angular momentum still has that value at the instant of collision, before the cube has had time to rotate much. (c) What is the rotational acceleration of the cube the instant after it hits the lip? (d) What maximum initial speed can the cube have so that it does not topple over the lip? •••

Figure P12.77

(a)   (b)

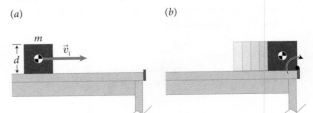

m

d

$\vec{v}_i$

12.8 The vector product

78. Which of the following expressions make sense: (a) $\vec{A} \cdot (\vec{B} \times \vec{C})$, (b) $\vec{A} \times (\vec{B} \cdot \vec{C})$, (c) $\vec{A} \times (\vec{B} \times \vec{C})$? •

79. When you want to turn a bicycle left, the first thing you do is lean left. Why is it important to do this? •

80. If the magnitude of the vector product of two vectors is the same as the magnitude of the scalar product of the same two vectors, what is the angle between them? •

81. A cyclist exerts a vertical force on a bike pedal. This pedal is at the end of a crank that is 0.20 m long and pivoted to rotate about the axle of the chain wheel. If the cyclist pushes downward with a force of 150 N, determine the magnitude of the torque caused by this force when the crank is (a) horizontal, (b) 30° from the horizontal, (c) 45° from the horizontal, (d) 60° from the horizontal, and (e) vertical. (f) Calculate the average magnitude of this torque exerted by one of the cyclist's feet during one cycle of the pedals, assuming there is no clip holding the cyclist's feet to the pedals. ••

82. The two arms of the L-shaped handle on the spigot of Figure P12.82 have a length ratio of $1:\sqrt{3}$. At what angle θ do you want to pull down on the end of the handle to maximize the torque your force causes? ••

Figure P12.82

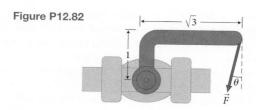

83. Show that, if vectors $\vec{A}$ and $\vec{B}$ are both in an xy plane, $\vec{A} \times \vec{B}$ is a vector that is perpendicular to the xy plane and has magnitude $|\vec{A} \times \vec{B}| = |A_x B_y - A_y B_x|$. ••

84. A 3.0-kg rod that is 1.5 m long is free to rotate in a vertical plane about an axle that runs through the rod's center, is perpendicular to the rod's length, and runs parallel to the floor. A 1.0-kg block is attached to one end of the rod, and a 2.0-kg block is attached to the other end. At some instant, the rod makes an angle of 30° with the horizontal so that the blocks are in the positions shown in Figure P12.84. (a) Determine the torque caused by the forces exerted on the system at this instant. (b) Determine the rotational acceleration of the system at this instant. Ignore friction and assume the blocks are small enough that any length they add to the rod can be ignored. ••

Figure P12.84

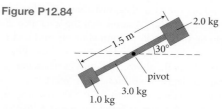

85. A motor drives a disk initially at rest through 23.9 rotations in 5.0 s. Assume the vector sum of the torques caused by the force exerted by the motor and the force of friction is constant. The rotational inertia of the disk is 4.0 kg·m². When the motor is switched off, the disk comes to rest in 12 s. (a) What is the torque created by the force of friction? (b) What is the torque caused by the force exerted by the motor? ••

86. The angular momentum of the propellers of a small airplane points directly forward from the plane. (a) In what direction do the propellers rotate as seen from the rear of the plane? (b) If the plane is flying horizontally and suddenly pulls upward, in which direction does the nose of the plane tend to move? Justify your answer. ••

87. A 40-kg sharpening wheel of radius 0.10 m is rotating at 3.3 revolutions per second. A 6.0-kg axe is pressed against the rim with a force of 40 N directed as shown in Figure P12.87. Treat the wheel as a disk, and assume a coefficient of kinetic friction between wheel and axe blade of $\mu_k = 0.35$. (a) If there is no power source keeping the wheel rotating, how long does it take to stop? (b) How many revolutions does the wheel make while it is slowing down? ••

Figure P12.87

88. A heavy-rimmed bicycle wheel is set spinning in the direction shown in Figure P12.88. A string is tied to one end of the axle, and someone is holding up the string. (a) Use torque arguments to explain why the wheel slowly revolves horizontally around the string end of the axle when the free end of the axle is released. (This effect is called *precession*.) (b) The precession speed increases as time passes. Explain why. •••

Figure P12.88

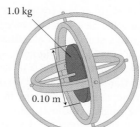

89. A gyroscope consists of a spinning wheel mounted inside a support structure (Figure P12.89). Because of conservation of angular momentum, the gyroscope keeps its orientation in space unless some outside force causes a torque on it. Suppose the 1.0-kg spinning wheel of a certain gyroscope has a radius of 0.10 m and is spinning at 14,200 revolutions per minute. What is the magnitude of the torque required to tilt the gyroscope 90° in 20 s such that the spin speed is still 14,200 revolutions per minute when the gyroscope is in its new orientation? •••

Figure P12.89   1.0 kg

PRACTICE

Additional Problems

90. Two identical boxes are placed on opposite ends of the board of a playground seesaw as in Figure P12.90a, so that the system is in mechanical equilibrium. (a) What happens to that equilibrium if you exert a small, brief force, either upward or downward, on either end of the board? (b) With the boxes attached to the underside of the board as in Figure P12.90b and the system again in mechanical equilibrium, what happens to the equilibrium if you exert such a force on either end? •

Figure P12.90

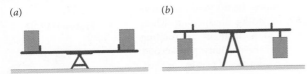

(a)   (b)

91. For the x and y axes of a right-handed Cartesian coordinate system, what is the direction of $\hat{\imath} \times \{\hat{\imath} \times [\hat{\imath} \times (\hat{\imath} \times \hat{\jmath})]\}$? •

92. Standing on a round raft floating on a pond, how do you turn the raft around 180°? •

93. Dragster drivers have to avoid supplying too much power to the vehicle because too much power causes the front end to rise in a "wheelie," compromising steering control. (a) Why does this happen? (b) What advantage comes from having front-wheel drive in a dragster? ••

94. If everyone on Earth simultaneously walked from west to east, by what fraction would the length of the day change? Would it lengthen or shorten? ••

95. A 51-kg box is suspended from the right end of a horizontal rod that has very small inertia. The left end of the rod is affixed to a wall by a pin. A wire connects the right end of the rod to the wall directly above the pin, making an angle of 40° with the rod. (a) Calculate the tension in the wire. (b) Determine the magnitude of the reaction force the pivot exerts on the rod. (c) Repeat parts a and b for a 10.2-kg rod. ••

96. A 35-kg child stands on the edge of a playground merry-go-round that has a radius of 2.0 m and a rotational inertia of 500 kg·m². The merry-go-round has a rotational speed of 0.20 s⁻¹ when the child is standing still. If the rotational speed increases to 0.25 s⁻¹ when the child starts walking along the edge, with what speed is he walking in the reference frame of the merry-go-round, and in which direction? ••

97. A neutron star is the compact remnant of a very large star that has exploded. Suppose that right after one such explosion, when much of the star's inertia has been blasted away, the star's remaining core has an inertia of 4×10^{30} kg, a radius of 13×10^8 m, and completes one revolution every five days. What is the new spin rate once the core gravitationally collapses into a neutron star of radius 20 km? Assume negligible loss of inertia during the collapse. ••

98. A horizontal 2.0-kg rod is 2.0 m long. An 8.0-kg block is suspended from its left end, and a 4.0-kg block is suspended from its right end. (a) Determine the magnitude and direction of the single extra force necessary to keep the rod in mechanical equilibrium. (b) At what distance from the left end of the rod must the point of application of this force be? ••

99. What maximum torque can a bicyclist deliver to the pedals? ••

100. A solid ball of inertia m rolls without slipping down a ramp that makes an angle θ with the horizontal. (a) What frictional force is exerted on the ball? (b) As a function of θ, what coefficient of friction is required to prevent slipping? ••

101. You make a lawn roller out of a 125-kg solid cylinder, set to rotate about a central axle. You rig up handles on either side of the axle so that you and a friend can pull on the handles horizontally to pull the roller in a direction perpendicular to its rotation axis (Figure P12.101). (a) If each of you pulls with a horizontal force of 500 N, what is the magnitude a of the roller's acceleration? (b) What is the minimum coefficient of static friction necessary to keep the roller from slipping? ••

Figure P12.101

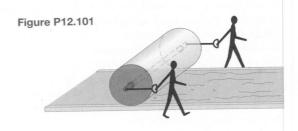

102. A marble is shot across a smooth wooden floor so that the marble both rotates about a horizontal axis and slides. If at a certain instant the marble's rotational kinetic energy equals the translational kinetic energy of its center of mass, what is the ratio of its center-of-mass speed to the speed of a point on its "equator" associated with rotation? ••

103. A 12-kg cylinder of radius 0.10 m starts at rest and rolls without slipping down a ramp that is 6.0 m long and inclined at 30° to the horizontal. When the cylinder leaves the end of the ramp, it drops 5.0 m to the ground. At what horizontal distance from the end of the ramp does it land? ••

104. A child runs in a straight line tangent to a playground carousel, jumps on, and holds tight as the carousel begins to rotate. Which of the following statements is correct for this collision? (a) The coefficient of restitution is greater than 1 because there is now rotational energy where there was none before. (b) The coefficient of restitution is 1 because the system's translational kinetic energy is completely converted to rotational kinetic energy. (c) The coefficient of restitution is less than 1 because the sum of the translational and rotational kinetic energies in the final state is less than the sum in the initial state. (d) The coefficient of restitution cannot be defined in a system in which there is both translational motion and rotational motion. (e) Explain why all the answers except the one you chose are wrong. ••

105. In a judo hip throw, you pull your 60-kg opponent onto your back to bring her center of mass just above your hip and then rotate her about your hip. Assume that the lever arm distance of your grip on her is 0.30 m from your hip and that her rotational inertia about your hip is 15 kg·m². (a) What is the magnitude of the pulling force you must exert on her to cause a rotational acceleration of 1.0 s⁻¹? (b) What is the magnitude of the pulling force if you throw her from a more upright position, so that the gravitational force has a lever arm of 0.12 m about your hip? •••

106. You've been asked to design a flywheel and the associated emergency friction braking system for an electric vehicle. In order to fit the available space, the flywheel must be cast in the form of a solid steel disk with a thickness of 50 mm and a radius that cannot exceed 0.20 m. For safety purposes, the emergency braking system must slow the flywheel from 6000 revolutions per minute to 0 in 12 revolutions. You know that typical braking systems utilize a pair of "shoes" mounted on calipers to squeeze the thickness of the disc symmetrically near its edge, but you are concerned about how much force the brake shoes might have to exert on the disk. ••• CR

107. A 3.0-kg block is attached to one end of a light, unstretchable string that is wrapped securely around a cylinder that has a 0.30-m radius and a rotational inertia of 0.80 kg · m². The cylinder is free to rotate about an axle aligned along its long axis. The block is held right alongside the cylinder (so there is essentially no unwound string) and then released from rest. As the block falls, it unwinds the string from the cylinder, with no slippage between string and cylinder. (*a*) Calculate the magnitude of the block's acceleration. (*b*) Calculate the rotational speed of the cylinder after the block has dropped 1.5 m. (Hint: First use the acceleration from part *a* to determine the block's speed.) (*c*) Use energy methods to obtain the cylinder's rotational speed after the block has dropped 1.5 m. (*d*) Calculate the tension in the string after the block has dropped this distance. (*e*) Calculate the instantaneous power delivered by the string and block to the cylinder after the block has dropped this distance. •••

108. You are building a pool table and want the design to be such that any ball that rolls into one of the side rails without slipping bounces off the rail and rolls away in another direction, again without slipping. After some experimenting, you decide that you must build the rails to a specific height that is some function of the radius of the balls. (Hint: It seems plausible to assume that the collision is elastic.) ••• CR

109. A yo-yo is composed of two disks of radius *a*, with many layers of string wrapped around the axle. The radial distance from the center of the axle to the top layer of string at any instant is *b*. With the end of the string wrapped around your finger and $b/a \approx 1$, you let the yo-yo drop, unwinding string and ultimately spinning pretty fast when it reaches the bottom of the string ($b/a \approx 0$). You begin to wonder how the motion of the yo-yo as the string unwinds depends on the value of *b*. ••• CR

Answers to Review Questions

1. Yes, because the torque magnitude depends not only on the magnitude of the force but also on the lever arm.

2. Not quite. The fact that the door moves the same distance in 2 s in both cases tells you that the rotational acceleration is the same in both cases, which means that the torque caused by your force must be the same as the torque caused by your friend's force. However, because the rotation axis runs along the hinged edge, the magnitude of your force must be greater than the magnitude of his force because the lever arm of your force is smaller than the lever arm of his force.

3. You may choose any reference point, though it is best to choose a reference point that simplifies the calculation by eliminating one or more unknowns from the torque calculation.

4. There are many examples. The normal force that the ground exerts on the rim of a bicycle tire (when the tire is vertical) is one example. The torque is zero because there is no component of the force perpendicular to the line of action. Similarly, if you pull on a door handle with a force whose line of action passes through the hinged edge, the door will not rotate.

5. No. Objects can be made to rotate or revolve about a specified point only by imposing a constraint, such as a hinge, pivot, or string. Free rotation is the absence of such constraints, so a freely rotating object will rotate about its center of mass.

6. The center of mass of an object or system moves as if all the inertia of the object or system were concentrated at that point. Hence, any free rotation occurs about an axis through the center of mass, allowing complicated motion to be expressed as two independent motions: the translational motion of the center of mass and the rotational motion around that point.

7. Primarily downward, with a small horizontal component toward the hand. One way to see this is to look at the stationary forearm. Being free to choose any pivot around which to sum torques, let's arbitrarily take the pivot to be at the hand. About this point, the torque caused by the force exerted by the biceps muscle on the forearm bone is clockwise from your perspective. To balance this clockwise torque exactly (remember that the arm is not rotating), the torque caused by the force exerted by the humerus on the forearm bone must be counterclockwise. To get this counterclockwise torque, the force causing it must be directed downward. Another viewpoint: If the biceps muscle pulls upward on the forearm, then the humerus must exert a downward force on the elbow-joint end of the forearm in order to prevent upward translation of that end.

8. An extended free-body diagram shows the location of the point of application of each force on the object in question, a necessity for computing torques. Torques are required in order to analyze rotational motion. A standard free-body diagram shows the forces as if the points of application of all of them were at a common point, which is fine for force computations.

9. It is possible to use the extended diagram for both forces and torques, but it is best to draw a standard free-body diagram first until you have gained a lot of experience. You are familiar with free-body diagrams, so just a glance at one will tell you whether the force components are capable of canceling or adding up to the required components of ma. The extended free-body diagram is less familiar and has the forces moved to remote locations rather than all tail to tail at one spot. If you accidentally leave one force off the extended free-body diagram, it will be much harder to tell at a glance that anything is wrong. Leaving out a force (or including an inapplicable force) leads to an erroneous result.

10. Curl the fingers of your right hand so that they follow the rotation of the object. Your thumb then points in the direction of the vector that specifies the direction of rotation.

11. The direction is into the wall, by the right-hand rule.

12. In three dimensions, rotations that appear clockwise from one perspective can appear counterclockwise from another.

13. The latter statement is correct. The former cannot be right because the units of force and torque aren't the same and so they can't be added.

14. Yes. As long as the single force is exerted, the momentum changes. As long as that force has a lever arm with respect to some pivot, the force causes a torque, which changes the angular momentum.

15. No. The momentum is not constant because a gravitational force is exerted on the baton. The baton is unconstrained and hence rotates about an axis through its center of mass. Because only a gravitational force is exerted on the baton while it is in the air, and this force is exerted straight downward on the center of mass, it causes no torque about that point. Thus the angular momentum is constant.

16. Rotational impulse represents the transfer of angular momentum between a system and its environment. It can be computed as the product of external torque and the time interval during which the torque is applied, quantified in the rotational impulse equation, $J_\vartheta = (\sum \tau_{\text{ext}\,\vartheta}) \Delta t$. The angular momentum law, by analogy to the momentum law from translational motion, states that any change in the angular momentum of a system must be due to the rotational impulse caused by external torque: $\Delta L_\vartheta = J_\vartheta$.

17. Angular momentum is always conserved, which means it cannot be created or destroyed. The angular momentum of a system remains constant if the system is in rotational equilibrium. This occurs when the system is isolated or when the rotational impulse acting on the system is zero.

18. They are related by a factor of the radius of the wheel: $|v_{\text{cm}\,x}| = |R\omega_\vartheta|$.

19. When gears mesh, they turn without slipping, so the tangential speeds of the teeth of the two gears must be identical. The rotational speeds are generally not the same because they depend on the radii of the gears.

20. The shape factor is the ratio of an object's rotational inertia to the rotational inertia of a hoop of the same inertia m and radius R: $c = I/mR^2$. This factor compares the rotational inertia of an object with the rotational inertia if all the object's inertia were concentrated on its rim, so it measures the distribution of inertia in the object. This distribution is loosely referred to as "shape."

21. They are related by a factor of the object's radius: $|a_{\text{cm}\,x}| = |R\alpha_\vartheta|$.

22. The work done is the product of the sum of the torques $\sum \tau$ and the change in the object's rotational coordinate $\Delta \vartheta$. It is necessary to first convert the number of rotations to the change in the rotational coordinate.

23. The kinetic energy of an object or system that is in both translational and rotational motion is equal to the sum of its center-of-mass and rotational kinetic energies.

24. The force of static friction does no work on the object and converts some of the object's translational kinetic energy to rotational kinetic energy.

25. The vector product vector is perpendicular to the desktop, pointing either up or down. Which direction is correct is determined by the right-hand rule. Other than that, the angle is independent of the orientation of the two vectors.

26. The magnitude of the vector product of two vectors is equal to the area of the parallelogram defined by the two vectors. It is also equal to the product of the magnitudes of each vector and the sine of the smaller angle between them when they are placed tail to tail.

27. Zero. The term in parentheses is perpendicular to $\vec{A}$, and so the scalar product is zero.

28. The sign. The magnitude of your answer is correct, but the direction is opposite the correct direction.

29. The angular momentum of an isolated particle about any reference point is equal to the vector product of the position vector that locates the particle relative to the reference point and the momentum of the particle: $\vec{L} = \vec{r} \times \vec{p}$.

Answers to Guided Problems

Guided Problem 12.2 $|\vec{\tau}| = 1.8 \, \text{N} \cdot \text{m}$

Guided Problem 12.4 (*a*) The disk wins; (*b*) $\Delta t = 0.4 \, \text{s}$

Guided Problem 12.6 (*a*) $a_x = \dfrac{\mathcal{T}}{m_Y(1 + c)}\left(\cos \phi - \dfrac{R_a}{R_Y} \right)$;

(*b*) $\phi_{\text{crit}} = \cos^{-1}\left(\dfrac{R_a}{R_Y} \right)$

Guided Problem 12.8 (*a*) beneath the edge of the base opposite you; (*b*) $h_{\text{max}} = d/2\mu$

13

PRACTICE
Gravity

Chapter Summary   233
Review Questions   234
Developing a Feel   235
Worked and Guided Problems   236
Questions and Problems   241
Answers to Review Questions   247
Answers to Guided Problems   247

PRACTICE

Chapter Summary

Universal gravity and the principle of equivalence (Sections 13.1, 13.3–13.5, 13.8)

Concepts The force of gravity is an attraction between all objects that have **mass,** which is a measure of the quantity of material in an object. Mass is the property of an object that determines the strength of its gravitational interaction with other objects.

At everyday speeds, the mass of an object is equal to its inertia.

According to the **principle of equivalence,** one cannot distinguish locally between the effects of a constant gravitational acceleration of magnitude g and the effects of an acceleration of the reference frame of magnitude $a = g$.

Quantitative Tools According to Newton's **law of gravity,** if two objects that have masses m_1 and m_2 are a distance r apart, the magnitude of the gravitational force that each object exerts on the other is

$$F_{12}^G = G\frac{m_1 m_2}{r_{12}^2}, \tag{13.1}$$

where $G = 6.6738 \times 10^{-11}$ N·m²/kg² is the **gravitational constant.**

The acceleration due to gravity near Earth's surface is

$$g = \frac{Gm_E}{R_E^2}, \tag{13.4}$$

where m_E is Earth's mass and R_E is its radius.

The gravitational force F_{sp}^G exerted by a uniform solid sphere on a particle of mass m is the same as if all the matter in the sphere were concentrated at its center:

$$F_{sp}^G = G\frac{mM_{sphere}}{r^2}, \tag{13.37}$$

where M_{sphere} is the mass of the sphere and r is the distance from its center to the particle.

Angular momentum and gravitational potential energy (Sections 13.2, 13.6)

Concepts The force of gravity is a **central force,** meaning that its line of action always lies along the line connecting the interacting objects.

In an isolated system of two objects that interact through a central force, each object has a constant angular momentum about the center of mass of the system.

Quantitative Tools The **gravitational potential energy** $U^G(x)$ of a system made up of two objects of masses m_1 and m_2 that are a distance x apart is

$$U^G(x) = -G\frac{m_1 m_2}{x}, \tag{13.11}$$

where we take $U^G(x)$ to be zero at $x = \infty$.

Celestial mechanics (Sections 13.2, 13.7)

Concepts **Kepler's first law of planetary motion:** All planets move in elliptical orbits with the Sun at one focus. This law is a consequence of the fact that the magnitude of the gravitational force has a $1/r^2$ dependence.

Kepler's second law: The straight line from any planet to the Sun sweeps out equal areas in equal time intervals. This law reflects the fact that the angular momentum of a planet is constant because the gravitational force is a central force.

Kepler's third law: The squares of the periods T of the planets are proportional to the cubes of the semimajor axes a of their orbits around the Sun ($T^2 \propto a^3$). This law is a consequence of the fact that the magnitude of the gravitational force has a $1/r^2$ dependence.

Quantitative Tools The energy of a closed ($\Delta E = 0$), isolated ($\Delta L = 0$) system made up of a small object of mass m orbiting a large object of mass $M \gg m$ is

$$E = K + U^G = \tfrac{1}{2}mv^2 - G\frac{Mm}{r}, \tag{13.19}$$

where v is the speed of the small object, r is the distance between the objects, and the large object does not move.

The magnitude of angular momentum of the small object is

$$L = r_\perp mv. \tag{13.20}$$

The value of E in Eq. 13.19 determines the orbit's shape:

$$E_{mech} < 0: \quad \text{elliptical (bound) orbit}$$
$$E_{mech} = 0: \quad \text{parabolic orbit}$$
$$E_{mech} > 0: \quad \text{hyperbolic (unbound) orbit.}$$

PRACTICE

Review Questions

Answers to these questions can be found at the end of this chapter.

13.1 Universal gravity

1. Which object exerts a greater pulling force on the other: Earth or the Moon? Which object experiences greater acceleration due to this mutual pull?
2. What is the evidence for Newton's $1/r^2$ law of universal gravitation?
3. What orbital shapes are possible with Newton's $1/r^2$ law of universal gravitation?
4. The length of a year on any planet is defined as the time interval the planet takes to make one revolution around the Sun. What is the length of a year on a planet whose orbital radius is four times Earth's orbital radius?
5. What is the distinction between mass and inertia?

13.2 Gravity and angular momentum

6. Explain how, in an isolated system of two objects interacting through a central force, each object has constant angular momentum about the system's center of mass.
7. State Kepler's three laws of planetary motion.
8. In *Principles* Figure 13.12*a* no force is exerted on the moving object, whereas in Figure 13.12*b* and *c*, there is (the force that keeps the object moving along a curved trajectory). Why is the conclusion about angular momentum the same in all three cases?
9. Which has greater acceleration in its orbit around Earth: the Moon or the International Space Station?

13.3 Weight

10. What is your weight measured by a spring scale as you fly through the air over a trampoline? Assume that you can keep the spring scale beneath your feet.
11. Suppose a balance is loaded with objects of equal mass on each side and then dropped in free fall. Does it remain balanced?
12. You stand on a spring scale placed on the ground and read your weight from the dial. You then take the scale into an elevator. Does the dial reading increase, decrease, or stay the same when the elevator accelerates downward as it moves upward?
13. Explain why astronauts in an orbiting space station are able to float around in the cabin.

13.4 Principle of equivalence

14. State the principle of equivalence, and explain why you must use the word *locally* in your statement.
15. Should we expect light to travel along a curved path when it passes near a massive object?

13.5 Gravitational constant

16. Gravitational forces decrease with a $1/r^2$ dependence, and yet the value of g is constant near Earth's surface. Is this a contradiction?
17. Does a planet that has a greater mass than Earth necessarily have a greater acceleration due to gravity near its surface? (Saturn, for example, has about a hundred times the mass of Earth.)

13.6 Gravitational potential energy

18. What choice causes the gravitational potential energy described by Eq. 13.14, $U(r) = -Gm_1m_2/r$, to be negative for any finite r?
19. What does it mean to say that the work done by the gravitational force exerted on an object is *path independent*?
20. Under what condition is it reasonable to approximate a $1/r$ potential energy curve by a straight line?

13.7 Celestial mechanics

21. When you throw a ball from the roof of a building, is the trajectory a parabola or an ellipse?
22. Where in the orbit of any planet in our solar system is the planet's speed smallest? Where is it greatest?
23. If $E = 0$ for a system that consists of an object orbiting a much more massive object, is the orbit bound or unbound?
24. Describe the orbital shapes available for motion under the gravitational force exerted by the Sun, and relate each to the energy of the system comprising the Sun and the orbiting object.

13.8 Gravitational force exerted by a sphere

25. What approach is used to show that the gravitational force exerted by a solid sphere on an external particle is the same as if all the mass of the sphere were concentrated at its center?
26. Can the process for computing the force exerted by a solid sphere on an external particle be extended to a particle located inside the sphere? To a hollow sphere? To a sphere of nonuniform density?

Developing a Feel

Make an order-of-magnitude estimate of each of the following quantities. Letters in parentheses refer to hints below. Use them as needed to guide your thinking.

1. The magnitude of the gravitational force exerted on you by your partner for a tango (B, N)
2. The magnitude of the gravitational force exerted on you by the population of Earth (B, Q, D, J)
3. The magnitude of the gravitational force exerted on you by a nearby large mountain (S, I, B)
4. The radius of the orbit in which your running speed would be sufficient to keep an object moving in a circular orbit around Earth (A, V)
5. The orbital speed required for a near-surface circular orbit of the Moon (V, C, R)
6. The maximum radius for a planet of Earth's density that you could jump from without falling back (F, K, T)
7. The radius to which Earth must be compressed in order for a light beam to orbit the planet in a circular path (V, G, O)
8. The energy required to remove all the planets from the solar system (E, P, L)
9. The maximum and minimum magnitudes of the vector sum of forces you experience during a bungee jump (H, M, U)
10. The magnitude of Earth's angular momentum as it orbits the Sun (W)

Hints

A. What is a typical human running speed for long distances?
B. What is the mass of a typical human?
C. What is the mass of the Moon?
D. How is the population of Earth distributed?
E. What is the gravitational potential energy of the system after the planets are removed?
F. What is the average density of Earth?
G. What principle provides the key to analyzing the gravitational behavior of light?
H. Where in the jump does the bungee cord begin to stretch?
I. What is the distance to the center of mass of a nearby mountain?
J. What is the radius of Earth?
K. How much can you raise your center of mass on Earth by running and then jumping up?
L. How many terms should you include in computing the initial gravitational potential energy of the solar system?
M. Where in the jump does the bungee cord exert maximum force on you?
N. What typical distance separates your center of mass from that of your partner as you tango?
O. What is the required orbital speed?
P. Do you need to add or remove energy?
Q. What is the population of Earth?
R. What is the radius of the Moon?
S. What is the mass of a large mountain?
T. How can you relate jumping height to vertical speed or planet radius?

U. How is the energy stored in a bungee cord related to the distance through which it has stretched?
V. How is the speed of a small object of mass m_s orbiting a large object of mass m_ℓ related to the radius of the orbit?
W. How is angular momentum related to radius and speed of orbit?

Key (all values approximate)

A. 4 m/s; B. 7×10^1 kg; C. 7×10^{22} kg; D. around Earth's surface—assume a uniform spherical-shell distribution, which is a rough approximation; E. zero; F. 6×10^3 kg/m³; G. the principle of equivalence; H. about one-third of the way down; I. 3×10^1 km is a reasonable choice; J. 6×10^6 m; K. 0.7 m for an average person; L. ideally, one term for each pair of objects (Mercury-Sun, Mercury-Venus, Mercury-Earth, and so on), but compare terms for the Sun and the three largest planets to get a sense of scale; M. the spring force kx is greatest where the stretch length x is greatest, at the bottom; N. 0.5 m; O. the speed of light, 3×10^8 m/s; P. for each planet in orbit, the gravitational potential energy of the Sun-planet system is negative, so you must add potential energy; Q. 7 billion; R. 2×10^6 m; S. 2×10^{13} kg; T. use energy methods: the initial kinetic energy is converted to gravitational potential energy during a jump; U. the energy stored is approximately $kx^2/2$, where x is the stretch and k is the cord's spring constant; V. the gravitational force is the only force in the vector sum for the small object, and the acceleration is centripetal for circular orbit, so $v_s = (Gm_\ell/r)^{1/2}$; W. for circular orbit, $L = rmv$

Worked and Guided Problems

Table 13.1 Solar system data (in SI units and relative to Earth)

	Mass (kg)	Mass (m_E)	Equatorial radius (m)	Equatorial radius (R_E)	semimajor axis (m)	semimajor axis (a_E)	Orbit[†] eccentricity	period (s)	period (years)
Sun	2.0×10^{30}	3.3×10^5	7×10^8	110	–	–	–	–	–
Mercury	3.30×10^{23}	0.06	2.440×10^6	0.38	5.79×10^{10}	0.39	0.206	7.60×10^6	0.24
Venus	4.87×10^{24}	0.81	6.052×10^6	0.95	1.082×10^{11}	0.72	0.007	1.94×10^7	0.62
Earth	5.97×10^{24}	1	6.378×10^6	1	1.496×10^{11}	1	0.017	3.16×10^7	1
Mars	6.42×10^{23}	0.11	3.396×10^6	0.53	2.279×10^{11}	1.52	0.09	5.94×10^7	1.88
Jupiter	1.90×10^{27}	318	7.149×10^7	11.2	7.783×10^{11}	5.20	0.05	3.74×10^8	11.86
Saturn	5.68×10^{26}	95.2	6.027×10^7	9.45	1.427×10^{12}	9.54	0.05	9.29×10^8	29.45
Uranus	8.68×10^{25}	14.5	2.556×10^7	4.01	2.871×10^{12}	19.2	0.05	2.65×10^9	84.02
Neptune	1.02×10^{26}	17.1	2.476×10^7	3.88	4.498×10^{12}	30.1	0.01	5.20×10^9	164.8
Pluto	1.31×10^{22}	0.002	1.151×10^6	0.18	5.906×10^{12}	39.5	0.25	7.82×10^9	247.9
Moon	7.3×10^{22}	0.012	1.737×10^6	0.27	3.84×10^8	0.0026	0.055	2.36×10^6	0.075

[†]The elliptical orbits of the planets and the Moon are specified by their *semimajor axis a* (half the major axis) and eccentricity *e*; see Figure 13.7. With the exception of Mercury and Pluto, the eccentricity is small and so the orbits are close to being circular.

These examples involve material from this chapter but are not associated with any particular section.
Some examples are worked out in detail; others you should work out by following the guidelines provided.

Worked Problem 13.1 Reliable 24/7 communications

In order to supply 24-h reception, a communications satellite is often placed in *geosynchronous orbit,* which means the satellite always appears in the same location relative to the receiver on the ground. This requires, among other things, that the rotational speed at which the satellite orbits Earth is the same as the rotational speed at which Earth spins on its axis. How far above Earth's surface is a satellite in geosynchronous orbit? What else is required?

❶ **GETTING STARTED** We seek an orbit whose period is $T = 24.0 \text{ h} = 8.64 \times 10^4 \text{ s}$. Determining the radius of such an orbit is important, but what else might be needed? Let's consider the goal in detail: In order for the satellite to always appear at the same location relative to a point on the ground, its orbital speed must be constant, matching the rotational speed of the Earth. This means the orbit must be circular. We draw a sketch of the system, labeling both the known mass of Earth m_E and the unknown mass of the satellite m_s (Figure WG13.1).

Kepler's third law says the square of the orbital period is proportional to the cube of the distance from the center of the object exerting the gravitational force. Thus a specified period implies a certain orbit radius r. We need to obtain an expression for r in terms of our given parameters. However, does just any orbit of this radius work? The goal is for the satellite to remain fixed relative to a specific point on Earth. Points on the surface of Earth move in circles parallel to the equator as Earth rotates. Thus it seems that our circular orbit must also be parallel to the plane of the equator in order to remain at the same location relative to a point on the ground.

❷ **DEVISE PLAN** As the satellite moves in a circular orbit, it is subject to the gravitational force of Earth only. We draw a free-body diagram for the satellite (Figure WG13.2), choosing the x axis to be in the direction of this force, toward the center of the orbit. Because the orbit is circular, we can use the relationship between radial (centripetal) acceleration and rotational speed ω from *Principles*

Figure WG13.1

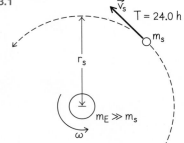

Figure WG13.2

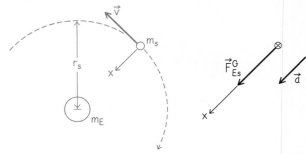

Eq. 11.15: $a_c = \omega^2 r$. The satellite's rotational speed is 2π rad in 24.0 h, making $\omega = 2\pi/T$. The centripetal acceleration is due to the force of gravity exerted by Earth. Because the satellite is not near Earth's surface, we cannot set a_c equal to g. Instead, we can solve for the acceleration by force analysis, using Newton's law of gravity, $F^G_{Es} = Gm_Em_s/r^2$. The only unknown in the equation of motion (*Principles* Eq. 8.8) will be r.

3 EXECUTE PLAN We begin by substituting known formulas into the component form of the equation of motion for the satellite:

$$\Sigma F_x = ma_x$$

$$F^G_{Esx} = m_s a_x$$

$$+F^G_{Es} = m_s(+a_c)$$

$$G\frac{m_E m_s}{r^2} = m_s(\omega^2 r)$$

$$Gm_E = \omega^2 r^3.$$

We want to solve for r in terms of T:

$$Gm_E = \left(\frac{2\pi}{T}\right)^2 r^3$$

$$r^3 = Gm_E\left(\frac{T}{2\pi}\right)^2.$$

This is Kepler's third law in mathematical form for a satellite orbiting Earth. Substituting the values given, we get

$$r^3 = (6.67 \times 10^{-11} \text{ N} \cdot \text{m}^2/\text{kg}^2)(5.97 \ 10^{24} \text{ kg})\left(\frac{8.64 \times 10^4 \text{ s}}{2\pi}\right)^2$$

$$r = 4.22 \times 10^7 \text{ m}.$$

Thus the satellite must follow a circular orbit whose plane is parallel to Earth's equator, and it must orbit at height

$$h = r - R_E = (4.22 \times 10^7 \text{ m}) - (6.38 \times 10^6 \text{ m})$$
$$= 3.58 \times 10^7 \text{ m} = 3.6 \times 10^4 \text{ km.} ✔$$

4 EVALUATE RESULT This is pretty far up, considering that the International Space Station (ISS) orbits at only a few hundred kilometers above ground. On the other hand, our value for r is an order of magnitude smaller than the distance to the Moon, so the answer is at least plausible. The ISS completes an orbit in a couple of hours, while the Moon requires a month. Our geosynchronous orbit is thus bracketed by these examples in both period and radius—which gives us some confidence in our results.

Note that the mathematical form of Kepler's third law does not depend on the mass of the satellite. (Neither does the version expressed purely in words!) Any geosynchronous satellite must therefore orbit in a circular, equatorial orbit at a common height. This raises a practical issue: If many countries and companies all want geosynchronous satellites, how do we find room up there for them?

Guided Problem 13.2 Disruptive spin

Objects on Earth's surface remain there because of gravitational attraction, despite the fact that objects near the equator move at about 1600 km/h because of Earth's rotation. If the Earth were to spin a lot faster, though, the gravitational force might not be strong enough to prevent objects from flying away from Earth's surface. This effect of rapid spin is especially acute for fluid objects like stars, particularly for the remnant cores of stars that have exploded in a supernova explosion. Some of these, called millisecond pulsars, rotate very rapidly (about once every millisecond) because their angular momentum remained constant as they collapsed. These stars typically have 2 times the mass of our Sun. What is the maximum radius a millisecond pulsar can have in order to avoid losing matter at its surface?

1 GETTING STARTED

1. Describe the problem in your own words. How is the radius involved?
2. What is the shape of the path followed by a small clump of material at the surface of a millisecond pulsar? What force keeps the clump on that path rather than traveling in a straight line?
3. Draw an appropriate diagram.

2 DEVISE PLAN

4. How is the magnitude of the relevant force related to the mass and radius of the pulsar?
5. What physics law allows you to relate this force to the rotational speed of the pulsar? How is the rotational speed related to the pulsar's rotational period?
6. Which unknown quantity do you need to determine?
7. Does your approach allow you to express the unknown quantity in terms of known ones?

3 EXECUTE PLAN

8. Work through the algebra to solve for the unknown quantity.

4 EVALUATE RESULT

9. Did you make any assumptions?
10. Compare your answer with the range of pulsar radii available online.

PRACTICE

Worked Problem 13.3 Escape at last

The Mars Colony wants to launch a deep-space probe, but they have no rocket engines. They decide to launch a probe with an electro-magnetic cannon, which means they must launch at escape speed. Determine this speed.

❶ GETTING STARTED Let us do a quick sketch to help our thinking (Figure WG13.3). We select the Mars-probe system for analysis. In order to reach "deep space," the probe must attain a very great distance from Mars. This will require a significant amount of initial kinetic energy, which the probe must acquire during launch. After launch, the kinetic energy immediately begins to decrease, and the potential energy of the Mars-probe system increases as the separation distance increases. We assume a reference frame where Mars is fixed and only the probe moves. When the probe is far enough away (infinity, really, but practically it doesn't need to go quite this far), the kinetic energy has its minimum value, which we can take to be zero because the colonists presumably do not want to supply any more energy than needed to get the probe out there. The gravitational potential energy has its maximum value, which is also zero. (Remember that universal gravitational potential energy is *negative*.) We also assume that the Sun and other planets have a negligible influence on our system, and we ignore the rotation of Mars.

Figure WG13.3

❷ DEVISE PLAN We can use conservation of energy because the probe has all of the needed kinetic energy at the beginning, as it is shot from a cannon. As the probe travels, this kinetic energy is converted to gravitational potential energy of the Mars-probe system. We want to know the initial speed of the probe acquired at launch. The initial potential energy is the value when the probe is still near the Martian surface. The final state of the probe is zero speed at an infinite distance from Mars. The *Principles* volume analyzes a similar situation in Section 13.7, leading to Eq. 13.23, so there is no need to derive this result again here. We begin with Eq. 13.23, solving this version of an energy conservation equation for $v_i = v_{esc}$ in terms of the known quantities.

❸ EXECUTE PLAN Let us use r_i for the initial Mars-probe radial center-to-center separation distance, $r_f = \infty$ for the final separation distance, R_M for the radius of Mars, and m_M and m_p for the two masses. We begin with Eq. 13.23:

$$E_{mech} = \tfrac{1}{2} m_p v_{esc}^2 - G\frac{m_M m_p}{R_M} = 0$$

$$\tfrac{1}{2} v_{esc}^2 - G\frac{m_M}{R_M} = 0$$

$$\tfrac{1}{2} v_{esc}^2 = G\frac{m_M}{R_M}$$

$$v_{esc} = \sqrt{2G\frac{m_M}{R_M}}$$

$$v_{esc} = \sqrt{2(6.67 \times 10^{-11}\,\text{N} \cdot \text{m}^2/\text{kg}^2)\frac{6.42 \times 10^{23}\,\text{kg}}{3.40 \times 10^6\,\text{m}}}$$

$$= 5.02 \times 10^3\,\text{m/s} = 5.02\,\text{km/s}. \; ✔$$

Notice that this speed does not depend on the mass of the probe. A probe of any other size shot from the cannon would need the same minimum speed to break free of Mars's gravitational pull.

❹ EVALUATE RESULT Our algebraic expression for the escape speed is plausible because it involves the mass of Mars, the initial center-to-center radial separation distance of our two objects (which is Mars's radius), and G. We expect v_{esc} to increase with m_M because the gravitational pull increases with increasing mass. We also expect v_{esc} to decrease as the distance between the launch position and Mars's center increases because the gravitational force exerted by the planet on the probe decreases with increasing separation distance. All this is just what our result predicts.

An escape speed of 18,000 km/h is smaller than (but on the order of) the escape speed from Earth, and so the answer is not unreasonable.

We assumed that the initial Mars-probe separation distance is equal to the planet's radius. Of course, the length of the cannon may be tens of meters, but this tiny difference would have no impact on the numerical answer. We ignored the rotation of Mars, which could supply a small amount of the needed kinetic energy. We also ignored the effect of the Sun, which is fine for getting away from the surface of Mars, but we would need to account for it if the destination was another star.

Guided Problem 13.4 Spring to the stars

Suppose that, instead of using chemical rockets, NASA decided to use a compressed spring to launch a spacecraft. If the spring constant is 100,000 N/m and the mass of the spacecraft is 10,000 kg, how far must the spring be compressed in order to launch the craft to a position outside Earth's gravitational influence?

❶ GETTING STARTED
1. Describe the problem in your own words. Are there similarities to Worked Problem 13.3?
2. Draw a diagram showing the initial and final states. What is the spacecraft's situation in the final state?
3. How does the spacecraft gain the necessary escape speed?

❷ DEVISE PLAN
4. What law of physics should you invoke?

5. As the spring is compressed, is the gravitational potential energy of the Earth-spacecraft system affected? If so, can you ignore this effect?
6. What equation allows you to relate the initial and final states?

❸ EXECUTE PLAN
7. What is your target unknown quantity? Algebraically isolate it on one side of your equation.
8. Substitute the numerical values you know to get a numerical answer.

❹ EVALUATE RESULT
9. Is your algebraic expression for the compression plausible for how the compression changes as the spring constant and Earth's mass and radius change?
10. If you were the head of a design team, would you recommend pursuing this launch method?

Worked Problem 13.5 Close encounter

Suppose astronomers discover a large asteroid, still far away, headed for Earth. Calculations indicate that it is traveling at $v_i = 754$ m/s and that a straight-line extension of its velocity would bring it within a distance $d = 3.30 \times 10^8$ m of Earth (which is about the radius of the Moon's orbit around Earth). However, Earth's gravitational attraction will cause its trajectory to be a conic section, as Figure WG13.4 shows. (*a*) What is the value of r_f, the radius of its orbit when the asteroid is closest to Earth? (*b*) What speed v_f does the asteroid have when it gets closest to Earth? Note from Figure WG13.4 that r_f is to be measured from Earth's center, not from Earth's surface.

Figure WG13.4

r_f

$d = 3.30 \times 10^8$ m

❶ **GETTING STARTED** The problem is one of celestial mechanics. We have an asteroid in orbit (it may be an unbound orbit), and we need to calculate the radius r_f of its orbit when it comes closest to Earth and its speed v_f when it is at that closest-approach position. We know both its speed and direction of motion a long way off, although we don't know its mass. Perhaps we don't need to know this mass. We assume that when the asteroid is first sighted, Earth's gravitational influence on it is negligible.

❷ **DEVISE PLAN** For celestial mechanics problems, we know that both the energy and the angular momentum of a system of objects interacting only gravitationally are constant: $E_i = E_f$ and $\vec{L}_i = \vec{L}_f$. In Earth's reference frame, the angular momentum of the asteroid (of unknown mass m_a) and the energy of the Earth-asteroid system do not change as the asteroid moves from far away to its position of closest approach, where the radius of its orbit is r_f. We can determine the magnitude of its angular momentum at closest approach by looking at its angular-momentum magnitude when it is far from Earth. Its velocity is always tangent to its trajectory, and when it is far away we have information about the perpendicular distance of this tangent line from Earth:

$$L_i = |\vec{r}_i \times \vec{p}_i| = m_a v_i d,$$

where $\vec{p}_i$ is the asteroid's momentum when it is traveling along its straight-line course. At the position of closest approach, the direction of velocity is also perpendicular to its radius of curvature (see *Principles* Example 13.5), so its angular momentum is $L_f = r_f m_a v_f$.

The energy of the Earth-asteroid system at the initial position of sighting, where the influence of Earth's gravitational pull is negligible, is purely kinetic. At the point of closest approach, however, it has both kinetic energy and (negative) gravitational potential energy. All that remains is to do the algebra to solve for r_f and v_f.

❸ **EXECUTE PLAN**
(*a*) Using r_i and r_f for the initial and final Earth-asteroid radial center-to-center distances, we have from conservation of energy:

$$K_i + U_i^G = K_f + U_f^G$$

$$\tfrac{1}{2}m_a v_i^2 + \left(-G\frac{m_E m_a}{r_i}\right) = \tfrac{1}{2}m_a v_f^2 + \left(-G\frac{m_E m_a}{r_f}\right)$$

$$\tfrac{1}{2}v_i^2 + \left(-G\frac{m_E}{\infty}\right) = \tfrac{1}{2}v_f^2 + \left(-G\frac{m_E}{r_f}\right)$$

$$\tfrac{1}{2}v_i^2 = \tfrac{1}{2}v_f^2 - G\frac{m_E}{r_f}.$$

This equation contains the two desired quantities: the orbit radius at closest approach r_f and the speed v_f when the asteroid is at that position. Next we apply conservation of angular momentum:

$$L_i = L_f$$

$$m_a v_i d = r_f m_a v_f$$

$$v_f = \frac{v_i d}{r_f}.$$

We note that the asteroid's mass m_a cancels in all our equations. Substituting this expression for v_f into our energy equation yields

$$\tfrac{1}{2}v_i^2 = \tfrac{1}{2}\left(\frac{v_i d}{r_f}\right)^2 - G\frac{m_E}{r_f} = \frac{1}{2}\frac{(v_i d)^2}{r_f^2} - G\frac{m_E}{r_f}.$$

Multiplying through by r_f^2 and then by $2/v_i^2$, we get a quadratic equation:

$$\tfrac{1}{2}v_i^2 r_f^2 = \tfrac{1}{2}(v_i d)^2 - G m_E r_f$$

$$r_f^2 + 2\frac{G m_E}{v_i^2}r_f - d^2 = 0$$

$$r_f = -\frac{G m_E}{v_i^2} \pm \tfrac{1}{2}\sqrt{\left(2\frac{G m_E}{v_i^2}\right)^2 + 4d^2}$$

$$= -\frac{G m_E}{v_i^2} \pm \sqrt{\left(\frac{G m_E}{v_i^2}\right)^2 + d^2}.$$

The positive result is the physical one we seek, yielding

$$r_f = 7.38 \times 10^7 \, \text{m} = 7.38 \times 10^4 \, \text{km.} \ ✔$$

(*b*) The speed at closest approach is

$$v_f = \frac{d v_i}{r_f} = \frac{(3.30 \times 10^8 \, \text{m})(754 \, \text{m/s})}{7.38 \times 10^7 \, \text{m}}$$

$$= 3.37 \times 10^3 \, \text{m/s} = 3.37 \, \text{km/s.} \ ✔$$

❹ **EVALUATE RESULT** Earth's radius is 6.38×10^3 km, making $r_f \approx 10 R_E$, which might generate a bit of anxiety among the planet's inhabitants when one considers possible observational errors. The distance of closest approach is about 74,000 km, roughly one-fourth of the Earth-Moon distance. Fortunately, the asteroid does not hit Earth because at closest approach it is moving a lot faster than it was at first sighting ($v_f \approx 5v_i$). You also might notice that the speed at closest approach is barely greater than the escape speed for an object at this distance r_f from Earth's center, $v_{esc} = \sqrt{2Gm_E/(7.4 \times 10^7 \, \text{m})} = 3.3 \times 10^3$ m/s. A near miss indeed!

The expression for r_f shows that a greater straight-path distance d gives a greater r_f value, as we expect. Because we are not given any information about how far away the asteroid is initially, we assumed that Earth has negligible gravitational influence at that instant. Without that assumption, we would not have been able to solve the problem. In any case, if our assumption is false, the decrease in potential energy would have been (somewhat) smaller and therefore the final velocity would have been smaller.

Guided Problem 13.6 More rocket launches

Three rockets are launched from Earth as shown in Figure WG13.5. One rocket is launched vertically (*a*), one horizontally (*b*), and one at 45° with respect to the ground (*c*). All three are launched with a speed $\frac{3}{4}$ of the escape speed from Earth's surface, $v_{esc} = 1.12 \times 10^4\,\text{m/s}$ (from *Principles* Checkpoint 13.22). Calculate the maximum distance from Earth each rocket achieves. Ignore Earth's rotation, and express your answer in terms of Earth's mass and radius.

Figure WG13.5

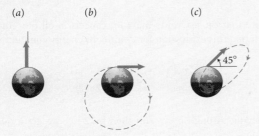

(*a*)   (*b*)   (*c*)

①GETTING STARTED

1. What are the physical quantities of interest in this problem?
2. Consider the shape of each trajectory. *Principles* Figure 13.37 may aid your thinking.

②DEVISE PLAN

3. What conservation law(s) should you use? How are the maximum and minimum orbital distances related to other physical quantities?
4. Is the information about the magnitude and direction of the velocities in each of the three cases useful?
5. Which unknown quantities do you need to determine in each case?

③EXECUTE PLAN

6. Does the same *general* approach work in all three cases?
7. Work through the algebra and obtain an expression for the maximum distance in terms of Earth's mass and radius. Is it possible to use the same expression in all three cases?

④EVALUATE RESULT

8. Does your expression (or expressions) show plausible algebraic behavior when you vary the launch condition and when you vary Earth's mass and radius?
9. Are your assumptions reasonable?

Worked Problem 13.7 Black hole

A black hole is the remnant of a massive star that, once its nuclear fuel burned out, has collapsed to essentially a point under its own gravitational attraction. The gravitational force exerted by a black hole is so great that not even light can escape from within its interior—hence the name: (black, for the absence of light). Suppose a certain black hole has a mass equal to ten times the mass of the Sun. What is the radius of closest approach for a passing light ray to avoid capture?

①GETTING STARTED The problem asks about the motion of a light ray in the vicinity of a black hole. We might consider the gravitational force exerted by the black hole on the ray or work from energy or angular momentum conservation. Light has no mass, however, and so we will likely need to invoke the equivalence principle. Conservation principles are usually easier to use than other laws, so let's begin there. It is possible that energy alone can provide the equation we need because only one unknown is sought: a radial distance from the center of the black hole. We know the speed of light, $c = 3.0 \times 10^8\,\text{m/s}$, and are given the mass of the black hole as 10 solar masses. The mass of the Sun is given in Table 13.1.

②DEVISE PLAN In any two-object system where one object moves relative to the other, angular momentum, gravitational force, and gravitational potential energy each depend on the masses of the objects. That seems to rule out any of our usual methods for solving this problem because one of our objects has no mass. By the equivalence principle, however, we might be able to treat the light ray as if it were an object subject to the usual laws. Thus we might imagine the ray as having an initial kinetic energy to use in attempting to escape the gravitational pull of the black hole. We used energy methods in Worked Problem 13.3 to derive an expression for the escape speed that depends *not* on the mass of the "escaping" object but only on the mass of the object exerting the gravitational force (which means m_h here) and on the radial distance r separating the objects at the instant of launch: $v_{esc} = \sqrt{2Gm_h/r}$. This expression was derived for an object escaping from a planet's surface, and so we must assume that the gravitational behavior of the black hole is similar enough to that of a planet to allow us to use this relationship.

We apply this approach here, with the light ray being the escaping "object" and with r being the "escape radius," the radius at which the light can just barely escape from the black hole, which is the radius of closest approach that we seek. We assume the ray obeys mechanical laws and is launched with speed c, which must be the escape speed at the radius of closest approach.

③EXECUTE PLAN Squaring our equation for the escape speed from Worked Problem 13.3, substituting $v_{esc} = c$, and solving for r, we get

$$r = 2G\frac{m_h}{c^2}.$$

The mass of the Sun is 2.0×10^{30} kg. Putting in numbers, we arrive at

$$r = 2(6.67 \times 10^{-11}\,\text{N}\cdot\text{m}^2/\text{kg}^2)\frac{10(2.0 \times 10^{30}\,\text{kg})}{(3 \times 10^8\,\text{m/s})^2}$$

$$= 3.0 \times 10^4\,\text{m} = 30\,\text{km.} ✔$$

④EVALUATE RESULT The gravitational force must be very strong to affect light significantly, and the inverse square dependence of this force (or the inverse first power dependence of the gravitational potential energy) on the radius therefore demands a very small radius. That black holes have much smaller radii than ordinary objects of the same mass is no surprise: Black holes must be very exotic or we would have noticed them in our everyday lives. In addition, light seems unaffected by gravity in ordinary life, so we would expect that a very intense gravitational pull would be required. If the mass is fixed, that leaves only the radius to adjust, and the smaller the radius, the greater the gravitational force.

Note: The assumption we made—that we can ignore the fact that light has no mass and simply apply the escape speed formula—is not technically valid. However, when this calculation is done correctly, using Einstein's general theory of relativity, our expression for the black hole radius of closest approach is correct. The value of this radius is known as the "event horizon" for the black hole because events that take place at radii smaller than this cannot be seen from the outside.

PRACTICE

Guided Problem 13.8 Heavenly rock

A planet that has a mass 25 times that of Earth is orbited by a moon that has a mass eight times that of Earth (Figure WG13.6). The orbital radius is equal to twelve Earth radii. A large rock located near this system is sixteen Earth radii away from the moon, with the line connecting the rock and moon perpendicular to the line connecting the planet and moon. What is the magnitude of the rock's acceleration in terms of the free-fall acceleration (g) it would have on Earth?

Figure WG13.6

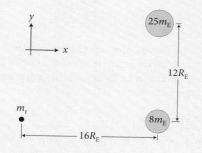

① GETTING STARTED

1. Given that the question asks for acceleration, what approach should you take?

2. Does it matter that the rock's mass is not specified? Even though this mass is not given, you should give it a symbolic value m_r in any diagrams you draw and any equations you use to see how it comes in and, perhaps, cancels out.

3. How many forces are exerted on the rock? What is the direction of each force? Draw a free-body diagram for the rock.

② DEVISE PLAN

4. What is the center-to-center distance between rock and moon? The center-to-center distance between rock and planet?

5. What is the magnitude of the gravitational force exerted by the moon on the rock? The magnitude of the gravitational force exerted by the planet on the rock?

③ EXECUTE PLAN

6. Determine the x component of the vector sum of forces exerted on the rock.

7. Determine the y component of the vector sum of forces exerted on the rock.

8. Remember that acceleration is a vector.

④ EVALUATE RESULT

9. Is your result greater or smaller than g? Given the masses and distances, is this reasonable?

Questions and Problems

Dots indicate difficulty level of problems: • = *easy,* •• = *intermediate,*
••• = *hard;* CR = *context-rich problem.*

[It is convenient for some problems to note that
$G = 6.67 \times 10^{-11} \text{ N} \cdot \text{m}^2/\text{kg}^2 = 6.67 \times 10^{-11} \text{ m}^3/(\text{kg} \cdot \text{s}^2).$]

13.1 Universal gravity

1. If the matter that makes up a planet is distributed uniformly so that the planet has a fixed, uniform density, how does the magnitude of the acceleration due to gravity at the planet surface depend on the planet radius? •

2. In Figure P13.2, suppose the mass of object 1 is three times the mass of object 2. (*a*) At Earth's surface, what ratio r_1/r_2 is needed to have the rod stay horizontal? (*b*) What is this ratio on the Moon, where the magnitude of the gravitational force is only one-sixth of the magnitude of Earth's gravitational force? •

Figure P13.2

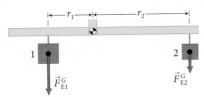

3. Suppose you're making a scale model of the solar system showing the eight planets (not including the dwarf planet Pluto) and having each planet at its farthest position from the Sun. If you use a marble of radius 10 mm for Mercury, how far from the Sun must you place your outermost planet? •

4. Four identical objects are placed at the four corners of a square, far from any star or planet. (*a*) Choose one object and draw to scale the vectors that represent the gravitational forces exerted on it by the other three objects. (*b*) Draw a vector that represents the vector sum of the forces exerted on your chosen object. ••

5. Two basketballs, each of mass 0.60 kg and radius 0.12 m, are placed on a floor so that they touch each other. Two golf balls, each of mass 0.045 kg and radius 22 mm, are placed on a table so that they touch each other. What is the ratio of the gravitational force exerted by one basketball on the other to that exerted by one golf ball on the other? ••

13.2 Gravity and angular momentum

6. An astronaut on the International Space Station gently releases a satellite that has a mass much smaller than the mass of the station. Describe the motion of the satellite after release. •

7. An isolated object of mass m can be split into two parts of masses m_1 and m_2. Suppose the centers of these parts are then separated by a distance r. What ratio of masses m_1/m_2 would produce the largest gravitational force on each part? ••

8. Particles of mass m, $2m$, and $3m$ are arranged as shown in Figure P13.8, far from any other objects. These three particles interact only gravitationally, so that each particle experiences a vector sum of forces due to the other two. Call these $\vec{F}_m, \vec{F}_{2m}$, and $\vec{F}_{3m}$. (*a*) Show that the lines of action of $\vec{F}_m, \vec{F}_{2m}$, and $\vec{F}_{3m}$ intersect at a common point. (*b*) Is this point of intersection the center of mass of the system? (*c*) Is the analysis of the motion of this system straightforward? ••

Figure P13.8

9. Angular momentum can be represented as a vector product and interpreted in terms of the rate of change of an area with respect to time. Give a similar interpretation of another vector product: torque. (Hint: Consider derivatives.) •••

13.3 Weight

10. Suppose the spring/balance device in Figure P13.10 is adjusted so that the spring is slightly stretched and the balance arm is level when the device is in an elevator at rest relative to the ground and a 1.0-kg object is placed as shown. If the elevator accelerates upward, does the balance arm tilt clockwise, tilt counterclockwise, or remain level? •

Figure P13.10

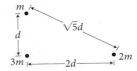

11. Step gently onto a bathroom scale, read the dial, and then jump from a chair onto the same scale. (*a*) Does the dial show different readings in the two cases? (*b*) Has the gravitational force exerted by Earth on you changed? •

12. A balance is purposefully prepared in an unbalanced condition, with too much mass on one side. It is carefully held with two hands, one on the base of the balance and one on the heavier end, so that the balance arm is level. Then both hands release the system from rest at the same instant. While it is in free fall, what happens to the balance arm? ••

13. (*a*) For what downward acceleration does the spring scale in Figure P13.13 read zero? (*b*) What would the answer in part *a* be if the experiment were done on the Moon? (*c*) How would your conclusion about the spring forces change if the brick were hung from the spring scale rather than supported by it? ••

Figure P13.13

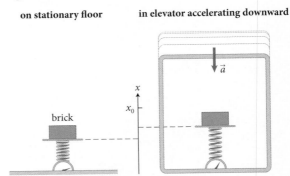

14. You ride your pogo stick (Figure P13.14) across a diving board and into the local swimming pool. Your pogo stick is a special model equipped with a sensor that records the force measured by the spring. Describe the values of force recorded during your ride, using the gravitational force exerted on you by Earth as a benchmark. ••

Figure P13.14

13.4 Principle of equivalence

15. A commercial airliner hits a pocket of turbulence and experiences a downward acceleration of about 2g. A person who is not fastened into a seat by a safety belt "falls" to the ceiling and suffers a broken neck. Explain how this injury can be so severe. •

16. How should the simulator in Figure P13.16 be tilted to simulate a left turn in a bobsled? Assume that the bobsled run has banked turns and that the sled exerts no tangential force on the ice. ••

Figure P13.16

17. A vertical accelerometer for measuring accelerations on a roller coaster can be constructed using a piece of clear plastic pipe, a cork, a rubber band, and a metal bob (Figure P13.17). Where the bob hangs down normally is where you make a mark for an acceleration of g. Without accelerating the pipe, explain how you can calibrate the accelerometer as you assemble it for accelerations of 2g, 3g, and so on. ••

Figure P13.17

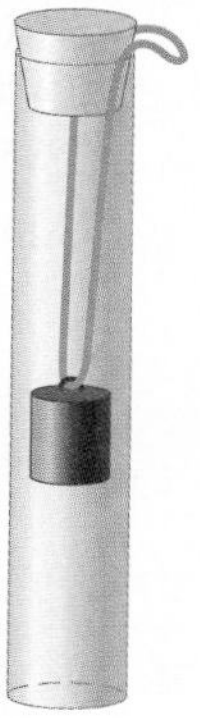

18. Consider two satellites simultaneously launched 180 km apart at the equator, each placed in a circular orbit that passes above the North Pole and its launch point. (a) Describe their relative motion with respect to Earth and with respect to each other. (b) Does this motion imply that each satellite exerts a force on the other? ••

19. Let us explore whether there is any way to distinguish acceleration due to rotation from acceleration due to a gravitational force. Imagine a deep bowl with a small volume of milk in it. (a) What happens to the milk when you spin the bowl about a perpendicular axis that runs through the center of the bowl bottom? (b) What is the shape of an object that exerts a gravitational force that could make the milk behave in a similar fashion? •••

13.5 Gravitational constant

20. What is the magnitude of the gravitational force exerted (a) by the Sun on Mars and (b) by Mars on the Sun? Approximate Mars's orbit as circular. •

21. What is the magnitude of the gravitational force between two 1.0-g marbles placed 100 mm apart? •

22. The gravitational force between two spherical celestial bodies, one of mass 2×10^{12} kg and the other of mass 5×10^{20} kg, has a magnitude of 3×10^7 N. How far apart are the two bodies? •

23. Mars has a mass that is about one-ninth of Earth's and a radius that is about half of Earth's. What is the ratio of the acceleration due to gravity on Mars to that on Earth? ••

24. Gravity is the weakest of the fundamental interactions (see *Principles* Section 7.6). (a) This being so, what makes it such an important force on Earth? (b) What makes it the dominant force in galaxies? ••

25. A neutron star has about two times the mass of our Sun but has collapsed to a radius of 10 km. What is the acceleration due to gravity on the surface of this star in terms of the free-fall acceleration at Earth's surface? ••

26. Which pulls harder on the Moon: Earth or the Sun? ••

27. How far above the surface of Earth do you have to go before the acceleration due to gravity drops by 0.10%, 1.0% and 10%? ••

28. (a) As a spacecraft travels along a straight line from Earth to the Moon, at what distance from Earth does the force of gravity exerted by Earth on the coasting spacecraft cancel the force of gravity exerted by the Moon on the spacecraft? (b) What do the passengers in the spacecraft notice, if anything, before and after passing this location? ••

29. From the picture of Saturn in Figure P13.29 and known values, estimate the orbital period of particles in Saturn's rings. ••

Figure P13.29

30. The Sun and Moon both exert a gravitational force on Earth. (a) Which force has greater magnitude? (b) What is the ratio of the two force magnitudes? ••

31. Given that the acceleration due to gravity is 0.0500 m/s² at the surface of a spherical asteroid that has a radius of 3.75×10^4 m, determine the asteroid's mass. ••

32. At what height above the surface of Pluto is the acceleration due to gravity half its surface value? ••

33. A test object of mass m_{test} is placed at the origin of a two-dimensional coordinate system (Figure P13.33). An object 1, of the same mass, is at $(d, 0)$, and an object 2, of mass $2m_{test}$, is at $(-d, \ell)$. What is the magnitude of the vector sum of the gravitational forces exerted on the test object by the other two objects? •••

Figure P13.33

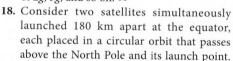

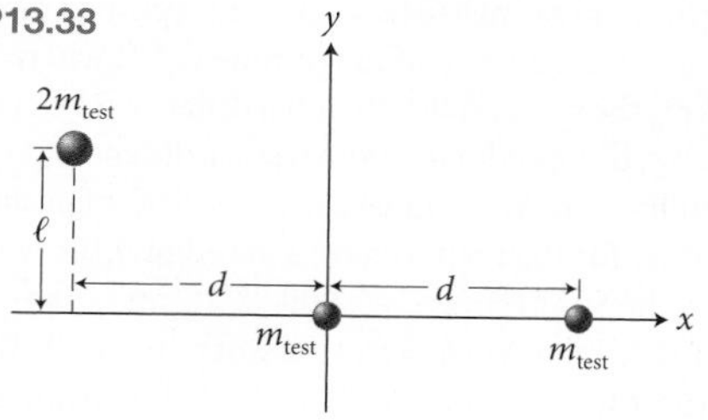

PRACTICE

34. Using g for the acceleration due to gravity is valid as long as the height h above Earth's surface is small. Derive an expression for a more accurate value of the acceleration due to gravity as a quadratic function of h. ●●●

35. Calculate the acceleration due to gravity *inside* Earth as a function of the radial distance r from the planet's center. (Hint: Imagine that a mine shaft has been drilled from the surface to Earth's center and an object of mass m has been dropped down the shaft to some radial position $r < R_E$ from the center, where R_E is the radius of Earth. What is the vector sum of the forces exerted on the object by all of Earth's mass that lies at any distance $d > r$ from the center? $d < r$ from the center?) ●●●

13.6 Gravitational potential energy

36. What is the gravitational potential energy of the Earth-Sun system? ●

37. The asteroid known as Toro has a radius of about 5.0 km and a mass of 2.0×10^{15} kg. Could a 70-kg person standing on its surface jump free of Toro? ●

38. (*a*) How much gravitational potential energy does a system comprising a 100-kg object and Earth have if the object is one Earth radius above the ground? (*b*) How fast would a 100-kg object have to be moving at this height to have zero energy? ●●

39. An object 1 of mass m_1 is separated by some distance d from an object 2 of mass $2m_1$. An object 3 of mass m_3 is to be placed between them. If the potential energy of the three-object system is to be a maximum (closest to zero), should object 3 be placed closer to object 1, closer to object 2, or halfway between them? ●●

40. Two particles, each of mass m, are initially at rest very far apart. Obtain an expression for their relative speed of approach at any instant as a function of their separation distance d if the only interaction is their gravitational attraction to each other. ●●

41. Consider an alternate universe in which the magnitude of the attractive force of gravity exerted by Earth on a meteor of mass m_m approaching Earth is given by $F = Cm_m m_E / r^3$, where C is some positive constant and r is the center-to-center distance between Earth and the meteor. (*a*) How much work is done by the gravitational force on the meteor as it falls from very far away (infinite distance) to some height h above Earth's surface? (*b*) If the meteor is moving very slowly when it is very far away, how fast is it moving when it gets to that height? ●●

42. Derive expressions for the orbital speed and energy for a satellite of mass m_s traveling in a circular orbit of radius a around a planet of mass $m_p \gg m_s$. ●●

43. What maximum height above the surface of Earth does an object attain if it is launched upward at 4.0 km/s from the surface? ●●

44. (*a*) Derive an expression for the energy needed to launch an object from the surface of Earth to a height h above the surface. (*b*) Ignoring Earth's rotation, how much energy is needed to get the same object into orbit at height h? ●●

45. A spacecraft propelled by a solar sail is pushed directly away from the Sun with a force of magnitude C/r^2, where C is a constant and r is the spacecraft-Sun radial distance. The craft has a mass of 5.0×10^4 kg and starts from rest at a distance $r_i = 1.0 \times 10^8$ km from the Sun. What must the numerical value and units of C be in order for the craft to have a speed of $0.10c$ when it is very far from the Sun? (Hint: $c = 3.00 \times 10^8$ m/s.) ●●

46. In 1865, Jules Verne wrote a story in which three men went to the Moon by means of a shell shot from a large cannon sunk in the ground. (*a*) What muzzle speed must the cannon have in order for the shell to reach the Moon? (*b*) If the acceleration was constant throughout the muzzle, how long would the muzzle need to be in order to achieve this speed without exposing the men to an acceleration greater than $6g$? (*c*) Argue whether this launch method is practical or not. ●●

47. The center-to-center distance between a 200-g lead sphere and an 800-g lead sphere is 0.120 m. A 1.00-g object is placed 0.0800 m from the center of the 800-g sphere along the line joining the centers of the two spheres. (*a*) Ignoring all sources of gravitational force except the two spheres, calculate the gravitational force exerted on the object. (*b*) Determine the gravitational potential energy per gram at the position of the object. (*c*) How much work is needed to bring the object 0.0400 m closer to the 800-g sphere? ●●

48. Derive an expression for the gravitational potential energy of a system consisting of Earth and a brick of mass m placed at Earth's center. Take the potential energy for the system with the brick placed at infinity to be zero. ●●●

49. Is approximating the gravitational potential energy near the surface of the Sun as $m g_S \Delta x$ accurate over a smaller or larger range of values than approximating the gravitational potential energy near the surface of Earth as $m g_E \Delta x$? ●●●

50. A uniform rod of mass m_{rod} and length ℓ_{rod} lies along an x axis far from any stars or planets, with the center of the rod at the origin (Figure P13.50). A ball of mass m_{ball} is located at position x_{ball} on the axis. (*a*) Write an expression for the gravitational potential energy of the system made up of the ball and a small element dm_{rod} of the rod located at position x_{dm}. (*b*) Integrate your expression over the length of the rod to determine the potential energy of the system for $x_{ball} > \ell_{rod}/2$. (*c*) It can be shown that for such a system the gravitational force exerted by the rod on an object located a distance x from the origin is $F_x = -dU/dx$. Using this relationship, compute the force exerted by the rod on the ball at any distance x from the rod center. ●●●

Figure P13.50

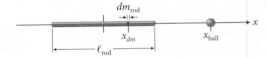

13.7 Celestial mechanics

51. The parabolic orbit of any asteroid around the Sun might be described as a collision between the two objects. Would it be better described as an elastic collision or an inelastic collision? ●

52. Assume a particle is located at the surface of the Sun. (*a*) If the particle has mass m, how fast must it be moving away from the Sun's center of mass to escape the gravitational influence of the Sun? (*b*) At what speed must a particle of mass $2m$ move in order to escape the Sun's gravitational influence? ●

53. What is the speed of a space probe when it is very far from Earth if it was launched from the surface of Earth at twice its escape speed? ●

54. Which trip takes more rocket fuel: from Earth to the Moon or from the Moon to Earth? ●●

55. A satellite moves at speed v in very low orbit about a moon that has no atmosphere. You launch a projectile vertically from the same moon's surface at the same speed v. To what height does it rise? ●●

56. A satellite in an elliptical orbit around Earth has a speed of 8032 m/s when it is at perigee, the position in the orbit closest to Earth. At this position, the satellite is 112 km above Earth's surface. How far above the ground is the satellite when it is at apogee, the position in its orbit farthest from Earth? ••

57. A certain comet comes in on a parabolic approach around the Sun. Its perihelion is about the same distance from the Sun as the orbit of Mercury. How much faster than Mercury (expressed as a ratio) is the comet traveling at perihelion? ••

58. Show that for circular orbits of two objects about their center of mass, $E = \frac{1}{2}U = -K$. This is a special case of the more general mathematical statement known as the *virial theorem*. ••

59. An attractive central force exerted on a particle of mass m causes the particle to travel in a bound orbit. The orbit is elliptical with semimajor axis a, and the force has magnitude Cm/r^2, where C is a constant and r is the orbit radius. If the particle's speed is $v = \sqrt{C/2a}$ when it is farthest from the source of the force, what is its speed when it is nearest the force source? ••

60. A particle is leaving the Moon in a direction that is radially outward from both the Moon and Earth. What speed must it have to escape the Moon's gravitational influence? ••

61. A meteoroid passes through a position in space where its speed is very small relative to Earth's and it is at a perpendicular distance of 19 Earth radii above Earth's surface. The meteoroid is moving in such a way that Earth captures it. What is the speed of the meteoroid when it is one Earth radius above the ground? ••

62. A space probe can be either launched at escape speed from Earth or transported to a "parking orbit" above Earth and then launched. (a) If the parking orbit is 180 km above Earth's surface, what is the escape speed from the orbit? (b) What launch speed is required to have the probe escape Earth if it is launched from the ground? ••

63. Two identical stars, each of mass 3.0×10^{30} kg, revolve about a common center of mass that is 1.0×10^{11} m from the center of either star. (a) What is the rotational speed of the stars? (b) If a meteoroid passes through the center of mass, perpendicular to the orbital plane of the stars, how fast must the meteoroid be going if it is to escape the gravitational attraction of the stars? ••

64. Kepler's third law says that the square of the period of any planet is proportional to the cube of its mean distance from the Sun. Show that this law holds for elliptical orbits if one uses the semimajor axis as the mean distance. (Said another way, orbits that have the same energy but different angular momentum have the same period.) •••

65. Two celestial bodies of masses m_1 and m_2 are orbiting their center of mass at a center-to-center distance of d. Assume each body travels in a circular orbit about the center of mass of the system, and derive the general Newtonian form of Kepler's third law: $T^2 = 4\pi^2 d^3 / G(m_1 + m_2)$. •••

66. To travel between Earth and any other planet requires consideration of such things as expenditure of fuel energy and travel time. To simplify the calculations, one chooses a path such that the position in Earth's orbit where the launch occurs and the position in the other planet's orbit when the spacecraft arrives define a line that passes through the Sun, as shown for an Earth-Mars transfer in Figure P13.66. The path the spacecraft travels is an ellipse that has the Sun at one focus. Such a path is called a Hohmann transfer orbit, and the major axis of the ellipse, $2a$, is the sum of the radii of the Earth's and the other planet's orbits around the Sun. Take $m_S = 1.99 \times 10^{30}$ kg, $m_M = 6.42 \times 10^{23}$ kg, $m_E = 5.97 \times 10^{24}$ kg, $a_E = 1.50 \times 10^{11}$ m, and $a_M = 2.28 \times 10^{11}$ m,

and assume the planets have circular orbits. (a) What is the energy of the system comprising the Sun and a 1000-kg space probe in a Hohmann transfer orbit? (b) What is the speed of the probe in this orbit, as a function of r, the probe's radial distance from the Sun? (c) What is the probe's speed relative to the Sun as the probe enters the transfer orbit? (d) What is its speed relative to the Sun as it approaches Mars? (e) Given that the orbital speed of Earth is about 2.98×10^4 m/s, how much additional speed does the probe need to begin the transfer orbit? (f) What is the required launch speed from Earth's surface for a probe traveling to Mars in a Hohmann transfer orbit? •••

Figure P13.66

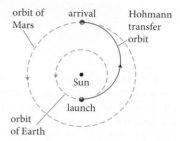

13.8 Gravitational force exerted by a sphere

67. A small disk of mass m is at the center of a larger disk that has an off-center hole drilled in it (Figure P13.67). What is the direction of the gravitational force exerted by the larger disk on the small one? •

Figure P13.67

68. Is Eq. 13.37 for the gravitational force exerted by a solid sphere of mass m_s on some object of mass m_o when the two are a radial distance r apart, $F_{so}^G = Gm_o m_s/r^2$, valid for a sphere whose mass is not distributed uniformly? •

69. A uniform ring of mass m_{ring} and radius R_{ring} is shown in Figure P13.69. A small object of mass m_{obj} sits a separation distance s from the ring on the line that is perpendicular to the plane of the ring and passes through the ring's center. Obtain an expression (in terms of the given variables and universal constants) for the magnitude of the gravitational force exerted by the ring on the object. ••

Figure P13.69

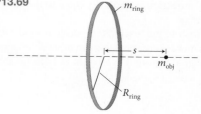

70. Two uniform spherical shells are located such that their centers are along the x axis in Figure P13.70. The inner shell has mass m_{inner} and radius R, and its center is at $x = 0.80R$. The outer shell has mass $3.0m_{inner}$ and radius $2.0R$, and its center is at $x = 0$. What is the magnitude of the vector sum of the gravitational forces exerted by these two shells on an object of mass m_{obj} placed (a) at $x = 3.0R$, (b) at $x = 1.9R$, and (c) at $x = 0.90R$? ●●

 Figure P13.70 $3.0m_{inner}$

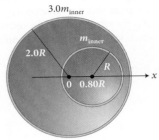

71. A thin disk of radius R has mass m_{disk} uniformly distributed over its area. (a) Derive an expression for the magnitude of the gravitational force this disk exerts on a particle of mass m_{part} located a distance y directly above the center of the disk. (b) If $y \gg R$, show that this expression reduces to that for two particles. ●●●

Additional Problems

72. In a system of N particles, how many terms are there in the expression for the gravitational potential energy of the system? ●

73. You equip a rocket with enough fuel to reach escape speed from Earth. If you plan to launch the rocket from just above ground level, does it matter whether you launch it vertically, at an angle, or horizontally? ●

74. What is the value of the acceleration g_{Venus} due to gravity at the surface of Venus? ●

75. Is there any position in the elliptical orbit of a satellite where the tangential component of the acceleration is greater than the component perpendicular to the tangential component? If so, what conditions on the orbit must there be for such a position to exist? ●●

76. The *gravitational vector field* $\vec{g}(\vec{r})$ is defined as the gravitational force exerted on a small object located at a particular position $\vec{r}$ divided by that object's mass. (a) Sketch the gravitational vector field produced by Earth at a sampling of positions around Earth. Make sure your vector lengths indicate the correct relative sizes. (b) Express the gravitational force exerted by Earth on an object of mass m_{obj} located a distance r_{obj} ($> r_E$) from Earth's center in terms of Earth's gravitational vector field. ●●

77. Some writers of fiction have placed human beings on asteroids or other "planetoids" whose radius, in the illustrations accompanying the stories, seems not too much greater than the height of the people themselves! What would the density of such a celestial body have to be for the human to feel a gravitational attraction numerically equal to the attraction we feel to Earth? ●●

78. You wish to put a satellite in orbit around Earth. (a) Which takes more energy per kilogram of satellite: launching the satellite from the ground to a height of 1600 km above the ground

or placing the satellite into orbit once it has reached that altitude? (b) Repeat for an orbit 3200 km above the ground. (c) Repeat for an orbit 4800 km above the ground. ●●

79. Unable to sleep, you crawl out of bed one evening and stare out your window at the night sky, finding Jupiter and sitting back to contemplate its grandeur. Remembering your physics, you imagine yourself at some height above the apparent surface of Jupiter, experiencing a gravitational force that has a magnitude equal to the magnitude of the gravitational force you experience at the surface of Earth. ●●● CR

80. After seeing an old movie, you become concerned about a large asteroid that could be gravitationally attracted to Earth from very far away. You realize that long ago such an asteroid's speed relative to Earth's might have been very small, and thus you are able to calculate the relative speed it has at any separation distance from Earth. In order to get a baseline, you assume that the only object exerting a gravitational force on the asteroid is Earth, and you ignore the speed of Earth in its path around the Sun. What worries you is that observers might not sight the asteroid until it passes the imaginary line defining the Moon's orbit around Earth, and how little time that would give humanity to react. ●●● CR

81. You are working on a project to plot the course of a spy satellite. The satellite has a polar orbit (which means it passes over both of Earth's poles during each orbit), and it is outfitted with a camera carrying a wide-angle lens that can "see" strips of land up to 2500 km wide. The enemy is believed to have chosen an elevation of orbit that just guarantees complete coverage of Earth each day by these surveillance strips. You cannot avoid or block the satellite unless you know where it is. ●●● CR

82. While helping an astronomy professor, you discover a binary star system in which the two stars are in circular orbits about the system's center of mass. From their color and brightness, you determine that each star has the same mass as our Sun. The orbital period of the pair is 24.3 days, based on the oscillation of brightness observed as one star occludes (hides) the other. From this information you are able to ascertain the distance between the stars. ●●● CR

83. Scientists still are not entirely sure whether the universe is open (which means it will keep expanding forever) or closed (which means it will eventually fall back in on itself in a "Big Crunch"). What determines whether it is open or closed is basically the same thing we calculate to determine escape speeds. Assume that the universe has a uniform density ρ. Then the mass contained in some spherical volume of space of radius R is $\frac{4}{3}\pi R^3 \rho$. Model a galaxy located at the edge of this spherical volume as a particle and consider its energy. According to Hubble's law, the speed of this galaxy is HR, where $H = 2.2 \times 10^{-18}\ \text{s}^{-1}$ is called the Hubble constant. (a) What is the maximum density of the universe such that it does not close? (b) The current estimate of the density of the universe is $10^{-26}\ \text{kg/m}^3$, including about 25% *dark matter,* which is matter we cannot observe directly, and 70% *dark energy,* another constituent that is not directly observable. Both dark matter and dark energy contribute to gravity, though, and so are included in the density estimate. Compare this value with your result and discuss what difference it would make if there were no dark matter in the universe. ●●●

Answers to Review Questions

1. They exert forces of equal magnitude because the two forces in an interaction pair are always of equal magnitude. However, the Moon has the smaller mass, and so, from $a = F/m$, its acceleration is greater.

2. Several bits of evidence are presented in this section. First, the centripetal acceleration of the Moon (required to produce its known orbital radius and speed) matches the expected $1/r^2$ strength of gravity at the Moon's orbital radius (a distance of 60 Earth radii). Second, the proportionality between period squared and radius cubed for all planets, first identified by Kepler, agrees with a $1/r^2$ force law. Finally, there is the elliptical shape of the planetary orbits, which can be derived from the inverse square law plus conservation of energy.

3. Ellipse, circle, parabola, and hyperbola are possible.

4. Because period squared is proportional to orbital radius cubed, if the radius quadruples, then the period increases by a factor of 8. Thus a "year" on the planet is equivalent to eight years on Earth.

5. Mass is the property of an object that determines the magnitude of the gravitational force it feels due to, or exerts on, another object. Inertia is the property of an object that determines its acceleration under the influence of any nonzero vector sum of forces. These are very different concepts, even though it turns out that mass and inertia have been measured to be identical to great accuracy under ordinary circumstances.

6. The torque about this axis is zero because the lever arm is always zero. This means that the angular momentum about this axis does not change.

7. (1) Planets move in elliptical orbits with the Sun at one focus. (2) As a planet orbits the Sun, a straight line connecting the two bodies sweeps out equal areas in equal time intervals. (3) The period squared for any planet is proportional to its orbital radius cubed.

8. The force exerted in (b) and (c) is a central force and so causes zero torque. For angular momentum to change, the vector sum of torques must be nonzero.

9. The ISS. The acceleration due to gravity is independent of the orbiting object's mass and is inversely proportional to (the square of) the distance from Earth. Because the station orbits closer to Earth, its acceleration is greater.

10. Zero. During your entire trajectory, the spring scale is pulled toward the trampoline right along with you, so you do not exert any force on it.

11. Yes. The balance compares the normal forces needed to support each mass against their respective gravitational forces, but in free fall no such normal forces are needed. Thus the normal force is zero on each side; that is, it is still balanced.

12. The reading decreases because the supporting force exerted by the scale has a smaller magnitude when you are accelerating downward.

13. While the astronauts are in orbit, the gravitational force exerted on them by Earth is exactly the amount needed to supply the centripetal acceleration that keeps them in orbit. Thus there is no support force required to satisfy Newton's second law.

14. The principle of equivalence states that there is no way to distinguish locally between gravitational effects and those due to acceleration of the reference frame. The qualification *locally* must be used because over large scales the inverse square nature of gravitational forces would become apparent, causing, for example, the gravitational acceleration experienced by an object at two widely separated locations to be different.

15. Yes. According to the principle of equivalence, anything—including light—that moves close to any object that has mass must move in a curved path.

16. The gravitational force exerted by Earth on an object in free fall near the ground does indeed vary as $1/r^2$. However, as the object moves, the change in its distance r from Earth's center is negligible. Therefore the force and hence the acceleration g are constant to a very good approximation.

17. Not necessarily. The larger planet might also have a larger radius, and the acceleration due to gravity decreases inversely with the radius squared. In fact, the acceleration due to gravity on Saturn is very close to that on Earth.

18. The choice that the potential energy is zero when two objects in a system are infinitely far apart makes the potential energy at all finite distances negative.

19. For any choice of initial and final positions, the work done by the gravitational force on an object is the same regardless of the particular path over which the object moves to get from initial to final position.

20. It is reasonable over any distance that is much smaller than the separation distance r.

21. Strictly speaking, the path is an ellipse. You cannot throw an object fast enough so that its kinetic energy exceeds the gravitational potential energy of the Earth-object system. Thus the mechanical energy is negative and the path is elliptical. However, the path is well approximated by a parabola.

22. The speed is smallest when the planet is at aphelion and greatest when the planet is at perihelion.

23. It is just barely unbound; the object can "escape" to an infinite distance from the more massive object.

24. The orbit is an ellipse if the mechanical energy of the object-Sun system is negative (negative potential energy has a greater magnitude than positive kinetic energy). The circle is also possible as a special case of an ellipse with zero eccentricity. The orbit is one branch of a hyperbola if the mechanical energy of the object-Sun system is positive (positive kinetic energy has a greater magnitude than negative potential energy). If the mechanical energy is zero, the orbit is a parabola, the limiting case between an ellipse and a hyperbola.

25. The sphere is divided into concentric thin shells, and then a shell is divided into rings such that each ring lies between two planes perpendicular to the line joining the particle and the center of the sphere. The force exerted by a ring on the particle is computed, and then this is integrated to obtain the force exerted by one spherical shell. The results for each shell are then superposed to obtain the result for the full sphere.

26. The particle may be located inside the sphere; the same process yields a result of zero for the force on a particle inside a spherical shell. The sphere may be hollow because the process involves superposing the results for spherical shells. The density does not have to be uniform across the solid sphere, but it is assumed to be spherically symmetrical for each shell. Hence the density could vary with radius without affecting the result, but the result would change if the density varied with some other component of the position vector.

Answers to Guided Problems

Guided Problem 13.2 $R_{max} = 19$ km

Guided Problem 13.4 $\Delta x = 3.5$ km

Guided Problem 13.6 vertical launch: $r_{max} = \frac{16}{7}R_E$ or $h_{max} = \frac{9}{7}R_E$;

horizontal launch: $r_{max} = \frac{9}{7}R_E$ or $h_{max} = \frac{2}{7}R_E$;

45° launch: $r_{max} = 1.96R_E$ or $h_{max} = 0.96R_E$

Guided Problem 13.8 $a_{rock} = 0.089g$

PRACTICE

14

PRACTICE
Special Relativity

Chapter Summary   249

Review Questions   251

Developing a Feel   252

Worked and Guided Problems   253

Questions and Problems   261

Answers to Review Questions   266

Answers to Guided Problems   266

Chapter Summary

Time measurements (Sections 14.1, 14.2)

Concepts An **event** is a physical occurrence that takes place at a specific location at a specific instant.

Two events are **simultaneous** when they are observed to occur at the same instant in a reference frame made up of a grid of synchronized clocks. Events that are simultaneous in one reference frame need not be simultaneous in another reference frame moving relative to the first one.

An **invariant** is a quantity whose value does not depend on the choice of reference frame.

The **speed of light** is independent of the motion of the light source relative to any moving observer. Therefore the speed of light is an invariant. It is the limiting speed of any object moving relative to any other object.

Quantitative Tools In SI units, the speed of light in vacuum is $c_0 = 3.00 \times 10^8$ m/s.

Space-time (Sections 14.3, 14.5, 14.6)

Concepts The **proper time interval,** Δt_{proper}, is the time interval between two events that occur at the same position in a particular reference frame but at different instants. An observer measuring the proper time interval between two events records the shortest time interval between those events.

Time dilation: A clock moving relative to an observer is measured to be running more slowly than an identical clock at rest relative to that observer.

The **proper length,** ℓ_{proper}, of an object is the length of that object measured by an observer who is at rest relative to the object.

Length contraction: When an object moves at speed v relative to an observer, the observer measures the length ℓ_v of the object along the direction of its motion to be shorter than the proper length of the object.

Quantitative Tools For an object moving at speed v, the **Lorentz factor** γ is a measure of the extent to which relativity affects our everyday notions of length and time intervals:

$$\gamma \equiv \frac{1}{\sqrt{1 - v^2/c_0^2}}. \tag{14.6}$$

At nonrelativistic speeds ($v \leq 0.1c_0$), $\gamma \approx 1$, but it approaches infinity as v approaches c_0.

Time dilation: If Δt_{proper} is the proper time interval between two events measured by an observer for whom the two events occur at the same position but at different instants, then the time interval Δt_v between those events measured by an observer moving at speed v relative to the observer is

$$\Delta t_v = \gamma \Delta t_{\text{proper}}. \tag{14.13}$$

The **space-time interval** s^2 between two events that are separated by a time interval Δt and a space interval Δx is given by

$$s^2 \equiv (c_0 \Delta t)^2 - (\Delta x)^2. \tag{14.18}$$

If an object that has proper length ℓ_{proper} moves at speed v relative to an observer, that observer measures the length ℓ_v of the object to be

$$\ell_v = \frac{\ell_{\text{proper}}}{\gamma}. \tag{14.28}$$

The **Lorentz transformation equations** are relationships between the coordinates of an event e measured in two reference frames A and B that coincide at $t_A = t_B = 0$ and are moving in the x direction at constant velocity $\vec{v}_{AB}$ relative to each other:

$$t_{Be} = \gamma\left(t_{Ae} - \frac{v_{ABx}}{c_0^2}x_{Ae}\right) \tag{14.29}$$

$$x_{Be} = \gamma(x_{Ae} - v_{ABx}t_{Ae}) \tag{14.30}$$

$$y_{Be} = y_{Ae} \tag{14.31}$$

$$z_{Be} = z_{Ae}. \tag{14.32}$$

Velocity addition: From the Lorentz transformation equations, we find that $v_{\text{A}ox}$, the x component of the velocity of an object o measured in reference frame A, is related to $v_{\text{B}ox}$, the x component of its velocity measured in reference frame B, by

$$v_{\text{B}ox} = \frac{v_{\text{A}ox} - v_{\text{AB}x}}{1 - \dfrac{v_{\text{AB}x}}{c_0^2} v_{\text{A}ox}}. \tag{14.33}$$

Energy and momentum (Sections 14.4, 14.7, 14.8)

Concepts The internal energy E_{int} of an object (energy not associated with the object's motion) is an invariant and is proportional to the object's mass m (the property that determines its gravitational interaction). Mass is also an invariant.

Unlike mass, inertia (a quantitative measure of an object's tendency to resist any change in its velocity) is not an invariant: What an observer measures for an object's inertia depends on the motion of the observer relative to the object.

Quantitative Tools For an object of mass m, its inertia m_v is

$$m_v = \gamma m, \tag{14.41}$$

its momentum $\vec{p}$ is

$$\vec{p} = \gamma m \vec{v}, \tag{14.42}$$

and its kinetic energy K is

$$K = (\gamma - 1)mc_0^2. \tag{14.51}$$

The energy E of an object of mass m and inertia m_v is

$$E = m_v c_0^2, \tag{14.53}$$

and its internal energy E_{int} is

$$E_{\text{int}} = mc_0^2. \tag{14.54}$$

Review Questions

Answers to these questions can be found at the end of this chapter.

14.1 Time measurements

1. What is an event?

2. Name two attributes of clocks that allow us to measure precisely the instants at which various events occur.

3. What must happen in order for an event to be observed?

4. What must you observe about the locations of two events in order to measure the proper time interval between them?

5. Describe what an observer must do to measure the instant at which an event occurs in a reference frame that consists of synchronized clocks placed at various equally spaced positions.

14.2 *Simultaneous* is a relative term

6. Where must a simultaneity detector be placed in order to determine whether two events are simultaneous? When placed at that location, how does the detector operate?

7. How does the speed of light in vacuum depend on the relative velocity of the source and the observer?

8. What can you say about the simultaneity of two events seen by two observers moving relative to each other along the line joining the events?

14.3 Space-time

9. Clock A is moving relative to identical clock B. To an observer at rest relative to B, how does the rate at which A runs compare with the rate at which B runs?

10. Cite two experiments that give direct evidence for time dilation.

11. What is the proper length of an object?

12. What is length contraction?

14.4 Matter and energy

13. What experimental evidence shows that the kinetic energy expression $K = \frac{1}{2}mv^2$ does not apply at relativistic speeds?

14. Which of the variables—inertia, mass, kinetic energy, and internal energy—are invariant, and which depend on the reference frame of the person measuring them?

15. If you change the internal energy of a system, does the system's inertia change, and if so, how?

16. What happens to an object's inertia when you decrease the object's kinetic energy?

14.5 Time dilation

17. What is the definition of the Lorentz factor, and at what speed do its effects on space-time properties become noticeable?

18. What is the relationship between the proper time interval between two events and the time interval between the events measured by an observer for whom the events occur at different locations?

19. What is the difference among *timelike*, *lightlike*, and *spacelike* values of the space-time interval $s^2 = (c_0\Delta t)^2 - (\Delta x)^2$ between two events?

20. How does the nature of the space-time interval between events (*timelike*, *lightlike*, or *spacelike*) affect the principle of causality?

14.6 Length contraction

21. What is the relationship between the proper length of an object and the length measured by an observer moving relative to the object?

22. How are time dilation and length contraction related to each other? Consider muons traveling from the top of Earth's atmosphere to the ground to support your answer.

23. What are the requirements for validity of the Lorentz transformation equations? How do they differ from the requirements for the time dilation and length contraction equations (Eqs. 14.13 and 14.28)?

14.7 Conservation of momentum

24. The momentum of a particle was defined in Chapter 4 as the product of its inertia and its velocity. What change, if any, is required so that momentum can be defined for a particle that is traveling at a relativistic speed?

25. Does the momentum of an isolated system remain constant when one or more components travel at a relativistic speed?

26. What is the relationship (*a*) between the mass m of a particle and the inertia m_v we measure as the particle moves at some speed v relative to us, and (*b*) between the particle's inertia and its momentum?

14.8 Conservation of energy

27. How does the mathematical expression for the kinetic energy of a particle moving at relativistic speeds differ from that for a particle moving at nonrelativistic speeds? Are the two expressions compatible?

28. For a system of mass m, what does the quantity mc_0^2 represent?

29. For a system of mass m, what does the quantity γmc_0^2 represent?

30. What is the difference between *conserved* and *invariant*, and how is this difference relevant when we measure energy and momentum?

Developing a Feel

Make an order-of-magnitude estimate of each of the following quantities. Letters in parentheses refer to hints below. Use them as needed to guide your thinking.

1. The difference between 1 and the Lorentz factor for a high-speed rifle bullet (E, T)
2. The difference between the time intervals needed to complete your walking trip, as measured by you and by your friend at rest in the Earth reference frame, beginning when you depart New York City and ending when you reach Los Angeles (Each of you measures only the time intervals you spend *walking*.) (V, J, M, B)
3. The difference between the time intervals, as measured by an astronaut on board the International Space Station (ISS) and by an observer at rest in the Earth reference frame, beginning when the ISS passes directly above Los Angeles and ending when the ISS is directly above New York City (F, J, U, C, P)
4. The length by which the distance from New York City to Los Angeles is contracted when measured by an observer moving east to west relative to Earth at the speed of the ISS (F, J, U, C, P)

5. The distance traveled by a pion during its lifetime, 26 ns in the pion rest reference frame, when the particle is moving at a speed of $0.995c_0$, as measured by an observer at rest in the Earth reference frame (H, Q)
6. The extra inertia in the Earth reference frame of a midsize car traveling at highway speed compared to the same car at rest (R, D, O)
7. The amount of mass converted to energy in the Sun every second (N, A, G)
8. The ratio of the energy of a proton moving at $0.99999c_0$ in the Earth reference frame to the *kinetic* energy of a snail in the same reference frame (S, L, I, K)

Hints

A. What is the distance from the Sun to Earth?
B. What is the Lorentz factor for your walking speed in the Earth reference frame?
C. What fraction of an orbital period does the ISS need to travel from Los Angeles to New York City in the Earth reference frame?
D. What is the Lorentz factor for this speed? (You calculated this in *Principles* Checkpoint 14.14.)
E. What is the speed of a high-speed bullet in the Earth reference frame?
F. At what height does the ISS orbit in the Earth reference frame?
G. How much power does the Sun generate?
H. What length of Earth moves past the pion during its lifetime in the pion reference frame?
I. What is the mass of a snail?
J. What is the distance from New York City to Los Angeles in the Earth reference frame?
K. What is the top speed of a snail in the Earth reference frame?
L. What is the Lorentz factor for this proton in the Earth reference frame?
M. What time interval (in the Earth reference frame) would you need to walk that distance nonstop?
N. What is the amount of average solar power that arrives at each square meter of Earth?

O. What is the mass of a midsize car?
P. What is the Lorentz factor for the ISS in the Earth reference frame?
Q. What is the Lorentz factor for the pion in the Earth reference frame?
R. What is a typical highway speed in the Earth reference frame?
S. What is the mass of a proton?
T. What is the Lorentz factor for this bullet in the Earth reference frame?
U. What is the period of orbit of the ISS in the Earth reference frame?
V. What is your average walking speed in the Earth reference frame?

Key (all values approximate)

A. 2×10^{11} m; B. $1 + 2 \times 10^{-17}$; C. 1×10^{-1}; D. $1 + 5 \times 10^{-15}$;
E. 1×10^3 m/s; F. 4×10^2 km; G. 5×10^{26} W; H. 8 m; I. 10^{-2} kg;
J. 4×10^6 m; K. 10^{-3} m/s; L. 2×10^2; M. 2×10^6 s; N. 1×10^3 W/m²;
O. 1×10^3 kg; P. $1 + 3 \times 10^{-10}$; Q. 1×10^1; R. 3×10^1 m/s;
S. 2×10^{-27} kg; T. $1 + 6 \times 10^{-12}$; U. 9×10^1 min; V. 2×10^0 m/s

Worked and Guided Problems

Procedure: Time and length measurements at very high speeds

At very high speeds, measurements of lengths and time intervals depend on the reference frame. Problems involving objects or observers moving at very high speeds therefore require extra care when we consider lengths and time intervals.

1. Identify each observer or reference frame with a letter and determine in which reference frame each numerical quantity given in the problem statement is measured.

2. If the problem involves any time intervals, identify the events that define that time interval. Determine

in which observer's reference frame, if any, the two events occur at the same position. This observer measures the proper time interval between the events (Figure 14.26).

3. Identify any distances or lengths mentioned in the problem. Determine for which observer, if any, the object defining that distance or length remains at rest. This observer measures the proper length (Figure 14.29).

4. Use Eqs. 14.13 and 14.28 to determine the time intervals and lengths in reference frames other than the ones in which the intervals and lengths are given.

These examples involve material from this chapter but are not associated with any particular section. Some examples are worked out in detail; others you should work out by following the guidelines provided.

Worked Problem 14.1 Space rescue

You are playing a rescue dispatcher on Earth in a video game involving interstellar travel. In the game, a spaceship that left Earth a few months ago, accelerating toward the star Vega, has suffered an engine failure and is now drifting toward Vega. A distress signal from the ship has been received on Earth, with the signal pulses arriving 2.00 s apart. You know that the distress beacon on the disabled ship emits pulses 1.00 s apart in the ship's reference frame. (*a*) In order to give instructions to a rescue crew being sent out, you must know the speed at which the disabled ship is drifting toward Vega. What is this speed in the Earth reference frame? (*b*) The standard reply signal telling a disabled ship that help is on the way consists of pulses arriving at the ship 1.00 s apart. If the disabled ship is to receive these reply pulses at 1.00-s intervals, what interval between pulses must you use in sending the reply signal?

❶ **GETTING STARTED** The problem asks us to relate the time difference between the pulses that are emitted in one reference frame and the pulses that are received in another reference frame, so let us consider two pairs of events: Event 1 is the emission of a pulse by the spaceship, and event 2 is the emission of the next pulse by the ship. Event 3 is the reception of the first pulse on Earth, and event 4 is the reception of the next pulse on Earth. Events 1 and 2 occur at the same location relative to the disabled ship, so crewmembers on the ship measure the proper time interval between those two events. Events 3 and 4 occur at the same location relative to Earth, so we measure the proper time interval between them. Because the ship is moving at constant speed with respect to Earth, we can use the formula for time dilation to determine the time interval between events 1 and 2 in the Earth reference frame. We need to remember, though, that we measure the instant at which an event takes place with a clock that is at rest in our reference frame *at the location of the event*. Because events 1 and 2 take place at different locations in the Earth reference frame, the time interval that we measure between them is not the same as the time interval that we measure between events 3 and 4. These two intervals are, however, related by the speed of the ship and the speed of the signal (that is, the speed of light, c_0).

We get oriented with a sketch showing the four events of interest in both the Earth and Ship reference frames (Figure WG14.1).

Figure WG14.1

Earth reference frame

$\Delta x_{E34} = 0 \quad \Delta t_{E34} = 2.00$ s

$\Delta x_{E12} = ? \quad \Delta t_{E12} = ?$

ship reference frame

$\Delta x_{S34} = ? \quad \Delta t_{S34} = ?$

$\Delta x_{S12} = 0 \quad \Delta t_{S12} = 1.00$ s

From the point of view of the crew on the disabled ship, they are at rest and Earth is moving away from them at the constant speed that we measure for the ship. Because the respective situations are symmetrical, we should be able to use similar reasoning to solve parts *a* and *b*.

❷ **DEVISE PLAN** Because the crewmembers on the disabled ship measure the proper time interval between events 1 and 2, we know that the interval as measured in the Earth reference frame is longer by the Lorentz factor, γ:

$$\Delta t_{E12} = \gamma \Delta t_{S12}. \tag{1}$$

In the Earth reference frame, the time interval it takes for the signals to reach Earth is the distance from the ship to Earth divided by the speed of light, so

$$\Delta t_{E34} = t_{E4} - t_{E3}$$

$$= (t_{E2} + x_{E2}/c_0) - (t_{E1} + x_{E1}/c_0)$$

$$= \Delta t_{E12} + \Delta x_{E12}/c_0. \tag{2}$$

The distance that the ship moves between events 1 and 2 is the ship's speed multiplied by the time interval:

$$\Delta x_{E12} = v_{ES}\Delta t_{E12}. \tag{3}$$

Substituting first Eq. 3 and then Eq. 1 into Eq. 2 gives

$$\Delta t_{E34} = \Delta t_{E12} + v_{ES}\Delta t_{E12}/c_0$$

$$= \Delta t_{E12}(1 + v_{ES}/c_0)$$

$$= \gamma\Delta t_{S12}(1 + v_{ES}/c_0),$$

which we can simplify by using the definition of the Lorentz factor:

$$\Delta t_{E34} = \frac{\Delta t_{S12}}{\sqrt{1 - (v_{ES}/c_0)^2}}(1 + v_{ES}/c_0)$$

$$= \Delta t_{S12}\sqrt{\frac{(1 + v_{ES}/c_0)^2}{1 - (v_{ES}/c_0)^2}}$$

$$= \Delta t_{S12}\sqrt{\frac{1 + v_{ES}/c_0}{1 - v_{ES}/c_0}}. \tag{4}$$

Finally, we can solve Eq. 4 for v_{ES}, the speed of the ship in the Earth reference frame:

$$\Delta t_{E34}^2(1 - v_{ES}/c_0) = \Delta t_{S12}^2(1 + v_{ES}/c_0)$$

$$\Delta t_{E34}^2 - \Delta t_{S12}^2 = \Delta t_{E34}^2 v_{ES}/c_0 + \Delta t_{S12}^2 v_{ES}/c_0$$

$$= v_{ES}(\Delta t_{E34}^2/c_0 + \Delta t_{S12}^2/c_0)$$

$$v_{ES} = c_0\frac{\Delta t_{E34}^2 - \Delta t_{S12}^2}{\Delta t_{E34}^2 + \Delta t_{S12}^2}.$$

❸ **EXECUTE PLAN** (a) Substituting the time intervals measured in each reference frame, we obtain

$$v_{ES} = c_0\frac{(2.00\text{ s})^2 - (1.00\text{ s})^2}{(2.00\text{ s})^2 + (1.00\text{ s})^2} = c_0\frac{3.00}{5.00} = 0.600c_0. ✔$$

(b) Because the relationship between the Earth reference frame and the ship's reference frames is symmetrical, we can use Eq. 4—but this time with E and S exchanged and labeling the two emissions on Earth as 5 and 6 and the two receptions on the ship as 7 and 8—to determine the ratio between the rate of pulses that are sent from Earth and the rate of pulses that are received on the ship:

$$\Delta t_{S78} = \Delta t_{E56}\sqrt{\frac{1 + v_{SE}/c_0}{1 - v_{SE}/c_0}}$$

$$\Delta t_{E56} = \Delta t_{S78}\sqrt{\frac{1 - v_{SE}/c_0}{1 + v_{SE}/c_0}}$$

$$\Delta t_{E56} = (1.00\text{ s})\sqrt{\frac{1 - 0.600}{1 + 0.600}} = 0.500\text{ s}. ✔$$

❹ **EVALUATE RESULT** The speed that we calculated in part a is substantial. An interstellar ship would have to move at great speed to travel between stars in a reasonable amount of time.

It makes sense that the ship's speed is relativistic because of the large factor by which the received pulse rate differs from the rate at which the pulses were emitted. It also makes sense that we got the same factor for the Earth-to-ship pulses as for the ship-to-Earth pulses because of the symmetry of their situations with respect to each other.

As a check on our math, we can compare the invariant interval between events 1 and 2 using numerical values from each reference frame:

$$s^2 = (c_0\Delta t^2) - \Delta x^2$$

For events 1 and 2 in the Earth frame we have (using Eqs. 1 and 3):

$$s_{12}^2 = (c_0\Delta t_{E12})^2 - \Delta x_{E12}^2 = (c_0\Delta t_{E12})^2 - (v_{ES}\Delta t_{E12})^2$$

$$= (c_0\gamma\Delta t_{S12})^2 - (v_{ES}\gamma\Delta t_{S12})^2$$

$$= [(1.25c_0\text{ m/s})(1.00\text{ s})]^2 - [(0.600c_0\text{ m/s})(1.25)(1.00\text{ s})]^2$$

$$= c_0^2\text{ m}^2$$

In the Ship frame

$$s_{12}^2 = (c_0\Delta t_{S12})^2 - \Delta x_{S12}^2$$

$$= [(c_0\text{ m/s})(1.00\text{ s})] - 0$$

$$= c_0^2\text{ m}^2$$

This is reassuring.

We are reminded that, when we consider objects that move at relativistic speeds, what we measure for distant events is not what we actually see when we look at them: When an object is moving at a significant fraction of the speed of light, we have to account for the finite time interval that it takes for light to travel from the object to our eyes (or another detector) because we define our measurements as being made locally to the events. In this situation, with the disabled spaceship moving away from Earth, the time interval between the pulses that we receive on Earth is even longer than the time-dilated interval that we measure for the emission of those pulses from the ship. If the ship had been approaching Earth, the time interval between the pulses that we received on Earth would have been shorter than the time interval that we measured for their emission from the ship. (The sign of Δx in Eq. 2 would be opposite, giving us the inverse ratio in Eq. 4.)

Guided Problem 14.2 Relativistic fly-by

Spaceship A makes a fly-by of space station B at very close distance. The duty officer on B determines that, during the fly-by, the length of A was 50.0 m in the reference frame of B. Records show that the measured length of A was 80.0 m when A was in dry-dock on B. (a) What does the duty officer on B measure for A's speed during the fly-by? (b) The crew on B measures the length of B to be 200 m. What length does an observer on A measure for the length of B during the fly-by?

❶ GETTING STARTED

1. Which reference frames should you compare?
2. For each reference frame you use, make a sketch of the objects involved and the information you know about them.

❷ DEVISE PLAN

3. How is the speed you must determine related to the two lengths given for the spaceship?
4. Is there any reference frame in which the station length you must determine is a proper length?

❸ EXECUTE PLAN

❹ EVALUATE RESULT

5. Are your proper lengths greater than their corresponding nonproper lengths?

Worked Problem 14.3 Galactic fireworks

A fast *nova* (from the Latin, *nova stella,* "new star") occurs when a white dwarf star experiences a runaway nuclear reaction, which increases its brightness dramatically during a short interval of time. The enhanced brightness then persists for about a month. Suppose that bright light from "new star" A first appears to observers at MacDonald Observatory on Earth at 11:00 p.m. on July 12. Measurements indicate that star A is moving almost directly away from Earth with a relative speed of 6.0×10^4 m/s and is first observed approximately 9.33×10^{16} m from Earth in the direction of Vega. While still studying star A, observers see another "new star," star B, also in the direction of Vega. Bright light from star B first arrives at the observatory at 2:00 a.m. on July 30, and measurements indicate that star B is moving almost directly away from Earth with a relative speed of 4.0×10^4 m/s and is first observed approximately 6.63×10^{16} m from Earth. (a) Determine the space-time interval between these two events: star A goes nova and star B goes nova. (b) Describe an inertial reference frame in which these two events are simultaneous, or argue that such a reference frame does not exist.

❶ GETTING STARTED The information we have is from the Earth reference frame, so we begin with a picture of the situation in that reference frame (Figure WG14.2). It is convenient to choose an *x* axis along the straight line that contains both stars and Earth. We arbitrarily choose the positive direction to be toward Vega (toward both new stars). We can compute the light travel time from each star from the given distance. It may be necessary to adjust for the stellar motion, but we have speed and direction information in any case. This should allow us to compute both the *x* component of displacement between the two events and the time interval between the two events in the Earth reference frame. The value of the space-time interval is the same in all inertial reference frames. It can be computed directly from these Δx and Δt values in the Earth reference frame. Depending on the sign of this result, we should be able to determine whether there is a reference frame in which the events are simultaneous. If there is, such a reference frame could be imagined as moving past Earth at some relative velocity in the *x* direction—that is, toward one new star and away from the other.

Figure WG14.2

❷ DEVISE PLAN For part *a,* the space-time interval between two events can be computed in any reference frame using Eq. 14.18:

$$s^2 = (c_0 \Delta t)^2 - (\Delta x)^2.$$

We have enough information to compute the displacement and time interval between the two specified events, but we must decide whether the velocities of the two stars relative to Earth should be taken into account. Both relative velocities are of order of magnitude 10^5 m/s, and the time interval between observations of "first" light from the two stars is of the order 10^6 s. This means that star A might have moved as much as $d = vt = (10^5 \text{ m/s})(10^6 \text{ s}) = 10^{11}$ m before star B was first observed. However, we are given stellar distance information to three significant digits, and both distances are of the order 10^{17} m. Thus any consideration of additional distance due to stellar motion is several digits beyond the precision of our data. In addition, relativistic effects depend on the Lorentz factor:

$$\gamma = \frac{1}{\sqrt{(1 - v^2/c_0^2)}}.$$

For the given relative velocities of these stars, the difference between γ and 1 is on the order of 10^{-8}, which is well beyond the precision of our distance information and, assuming we know the measured times to within 1 minute, also several digits beyond the precision of our time data. Thus a simple calculation of light travel time is appropriate in determining the space-time interval.

For part *b,* the space-time interval can be positive, negative, or zero. If it is positive, the interval is timelike, and it is possible to establish a definite order of events. Thus one of the two events occurs first in all inertial reference frames, and there is no inertial reference frame in which the two events are simultaneous. If the space-time interval is negative, the interval is spacelike. In this case, there is a reference frame in which the events are simultaneous, and we can obtain it by using a simultaneity detector—that is, by devising a reference frame that receives light from each event at an instant when it is equidistant in space between the two events. If the space-time interval is zero, it is lightlike in all inertial reference frames, so a ray of light could travel from one event to the other. Because the speed of light is the same in all reference frames, the events cannot be simultaneous unless they occur at exactly the same place.

❸ **EXECUTE PLAN**

(a) The time interval between the two events must be computed using the travel time of light. We observe star A at the location where its light originated—that is, where the star was when it "went nova." Light would require a time interval of

$$\Delta t_{EA} = \frac{\Delta x_{EA}}{c_0}$$

to travel from star A to Earth.

Similarly, the travel time for light from the second event to reach Earth is

$$\Delta t_{EB} = \frac{\Delta x_{EB}}{c_0}.$$

However, these time intervals overlap because star B does not wait for light from star A to reach Earth before "going nova." Thus the time interval between the two events is the difference in the two light travel times plus the delay in reception of the signal from star B compared to star A. The result is a time interval, in the Earth reference frame, between the two events of

$$\Delta t_E = \Delta t_{EA} - \Delta t_{EB} + \Delta t_{E\ \text{detection}}$$

$$= \frac{\Delta x_{EA}}{c_0} - \frac{\Delta x_{EA}}{c_0} + \Delta t_{E\ \text{detection}}$$

$$= \frac{9.33 \times 10^{16}\ \text{m}}{3.00 \times 10^8\ \text{m/s}} - \frac{6.63 \times 10^{16}\ \text{m}}{3.00 \times 10^8\ \text{m/s}}$$

$$+ \left[17\ \text{days} \left(\frac{24\ \text{h}}{1\ \text{day}} \right) + 3\ \text{h} \right] \left(\frac{3600\ \text{s}}{1\ \text{h}} \right)$$

$$= (3.11 \times 10^8\ \text{s}) - (2.21 \times 10^8\ \text{s}) + (1.48 \times 10^6\ \text{s})$$

$$= 9.15 \times 10^7\ \text{s}.$$

The displacement between the two events is, to the precision available, simply the difference in their apparent positions when they went nova:

$$\Delta x_{EA} = 9.33 \times 10^{16}\ \text{m}$$

$$\Delta x_{EB} = 6.63 \times 10^{16}\ \text{m}$$

$$\Delta x_E = \Delta x_{EA} - \Delta x_{EB} = (9.33 \times 10^{16}\ \text{m}) - (6.63 \times 10^{16}\ \text{m})$$

$$= 2.70 \times 10^{16}\ \text{m}.$$

The space-time interval is then

$$s^2 = (c_0 \Delta t_E)^2 - (\Delta x_E)^2$$

$$= [(3.00 \times 10^8\ \text{m/s})(9.15 \times 10^7\ \text{s})]^2 - (2.70 \times 10^{16}\ \text{m})^2$$

$$= 2.4 \times 10^{31}\ \text{m}^2. ✔$$

(b) The invariant space-time interval is positive, or timelike, so it must be possible to order the events in time. Indeed, star A must go nova first; otherwise, the light from star B, which is closer to Earth, would arrive first. This is true in any inertial reference frame. It is therefore not possible to construct an inertial reference frame in which the events are simultaneous. ✔

❹ **EVALUATE RESULT** Distances between stars are typically very large. The stars in this problem are at distances on the order of 10^{17} m, which is about the distance light travels in 10 years.

For part *a*, the space-time interval is less than $(c_0 \Delta t_E)^2$, which is consistent for events that do not occur at the same location. The time required for light to travel between stars A and B is slightly smaller than the time interval we computed between the events, so in principle a signal from one star could have triggered the other to explode. This is also consistent with a timelike interval between events.

For part *b*, we know that events separated by a timelike space-time interval in one reference frame are also separated by a timelike interval in any other inertial reference frame; hence, they cannot be simultaneous in any reference frame.

Guided Problem 14.4 Lorentz would love it

Observers Carol and Diane move at constant velocity relative to each other along a straight line. They agree to align the positive *x* axis of their respective reference frames along this line. They set their respective clocks to zero just as the origins of the two *x* axes coincide. Diane then observes event 1 at $x_{D1} = +400$ m, $t_{D1} = 1.00 \times 10^{-6}$ s, and event 2 at $x_{D2} = +900$ m, $t_{D2} = 5.00 \times 10^{-7}$ s. Carol observes these two events as occurring simultaneously. (a) What is the velocity $\bar{v}_{DC}$ of Carol relative to Diane? (b) At what instant does Carol observe the events?

❶ **GETTING STARTED**

1. For each reference frame, make a sketch of the objects involved and the information you know about them.
2. Is any length a proper length in one of the reference frames? Is any time interval a proper time interval in one of the reference frames?

❷ **DEVISE PLAN**

3. What general transformation equation relates time and space coordinates in different reference frames?
4. How can you express Carol's observations of the time coordinates of events 1 and 2 in terms of Diane's observations of the same events?
5. How can you express the difference in time between events 1 and 2 as measured by Carol in terms of Diane's coordinates?
6. Can the Lorentz factor that relates Carol's reference frame to Diane's reference frame be zero?

❸ **EXECUTE PLAN**

7. Can you rearrange the expression for the time difference as measured by Carol to get Carol's velocity as measured by Diane?

❹ **EVALUATE RESULT**

8. Can you check your result for consistency by using the space-time interval between events?

Worked Problem 14.5 To the stars!

The first interstellar exploration spaceship passes remote science post P and has already turned off its engines to coast at a constant cruising speed for its journey beyond the star Alpha Centauri. The scientists on post P measure the ship's speed to be $0.600c_0$. Because Alpha Centauri is known to be 4.00×10^{16} m distant from post P, according to measurements made on post P, the scientists have some time to plan a celebration of humans moving past Alpha Centauri. Cleverly, they schedule the festivities so that astronauts aboard the ship will observe (through their telescopes) the beginning of the scientists' celebration at the same instant the ship speeds past Alpha Centauri. (*a*) What time interval is required for the trip according to spaceship clocks? (*b*) How much time will have elapsed on post P between the instant the ship passes their post and the start of the celebration?

❶ GETTING STARTED Because we need to compare information from post P and the ship, we need to analyze the problem from the perspective of both the scientists at post P and the astronauts on the ship. We define two reference frames: reference frame P attached to the post and reference frame S attached to the spaceship. Thus the ship's speed relative to the post is $v_{PS} = 0.600c_0$. We are told that in reference frame P, the distance to Alpha Centauri is 4.00×10^{16} m. We define event 0 to occur when the ship passes the post, event 1 to be when the ship reaches Alpha Centauri, event 2 to occur when the party starts at post P, and event 3 to be when the astronauts on the ship as it passes Alpha Centauri observe the party starting. It is useful to note that, in order for the astronauts to observe the party starting, the light from that event must travel from post P to the ship. We can think of this light representing the start of the party as a signal sent (event 2) and received (event 3). Table WG14.1 organizes the information we know about these events. For convenience, we choose the origin of both time and position in both reference frames at event 0, with the positive x axis pointing toward Alpha Centauri in reference frame P and toward post P in reference frame S. This allows all x components to be positive for each event in each reference frame.

❷ DEVISE PLAN We can glean significant information about the trip from the time dilation equation (Eq. 14.13). In order to determine the time interval for the trip according to scientists at post P ($\Delta t_{P01} = t_{P1} - t_{P0}$), we divide the distance traveled by the spaceship as measured in reference frame P by the ship's speed. Because in reference frame S events 0 and 1 both take place at the location of the ship, the time interval between these two events as measured by the ship's clocks ($\Delta t_{S01} = t_{S1} - t_{S0}$) is a proper time interval in reference frame S. Therefore we can use Eq. 14.13 to relate Δt_{S01} and Δt_{P01}, using the appropriate Lorentz factor given by Eq. 14.6. The v in the Lorentz factor is equal to the speed of the ship relative to the monitoring post: $v = v_{PS} = 0.600c_0$.

Because events 1 and 3 are simultaneous in reference frame S, we know that $\Delta t_{S03} = \Delta t_{S01}$. Because event 2 takes place at the post, we know that Δt_{P02} is a proper time interval in reference frame P. Thus we can add some information to Table WG14.1 by using Eq. 14.13. Sorting events 2 and 3 requires something more general, however. We use the invariant (reference-frame-independent) speed of light plus the invariant space-time interval s^2 to compare information between reference frames and then calculate the desired temporal information.

❸ EXECUTE PLAN The time interval Δt_{P01} needed for the spaceship to reach Alpha Centauri as measured by clocks at the monitoring post is

$$\Delta t_{P01} = \frac{d_{P01}}{v_{PS}} = \frac{4.00 \times 10^{16} \text{ m}}{0.600(3.00 \times 10^8 \text{ m/s})} = 2.22 \times 10^8 \text{ s}.$$

We then use Eq. 14.6 to determine the Lorentz factor:

$$\gamma = \frac{1}{\sqrt{1 - v_{PS}^2/c_0^2}} = \frac{1}{\sqrt{1 - (0.600)^2}} = \frac{1}{0.800} = 1.25.$$

(*a*) Now we can use Eq. 14.13 to determine what clocks on the spaceship measure for the proper time interval needed for the ship to travel from the monitoring post to Alpha Centauri:

$$\Delta t_{S01} = \frac{\Delta t_{P01}}{\gamma_{PS}} = \frac{2.22 \times 10^8 \text{ s}}{1.25} = 1.78 \times 10^8 \text{ s.} ✔$$

Both Δt_{S01} and Δt_{P01} begin at $t = 0$, and so we now have full information about events 0 and 1 in Table WG14.1:

$$t_{S1} = 1.78 \times 10^8 \text{ s}, \quad t_{P1} = 2.22 \times 10^8 \text{ s}.$$

Note that we also have most of the information about event 3, missing only a value for t_{P3}. In reference frame S, events 1 and 3 are simultaneous and occur at the same position. This means that in this reference frame the space-time interval s^2 between events 1 and 3 is zero. Thus these two events are separated by a lightlike space-time interval. Events that are simultaneous *and* occur at the same location in one reference frame must also be simultaneous and occur at some common location in *any* inertial reference frame. This is not immediately obvious because it might be possible to calculate a number of different time and space intervals between events 1 and 3 that, when combined, give a space-time interval $s^2 = 0$.

Table WG14.1 Events for Worked Problem 14.5

Event	t_S	x_S	t_P	x_P
0: ship passes post P	$t_{S0} = 0$	$x_{S0} = 0$	$t_{P0} = 0$	$x_{P0} = 0$
1: ship arrives at AC	t_{S1}	$x_{S1} = 0$	t_{P1}	$x_{P1} = +4.00 \times 10^{16}$ m
2: signal that party starts sent from post P	t_{S2}	x_{S2}	t_{P2}	$x_{P2} = 0$
3: signal that party starts received by ship at AC	$t_{S3} = t_{S1}$	$x_{S3} = 0$	t_{P3}	$x_{P3} = x_{P1} = 4.00 \times 10^{16}$ m

Let us try it: Suppose scientists at the monitoring post see the spaceship reach Alpha Centauri (event 1) at some instant and then see the light signal indicating the start of the party received at the star (event 3) at a different instant. Then the signal must be received at a position that makes the space-time interval between events 3 and 1 equal to zero, just as in reference frame S. However, that would mean that the ship travels from the position of event 3 to the position of event 1 at the speed of light because only then does $s^2 = 0 = (c_0 \Delta t)^2 - (\Delta x)^2$. This is not possible! The result is that the scientists must also see the signal as being received at Alpha Centauri just as the ship reaches the star. Thus we add to our table of values:

$$t_{P3} = t_{P1} = 2.22 \times 10^8 \text{ s}.$$

(b) Now we must determine what scientists at the post measure for the time interval between the instant the ship passes the post (event 0) and the instant the scientists begin their party (event 2). The scientists know that both the ship and the light signal indicating the start of the party must arrive at Alpha Centauri at instant $t_{P3} = t_{P1} = 2.22 \times 10^8$ s. They also know the speed at which that signal travels and the distance it travels. Thus from the instant the signal is sent out (event 2) to the instant it is received at Alpha Centauri (event 3) is a time interval of

$$\Delta t_{P23} = t_{P3} - t_{P2} = \frac{d_{P23}}{c_0} = \frac{d_{P01}}{c_0} = \frac{4.00 \times 10^{16} \text{ m}}{3.00 \times 10^8 \text{ m/s}}$$

$$= 1.33 \times 10^8 \text{ s}.$$

We know t_{P3} and so

$$t_{P2} = t_{P3} - \Delta t_{P23} = (2.22 \times 10^8 \text{ s}) - (1.33 \times 10^8 \text{ s})$$

$$= 0.89 \times 10^8 \text{ s}.$$

We have set $t_{P0} = 0$, so by the clocks at post P the party must start

$$\Delta t_{P02} = t_{P2} - t_{P0} = t_{P2} - 0 = 8.9 \times 10^7 \text{ s} ✔$$

after the ship passes.

④ EVALUATE RESULT The light signal indicating the start of the party travels at speed c_0, and the spaceship travels at 60% of that value. If the two must travel the same distance in reference frame P and the signal moves 1.67 times faster than the ship in that reference frame, it seems reasonable that the post scientists should start their party after about 30% to 40% of the ship travel time interval has elapsed. The astronauts, of course, measure a different value for the instant at which the party signal is sent. Because the time interval Δt_{P02} is proper in reference frame P, we can use Eq. 14.13 to calculate the time interval in reference frame S:

$$\Delta t_{S02} = \gamma \Delta t_{P02} = 1.25(8.9 \times 10^7 \text{ s}) = 1.1 \times 10^8 \text{ s}.$$

In reference frame P, the space-time interval between events 0 and 2 is

$$s_{02}^2 = (c_0 \Delta t_{P02})^2 - 0 = 7.1 \times 10^{32} \text{ m}^2.$$

The space-time interval must have the same value in reference frame S, allowing us to compute the displacement of post P for the ship when the signal was sent in reference frame S:

$$s_{02}^2 = 7.1 \times 10^{32} \text{ m}^2 = (c_0 \Delta t_{S02})^2 - (\Delta x_{S02})^2$$

$$\Delta x_{S02} = \sqrt{[c_0(1.1 \times 10^8 \text{ s})]^2 - (7.1 \times 10^{32} \text{ m}^2)} = 1.9 \times 10^{16} \text{ m}.$$

Thus in reference frame S the party signal travels this distance to arrive at the ship (at Alpha Centauri) for event 3. The time interval the astronauts measure for how long the signal takes to travel this distance is

$$\Delta t_{S23} = \frac{1.9 \times 10^{16} \text{ m}}{3.00 \times 10^8 \text{ m/s}} = 6.3 \times 10^7 \text{ s},$$

so that the time interval from passing post P to the arrival of the party signal is

$$\Delta t_{S03} = \Delta t_{S02} + \Delta t_{S23} = (1.1 \times 10^8 \text{ s}) + (6.3 \times 10^7 \text{ s})$$

$$= 1.7 \times 10^8 \text{ s}.$$

The results are all consistent within the two significant digits allowed by the subtraction we used in calculating t_{P2}.

Guided Problem 14.6 Space-time interval

An unstable high-energy particle enters a detector and, before decaying, leaves a track 1.05 mm long. The speed of the particle relative to the detector is $0.992c_0$. Show that the value of the space-time interval s^2 for the interval from the instant the particle enters the detector to the instant the particle decays is the same when computed in a reference frame D at rest relative to the detector and in a reference frame P at rest relative to the particle.

① GETTING STARTED

1. Identify the two events that bracket the desired space-time interval.
2. Which time or space intervals given in the problem statement are useful in computing the desired space-time interval? In which reference frame(s)?

② DEVISE PLAN

3. How can you determine the time interval in reference frame D?
4. Is Eq. 14.18 relevant?
5. Is a proper time interval involved in either reference frame? What about a proper length? How might either of these allow you to obtain information in the second reference frame?

③ EXECUTE PLAN

④ EVALUATE RESULT

6. Do your answers for the space-time interval in the two reference frames agree?

Worked Problem 14.7 Pion production

Protons p_1 and p_2, initially moving at the same speed in the Earth reference frame, collide head-on, and the products of the collision are a proton p_3, a neutron n, and a pion π^+ that are at rest in the Earth reference frame. The masses of these particles, in terms of the proton mass m_p, are: neutron mass $m_n = 1.0014\, m_p$ and pion mass $m_{\pi^+} = 0.1488 m_p$. In a reference frame in which proton p_2 is at rest, what is the ratio of the kinetic energy of proton p_1 to its internal energy?

❶ GETTING STARTED We are given a description of two protons p_1 and p_2 colliding to create a proton p_3, a neutron, and a pion, and we are told the mass ratios of the products: $m_n/m_p = 1.0014$ and $m_{\pi^+}/m_p = 0.1488$. Our task is to determine the value of the ratio $K_{p1}/E_{int,p1}$ in a reference frame in which p_2 is at rest. We begin by making a sketch of the collision in the Earth reference frame, showing an unknown common speed v for p_1 and p_2 (Figure WG14.3). Note that the Earth reference frame is the zero-momentum reference frame because the protons are moving with opposite velocities and hence the sum of their momenta is zero. We choose to label this reference frame Z because this label will remind us of important momentum information.

Figure WG14.3   zero-momentum reference frame

We must also consider the collision from a reference frame in which p_2 is at rest. To keep things simple, let's call this the target reference frame T (because in this reference frame p_2 behaves like a stationary target waiting to be hit by p_1). In reference frame T, the three product particles are all moving to the right at speed v (Figure WG14.4).

Figure WG14.4   target reference frame

❷ DEVISE PLAN We can consider the system (which initially is the protons p_1 and p_2 and finally is the three product particles p_3, n, and π^+) as being both isolated and closed. Therefore the system's initial energy equals its final energy, and its initial momentum equals its final momentum. We know that the energy of a particle is the sum of its kinetic energy and internal energy (Eq. 5.21) and that its internal energy is $E_{int} = mc_0^2$ (Eq. 14.54).

Determining E_{int} for p_1. We can set the initial energy equal to the final energy in both reference frames. This gives us two equations in five quantities: $K_{Z,i}$, $E_{int,i}$, $E_{int,f}$, $K_{T,i}$, and $K_{T,f}$. We can express the internal energies in terms of c_0^2 and the masses m_p, m_n, and m_{π^+}. We can also express the internal energies in terms of the internal energy of a proton, $E_{int,p}$.

Determining K for p_1: We know that in the zero-momentum reference frame Z protons p_1 and p_2 are moving with speed v, and

so we can calculate $K_{Z,i}$ using Eq. 14.51 and the Lorentz factor γ_{Zv} (Eq. 14.6). In order to calculate γ_{Zv}, we use Eqs. 14.53 and 14.54 to calculate the initial and final energies of p_1 and p_2 in reference frame Z and then set the initial energy equal to the final energy in order to determine the inertia m_{vp} of p_1 and p_2. We can then use Eq. 14.41 to determine γ_{Zv} for p_1 and p_2. We know that in reference frame T the three product particles are moving with speed v, and so we can calculate $K_{T,f}$ using the same Lorentz factor γ_{Zv}. We can then solve for the kinetic energy $K_{T,i}$ of p_1 in reference frame T.

❸ EXECUTE PLAN In the zero-momentum reference frame, the three product particles are created at rest. Therefore none of the energy initially contained in p_1 and p_2 goes into kinetic energy in the product particles. All of the initial energy of the system thus is converted to internal energy in the product particles. Setting the system initial energy (kinetic energy of p_1 and p_2, internal energy in p_1 and p_2) equal to the system final energy (internal energy only, in p_3, n, π^+) yields

$$E_Z = K_{Z,i} + E_{int,i} = E_{int,f},$$

which tells us that the initial kinetic energy is equal to the increase in internal energy:

$$K_{Z,i} = E_{int,f} - E_{int,i}. \tag{1}$$

In reference frame T, the initial momentum is nonzero, which means that the final momentum must also be nonzero. Thus the proton, neutron, and pion are moving at speed v, and the kinetic energy $K_{T,f}$ is also nonzero. Setting the initial energy equal to the final energy yields

$$E_T = K_{T,i} + E_{int,i} = K_{T,f} + E_{int,f},$$

and so we have an expression for the initial kinetic energy in the target reference frame T:

$$K_{T,i} = (E_{int,f} - E_{int,i}) + K_{T,f}. \tag{2}$$

This equation tells us that in reference frame T, the initial kinetic energy is equal to the increase in internal energy plus the final kinetic energy of the three product particles.

Substituting Eq. 1 into Eq. 2 tells us that the initial kinetic energy in reference frame T is

$$K_{T,i} = K_{Z,i} + K_{T,f}. \tag{3}$$

In the zero-momentum reference frame, we set the initial energies of p_1 and p_2 (in the form $m_v c_0^2$ from Eq. 14.53) equal to the final internal energy of the product particles (in the form mc_0^2 from Eq. 14.54) to determine the inertia m_{vp} of p_1 and p_2:

$$2m_{vp}c_0^2 = 2\gamma_{Zv}m_p c_0^2 = (m_p + m_n + m_{\pi^+})c_0^2.$$

Now we use Eq. 14.41 to determine the Lorentz factor γ_{Zv} associated with p_1 and p_2:

$$m_v = \gamma_{Zv}m$$

$$\gamma_{Zv} = \frac{m_v}{m} = \frac{m_p + m_n + m_{\pi^+}}{2m_p} = \frac{1 + (m_n/m_p) + (m_{\pi^+}/m_p)}{2}$$

$$= \frac{1 + 1.0014 + 0.1488}{2} = 1.0751.$$

Because p_1 and p_2 move at speed v in the zero-momentum reference frame, the incoming kinetic energy is (Eq. 14.51)

$$K_{Z,i} = 2(\gamma_{Zv} - 1)m_p c_0^2 = 0.1502 m_p c_0^2.$$

Because the three particles are moving with speed v in reference frame T, we can use the same Lorentz factor to determine the final kinetic energy in this reference frame:

$$K_{T,f} = (\gamma_{Zv} - 1)m_p c_0^2 + (\gamma_{Zv} - 1)m_n c_0^2 + (\gamma_{Zv} - 1)m_{\pi^+} c_0^2$$

$$= (0.0751)(m_p + m_n + m_{\pi^+})c_0^2$$

$$= (0.0751)(1 + 1.0014 + 0.1488)m_p c_0^2$$

$$= 0.162 m_p c_0^2.$$

We now substitute these values for $K_{Z,i}$ and $K_{T,f}$ into Eq. 3 to determine the initial kinetic energy measured in reference frame T. Because we have specified that p_2 is at rest in this reference frame, this initial kinetic energy all belongs to p_1:

$$K_{T,i} = K_{p_1} = K_{Z,i} + K_{T,f}$$

$$= 0.1502 m_p c_0^2 + 0.162 m_p c_0^2 = 0.312 m_p c_0^2.$$

From Eq. 14.54 we know that $m_p c_0^2$ is the internal energy of p_1. Thus the ratio we are asked for, $K_{p_1}/E_{int, p_1}$ in a reference frame in which p_2 is at rest, is

$$\frac{K_{p_1}}{E_{int, p_1}} = \frac{K_{T,i}}{m_p c_0^2} = 0.312. ✔$$

❹ **EVALUATE RESULT** We see that $K_{T,i} > 2K_{Zi}$. If we ignore relativistic effects, the speed of p_1 in the target reference frame is $2v$, and so the kinetic energies in the two reference frames are $K_{T,i} = \frac{1}{2}m(2v)^2 = 2mv^2$ and $K_{Z,i} = 2(\frac{1}{2}mv^2) = mv^2$. Thus $K_{T,i} = 2K_{Z,i}$. We would expect that the relativistic increase of inertia with speed should provide extra initial kinetic energy in reference frame T, so our result makes sense.

We can also evaluate our result by using an alternative method to calculate the ratio. First we calculate the speed v of p_1 and p_2 in the zero-momentum reference frame from our calculation for γ_{Zv}:

$$\gamma_{Zv} = \frac{1}{\sqrt{1 - (v/c_0)^2}}$$

$$\gamma_{Zv}^2 = \frac{1}{1 - (v/c_0)^2}$$

$$1 - (v/c_0)^2 = \frac{1}{\gamma_{Zv}^2}$$

$$(v/c_0)^2 = 1 - \frac{1}{\gamma_{Zv}^2}$$

$$v = \left(\sqrt{1 - \frac{1}{\gamma_{Zv}^2}}\right)c_0 = \left(\sqrt{1 - \frac{1}{(1.0751)^2}}\right)c_0 = 0.367 c_0.$$

In order to determine the initial kinetic energy of p_1 in the target reference frame, we need to know the x component of the velocity $v_{Tx,i}$ of p_1 (we choose the positive x direction to be the direction in which p_1 moves). We use Eq. 14.33 with $v_{ZTx} = -v$ and $v_{Zx,i} = v$:

$$v_{Tx,i} = \frac{v_{Zx,i} - v_{ZTx}}{1 - \frac{v_{ZTx} v_{Zx,i}}{c_0}} = \frac{v + v}{1 + \frac{v^2}{c_0^2}} = \frac{2(0.367)c_0}{1 + (0.367)^2} = 0.647 c_0.$$

Because the kinetic energy of p_1 in reference frame T depends on the Lorentz factor associated with its motion (Eq. 14.51), we first calculate that factor:

$$\gamma_{Tv} = \frac{1}{\sqrt{1 - \left(\frac{v_{Tx,i}}{c_0}\right)^2}} = \frac{1}{\sqrt{1 - (0.647)^2}} = 1.311.$$

The kinetic energy of p_1 in the target reference frame is thus

$$K_{T,i} = (\gamma_{Tv} - 1)m_p c_0^2 = 0.311 m_p c_0^2$$

$$\frac{K_{T,i}}{m_p c_0^2} = 0.311,$$

in agreement with our result above.

Guided Problem 14.8 Particle production

A pion π^- of mass m_{π^-} collides with a proton p of mass m_p that is at rest relative to an inertial reference frame R. The pion has the minimum kinetic energy necessary such that the collision destroys the proton and pion and produces two particles: a neutral kaon K of mass m_K and a lambda particle Λ of mass m_Λ. Derive an expression that gives, in reference frame R, the kinetic energy of the pion in terms of c_0 and the masses of the four particles.

❶ **GETTING STARTED**

1. If possible, choose an isolated and closed system.
2. Decide how many reference frames you must analyze, and sketch the initial and final states in each reference frame.
3. In the zero-momentum reference frame, at the minimum kinetic energy necessary to produce the kaon and lambda particle, are these two particles at rest or moving? Is the reference frame R the zero-momentum reference frame?
4. Assign symbols for the velocity of the kaon and lambda particle in the reference frame R in which the proton is at rest. It is generally good to include symbols in your sketches for any

quantities that you may need for the energy equations and the momentum equations.

❷ **DEVISE PLAN**

5. Can you write invariant expressions for the energy and momentum for the pion and proton in reference frame R?
6. Is there a reference frame in which the magnitude of the momentum of the kaon equals that of the lambda particle? What is the invariant expression for the energy and momentum of the kaon-lambda particle combination?

❸ **EXECUTE PLAN**

7. Determine the kinetic energy of the pion in the reference frame R in which the proton is at rest.

❹ **EVALUATE RESULT**

8. In the reference frame in which the proton is at rest, do you expect the sum of the energy of the pion and the internal energy of the proton to be greater than, equal to, or less than the sum of the internal energies of the kaon and lambda particle?

Questions and Problems

For instructor-assigned homework, go to MasteringPhysics® (MP)

Dots indicate difficulty level of problems: • = *easy,* •• = *intermediate,*
••• = *hard;* CR = *context-rich problem.*

14.1 Time measurements

1. Which of the following are events? (*a*) The local television station's newscast runs from 11:00 to 11:30 p.m. (*b*) The ceremonial first pitch is thrown at a baseball game in Omaha, Nebraska. (*c*) The Perseid meteor shower is seen across the eastern coast of North America. (*d*) A rainstorm lasts all day in Kyoto, Japan. (*e*) Two cars crash in front of your home. (*f*) An all-night concert starts in a local park. (*g*) Light emitted by a supernova reaches a spherical shell of radius 40,000 light years relative to its source. (One light year is the distance light travels in one year.) •

2. You are in a rocket moving away from Earth at one-third the speed of light relative to Earth. A friend is on Earth, and an astronaut in another rocket is moving toward Earth at one-third the speed of light (in the Earth reference frame), on a path collinear with your path. If each of you records the duration of your journey in your three respective reference frames, which of you records the proper time interval? •

3. A set of atomic clocks is placed on a square grid as shown in Figure P14.3. In order to synchronize these clocks, you set the time on each to a certain reading before they are allowed to run. Clock *O* starts first, and at precisely 12:00 noon, it sends out a light pulse in all directions. Each of the other clocks starts running as soon as the pulse reaches it. What initial time should you use for clocks A, B, C, and D to synchronize them with clock *O*? Express your answer to the nearest microsecond before or after 12:00 noon. ••

Figure P14.3

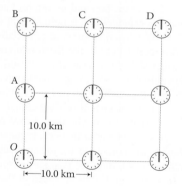

4. Events 1 and 2 are exploding firecrackers that each emit light pulses. In the reference frame of the detector, event 1 leaves a char mark at a distance 3.40 m from the detector, and event 2 leaves a similar mark at a distance 2.10 m from the detector. If the two events are simultaneous in the reference frame of the detector and occur at instant *t* = 0, at what instant of time will each light pulse be detected? ••

5. Consider the grid of clocks in *Principles* Figure 14.6. Do observers stationed at clocks far from the reference clock read earlier, later, or the same time as the time on the reference clock (*a*) if the clocks are to be synchronized after a pulse is sent out activating them but this pulse has not yet been sent, and (*b*) if the pulse has been sent out and all clocks are running? (*c*) Suppose the pulse has been sent and all clocks are running.

If you stand at the reference clock and use a telescope to read the time on distant clocks, do you read a time earlier, later, or the same as the time on the reference clock? ••

6. Observers A and B are both at rest in the Earth reference frame, in different parts of a large city, awaiting the start of a fireworks show. They want to synchronize their atomic clocks for some experiments to be conducted later that evening. A and B plan to start their clock displays at an agreed signal, so each clock display is preset to exactly 10:00 p.m., but they will not record time until the "start" buttons are pressed. Observer A suggests that each of them start upon *seeing* the first firework, and observer B suggests that each of them start upon *hearing* the first firework. (*a*) Which plan leads to the clocks being more closely synchronized, and why? (*b*) Is it possible for the two methods to give the same agreement between clocks? •••

14.2 *Simultaneous* is a relative term

7. Observer A is stationary in the Earth reference frame, and pilot B is flying toward pilot A at relative speed v_{AB}. What speed does observer A measure for the speed of the light emitted by a signal light on pilot B's ship? •

8. Alarm clock 1 is positioned at $x = -d$ and alarm clock 2 at $x = +d$ in reference frame A. According to a clock in reference frame A, both go off at $t = 0$. An observer in reference frame B, which is moving relative to A, measures alarm clock 2 as going off first. In what direction is B moving relative to A? •

9. You are traveling toward a large fixed mirror at a constant relative speed $0.20c_0$. At $t = 0$, when the mirror is a distance d from you (measured in your reference frame), you emit a light pulse from a lantern and then detect the reflected light pulse 0.80 μs later. What is the value of d? ••

10. You are in a jet helicopter traveling horizontally at 180 m/s relative to Earth. At one instant you observe the light pulses from two lightning strikes, one directly ahead of the helicopter and one directly behind it. (*a*) Can you say that the strikes occurred simultaneously in your reference frame? (*b*) How would your answer change if you were instead hovering at rest relative to the ground when you observed the two pulses? (*c*) How would your answer change if the pilot told you that the two strikes occurred at equal distances from the helicopter in the helicopter's reference frame? (*d*) Would an observer on the ground agree with your answers to parts *a*, *b*, and *c*? ••

11. Consider a lightweight, straight, rigid rod that has a proper length of 1×10^6 m. If you rotate the rod at 100 turns/s about a perpendicular axis through its center, at what speed do the ends of the rod move? What prevents such a case from occurring? ••

12. Observer A on Earth sees spaceship B moving at speed $v = 0.600c_0$ away from Earth and spaceship C moving at the same speed in the opposite direction. Observer A determines that both ships simultaneously sent out a radio signal at 12:00:00 noon and another signal at 12:05:00 p.m., making the time interval A measures between the emission of the signals from each ship $\Delta t_A = 5.00$ min. Let Δt_{BB} denote the time interval measured on ship B between the emission of signals on ship B, Δt_{BC} denote the time interval measured on ship B between the emission of signals on ship C, and Δt_{CB} denote the time interval measured on ship C between emission of signals on ship B. (a) Is Δt_{BB} greater than, less than, or equal to Δt_A? (b) Is Δt_{BC} greater than, less than, or equal to Δt_{CB}? ••

14.3 Space-time

13. An astronaut takes what he measures to be a 10-min nap in a space station orbiting Earth at 8000 m/s. A signal is sent from the station to Earth at the instant he falls asleep, and a second signal is sent to Earth the instant he wakes. Does an observer on Earth measure the astronaut's nap as being shorter than, longer than, or equal to 10 min? •

14. Suppose you could move at nearly the speed of light. If you were to move at this speed for a large portion of your life, would it be possible for you to live long enough to see a later calendar year on Earth than you otherwise would have? Compare your answer with that for Checkpoint 14.8. •

15. Standing somewhere between two vertical mirrors, you hold a lantern and at $t = 0$ emit a light pulse that travels in all directions. You observe the pulse reflected from the mirror on your right at $t = 2.5 \ \mu s$ and from the mirror on your left at $t = 6.5 \ \mu s$. What is the distance between the two mirrors in your reference frame? •

16. A spaceship of proper length $\ell = 100$ m travels in the positive x direction at a speed of $0.800c_0$ relative to Earth. An identical spaceship travels in the negative x direction along a parallel course at the same speed relative to Earth. At $t = 0$, an observer on Earth measures a distance $d = 50,000$ km separating the two ships. At what instant does this observer see the leading edges of the two ships pass each other? ••

17. A spacecraft traveling away from Earth at relative speed $0.850c_0$ sends a radio message when it is 65,000,000 km away from Earth in the Earth reference frame. Earth-based engineers immediately send a reply to the spacecraft. What do these engineers measure for the time interval from the instant the spacecraft sends the original message to the instant the spacecraft receives the reply? ••

18. Spaceships A, B, and C in Figure P14.18 all have the same proper length and all fly with the same speed according to an observer on Earth. Rank the lengths of the three ships as measured by an observer in ship A, longest first. ••

Figure P14.18

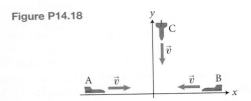

19. The boundary of a lunar base is a square 1 km wide and 1 km long in the Moon reference frame. Spaceship A flies at high relative speed close to the Moon parallel to the edge that we shall call the length of the field. Crew A measures the length and width, and reports that the area of the field is 0.5 km² in reference frame A. Spaceship B flies with the same relative speed as spaceship A, but parallel to one diagonal of the field. Crew B measures both diagonals. What surface area does crew B calculate in reference frame B? •••

14.4 Matter and energy

20. How fast must a pin move relative to your reference frame so that you measure its inertia to be three times its mass? •

21. A golf ball hit from a tee accelerates from rest to a speed of 40.0 m/s relative to the Earth reference frame. By what percent does the mass of the ball increase as it is hit? •

22. One end of a vertical spring of spring constant $k = 1500$ N/m is attached to the floor. You compress the spring so that it is 2.40 m shorter than its relaxed length, place a 1.00-kg ball on top of the free end, and then release the system at $t = 0$. (All values are measured in the Earth reference frame.) (a) By how much does the mass of the spring change during the time interval from $t = 0$ to the instant the ball leaves the spring? (b) By how much does the mass of the Earth-ball-spring system change during the time interval from $t = 0$ to the instant the ball reaches its maximum height? ••

23. You pick up your physics book from the floor and put it on your desk. In the Earth reference frame, which of the following quantities have changed for the system made up of Earth and the book: mass, energy, inertia, kinetic energy? ••

24. Observer A is on Earth looking at the Moon, and observer B is on the Moon looking at Earth. Which of the following quantities do the observers agree on for the Earth-Moon system: inertia, energy, mass, kinetic energy? ••

25. How far above Earth's surface must a 10,000-kg boulder be moved to increase the mass of the Earth-boulder system by 2.50 mg? Assume the same ratio of energy change to mass change as in *Principles* Exercise 14.5(b): 8.98×10^{16} J/kg. ••

14.5 Time dilation

26. A bullet is fired from a rifle (event 1) and then strikes a soda bottle, shattering it (event 2). Is there some inertial reference frame in which event 2 precedes event 1? If so, does the existence of this reference frame violate causality? •

27. Classify the space-time interval between each pair of events as timelike, spacelike, or lightlike: (a) A sunspot erupts on the surface of the Sun and is seen on Earth 500 s later. (b) A supernova explodes and is noticed 8.00×10^{20} m away after 8.45×10^4 y by an observer at rest relative to the supernova. (c) You receive light in the Earth reference frame emitted by a satellite 60,000 km away at $t = 0$, and 233 ms later receive light emitted by a spaceship 180,000 km away in the same direction as the satellite. (d) At $t = 0$, a pilot traveling at $0.75c_0$ relative to a star system receives signals from two planets A and B in that system. She then passes planet A (event 1) at $t = 450$ s and 90 s after that passes planet B (event 2). •

28. Calculate the Lorentz factor for an object moving relative to Earth at (a) 60 mi/h, (b) 0.34 km/s (speed of sound in air), (c) 28×10^3 km/h (orbital speed of the International Space Station), (d) $0.1c_0$, (e) $0.3c_0$, and (f) $0.995c_0$. Express your answers as "1 + correction." ••

29. You measure the period of light clock A to be T as it moves relative to you with a speed $0 < v_A < 0.5c_0$. You measure the period of an identical light clock (clock B) to be $3T$ as it moves relative to you with speed $v_B = 2v_A$. Determine numerical values for the speeds of each clock in your reference frame. ••

30. A space station sounds an alert signal at time intervals of 1.00 h. Spaceships A and B pass the station, both moving at $0.400c_0$ relative to the station but in opposite directions. How long is the time interval between signals (a) according to an observer on A and (b) according to an observer on B? (c) At what speed must A move relative to the station in order to measure a time interval of 2.00 h between signals? ••

31. A certain muon detector counts 600 muons per hour at an altitude of 1900 m and 380 muons per hour at sea level. Given that the muon half-life at rest is 1.5×10^{-6} s, determine the speed of the muons relative to Earth, assuming they all have the same speed. ••

32. Planets A and B are 10 light years apart in the reference frame of planet A. (One light year is the distance light travels in one year.) A deep-space probe is launched from A, and 5 y later (in reference frame A) a similar probe is launched from B. Does a reference frame exist in which these two events (*a*) are simultaneous and (*b*) occur at the same position? (*c*) Compute, if possible, the proper time interval between the events? (*d*) How do your answers to parts *a–c* change if the planets are (in reference frame A) 5 light years apart and the time interval between launch from A and launch from B is 10 y? ••

33. *Principles* Section 14.3 describes how the number of muons reaching Earth's surface is greater than the number expected based on the muon half-life of 1.5×10^{-6} s. How fast, relative to Earth, must muons be moving in order for one of every million muons to pass through a distance equal to Earth's diameter without decaying? ••

34. One cosmonaut orbited Earth for 437 days, as measured by Earth clocks. His speed of orbit was 7700 m/s relative to Earth during this time interval. Assume two clocks were synchronized on Earth, and one went into space with the cosmonaut while the other remained on Earth. You are interested in knowing by how much the clock readings disagree at the end of the 437 days. What are some problems with applying Eq. 14.13 to this situation? What information other than the cosmonaut's speed relative to Earth might you need to know? If you used Eq. 14.13, would you expect the clock readings to be very different from each other? •••

14.6 Length contraction

35. A cosmic ray traveling at $0.400c_0$ relative to Earth passes observer A (event 1) and then a short time interval later passes observer B (event 2). The observers are at rest, 8.00 km apart in the Earth reference frame. What are (*a*) the proper length and (*b*) the proper time interval between these events? •

36. An elementary particle is launched from Earth toward the Regulus system, 77.5 light years distant. At what speed relative to Earth must this particle travel to make this trip in 10 y in the reference frame of the particle? •

37. Spaceships A and B pass a space station at the same instant, the two ships moving in opposite directions at the same speed of $0.600c_0$ relative to the station. What is the Lorentz factor associated with the relative motion of the ships? •

38. Two elementary particles pass each other moving in opposite directions, each moving at speed $0.800c_0$ relative to Earth. What is the speed of one particle relative to the other? •

39. The radius of Earth is 6370 km in the Earth reference frame. In the reference frame of a cosmic ray moving at $0.800c_0$ relative to Earth, how wide does Earth seem (*a*) along the flight direction and (*b*) perpendicular to the flight direction? •

40. Show that the Lorentz transformation equations (Eqs. 14.29–14.32) reduce to the Galilean transformation equations in the limit $v_{AB} \ll c_0$. ••

41. An elementary particle moving away from Earth reaches its destination after two weeks (in the reference frame of the particle) traveling at $0.9990c_0$ relative to Earth. How far has it traveled in (*a*) its reference frame and (*b*) the Earth reference frame? ••

42. Describe the shape of the Moon as measured by an observer in a reference frame traveling past the Moon at a relative speed of (*a*) 1000 m/s, (*b*) $0.50c_0$, and (*c*) $0.95c_0$. ••

43. When at rest in the Earth reference frame, the delta-wing spaceship in Figure P14.43 is 7.00 m long and has a wingspan of 8.00 m. (*a*) What is the opening angle α of its wings? (*b*) If the same ship is moving forward at $0.700c_0$ relative to Earth, what does an observer on Earth measure for the plane length, wingspan, and opening angle α? ••

Figure P14.43

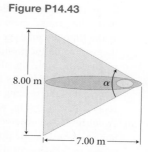

44. Standing at rest in the Earth reference frame, you observe two events, also at rest in that reference frame, occurring at these coordinates: event 1: $x_1 = 10$ km, $y_1 = 5.0$ km, $z_1 = 0$, $t_1 = 20$ μs; event 2: $x_2 = 30$ km, $y_2 = 5.0$ km, $z_2 = 0$, $t_2 = 30$ μs. At what speed and in what direction must your friend travel relative to Earth so as to observe these events as being simultaneous? ••

45. Use the Lorentz transformation equations (Eq. 14.29–14.32) to prove that the space-time interval s^2 between two events is *Lorentz-invariant*, which means it has the same value for all possible inertial reference frames. ••

46. An elementary particle travels at $0.840c_0$ across a solar system that has a diameter of 8.14×10^{12} m (both measurements in the reference frame S of the solar system). (*a*) What value does an observer in reference frame S measure for the time interval needed for the particle to cross the solar system? (*b*) What time interval is needed in the reference frame P of the particle? (*c*) What is the proper length of the diameter? (*d*) What is the proper time interval for the particle's trip across the system? (*e*) What is the diameter of the solar system in reference frame P? ••

47. A space colony defines the origin of a coordinate system in reference frame C, and you are at rest in this reference frame at position $x = +3.000 \times 10^5$ km. A ship passes the colony moving in the positive x direction at $0.482c_0$ in reference frame C. By prearrangement, observers on the ship (reference frame S) and on the colony use this event (event 0) to define $x = 0$ and $t = 0$ in both reference frames. Just as the ship passes, the colonists send out a light pulse toward you (event 1), and you transmit a reply pulse (event 2) immediately upon receipt. (*a*) At what position in reference frame S do observers see the colonists' pulse originate? (*b*) At what instant does the pulse originate according to these observers? (*c*) At what position in reference frame S do the observers see your reply pulse originate? (*d*) At what instant does the reply pulse originate according to observers on the ship? ••

48. Observer O at the origin of a coordinate system is at rest relative to two equidistant space stations located at $x = +3.00 \times 10^6$ km (station A) and $x = -3.00 \times 10^6$ km (station B) on the x axis. In reference frame O, station A sends out a light pulse at $t = 0$ (event 1) and station B also sends out a light pulse at $t = 0$ (event 2). Observer C moves relative to O with velocity $0.600c_0$ in the positive x direction, and observer D moves relative to O with velocity $0.600c_0$ in the negative x direction. What are the displacement from event 1 to event 2, and the time interval between events 1 and 2, according to (*a*) observer C and (*b*) observer D? ••

49. A spaceship travels at $0.610c_0$ away from Earth. When the ship is 3.47×10^{11} m away in the Earth reference frame, the ship clock is set to $t = 0$ and a message (light signal) is sent to its ground crew on Earth. When this message is received, the ground crew immediately sends a reply (light signal) to the ship. The ground crew adjust their clocks to specify $t = 0$ as the time the message was sent from the ship. (a) What is the reading on the ground crew's clock at the instant the ship receives the reply signal? (b) According to the ground crew, how far from Earth is the ship when it receives the reply? (c) What is the reading on the ship's clock at the instant the reply reaches the ship? •••

50. According to an observer at rest in the Earth reference frame, ship A travels in one direction at speed $0.862c_0$ while ship B moves in the opposite direction at $0.717c_0$. Ship A sends out a shuttle module M that travels to ship B at speed $0.525c_0$ relative to A. For every 1.00 s that passes on M, how many seconds pass according to (a) a clock on A, (b) a clock on B, and (c) a clock held by the observer on Earth? •••

51. Two asteroid clusters at rest with respect to each other are 3.00×10^9 m apart, and both measure 6.00×10^8 m across in the reference frame of the clusters. A pilot in a spaceship moving at $0.900c_0$ relative to the cluster reference frame is trying to pass between the two clusters. To an observer who is stationary in the reference frame of the clusters, the ship just barely misses the corners of the clusters as shown in Figure P14.51. If a line were drawn between the centers of the two clusters, what would be the angle between that line and the ship's path in (a) the reference frame of the clusters and (b) the reference frame of the ship? •••

Figure P14.51

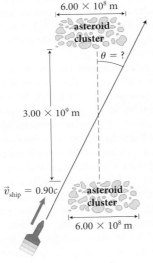

14.7 Conservation of momentum

52. Determine the magnitude of the momentum of a muon in a reference frame in which the muon moves with speed $0.500c_0$. The mass of a muon is 207 times the mass of an electron. •

53. An electron accelerates from $0.700c_0$ to $0.900c_0$ in the Earth reference frame. In this reference frame, by what factor do (a) its mass and (b) its inertia increase? •

54. (a) What is the magnitude of the momentum of an electron in a reference frame in which it is moving at $0.500c_0$? (b) At what speed must the electron move in a reference frame in which it has twice the momentum you calculated in part a? ••

55. Object A, mass 4.24×10^5 kg, is at rest in reference frame A, and object B, mass 7.71×10^4 kg, moves at $0.875c_0$ in this reference frame. (a) What does an observer at rest in A measure for the momentum of B? (b) What does an observer at rest with respect to object B measure for the momentum of A? (c) A reference frame C moves at $0.300c_0$ relative to reference frame A and moves in the same direction as object B. What are the momenta of objects A and B as measured by an observer at rest in reference frame C? ••

56. A uniform 200-kg cube that has a volume of 8.00 m^3 (measured when the cube is at rest in the Earth reference frame) travels perpendicular to a pair of its faces at $0.672c_0$ relative to Earth. (a) What is the mass density of the cube according to an observer at rest on Earth? (b) How fast (in the Earth reference frame) must a 100-kg object travel in the opposite direction to stop this cube in a collision? ••

57. Muons are formed 10.0 km above Earth's surface. What magnitude of momentum (in the Earth reference frame) must they have immediately after formation in order for only one of every 10,000 of them to reach the ground? The mass of a muon is 207 times the mass of an electron. ••

58. Two chunks of rock, each having a mass of 1.00 kg, collide in space. Just before the collision, an observer at rest in the reference frame of a nearby star determines that rock A is moving toward the star at $0.500c_0$ and rock B is moving away from the star at $0.300c_0$. If the rocks stick together after the collision, what are the direction and magnitude of their velocity (in the reference frame of the star) immediately after the collision? ••

59. In the Earth reference frame, a 150-kg probe travels at $0.860c_0$ while a 250-kg probe travels at $0.355c_0$ in the opposite direction. What are the velocities of the probes in the zero-momentum reference frame? •••

14.8 Conservation of energy

60. At what speed must a particle move in your reference frame so that its kinetic energy is equal to its internal energy? •

61. Particles A, B, and C each have mass m, and their energies are $E_A = E$, $E_B = 2E$, and $E_C = 3E$. Rank the particles, largest first, in order of (a) Lorentz factor, (b) kinetic energy, (c) speed, and (d) magnitude of momentum.

62. What is the internal energy of an electron moving at $0.750c_0$ in the Earth reference frame? •

63. Uranium-238 decays to thorium and helium according to the reaction $^{238}\text{U} \rightarrow {}^{234}\text{Th} + {}^4\text{He}$. How much energy is released when 1.00 kg of uranium-238 decays? Compare this amount with the energy released when 1.00 kg of coal is burned (approximately 30 MJ). The atomic masses are $m_U = 395.292599 \times 10^{-27}$ kg, $m_{Th} = 388.638509 \times 10^{-27}$ kg, and $m_{He} = 6.646478 \times 10^{-27}$ kg. ••

64. Assume that 437 days is a reasonable limit for how long a human can endure constant-velocity space travel. Proxima Centauri, the star closest to our Sun, is 4.24 light years away from Earth. If you wanted to fly to Proxima Centauri within the 437-day limit in a rocket of mass 2.00×10^6 kg, how much energy would be required to accelerate the rocket to the necessary speed in the Earth reference frame? For this rough estimate, ignore the energy required to stop, and disregard the hundreds of days required at each end of the journey for acceleration. Compare your answer with 0.6×10^{21} J, the yearly worldwide energy consumption predicted for 2015. ••

65. A particle of mass m_{orig}, initially at rest in the Earth reference frame, decays into two particles 1 and 2 that have masses m_1 and m_2, respectively. Use conservation of energy, conservation of momentum, and Eq. 14.57, $E^2 - (c_0 p)^2 = (mc_0^2)^2$, to calculate the energy and momentum magnitude of particle 1 in the Earth reference frame in terms of m_{orig}, m_1, m_2, and c_0. ••

66. Which of the following forms of energy contribute to the mass of a gas molecule: (a) energy due to the molecule rotating about its center of mass, (b) energy due to compression of atoms in the molecule, (c) energy due to the molecule's translational motion relative to Earth? ••

67. At the Large Hadron Collider in Switzerland, two high-energy protons collide to create new particles. Prior to collision, each proton is accelerated to an energy of 7000 GeV in the Earth reference frame. (a) What is the speed of each proton? (b) What is the maximum mass possible for any particle created in the collision? The proton mass is 938 MeV$/c_0^2$; 1 GeV = 1000 MeV. ••

68. An electron e^- and positron e^+ moving at the same speed in the Earth reference frame collide head-on and produce a proton p and an antiproton $\bar{p}$. The electron and positron have the same mass. The proton and antiproton also have the same mass. The mass of the proton is 1836.15 times the mass of the electron. Calculate, in the Earth reference frame, the minimum value possible for the ratio of the electron's kinetic energy to its internal energy in order to have the reaction $e^- + e^+ \rightarrow p + \bar{p}$ take place. ••

69. Antihydrogen is the only antimatter element that has been produced in the laboratory, albeit just a few atoms at a time. Each antihydrogen atom consists of a positron in orbit around an antiproton and has the same atomic mass as hydrogen. If an antihydrogen atom collides with a hydrogen atom, they annihilate each other and create gamma radiation. (a) What minimum amount of energy is released in this process? (b) If this energy could be harnessed in a matter-antimatter automobile engine, how far could a car travel on 10 mg each of antihydrogen and hydrogen? At highway speeds, a typical automobile expends about 2.5×10^3 J per meter. •••

Additional Problems

70. Galaxy A moves away from galaxy B at $0.600c_0$ relative to B. A spaceship leaves a planet in galaxy A traveling at $0.500c_0$ relative to galaxy A. If the direction in which the ship travels is the same as the direction in which A is moving away from B, what do observers in B measure for the ship's speed? •

71. A tree that is 32.6 m tall leans 26.0° from the vertical in the Earth reference frame. How tall is the tree, and at what angle does it lean, in the reference frame of a cosmic ray muon moving parallel to the tree at a speed of $0.382c_0$ relative to Earth? •

72. Consider a searchlight on the ground that casts a spot on a cloud 1000 m overhead. If the searchlight is rotated rapidly— say, 30° in 1 μs—how fast does the spot move in the Earth reference frame? Is this a violation of special relativity? ••

73. Two events occur at different locations on Earth and are simultaneous in the reference frame of an observer standing on Earth halfway between the events. Is there any other inertial reference frame in which the two events are simultaneous? ••

74. Pilot A is seated in the middle of a ship that has a proper length of 250 m, and pilot B is seated in the middle of an identical ship. Pilot B passes A at a speed that A measures to be $0.580c_0$. How long does it take B's ship to pass pilot A (a) according to A and (b) according to B? How long does it take A's ship to pass pilot B (c) according to A and (d) according to B? ••

75. Relative to Earth, spaceship A travels at $0.732c_0$ away from Earth, and spaceship B travels at $0.914c_0$ toward Earth along the same straight line. (a) How fast does A move according to an observer aboard B? (b) At $t = 0$ the two ships are separated by 4.5×10^{10} m in the Earth reference frame. At what instant t do they pass each other according to an observer at rest on Earth? ••

76. A proton p_1 moving in the Earth reference frame collides with a proton p_2 that is at rest in that reference frame. The collision creates a particle that has mass m and an internal energy 40 times the internal energy of a proton: $mc_0^2 = 40m_p c_0^2$. The velocity of this new particle in the Earth reference frame is not given. What is the ratio of the energy of p_1 to its internal energy in the Earth reference frame? (Hint: First analyze the collision in the zero-momentum reference frame in which p_1 and p_2 are both in motion, moving in opposite directions at the same initial speed.) ••

77. You are consulting for an episode of *Stars Wars, Part XVI*. As this episode opens, starship Orion zooms past a space station at $t = 0$ ship's time, traveling at a constant $0.6000c_0$ relative to the station. When the Orion's clock reads $t = 1000$ s, her sensors detect a hostile battle cruiser 8.00×10^7 km dead ahead and closing in, traveling on a direct line toward the Orion and the space station at a speed of $0.8000c_0$ relative to the Orion. The Orion immediately sends a light signal to the station warning the crew to evacuate. Your boss says to assume that evacuating the station requires 45.0 min measured in the reference frame of the station, and wants to know if a rewrite is required. ••• CR

78. Your boss wants you to construct a spring that, when compressed, experiences a mass increase of 1/10,000 of 1%. He has no idea whether this is possible, but you decide to write a report listing desirable parameters (spring mass, spring constant, load mass), discussing problems you might encounter in attempting to make such a spring, and considering whether or not the conventional spring equations apply. ••• CR

79. The Lorentz Cup spaceship race is won by the pilot who can travel 1.00×10^9 m in the shortest time interval. The rules state that the start and finish lines must be 10^9 m apart in the ships' reference frame, but the time interval is measured by an observer at rest in the Earth reference frame. There are many racing speeds you might try, such as $0.420c_0$ and $0.980c_0$ relative to the Earth reference frame, but trying a few combinations randomly makes you wonder if there is a better way than trial and error to decide on a strategy for this race. ••• CR

PRACTICE

Answers to Review Questions

1. An event is something that happens at a specific location at a specific instant.
2. Clocks measure equal known time intervals, and they can be synchronized so that their readings agree.
3. A signal of some sort must travel from the event to the observer.
4. You must see the two events as occurring at the same location. If you see them taking place at different locations, you cannot measure the proper time interval.
5. The observer must note, at the instant the event occurs, the reading on the clock nearest the event location.
6. The detector must be placed equidistant from the two locations at which the events take place. If the signals from the two events reach it at the same instant, the detector activates; if the signals arrive at different instants, the detector stays unchanged.
7. The speed is always $c_0 = 3.00 \times 10^8$ m/s, independent of the relative velocity of the source and the observer.
8. If the relative velocity of the observers is sufficiently large, they do not both report the two events as being simultaneous.
9. The observer sees time passing more slowly (time *dilates*) on clock A.
10. Time dilation has been confirmed by measurements on atomic clocks flown in airplanes around the world and by observations of cosmic-ray muons.
11. An object's proper length is its length measured by an observer who is at rest relative to the object.
12. Length contraction is the phenomenon whereby an observer moving relative to an object measures the object's length in the direction of motion as being shorter than the object's proper length.
13. This expression solved for v says that quadrupling K doubles v, but experiments on objects initially moving at relativistic speeds measure an increase in speed much less than the predicted doubling.
14. Mass and internal energy are invariant; inertia and kinetic energy depend on the measurer's reference frame.
15. The inertia changes in the same direction as the internal energy: Increasing the internal energy increases the inertia, and decreasing the internal energy decreases the inertia.
16. The inertia decreases.
17. The Lorentz factor is $\gamma = 1/\sqrt{1 - (v^2/c_0^2)}$ (Eq. 14.6). Its effects become noticeable at speeds greater than $0.1c_0$.
18. The time interval the observer measures is equal to the product of the Lorentz factor and the proper time interval (Eq. 14.13.)
19. The space-time interval is spacelike if less than zero, timelike if greater than zero, and lightlike if zero. When the interval between the events is timelike, there is an inertial reference frame in which the events occur at the same location but at different instants and are hence separated by only a time interval. When the interval is spacelike, there is an inertial reference frame in which the events occur at different locations but at the same instant and hence are separated by only a distance in space. When the interval is lightlike, the two events are separated spatially by exactly the distance light travels in the time interval between the events.
20. To distinguish cause and effect, it is necessary to determine the sequence of events. For a pair of events that have a timelike space-time interval ($s^2 > 0$), we can establish a reference frame in which we can study both events at the same position, and so we can determine their sequence with certainty and establish their causal relationship. For a pair of events that have a spacelike space-time interval ($s^2 < 0$), we cannot uniquely determine the time order of events and therefore cannot determine any causal relationship. However, for two events to be causally related, they must be able to interact physically. When $s^2 < 0$, the events occur in such rapid

succession that a light signal cannot travel from one event to the other in the time interval between them. Because no information can travel faster than the speed of light, such events cannot interact and so cannot be causally related. Only events that have a timelike space-time interval can be causally related, and for such events the time sequence can be uniquely determined.

21. The length the observer measures is equal to the proper length divided by the Lorentz factor (Eq. 14.28).
22. Time dilation and length contraction are the same effect seen by different observers. Muons can travel the proper distance through the atmosphere to reach Earth's surface because according to an observer at the surface their lifetime is lengthened by time dilation and according to an observer traveling with the muons their proper lifetime is sufficient for them cover the contracted length of the atmosphere.
23. The Lorentz transformation equations allow observers in any two inertial reference frames to compare measurements, whereas the length contraction and time dilation equations allow comparisons only between specific reference frames in which a proper time interval or proper length cannot be measured.
24. The momentum for a particle traveling at a relativistic speed is still defined as the product of its inertia and its velocity.
25. Yes. The system momentum remains constant because even with components moving at relativistic speeds, the definition of momentum is still the product of each component's inertia and its velocity.
26. (a) The inertia m_v is equal to the product of the particle's mass and the particle's Lorentz factor: $m_v = \gamma m$ (Eq. 14.41). (b) The momentum is the product of the particle's inertia and its velocity: $\vec{p} = \gamma m \vec{v} = m_v \vec{v}$.
27. The nonrelativistic form of kinetic energy is $K = \frac{1}{2}mv^2$. The relativistic expression is $K = (\gamma - 1)mc_0^2 = m_v c_0^2 - mc_0^2$ (Eq. 14.51). The two expressions are compatible; the latter reduces to the former at speeds much smaller than the speed of light in vacuum.
28. The quantity mc_0^2 represents the internal energy of the system (Eq. 14.54).
29. The quantity γmc_0^2 represents the energy of the system (Eq. 14.53), which equals the sum of the kinetic and internal energies (Eq. 14.52).
30. A conserved quantity is one that cannot be created or destroyed, but observers in different reference frames can measure different values for a given conserved quantity. An invariant quantity is one that has the same value for all observers, regardless of any difference in the reference frames of the observers. Because they are conserved quantities, the energy and momentum of an isolated, closed system have constant values for a given observer but can have a different set of values for an observer measuring from a different reference frame. In order to obtain, for a given system, a parameter value that is the same for all observers in all reference frames, the invariant combination $E^2 - (c_0 p)^2 = (mc_0^2)^2$ must be used.

Answers to Guided Problems

Guided Problem 14.2 (a) $0.781c_0$; (b) 125 m

Guided Problem 14.4 (a) $\vec{v}_{DC} = -0.300c_0\,\hat{i}$;
(b) $t_{C1} = t_{C2} = 1.47 \times 10^{-6}$ s

Guided Problem 14.6 $s^2 = 1.79 \times 10^{-8}$ m^2

Guided Problem 14.8 $K_{\pi^-} = \dfrac{(m_K + m_\Lambda)^2 c_0^4 - m_\pi^2 c_0^4 - m_p^2 c_0^4}{2m_p c_0^2} - m_{\pi^-} c_0^2$

PRACTICE
Periodic Motion

Chapter Summary 268

Review Questions 271

Developing a Feel 272

Worked and Guided Problems 273

Questions and Problems 278

Answers to Review Questions 285

Answers to Guided Problems 285

PRACTICE

Chapter Summary

Fundamental characteristics of periodic motion (Sections 15.1, 15.5)

Concepts **Periodic motion** is any motion that repeats itself at regular time intervals. **Oscillation** (or **vibration**) is back-and-forth periodic motion.

The *period T* is the minimum time interval in which periodic motion repeats, and the **amplitude** *A* of the motion is the magnitude of the maximum displacement of the moving object from its equilibrium position.

Quantitative Tools The **frequency** *f* of periodic motion is the number of cycles per second and is defined as

$$f \equiv \frac{1}{T}. \tag{15.2}$$

The SI unit of frequency is the **hertz** (Hz):

$$1 \text{ Hz} \equiv 1 \text{ s}^{-1}. \tag{15.3}$$

Simple harmonic motion (Sections 15.2, 15.4, 15.5)

Concepts **Simple harmonic motion** is periodic motion in which the displacement of a system from its equilibrium position varies sinusoidally with time. A system moving in this way is called a *simple harmonic oscillator*.

A restoring force that is linearly proportional to displacement tends to return a simple harmonic oscillator to its equilibrium position. For small displacements, restoring forces are generally proportional to the displacement and therefore cause objects to execute simple harmonic motion about any stable equilibrium position.

A **phasor** is a rotating arrow whose component on a vertical axis traces out simple harmonic motion. The **reference circle** is the circle traced out by the tip of the phasor, and the length of the phasor is equal to the amplitude *A* of the simple harmonic motion.

Quantitative Tools The **angular frequency** ω of oscillation is equal to the rotational speed of a rotating phasor whose vertical component oscillates with a frequency *f*. Angular frequency is measured in s^{-1} and is related to the frequency (measured in Hz) by

$$\omega = 2\pi f. \tag{15.4}$$

For a simple harmonic oscillator of amplitude *A*, the displacement *x* as a function of time is

$$x(t) = A \sin(\omega t + \phi_i), \tag{15.6}$$

where the sine argument is the **phase** $\phi(t)$ of the periodic motion,

$$\phi(t) = \omega t + \phi_i, \tag{15.5}$$

and ϕ_i is the *initial phase* at $t = 0$.

The *x* components of the velocity and acceleration of a simple harmonic oscillator are

$$v_x \equiv \frac{dx}{dt} = \omega A \cos(\omega t + \phi_i) \tag{15.7}$$

$$a_x \equiv \frac{d^2x}{dt^2} = -\omega^2 A \sin(\omega t + \phi_i). \tag{15.8}$$

Any object that undergoes simple harmonic motion obeys the **simple harmonic oscillator equation**:

$$\frac{d^2x}{dt^2} = -\omega^2 x, \tag{15.10}$$

where *x* is the object's displacement from its equilibrium position.

The mechanical energy *E* of an object of mass *m* undergoing simple harmonic motion is

$$E = \frac{1}{2}m\omega^2 A^2. \tag{15.17}$$

Fourier series (Section 15.3)

Concepts **Fourier's theorem** says that any periodic function with period T can be written as a sum of sinusoidal simple harmonic functions of frequency $f_n = n/T$, where n is an integer. The $n = 1$ term is the **fundamental frequency** or **first harmonic**, and the other components are **higher harmonics**.

Oscillating springs (Section 15.6)

Quantitative Tools For an object attached to a light spring of spring constant k, the simple harmonic oscillator equation takes the form

$$\frac{d^2x}{dt^2} = -\frac{k}{m}x, \tag{15.21}$$

and the angular frequency of the oscillation is

$$\omega = +\sqrt{\frac{k}{m}}. \tag{15.22}$$

The motion of the object is described by

$$x(t) = A\sin\left(\sqrt{\frac{k}{m}}t + \phi_i\right). \tag{15.23}$$

Rotational oscillations (Section 15.7)

Concepts A horizontal disk suspended at its center by a thin fiber forms a type of *torsional oscillator*.

A pendulum is any object that swings about a pivot. A *simple pendulum* consists of a small object (the bob) attached to a very light wire or rod.

Quantitative Tools If a torsional oscillator of rotational inertia I is twisted through a small angle from its equilibrium position ϑ_0 to position ϑ, the restoring torque τ_ϑ is

$$\tau_\vartheta = -\kappa(\vartheta - \vartheta_0), \tag{15.25}$$

where κ is the *torsional constant*. When $\vartheta_0 = 0$, the simple harmonic oscillator equation for the torsional oscillator is

$$\frac{d^2\vartheta}{dt^2} = -\frac{\kappa}{I}\vartheta. \tag{15.27}$$

The rotational position ϑ of the torsional oscillator at instant t is given by

$$\vartheta = \vartheta_{max}\sin(\omega t + \phi_i), \tag{15.28}$$

where ϑ_{max} is the maximum rotational displacement and

$$\omega = \sqrt{\frac{\kappa}{I}}. \tag{15.29}$$

For small rotational displacements, the simple harmonic oscillator equation of a pendulum is

$$\frac{d^2\vartheta}{dt^2} = -\frac{m\ell_{cm}g}{I}\vartheta \tag{15.32}$$

and its angular frequency is

$$\omega = \sqrt{\frac{m\ell_{cm}g}{I}}, \tag{15.33}$$

where ℓ_{cm} is the distance from the center of mass of the pendulum to the pivot.

The period of a simple pendulum is $T = 2\pi\sqrt{\dfrac{\ell}{g}}$.

Damped oscillations (Section 15.8)

Concepts In **damped oscillation,** the amplitude decreases over time due to energy dissipation. The cause of the dissipation is a *damping force* due to friction, air drag, or water drag.

A damped oscillator that has a large *quality factor Q* keeps oscillating for many periods.

Quantitative Tools At low speeds, the damping force $\vec{F}_{ao}^d$ tends to be proportional to the velocity of the object:

$$\vec{F}_{ao}^d = -b\vec{v}, \tag{15.34}$$

where b, the *damping coefficient,* has units of kilograms per second.

For small damping, the position $x(t)$ of a damped spring is

$$x(t) = Ae^{-bt/2m} \sin(\omega_d t + \phi_i), \tag{15.37}$$

and its angular frequency ω_d is

$$\omega_d = \sqrt{\frac{k}{m} - \frac{b^2}{4m^2}} = \sqrt{\omega^2 - \left(\frac{b}{2m}\right)^2}. \tag{15.38}$$

The **time constant** τ for a damped system is $\tau = m/b$. In one time constant, the energy of a damped simple harmonic oscillator is reduced by a factor of $1/e$.

The amplitude $x_{max}(t)$ and energy $E(t)$ of a damped oscillation of initial amplitude A and initial energy E_0 decrease exponentially with time:

$$x_{max}(t) = Ae^{-t/2\tau} \tag{15.39}$$

$$E(t) = E_0 e^{-t/\tau}. \tag{15.40}$$

The *quality factor Q* of a damped system is

$$Q = 2\pi \frac{\tau}{T}. \tag{15.41}$$

Review Questions

Answers to these questions can be found at the end of this chapter.

15.1 Periodic motion and energy

1. What is required in order that the motion of an object be called periodic?
2. The terms *amplitude* and *frequency* are often applied to periodic motion. What is the meaning of each?
3. It requires 5.00 s for an oscillator to complete four cycles. What is the period T? What is the frequency f?
4. In periodic motion, where is the kinetic energy of the moving object greatest? Where is the magnitude of its potential energy greatest? Where is each of these variables zero?

15.2 Simple harmonic motion

5. A playground swing moves in simple harmonic motion. If you don't pump your legs on the swing, the amplitude of your motion decreases. What happens to the period?
6. Explain why the restoring force exerted on an object in simple harmonic motion must be a linear function of the object's displacement.
7. What is a turning point in simple harmonic motion?
8. How is simple harmonic motion related to circular motion at constant speed?

15.3 Fourier's theorem

9. State Fourier's theorem and explain why it is important in the study of periodic motion.
10. How can the fundamental frequency of a Fourier series be determined from the time dependence of the original periodic function?
11. What is meant by the *spectrum* of a periodic function?
12. What is Fourier analysis, and what is Fourier synthesis?

15.4 Restoring forces in simple harmonic motion

13. Which of the three types of equilibrium—stable, unstable, neutral—allow periodic motion?
14. Two pendulums have identical shape and length, but one of them has twice the mass of the other. Which has the greater period, if either?
15. What common feature of all restoring forces is crucial for ensuring simple harmonic motion for small displacements from equilibrium?
16. A child and an adult are on adjacent identical swings at the playground. Is the adult able to swing in synchrony with the child?

15.5 Energy of a simple harmonic oscillator

17. Is there anything actually going around in a circle as a spring oscillates? If not, what purpose does the reference circle serve?

18. Describe the simple harmonic oscillator equation in words.
19. If you know the initial position of an oscillator, what else do you need to know in order to determine the initial phase of the oscillation?
20. How do we distinguish frequency f from angular frequency ω if both have units of $(\text{time})^{-1}$?

15.6 Simple harmonic motion and springs

21. If you know the mass of an object hanging from a spring in an oscillating system, what else do you need to know to determine the period of the motion?
22. You stop your car to pick up a member of your car pool. After she gets in, does the angular frequency ω of oscillation due to the car's suspension increase, decrease, or stay the same?
23. Given an object suspended by a spring, which of these variables of the motion can you control by varying the initial conditions: period, amplitude, energy of the system, frequency, phase, maximum velocity, maximum acceleration?
24. How does the frequency of a vertically oscillating block-spring system compare with the frequency of an identical block-spring system oscillating horizontally?

15.7 Restoring torques

25. Two pendulums have identical shape and mass, but pendulum B is twice as long as pendulum A. Which has the greater period, if either?
26. The pendulum on a grandfather clock consists of a heavy disk fastened to a rod in such a way that the position of the disk along the rod is adjustable. If the clock runs slow, how should you adjust the pendulum?
27. What approximation is necessary in order for a pendulum to execute simple harmonic motion?
28. What are the units of the torsional constant in Eq. 15.25, $\tau_\vartheta = -\kappa(\vartheta - \vartheta_0)$?

15.8 Damped oscillations

29. Does an overdamped oscillator oscillate?
30. Describe the changes in the periodic motion of an oscillator as the damping coefficient is gradually increased from $b = 0$ to $b = 0.5m\omega_0$.
31. During a time interval of one time constant, by what factor does the mechanical energy of an oscillating system decrease?

Developing a Feel

Make an order-of-magnitude estimate of each of the following quantities. Letters in parentheses refer to hints below. Use them as needed to guide your thinking.

1. The energy involved in the oscillation of water sloshing in a bathtub (E, J, P, Z)
2. The energy involved in the oscillation of a child on a playground swing (O, A, F, U)
3. The maximum frequency at which you can continuously shake a bowling ball horizontally at chest height for 1 minute, fully extending your arms during each cycle (N, G, C, K, Q)
4. The length of a grandfather-clock pendulum for which the period of oscillation is 2 s (O, W)
5. The angular frequency ω of your leg oscillating as a pendulum (D, I, M)
6. The angular frequency ω of each piston in a four-cylinder gasoline engine as you drive your car, and the initial phase difference $\Delta\phi_i$ between each of the four oscillating pistons (R, X, L, T)
7. The maximum restoring torque caused by the forces exerted on a child on a playground swing (J, A, F, U)
8. The damping coefficient for a shock absorber on a midsize car (S, Y, B, H, V)

Hints

A. What is the maximum angle of oscillation?
B. What spring constant is needed for each wheel?
C. How many times during each cycle must you provide the full oscillation energy to move or stop the bowling ball?
D. What are the mass and length of your leg?
E. What maximum height above the equilibrium level does the sloshing water reach?
F. What is the mass of the child plus the swing seat? (You can ignore the mass of the chains supporting the swing.)
G. What energy is required for these pushups?
H. What is the frequency of the vertical oscillation of your car?
I. What is the rotational inertia of your leg pivoted at your hip?
J. What type of restoring force is involved?
K. What is the mass of a bowling ball?
L. Does each piston fire simultaneously?
M. What is the length from the pivot to your leg's center of mass?
N. How many pushups can you do in 1 minute?
O. Is it reasonable to approximate this as a simple pendulum?
P. How is the water's energy related to the maximum height the water reaches?
Q. What is the amplitude of oscillation?
R. What is a typical cruising-speed angular frequency ω, in revolutions per minute, for the crankshaft (the shaft to which the pistons are attached)?
S. What is the mass of a midsize car?

T. What is the initial phase between two pistons adjacent in the firing order?
U. What is the length of the pendulum?
V. How many oscillations occur before the shock absorber damps out the motion?
W. How is period related to length for a simple pendulum?
X. What is the angular frequency ω of each piston?
Y. Through what distance does a fender deflect when you sit on it?
Z. At maximum height, where is the center of mass of the volume of water above the equilibrium level?

Key (all values approximate)

A. $45°$; B. 2×10^4 N/m; C. four; D. 1×10^1 kg, 1 m; E. 0.2 m; F. 3×10^1 kg; G. $30mgh$, where $h \approx 0.3$ m, $m \approx 7 \times 10^1$ kg, gives 6×10^3 J; H. 1 Hz; I. 3 kg·m² because it is roughly a uniform rod; J. the gravitational force exerted by Earth; K. 7 kg; L. no, they fire one at a time; M. 0.4 m; N. if you are fit, 3×10^1; O. yes; P. at maximum height, there is no kinetic energy; Q. 0.3 m; R. 2×10^3 rev/min; S. 1×10^3 kg; T. π rad; U. 3 m; V. a well-adjusted system allows less than one full oscillation; W. $T = 2\pi\sqrt{\ell/g}$; X. the same as for the crankshaft, 2×10^2 s^{-1}; Y. 5×10^1 mm; Z. this volume has a triangular shape, so its center of mass is at about one-third the maximum height

Worked and Guided Problems

These examples involve material from this chapter but are not associated with any particular section.
Some examples are worked out in detail; others you should work out by following the guidelines provided.

Worked Problem 15.1 Swayin' in the breeze

A treetop sways back and forth with a period of 12 s, and you estimate the amplitude of the motion to be 1.2 m (Figure WG15.1). You begin timing at some instant after the treetop passes through the upright position, and after you have been watching this motion for 36 s, the treetop is 0.60 m to the left of the upright position. (*a*) Write the equation for the position of the treetop as a function of time. (*b*) What is the maximum speed of the treetop? (*c*) At what instants does the treetop have this speed? (*d*) What is the maximum magnitude of the treetop's acceleration? (*e*) What is the magnitude of its acceleration at the instant $t = 36$ s?

Figure WG15.1

❶ **GETTING STARTED** This treetop is executing periodic motion, and we assume the motion can be described with a sine function. We are given several bits of information from which to construct the position as a function of time. Two of these bits are the period $T = 12$ s and the amplitude $A = 1.2$ m. We also know that the treetop is 0.60 m to the left of vertical at $t = 36$ s. Parts *b*–*e* all require the general equation of motion, and so once we have that, we're set.

❷ **DEVISE PLAN** We begin with the general equation for position in simple harmonic motion, Eq. 15.6:

$$x(t) = A \sin(\omega t + \phi_i).$$

With the data we have, we can determine the amplitude A, angular frequency ω, and initial phase ϕ_i. Then we can obtain the maximum speed and the acceleration magnitude from $v_x = dx/dt$ and $a_x = dv_x/dt$.

❸ **EXECUTE PLAN** (*a*) The angular frequency is $\omega = 2\pi/T = 2\pi/(12$ s$) = (\pi/6)$ s^{-1}. (Generally you should leave π in expressions because it often either cancels or produces an easily evaluated trig function.) The initial phase ϕ_i can be found by using the information we have about the position at 36 s:

$$x(t = 36 \text{ s}) = -0.60 \text{ m} = (1.2 \text{ m}) \sin\left[(\tfrac{1}{6}\pi \text{ s}^{-1})(36 \text{ s}) + \phi_i\right]$$

$$-0.50 = \sin(6\pi + \phi_i) = \sin(\phi_i)$$

$$\phi_i = \sin^{-1}(-0.50) = -\tfrac{1}{6}\pi.$$

With this value for ϕ_i, $x(t) = A \sin(\omega t + \phi_i)$ for this particular motion becomes

$$x(t) = (1.2 \text{ m}) \sin\left[(\tfrac{1}{6}\pi \text{ s}^{-1})t - \tfrac{1}{6}\pi\right]. ✔$$

(*b*) The velocity is given by Eq. 15.7:

$$v_x(t) = \frac{dx}{dt} = \frac{d}{dt}A \sin(\omega t + \phi_i) = \omega A \cos(\omega t + \phi_i)$$

$$= (\tfrac{1}{6}\pi \text{ s}^{-1})(1.2 \text{ m}) \cos\left[(\tfrac{1}{6}\pi \text{ s}^{-1})t - \tfrac{1}{6}\pi\right]$$

$$= (0.20\pi \text{ m/s}) \cos\left[(\tfrac{1}{6}\pi \text{ s}^{-1})t - \tfrac{1}{6}\pi\right].$$

The maximum speed occurs when the cosine function is 1, which means that $v_{max} = 0.20\pi$ m/s $= 0.63$ m/s. ✔

(*c*) The tree has this maximum speed whenever it passes through the position $x = 0$, which is when the tree is vertical. This occurs when the cosine function is ± 1. Solving first for the positive values, we have

$$1 = \cos\left[(\tfrac{1}{6}\pi \text{ s}^{-1})t - \tfrac{1}{6}\pi\right]$$

$$\Rightarrow (\tfrac{1}{6}\pi \text{ s}^{-1})t_n - \tfrac{1}{6}\pi = 2n\pi, \qquad n = 0,1,2,\ldots$$

$$t_n - \tfrac{1}{6}\pi\left(\frac{6}{\pi}\text{ s}\right) = 2n\pi\left(\frac{6}{\pi}\text{ s}\right), \quad n = 0,1,2,\ldots$$

$$t_n = (1 + 12n)\text{ s}, \quad n = 0,1,2,\ldots$$

$$t = 1 \text{ s, } 13 \text{ s, } 25 \text{ s,} \ldots. ✔$$

We repeat the calculation for the negative values:

$$-1 = \cos\left[(\tfrac{1}{6}\pi \text{ s}^{-1})t - \tfrac{1}{6}\pi\right]$$

$$\Rightarrow (\tfrac{1}{6}\pi \text{ s}^{-1})t_n - \tfrac{1}{6}\pi = n\pi, \qquad n = 1,3,5,\ldots$$

$$t_n - \tfrac{1}{6}\pi\left(\frac{6}{\pi}\text{ s}\right) = n\pi\left(\frac{6}{\pi}\text{ s}\right), \quad n = 1,3,5,\ldots$$

$$t_n = (1 + 6n)\text{ s}, \quad n = 1,3,5,\ldots$$

$$t = 7 \text{ s, } 19 \text{ s, } 31 \text{ s,} \ldots. ✔$$

(d) The acceleration is given by Eq. 15.8:

$$a_x(t) = \frac{dv_x}{dt} = \frac{d}{dt}[\omega A \cos(\omega t + \phi_i)] = -\omega^2 A \sin(\omega t + \phi_i)$$

$$= -(0.33 \text{ m/s}^2) \sin\left[\left(\frac{\pi}{6} \text{ s}^{-1}\right)t - \frac{\pi}{6}\right].$$

The maximum acceleration occurs when the sine function is 1, which makes $a_{max} = 0.33 \text{ m/s}^2$. ✔

(e) At $t = 36$ s, the acceleration magnitude is $a_x(36 \text{ s}) = 0.16 \text{ m/s}^2$. ✔

④ EVALUATE RESULT Given that the amplitude is not very large relative to the tree height, the acceleration maximum indicates gentle swaying in the breeze rather than a hurricane. The value we obtained for the maximum speed confirms this. The answers are the right order of magnitude for a swaying tree.

Guided Problem 15.2 Archaic music medium

A long-playing phonograph record is 12.0 in. (305 mm) in diameter and rotates at $33\frac{1}{3}$ rev/min. Suppose a bug is at the edge of one such record, as in Figure WG15.2, at instant $t = 0$. Imagine that a distant light source casts a shadow of the bug along an x axis that runs tangent to the turntable at the rim position opposite the bug's initial position. (a) Write an equation that describes the motion of the shadow as a function of time. (b) What is the shadow's maximum speed along the x axis? (c) What is the maximum magnitude of its acceleration along the x axis?

Figure WG15.2

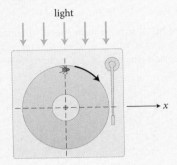

① GETTING STARTED

1. What does it mean to have the bug's motion projected onto the x axis? Make sketches showing the shadow's location at several instants as the turntable and bug rotate.

2. How can the properties of a right triangle and the sine and cosine functions help you in this problem?
3. Describe the relationship between circular motion and oscillation.

② DEVISE PLAN

4. What unknown quantities do you need to determine?
5. What equations allow you to express the unknown quantities in terms of known ones?
6. Write the generic equation for part a. Do you have to take derivatives of an expression?

③ EXECUTE PLAN

7. Manipulate your equations to obtain an expression for the position of the shadow along the x axis as a function of time.
8. How do you get the shadow's speed and the magnitude of its acceleration from the expression you found in step 7?
9. Which trigonometric function must be a maximum when the speed is a maximum, and what is that function's maximum value?
10. Which trigonometric function must be a maximum when the acceleration magnitude is a maximum, and what is that function's maximum value?

④ EVALUATE RESULT

11. Are the maximum speed and acceleration of the shadow of the right order of magnitude? Compare these values with the speed and maximum acceleration of the bug.

Worked Problem 15.3 Keeping lunar time

(a) What is the period of a simple pendulum of length 0.50 m on the Moon? (b) How does this period compare with the period of the same pendulum on Earth?

① GETTING STARTED We have to think about the factors that affect the period and about how these factors change when we move from Earth to the Moon. From videos of astronauts on the Moon, we are aware that things drop more slowly there, which implies that the free-fall acceleration is smaller there than on Earth.

② DEVISE PLAN From *Principles* Example 15.6, we know that the angular frequency of a simple pendulum is $\omega = \sqrt{g/\ell}$ and its period of oscillation is

$$T = \frac{2\pi}{\omega} = 2\pi\sqrt{\frac{\ell}{g}}.$$

The derivation of this relationship in the *Principles* volume does not require that the pendulum be oscillating on Earth, and so the result is general. Because Equation 13.4 in the generic form $g = Gm/R^2$ tells us the free-fall acceleration g at the surface of *any* celestial body of radius R, we can apply it to the Moon:

$$g_M = G\frac{m_M}{R_M^2}.$$

We get the mass and radius of the Moon from *Principles* Table 13.1.

③ EXECUTE PLAN (a) The magnitude of G is the same no matter where we are (that's why Newton's gravitational law is called the law of *universal* gravitation), so Eq. 13.4 gives

$$g_M = (6.674 \times 10^{-11} \text{ N} \cdot \text{m}^2/\text{kg}^2)\frac{7.3 \times 10^{22} \text{ kg}}{(1.74 \times 10^6 \text{ m})^2} = 1.6 \text{ m/s}^2.$$

The period of a 0.50-m lunar pendulum is therefore

$$T_M = 2\pi\sqrt{\frac{0.50\text{ m}}{1.6\text{ m/s}^2}} = 3.5\text{ s.} ✔$$

(*b*) We can use the general expression for period and our knowledge of the lunar and earthly free-fall accelerations to get the desired ratio of periods without having to calculate the pendulum's period on Earth:

$$\frac{T_M}{T_E} = \sqrt{\frac{g_E}{g_M}} = \sqrt{\frac{9.8\text{ m/s}^2}{1.6\text{ m/s}^2}} = 2.5. ✔$$

❹ **EVALUATE RESULT** The longer period on the Moon makes sense because, as we found in part *a*, the gravitational acceleration on the Moon, 1.6 m/s², is about one-sixth that on Earth.

Guided Problem 15.4 In sync

A block of mass *m* is hung from a vertical spring, stretching the spring a distance *h* beyond its relaxed length. The block is pulled down a bit and released, resulting in a vertical oscillation. The block is then removed from the spring and used as the bob on a simple pendulum. If the period of the oscillating pendulum is the same as the period of the oscillating block-spring system, what is the length of the pendulum?

❶ **GETTING STARTED**

1. How are the motions of the spring and pendulum analogous to each other?
2. Sketch both oscillating systems at an arbitrary instant, labeling key variables.
3. What assumptions do you need to make in order to solve this problem?

❷ **DEVISE PLAN**

4. What general expressions give you the periods of the two systems?
5. What unknown quantities must you determine before setting the two periods equal? What information can help you determine these quantities? For example, how can you calculate the spring constant *k*? Does calculating this constant help?

❸ **EXECUTE PLAN**

6. Work through the algebra and solve for the pendulum length.

❹ **EVALUATE RESULT**

7. Is the answer plausible, that is, does your expression behave as you expect it to with variations in *m* and *h*?

Worked Problem 15.5 A very tall grandfather clock

Suppose you could construct a pendulum consisting of a very long string (nearly infinitely long) and a bob that swings just above Earth's surface with an amplitude that is small compared to Earth's diameter. What would the period of its oscillation be?

❶ **GETTING STARTED** The problem statement describes a simple pendulum. From Worked Problem 15.3, we know that the period of a pendulum increases with the square root of its length: $T = 2\pi\sqrt{\ell/g}$. This relationship implies that the period of our very long pendulum should be really long, increasing without limit to infinity as the length increases. So, what is at issue? Why do we bother with this problem if the solution is that easy? To answer this question, we need to derive an expression for the period. Let us start with a sketch showing how this infinite pendulum might differ from a normal pendulum (Figure WG15.3).

Figure WG15.3

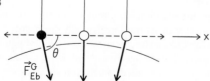

We note that an infinite string length means that the string is always vertical and the arc described by the bob is very nearly a straight line. We choose this line of motion as the *x* axis, with the origin at the equilibrium point and the positive direction to the right in the figure. Next we notice that the angle of the gravitational force exerted on the pendulum bob changes direction, always pointing toward Earth's center rather than straight down in the figure. From this changing direction, we conclude that although gravity provides a restoring force that pulls the bob toward the center of the motion, it is not the same as the constant gravitational force used in the analysis of a normal pendulum.

Because this is a hypothetical ("Suppose you could…") problem, we assume that practical considerations of pendulum construction and air drag should be ignored.

❷ **DEVISE PLAN** Because the pendulum's motion is nearly linear, we will use an analysis similar to that in *Principles* Example 15.8 to derive the angular frequency, beginning with Newton's second law, $\Sigma\vec{F} = m\vec{a}$, and casting it in a form that looks like a harmonic oscillator equation (Eq. 15.9):

$$a_x = \frac{d^2x}{dt^2} = -\omega^2 x. \tag{1}$$

We use a linear variable *x* to describe the position of the bob along its essentially straight-line horizontal path (straight and horizontal because we are taking our infinitely long string to be always vertical).

❸ **EXECUTE PLAN** From the free-body diagram (Figure WG15.4), we see that two forces are exerted on the bob: the tensile force exerted by the string and the force of gravity exerted by Earth. Because the pendulum is near Earth's surface at all times, we can use *g* for the acceleration magnitude.

Figure WG15.4

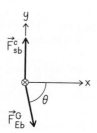

The horizontal component of the gravitational force, which points to the right in Figure WG15.4, is unbalanced and so produces an acceleration, causing the oscillations. The positive x direction is, for consistency with our earlier choice, to the right in Figure WG15.4. Thus the x component of force is

$$F_{\mathrm{Eb}x}^{G} = ma_x = mg\cos\theta. \qquad (2)$$

The gravitational-force vector in Figure WG15.4 is directed from the pendulum bob to the center of Earth. We can construct a right triangle (Figure WG15.5) that contains this same angle θ using the distance from the bob to Earth's center, the displacement x, and the radius of Earth. Note that the height above Earth's surface of the pendulum bob at the midpoint of its motion is negligible compared to the radius of Earth. Using this triangle, we can rewrite $\cos\theta$ in terms of the bob's displacement x and Earth's radius:

$$\cos\theta = \frac{-x}{\sqrt{R_{\mathrm{E}}^2 + x^2}} \approx \frac{-x}{R_{\mathrm{E}}}.$$

Using Eq. 2 in the form $a_x = g\cos\theta$ and then writing the acceleration a_x as the second derivative of position with respect to time give

Figure WG15.5

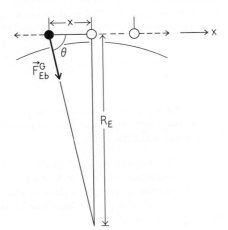

$$\frac{d^2x}{dt^2} = g\cos\theta = -\frac{g}{R_{\mathrm{E}}}x, \qquad (3)$$

our desired equation of motion for the oscillator.

We obtain the period of oscillation by comparing Eqs. 1 and 3:

$$a_x = -\omega^2 x$$

$$\omega^2 = -\frac{a_x}{x} = -\left(\frac{-gx}{R_{\mathrm{E}}}\right)\left(\frac{1}{x}\right) = \frac{g}{R_{\mathrm{E}}}.$$

Equation 15.1, $\omega = 2\pi/T$, gives us

$$\omega^2 = \frac{g}{R_{\mathrm{E}}} = \left(\frac{2\pi}{T}\right)^2, \quad \text{and so} \quad T = 2\pi\sqrt{\frac{R_{\mathrm{E}}}{g}}.$$

Notice that the period does not depend on the mass of the bob. Putting in numbers, we get

$$T = 2\pi\sqrt{\frac{6.38\times 10^6~\mathrm{m}}{9.8~\mathrm{m/s^2}}} = 5070~\mathrm{s} = 84~\mathrm{min}. \checkmark$$

❹ EVALUATE RESULT This is a large oscillation period, but long pendulums have large periods. Even if we didn't work through the calculation, it would be surprising that the period is not infinite, as the result of *Principles* Example 15.6, $T = 2\pi(\ell/g)^{1/2}$, implies it should be. The derivation of that equation assumed that the gravitational force always points vertically downward, whereas in this case we used the fact that the gravitational force is directed radially toward the center of Earth.

We assumed that the magnitude of the gravitational acceleration and the string direction do not change over the path of the swing, which is consistent with the requirement that the amplitude of the motion be small compared to the diameter of Earth. These assumptions do not undermine the main element of the derivation: The direction of the gravitational force changes even though we have assumed its magnitude does not.

Guided Problem 15.6 One deep hole

In another hypothetical exercise, suppose you could dig a tunnel that begins where you are standing now, passes through Earth's center, and comes out on the other side of the globe. What would happen to a rock you drop into this tunnel? How long would it take the rock to return to you? (For simplicity, assume Earth does not rotate and ignore air drag.)

❶ GETTING STARTED

1. Describe the physical situation and posed task in your own words. Draw a sketch consistent with your description.
2. What do you expect the rock to do after you release it?

❷ DEVISE PLAN

3. Recall from *Principles* Section 13.8 that a solid sphere exerts a gravitational force as if all the mass of the sphere were concentrated at its center.
4. To determine the magnitude of the gravitational force exerted by Earth on the rock after it has been dropped into the tunnel and is a radial distance $r < R_E$ from Earth's center, imagine the portion of Earth that has radius r (at the instant you are analyzing, your rock is on the surface of this imaginary sphere).

5. Refer to Checkpoint 13.23 if necessary.
6. What is the rock's acceleration as it passes through Earth's center?
7. Write the equation of motion for the rock as it passes a specific position x in its motion through the tunnel. Can you make this look like a simple harmonic oscillator equation?
8. What is the average density of Earth? Add information to your sketch as needed to help you visualize your approach.

❸ EXECUTE PLAN

9. Write an equation for the position of the rock as a function of time.
10. What is the angular frequency ω of the oscillation? What is the period T?

❹ EVALUATE RESULT

11. Compare your period with that of the "infinite" pendulum of Worked Problem 15.5. What insight do you gain from this comparison?

PRACTICE

Worked Problem 15.7 A damping influence

You decide to test the effect of a viscous liquid on the motion of a simple pendulum consisting of a thin rod 9.0 m long, a 2.0-kg spherical bob, and pivot bearings at the end of the rod opposite the bob. You place the pendulum in a huge vat of maple syrup. The rod is so thin that it experiences very little drag, so essentially the entire drag force is exerted on the bob. You start the pendulum swinging with a small amplitude, and 10 s later the motion is indiscernible. (*a*) Derive an equation that shows how the angle the pendulum makes with the vertical depends on time. (*b*) What is the approximate period of oscillation of this pendulum? Ignore any friction in the bearings.

❶ **GETTING STARTED** This problem involves damped oscillations but for a pendulum rather than the block-spring system examined in *Principles* Section 15.8. A drag force exerted by the syrup opposes the motion of the pendulum. We do not know anything about the amplitude of the motion except that after 10 s the motion is undetectable. So we have to use our judgment here. Damped oscillations die out exponentially, which means that they never truly reach zero amplitude (the motion just becomes too small to notice). A reasonable guess is that the final amplitude is 1% of the initial value.

❷ **DEVISE PLAN** To make use of the solution worked out in *Principles* Section 15.8, we need to derive an equation in the form of Eq. 15.36,

$$m\frac{d^2x}{dt^2} + b\frac{dx}{dt} + kx = 0,$$

and make the appropriate mathematical correspondence with our rotational quantities (for instance, x becomes ϑ). As with the block-spring system of *Principles* Section 15.8, we assume that the drag force exerted by the syrup on the pendulum is proportional to the pendulum's velocity: $\vec{F}^d_{sp} = -b\vec{v}_p$ (Eq. 15.34). This force produces a torque on the pendulum in addition to the gravitational torque produced by Earth. Therefore we can use Eq. 15.24, $\Sigma\tau_\vartheta = I\alpha_\vartheta$, in the form

$$I\frac{d^2\vartheta}{dt^2} = \tau_{Ep\,\vartheta} + \tau_{sp\,\vartheta}. \tag{1}$$

We use Eq. 15.31, $\tau_\vartheta = -(m\ell_{cm}g)\vartheta$, to simplify the $\tau_{Ep\,\vartheta}$ term. The torque $\tau_{sp\,\vartheta}$ is proportional to the pendulum's translational speed dx/dt and therefore to its rotational speed $d\vartheta/dt$. This information should give us the derivatives we need in the equation of motion for the damped oscillator in part *a*.

In order to get the period of the motion, we need to know the number of oscillations in a given time interval, which means we need to know the angular frequency ω_d of the damped motion. Because damping is present, the angular frequency ω_d is given by Eq. 15.38. Once we have ω_d, we obtain the period of the motion from the relationship $T_d \approx 2\pi/\omega_d$.

❸ **EXECUTE PLAN**

(*a*) Let ℓ be the length of the rod and m the mass of the pendulum. The torque caused by the gravitational force is given by Eq. 15.31, as noted above. The torque caused by the drag force exerted by the syrup is

$$\tau_{sp\,\vartheta} = F^d_{sp\perp}\ell = -(bv_{p\perp})\ell = -[b(\ell\omega_\vartheta)]\ell$$

$$= -b\ell^2\frac{d\vartheta}{dt},$$

where we have used Eq. 15.34 to substitute for the factor $F^d_{sp\perp}$ and the relationship between speed and rotational speed for motion along a circular arc, $v_{p\perp} = \ell\omega_\vartheta$, to substitute for the factor $v_{p\perp}$. Inserting these values into Eq. 1 yields

$$I\frac{d^2\vartheta}{dt^2} = [-(m\ell g)\vartheta] + \left[-b\ell^2\frac{d\vartheta}{dt}\right]$$

$$I\frac{d^2\vartheta}{dt^2} + (b\ell^2)\frac{d\vartheta}{dt} + (m\ell g)\vartheta = 0.$$

This result has the same form as Eq. 15.36,

$$m\frac{d^2x}{dt^2} + b\frac{dx}{dt} + kx = 0,$$

with the substitutions I for m, ϑ for x, $b\ell^2$ for b, and $m\ell g$ for k. We infer from Eq. 15.37 that the equation for the rotational position as a function of time for the damped pendulum is

$$\vartheta(t) = Ae^{-b\ell^2 t/2I}\sin(\omega_d t + \phi_i).$$

(Recall from *Principles* Section 15.5 that ϕ_i is the initial phase of the motion.) The rotational inertia for a simple pendulum is $I = m\ell^2$. Substituting this rotational inertia into our rotational position equation yields

$$\vartheta(t) = Ae^{-b\ell^2 t/2(m\ell^2)}\sin(\omega_d t + \phi_i),$$

where
$$\omega_d = \sqrt{\frac{m\ell g}{m\ell^2} - \left(\frac{b\ell^2}{2m\ell^2}\right)^2}$$

$$\vartheta(t) = Ae^{-bt/2m}\sin(\omega_d t + \phi_i),$$

where
$$\omega_d = \sqrt{\frac{g}{\ell} - \left(\frac{b}{2m}\right)^2}. \checkmark \tag{2}$$

As with the block-spring case, the angular frequency ω is modified by a term, $(b/2m)^2$, that depends on the damping, telling us that the oscillations decrease exponentially. The ratio $b/2m$ determines the damping.

(*b*) We assumed that the oscillation amplitude decreases to 1% of its initial value in 10 s: $(0.010)A = Ae^{-b(10\,s)/2m}$. Solving for $b/2m$ yields

$$\ln 0.010 = -\frac{b}{2m}(10\text{ s})$$

$$\frac{b}{2m} = \frac{4.61}{10\text{ s}} = 0.46\text{ s}^{-1}.$$

The estimated angular frequency and period are therefore

$$\omega_d = \sqrt{\frac{9.8\text{ m/s}^2}{9.0\text{ m}} - (0.46\text{ s}^{-1})^2}$$

$$= \sqrt{(1.09 - 0.21)\text{ s}^{-2}} = 0.94\text{ s}^{-1}$$

$$T_d \approx \frac{2\pi}{0.94\text{ s}^{-1}} = 6.7\text{ s.} \checkmark$$

④ EVALUATE RESULT If we remove the pendulum from the syrup, we have an undamped simple pendulum, for which the angular frequency is $\omega = \sqrt{g/\ell}$, which is what our expression for ω_d reduces to for $b = 0$.

We can gain further confidence by seeing whether our algebraic expression for ω_d behaves the way we expect as we change the value of the damping coefficient b. If b increases, we expect ω_d to decrease because the pendulum must overcome a greater resistive force as it swings. We see that the term $b/2m$ is subtracted in the square-root term in Eq. 2, causing ω_d to decrease with increasing b.

Note that the pendulum's angular frequency without syrup would be $\omega = 1.04$ s^{-1}, so under our assumption the syrup has a 10% effect on the angular frequency and period. Our decision to make 1% of the original amplitude the state of undetectable motion

is of course arbitrary. What if we chose 0.10% instead? The value of $b/2m$ becomes 0.69 s^{-1}, which makes $\omega_d = 0.78$ s^{-1}, and now we are closer to a 25% effect of damping on the angular frequency. The size of the effect clearly depends on our assumption. However, under this revised assumption, the period would be about 8 s, so the sensitivity is not dramatic.

One last thing to consider is why we must say that the period is only approximately equal to $2\pi/\omega_d$ instead of using the equality $T = 2\pi/\omega$ (from Eq. 15.1, $\omega = 2\pi/T$). The reason we must use the approximation is that the motion is not strictly periodic because of the exponential time dependence of the amplitude. The motion does not simply repeat itself, which makes the definition of *period* a bit ambiguous. Thus $T_d \approx 2\pi/\omega_d$ is the best estimate we can make.

Guided Problem 15.8 Prevent oscillation

A typical engineering design element in systems susceptible to unwanted oscillation is *critical damping*. The idea is that a disturbance displaces some moving part, but rather than oscillating around its equilibrium position, the part returns as quickly as possible to its equilibrium position without overshooting it. This requires that the damped angular frequency be zero: $\omega_d = 0$. (*a*) For the pendulum in Worked Problem 15.7, what must be the value of the damping coefficient b if critical damping is desired? (*b*) When the system is critically damped, how long does it take for the amplitude to decrease to 1% of its initial value?

① GETTING STARTED

1. What is the physical significance of the damping coefficient b?
2. Should the value of $b/2m$ be greater or smaller than it is in Worked Problem 15.7?

② DEVISE PLAN

3. Is the condition for critical damping given in the problem statement, $\omega_d = 0$, useful in determining the desired value for part *a*?
4. What equation or expression can be used to solve part *b*? Can you save a bit of derivation by using results from Worked Problem 15.7?

③ EXECUTE PLAN

5. Solve algebraically for the desired quantities, doing part *a* first and then part *b*. Are there any other unknown quantities in your expression that must be eliminated?
6. Substitute known values and compute your answers.

④ EVALUATE RESULT

7. How does your algebraic expression for part *b* behave if the value of b is twice the critical value? If it is half the critical value?

Questions and Problems

For instructor-assigned homework, go to MasteringPhysics® (MP)

Dots indicate difficulty level of problems: • = *easy,* •• = *intermediate,* ••• = *hard;* CR = *context-rich problem.*

15.1 Periodic motion and energy

1. If you photograph a flock of birds taking off from ground level, some of the wings will be blurred in the photograph even though the bodies are in focus. In which wing positions are the wings least blurry? •
2. Plot kinetic energy and potential energy as a function of time for the cart in *Principles* Figure 15.2. •
3. Based on the energy diagrams in *Principles* Figure 15.2, sketch a graph showing velocity as a function of position for the cart represented in the figure. ••
4. (*a*) At what displacement (expressed as a fraction of the amplitude of the motion) is the kinetic energy of the cart in *Principles* Figure 15.2 half its maximum value? (*b*) What is the velocity (expressed as a fraction of the maximum velocity) at this displacement? ••

5. Figure P15.5 shows a graph of the potential energy U of a moving object as a function of its position x. What is the maximum range of x values possible for periodic motion in this system? ••

Figure P15.5

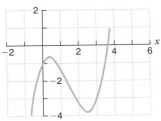

15.2 Simple harmonic motion

6. What distance does an oscillator of amplitude A travel in 2.5 periods? •

7. What is the period of the motion represented by the curve in Figure P15.7? •

Figure P15.7

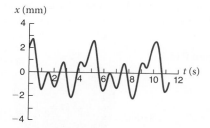

8. (*a*) Explain why the motion of the engine piston in Figure P15.8 is only approximately simple harmonic motion. (*b*) What must you do to the length of the connecting rod if you want to make the motion more nearly simple harmonic motion? ••

Figure P15.8

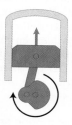

9. Is the typical up-and-down motion of a yo-yo periodic motion, assuming no energy is dissipated? Is it simple harmonic motion? ••

10. A highly elastic ball is dropped from a height of 2.0 m onto a hard surface. Assume the collision is elastic and no energy is lost to air friction. (*a*) Show that the ball's motion after it hits the surface is periodic. (*b*) Determine the period of the motion. (*c*) Is it simple harmonic motion? Why or why not? ••

15.3 Fourier's theorem

11. Fourier analysis of a particular spectrum includes the harmonic frequencies 889 Hz, 1143 Hz, and 1270 Hz. What is the fundamental frequency f_1, assuming that $f_1 > 100$ Hz? •

12. Even when a piano and a trumpet play the same note (which means the sounds have the same frequency f), the two instruments sound completely different. What is it about the sound from the instruments that accounts for this difference? ••

13. Estimate the number of harmonics (counting the fundamental frequency) that contribute significantly in the Fourier analysis of the spectrum in Figure P15.7. ••

14. Suppose you create a periodic function by adding together only sine functions whose amplitudes decrease like $1/n^2$, where n is any odd integer multiple of the fundamental frequency, and where every other term is subtracted rather than added. (*a*) What general properties would this periodic function exhibit? (*b*) Describe the shape of the function obtained by adding only the four terms of frequencies f, $3f$, $5f$, and $7f$. (Hint: Construct the graph using superposition.) •••

15. What harmonic series is required to produce a "square wave," as illustrated in Figure P15.15? •••

Figure P15.15

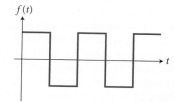

15.4 Restoring forces in simple harmonic motion

16. Which of these forces could result in simple harmonic motion: (*a*) $F(x) = 2x$, (*b*) $F(x) = -2x$, (*c*) $F(x) = -2x^2$, (*d*) $F(x) = 2x^2$, (*e*) $F(x) = -2x + 2$, (*f*) $F(x) = -2(x - 2)^2$? •

17. When you dive off a diving board, the board oscillates while you are flying toward the swimming pool. (These oscillations are significantly damped for a good diving board.) Identify the restoring force (or torque) responsible for these oscillations as well as the relevant mass. ••

18. A spring-cart system and a simple pendulum both have the same period near Earth's surface. What happens to the period of each motion when both systems are in orbit inside the International Space Station? ••

19. *Principles* Figure 15.12 illustrates the x component of the vector sum of forces on an object for conditions of stable, unstable and neutral equilibrium. (*a*) Examine this graph carefully. What visible feature differs in the three types of equilibrium? (*b*) Express your result as a set of mathematical relationships that can be used to discriminate stable from unstable or neutral equilibrium. (*c*) Are these relationships general? That is, can you construct a graph showing the x component of the vector sum of forces versus position that meets the requirements for stable equilibrium but violates your mathematical expression? ••

20. (*a*) When air resistance is ignored, does a pendulum clock run faster or slower at higher altitude? (*b*) Does your answer change if you don't ignore the effects of air resistance? •••

15.5 Energy of a simple harmonic oscillator

21. Write Eqs. 15.6, 15.7, and 15.8 in terms of the period T of the motion. •

22. Is the initial phase ϕ_i positive, zero, or negative for each oscillation graphed in Figure P15.22? •

Figure P15.22

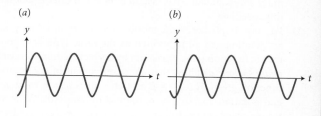

(*a*)

(*b*)

23. In *Principles* Figure 15.22, the vertical component of the phasor moves in simple harmonic motion. (*a*) Does the horizontal component also move in simple harmonic motion? (*b*) What is the phase difference $\Delta\phi = \phi_v - \phi_h$ between the vertical and horizontal components? ••

PRACTICE

24. Consider the nearly circular orbit of Earth around the Sun as seen by a distant observer standing in the plane of the orbit. What is the effective "spring constant" of this simple harmonic motion? ●●

25. Consider the reference circle shown in *Principles* Figure 15.22. (*a*) In terms of the rotational speed ω of the phasor, what is the magnitude of the tangential velocity $\vec{v}$ of the phasor tip? What is the vertical component of this velocity? How does this phasor velocity compare with the velocity of an object moving in simple harmonic motion? (*b*) In terms of the rotational speed ω of the phasor, what is the magnitude of the centripetal acceleration of the phasor tip? What is the vertical component of this acceleration? How does this phasor acceleration compare with the acceleration of an object moving in simple harmonic motion? ●●

26. The position of a particle undergoing simple harmonic motion is given by $x(t) = 20 \cos(8\pi t)$, where x is in millimeters and t is in seconds. For this motion, what are the (*a*) amplitude, (*b*) frequency f, and (*c*) period? (*d*) What are the first three instants at which the particle is at $x = 0$? (*e*) Determine the x components of position, velocity, and acceleration of the particle at $t = 0.75$ s. ●●

27. An object undergoes simple harmonic motion along an x axis with a period of 0.50 s and amplitude of 25 mm. Its position is $x = 14$ mm when $t = 0$. (*a*) Write an equation of motion with all variables identified. (*b*) Draw a position-versus-time graph for this motion. ●●

28. You are standing next to a table and looking down on a record player sitting on the table. Take the spindle (axis of rotation) to be the center of your coordinate system and the y axis to be perpendicular to the side of the player you are standing next to. Long-playing records revolve $33\frac{1}{3}$ times per minute. You put a small blob of clay at the edge of a record that has a radius of 0.15 m, positioning the clay such that it is at its greatest value of y at $t = 0$. (*a*) What is the rotational speed of the clay? (*b*) Write the equation of motion for the y component of the clay's position. ●●

29. A 1.0-kg object undergoes simple harmonic motion with an amplitude of 0.12 m and a maximum acceleration of 5.0 m/s². What is its energy? ●●

30. The position of a particle undergoing simple harmonic motion is given by $x(t) = a \cos(bt + \pi/3)$, where $a = 8.00$ m and $b = 2.00$ s^{-1}. What are the (*a*) amplitude, (*b*) frequency f, and (*c*) period? (*d*) What are the speed and acceleration magnitude of the particle at $t = \pi/2$ s? (*e*) What is its maximum acceleration magnitude, and at what instant after $t = 0$ does the particle first have this acceleration? (*f*) What is its maximum speed, and at what instant after $t = 0$ does it first have this speed? ●●

31. For any simple harmonic motion, the position, velocity, and acceleration can all be written as the same type of trigonometric function (sine or cosine) by correctly adding a *relative phase factor*. (*a*) Show that an x component of the position function given by $x(t) = \beta \cos(\omega t + \delta)$ is also a valid general solution for simple harmonic motion. (*b*) Write the x components of the velocity and acceleration associated with this position function. (*c*) Draw the reference circles for the x components of position, velocity, and acceleration, with the phasors correctly drawn for $t = 0$ and the amplitude clearly indicated. ●●●

32. You get a crazy idea to dig a tunnel through Earth from Boston to Paris, a surface distance of 5850 km, and run a passenger train between the two cities (Figure P15.32). The train would move down the first part of the tunnel under the pull of gravity and then coast upward against gravity. If you assume that the motion will be simple harmonic motion, how long will a one-way trip take? ●●●

Figure P15.32

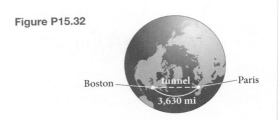

33. You have developed a method in which a paint shaker is used to measure the coefficient of static friction between various objects and a known surface. The shaker oscillates with a fixed amplitude of 50 mm, but you can adjust the frequency of the motion. You have affixed a horizontal tabletop (the known surface) to the shaker so that the tabletop oscillates with it. Then you put an object on the tabletop and increase the frequency until the object begins to slip on the surface. If a frequency $f = 1.85$ Hz is required before a penny positioned on the tabletop starts to slide, what is the coefficient of static friction between penny and tabletop? ●●●

15.6 Simple harmonic motion and springs

34. Even a spring without a block hanging on the end of it has an oscillation period. Why isn't the period zero? ●

35. A vertical spring on which is hung a block of mass m_1 oscillates with frequency f. With an additional block of mass $m_2 \neq m_1$ added to the spring, the frequency is $f/2$. What is the ratio m_1/m_2? ●

36. Two vertical springs have identical spring constants, but one has a ball of mass m hanging from it and the other has a ball of mass $2m$ hanging from it. If the energies of the two systems are the same, what is the ratio of the oscillation amplitudes? ●

37. A horizontal spring-block system made up of one block and one spring has oscillation frequency f. A second spring, identical to the first, is to be added to the system. (*a*) Does f increase, decrease, or stay the same when the two springs are connected as shown in Figure P15.37a? (*b*) What happens to f when the springs are arranged as in Figure P15.37b? ●●

Figure P15.37

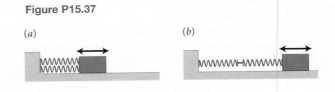

(*a*)   (*b*)

38. Two balls of unequal mass are hung from two springs that are not identical. The springs stretch the same distance as the two systems reach equilibrium. Then both springs are compressed and released. Which one oscillates faster? ●●

39. A table outfitted with springs on its feet bounces vertically in simple harmonic motion. A cup of coffee is sitting on the table. Discuss whether any of the following changes to this system could cause the cup to lose contact with the table: (*a*) increase the amplitude, (*b*) increase the period, (*c*) decrease the period, (*d*) decrease the mass of the table, (*e*) increase the mass of the table, (*f*) change the phase of the motion. (*g*) Would your answers be different if you considered not the cup bottom losing contact with the table but rather the coffee flying out of the cup? ••

40. A 2.0-kg cart is attached to a horizontal spring for which the spring constant is 50 N/m. The system is set in motion when the cart is 0.24 m from its equilibrium position, and the initial velocity is 2.0 m/s directed away from the equilibrium position. (*a*) What is the amplitude of the oscillation? (*b*) What is the speed of the cart at its equilibrium position? (*c*) How would the answers to parts *a* and *b* change if the initial direction of motion were toward the cart's equilibrium position? ••

41. A vertical spring-block system with a period of 2.3 s and a mass of 0.35 kg is released 40 mm below its equilibrium position with an initial upward velocity of 0.12 m/s. For this system, determine the (*a*) amplitude, (*b*) angular frequency ω, (*c*) energy, (*d*) spring constant, (*e*) initial phase angle, and (*f*) equation of motion. ••

42. A 5.0-kg object is suspended from the ceiling by a strong spring, which stretches 0.10 m when the object is attached. The object is lifted 0.050 m from this equilibrium position and released. Determine the amplitude and period of the resulting simple harmonic motion. ••

43. A 5.0-kg block is suspended from the ceiling by a strong spring and released to perform simple harmonic motion with a period of 0.50 s. The block is brought to rest, and the length of the spring with the block attached is measured. By how much is this length reduced when the block is removed? ••

44. A 4.0-kg object is suspended from the ceiling by a spring and undergoes simple harmonic motion with an amplitude of 0.50 m. At the highest position in the motion, the spring is at the length that would be its relaxed length if no object were attached. (*a*) Calculate the energy of the system. For the object at its lowest position, its equilibrium position, and its highest position, calculate (*b*) the elastic potential energy of the spring, (*c*) the kinetic energy of the object, and (*d*) the gravitational potential energy of the Earth-object system. ••

45. A 6.0-kg block free to slide on a horizontal surface is anchored to two facing walls by springs (Figure P15.45). Both springs are initially at their relaxed length. Then the block is displaced 20 mm to the right and released. (*a*) What is the effective spring constant of the system? (*b*) What is the equation for the position of the block as a function of time, assuming $+x$ to be to the right? Ignore friction. ••

Figure P15.45

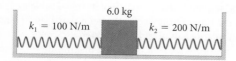

46. Block B in Figure P15.46 is free to slide on the horizontal surface. With block C placed on top of B, the system undergoes simple harmonic motion with an amplitude of 0.10 m. Block B has a speed of 0.24 m/s at a displacement of 0.060 m from its equilibrium position. (*a*) Determine the period of the motion. (*b*) What minimum value for the coefficient of static friction μ_s between B and C is needed if C is never to slip? Ignore any friction between B and the horizontal surface. •••

Figure P15.46

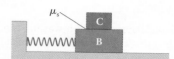

47. After gluing two 0.50-kg blocks together, you determine that you can pull the blocks apart with a force of 20 N. You attach the top block to a vertical spring for which the spring constant is 500 N/m. What is the maximum possible amplitude of oscillation for which the glue can keep the blocks together? •••

48. Two blocks, of masses m_1 and m_2, are placed on a horizontal surface and attached to opposite ends of a spring as in Figure P15.48. The blocks are then pushed toward each other to compress the spring. When the blocks are released, (*a*) describe the motion of the center of mass of the system and (*b*) derive an expression for the angular frequency ω of the motion. (*c*) What is ω in the limit $m_2 \gg m_1$? (*d*) What is ω when $m_1 = m_2$? Ignore friction and the mass of the spring. •••

Figure P15.48

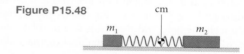

15.7 Restoring torques

49. You want to build a pendulum clock in which the time interval during which the "tick" sound is made (pendulum swinging one way) and the time interval during which the "tock" sound is made (pendulum swinging the other way) are each 0.50 s. If we assume the pendulum is a simple one, what should its length be? •

50. What happens to the period of the pendulum shown in Figure P15.50*a* when the stand supporting the pendulum is tipped backward as shown in Figure P15.50*b*? •

Figure P15.50

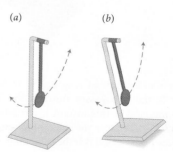

(*a*)   (*b*)

51. Which pendulum in Figure P15.51 has the greater period? ••

Figure P15.51

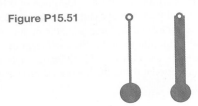

52. A small-angle approximation was used to derive Eq. 15.31, $\tau = -(m\ell g)\vartheta$. (*a*) What constitutes *small* in this context? In other words, how large can ϑ be before it can no longer be called small? (*b*) As a quantitative benchmark, how large does ϑ have to be before $(m\ell g)\vartheta$ deviates by more than 1% from the true value of τ? ••

53. The maximum angle from the vertical reached by a simple pendulum of length ℓ is ϑ_{max}. What is the linear speed v of the bob for any angle ϑ from the vertical? ••

54. The 800-mg balance wheel of a certain clock is made up of a thin metal ring of radius 15 mm connected by spokes of negligible mass to a fine suspension fiber as in *Principles* Figure 15.31. The back-and-forth twisting of the fiber causes the wheel to move in simple harmonic motion with period T. The clock ticks four times each second, with the interval between ticks equal to $T/2$. (*a*) Determine the rotational inertia I of the balance wheel. (*b*) What is the torsional constant of the fiber that drives the balance wheel? ••

55. A thin 0.100-kg rod that is 250 mm long has a small hole drilled through it 62.5 mm from one end. A metal wire is strung through the hole, and the rod is free to rotate about the wire. (*a*) Determine the rod's rotational inertia I about this axis. (*b*) Determine the period of the rod's oscillation. ••

56. A rod of mass m and length ℓ is rigidly connected at a right angle to the midpoint of one face of a uniform cube of mass m and side length ℓ (Figure P15.56). The other end of the rod is connected to a pivot so that the system can oscillate freely. What is the period of small oscillations about the pivot? ••

Figure P15.56

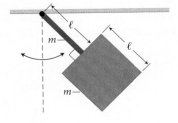

57. A simple pendulum of length 0.30 m has a 0.30-kg bob. At $t = 0$, the bob passes through the lowest position in its motion, and at this instant it has a horizontal speed of 0.25 m/s. (*a*) What is the maximum angular displacement ϑ_{max} away from vertical that the pendulum reaches? (*b*) What is the bob's linear speed v when it has angular displacement $\vartheta_{max}/2$? ••

58. A uniform disk of mass m and radius R lies in a vertical plane and is pivoted about a point a distance ℓ_{cm} from its center of mass (Figure P15.58). When given a small rotational displacement about the pivot, the disk undergoes simple harmonic motion. Determine the period of this motion. ••

Figure P15.58

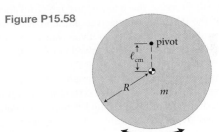

59. A thin 0.50-kg ring of radius $R = 0.10$ m hangs vertically from a horizontal knife-edge pivot about which the ring can oscillate freely. If the amplitude of the motion is kept small, what is the period? ••

60. If a person is 20% shorter than average, what is the ratio of his walking pace (that is, the frequency f of his motion) to the walking pace of a person of average height? Assume that a person's leg swings like a pendulum and that the angular amplitude of everybody's stride is about the same. •••

61. A pendulum consists of a bob of mass m_{bob} hanging at the end of a light rod of mass m_{rod} and length ℓ. If you ignore the mass of the rod, the period is T_0, and if you use m_{rod} in your calculation, the period is T. How far from the true value T is T_0? Express the error $\Delta T = T - T_0$ in terms of some power of the ratio m_{rod}/m_{bob}. (*Hint:* Some sort of series expansion is useful here.) •••

15.8 Damped oscillations

62. Verify that the time constant $\tau = m/b$ has units of time for damped oscillations. •

63. As the quality factor Q for a damped oscillation increases but the period T of undamped motion is held constant, what happens to the angular frequency ω_d of the damped motion? •

64. Express, as a function of τ, the time interval needed for the amplitude of an oscillation to reduce to half of its initial value. •

65. A 0.400-kg object is oscillating on a spring for which the force constant is 300 N/m. A damping force that is linearly proportional to the object's velocity is exerted on the system, and the damping coefficient is $b = 5.00$ kg/s. (*a*) Verify that the units given for b are correct. (*b*) What is the frequency f of the oscillation? (*c*) What is the time constant τ? (*d*) What is the quality factor Q? ••

66. A simple pendulum consists of a 1.00-kg bob on a string 1.00 m long. During a time interval of 27.0 s, the maximum angle this pendulum makes with the vertical is found to decrease from 6.00° to 5.40°. Determine the numerical values of the damping coefficient b and the time constant τ. ••

67. A 0.25-kg bob is suspended from a string that is 0.60 m long. When the pendulum is set into small-amplitude oscillation, the amplitude decays to half of its initial value in 35 s. (*a*) What is the time constant τ for the pendulum? (*b*) At what instant is the energy half of its initial value? ••

68. A restoring force of unknown magnitude is exerted on a object that oscillates with a period of 0.50 s. When the object is in an evacuated container, the motion is simple harmonic motion, with an amplitude of 0.10 m. When air is allowed into the container, the amplitude decreases by 2.0% with each cycle of the oscillation. (*a*) What is the amplitude after 25 cycles? (*b*) What fraction of the initial energy is left 6.3 s after air is admitted? ••

69. A 0.500-kg block oscillates up and down on a vertical spring for which $k = 12.5$ N/m. The initial amplitude of the motion is 0.100 m. (*a*) What is the natural angular frequency ω? (*b*) If the observed angular frequency is $\omega_d = 4.58$ s^{-1}, what is *b*? (*c*) Write an equation for the position as a function of time. ••

70. A damped oscillator has a quality factor of 20. (*a*) By what fraction does the energy decrease during each cycle? (*b*) By what percentage does the damped angular frequency ω_d differ from the undamped angular frequency? ••

71. One reference book states that, after a large earthquake, Earth can be treated as an oscillator that has a period of 54 min and a quality factor of 400. (*a*) What percentage of the energy of oscillation is lost to damping forces during each cycle? (*b*) What fraction of the original energy is left after 2.0 days? ••

72. A gong makes a loud noise when struck. The noise gradually gets less and less loud until it fades below the sensitivity of the human ear. The simplest model of how the gong produces the sound we hear treats the gong as a damped harmonic oscillator. The tone we hear is related to the frequency *f* of the oscillation, and its loudness is proportional to the energy of the oscillation. (*a*) If the loudness drops to 85% of its original value in 4.0 s, what is the time constant of the damped oscillation? (*b*) How long does it take for the sound to be 25% as loud as it was at the start? (*c*) What fraction of the original loudness remains after 1.0 min? ••

73. *Critical damping* is a term used to describe a situation where the damping coefficient *b* is just great enough that no oscillation occurs (See Guided Problem 15.8. This is how you want the springs in your car to behave, so that when they flex to take you smoothly over a bump, they don't cause the car to jump up and down, reducing contact with the road.) (*a*) For an object of mass *m* attached to a spring of spring constant *k*, what is the value b_{crit} needed for critical damping? (*b*) Describe the motion if $b > b_{crit}$. ••

74. A small 3.0-kg object dropped from the roof of a tall building acquires a terminal speed of 25 m/s. Assume the drag force exerted on the object has the same form as the damping force exerted on a damped oscillator; that is, the force is opposed to the motion, and its magnitude is linearly proportional to the object's speed. An object identical to the dropped one is attached to a vertical spring ($k = 230$ N/m) and set into oscillation in the air with an initial amplitude of 0.20 m. (*a*) What is the quality factor for this oscillation? (*b*) How long does it take for the amplitude to drop to half its initial value? (*c*) How much energy is lost in this time interval? •••

75. An object resting on a table oscillates on the end of a horizontal spring. Because the table is covered with a viscous substance, the motion is damped and the amplitude gets smaller with each cycle. At $t = 1.5$ s, the object is 60 mm from its equilibrium position, and this is the farthest from equilibrium it reaches. The amplitude of the next cycle of the motion is 56 mm, and the object reaches this maximum *x* value at $t = 2.5$ s. Write the equation for the object's *x* component of position as a function of time, using numerical values for all constants. •••

Additional Problems

76. Do the turning points in simple harmonic motion have to be equidistant from the equilibrium position? •

77. If the amplitude of a simple harmonic motion doubles, what happens to (*a*) the energy of the system, (*b*) the maximum speed of the moving object, and (*c*) the period of the motion? •

78. A pendulum is swinging in a stationary elevator. What happens to the period when the elevator (*a*) accelerates upward, (*b*) travels downward at constant speed, and (*c*) travels downward and gradually slows to a stop? •

79. You measure the oscillation frequency f_{whole} of a vertical block-spring system. You then cut the spring in half, hang the same block from one of the halves, and measure the frequency f_{half}. What is the ratio f_{half}/f_{whole}? ••

80. The hand in Figure P15.80*a* holds a vertical block-spring system so that the spring is compressed. When the hand lets go, the block, of mass *m*, descends a distance *d* in a time interval Δt before reversing direction (Figure P15.80*b*). Once the system comes to rest, the block is removed, a block of mass $2m$ is hung in its place, and the experiment is repeated. To what maximum distance does the block descend? ••

Figure P15.80

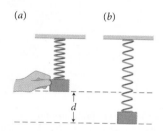

81. Two oscillatory motions are given by $x(t) = A \sin(m\omega t)$ and $y(t) = A \sin(n\omega t + \phi)$, where *m* and *n* are positive integers. Consider displaying these motions on a single graph with *x* and *y* axes oriented perpendicular to each other. What restrictions are necessary on *m*, *n*, and ϕ if the trajectory is to be a closed curve? These curves are called *Lissajous figures*. Plot these curves for (*a*) $m = n = 1$, $\phi = 0$; (*b*) $m = n = 1$, $\phi = \pi/4$; and (*c*) $m = 2$, $n = 3$, $\phi = 0$. ••

82. What would be the period of oscillation of this book if you held it at one corner between index finger and thumb and allowed it to swing slightly? To first order, ignore any effect on the oscillation by friction between your fingers and the book. ••

83. You have a teardrop-shaped 2.00-kg object that has a hook at one end and is 0.28 m long along its longest axis (Figure P15.83). When you try to balance the object on your fingers, it balances when your fingers are 0.20 m from the hook end. When you hang the object by the hook and set it into simple harmonic motion, it completes 10 cycles in 11 s. What is its rotational inertia *I*? ••

Figure P15.83

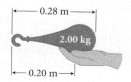

84. A 2.00-kg object is free to slide on a horizontal surface. The object is attached to a spring of spring constant 200 N/m, and the other end of the spring is attached to a wall. The object is pulled in the direction away from the wall until the spring is stretched 50.0 mm from its relaxed position. The object is not released from rest, but is instead given an initial velocity of 2.00 m/s away from the wall. (a) Determine the energy of the oscillating system. (b) Write the equation for the object's position as a function of time. Ignore friction. ••

85. A hole is drilled through the narrow end of a 0.960-kg baseball bat, and the bat is hung on a nail so that it can swing freely (Figure P15.85). The bat is 0.860 m long, the hole is drilled 0.0300 m from the end, and the center of mass is located 0.670 m from the end. If the period of oscillation is 1.85 s, what is the bat's rotational inertia (a) about the pivot and (b) about the center of mass? (c) If the hole had been drilled 0.200 m from the end, what would the period be? ••

Figure P15.85

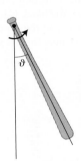

86. A meter stick is free to pivot around a position located a distance x below its top end, where $0 < x < 0.50$ m (Figure P15.86). (a) What is the frequency f of its oscillation if it moves as a pendulum? (b) To what position should you move the pivot if you want to minimize the period? ••

Figure P15.86

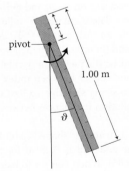

pivot

x

1.00 m

ϑ

87. To determine what effect a spring's mass has on simple harmonic motion, consider a spring of mass m and relaxed length ℓ_{spring}. The spring is oriented horizontally, and one end is attached to a vertical surface. When the spring is stretched a distance x, the potential energy of the system is $\frac{1}{2}kx^2$. If the free end of the spring is moving with speed v when the free end is at x, calculate the kinetic energy of the spring in terms of m and v. (Hint: Divide the spring into many infinitesimal elements each of length $d\ell_{\text{segment}}$ and located at position

ℓ_{segment}; then determine the speed of each piece in terms of ℓ_{segment}, $d\ell_{\text{segment}}$, m, v, and x and integrate from $\ell_{\text{segment}} = 0$ to $\ell_{\text{segment}} = \ell_{\text{spring}} + x$.) •••

88. (a) Show that for an object moving in simple harmonic motion, the speed of the object as a function of position is given by $v(x) = \omega\sqrt{A^2 - x^2}$. (b) Noting that $v_x = dx/dt$, isolate dt and integrate to determine how long it takes for an oscillator to move from its equilibrium position to some arbitrary position $x < A$. Compare this result with the value of $\Delta t = t_{x<A} - t_{x=A}$ calculated from Eq. 15.6, $x(t) = A\sin(\omega t + \phi_i)$. •••

89. The King's clock fails to keep time, and because you designed it, he's holding you responsible. Examining the spring-ball system you used as the central timekeeping oscillator, you realize that a simple pendulum might be more reliable, and there happens to be a bit of thin, strong wire handy. Unfortunately, you don't have any way of timing the oscillations, nor can you remember what the period is supposed to be. You don't even have a ruler to measure the wire for the pendulum, and there is not enough wire on hand to rely on trial and error. You suddenly realize that all you need to do is observe how far the ball stretches the spring when the spring is held vertically! You calm the King's wrath by putting your new design for the pendulum timekeeping mechanism in operation. ••• CR

90. In physics lab, you are measuring the period of a vertically hung spring-ball system in which the mass of the ball is 0.50 kg. As the system oscillates up and down, it also swings from side to side, a motion that interferes with your measurements of the vertical motion. After running several trials and calculating the period for the system set up this way, you decide to eliminate the side-to-side motion by rigging up the spring horizontally with a string draped over a pulley of radius 0.035 m, as in Figure P15.90. With this arrangement, your measured period is 10% greater than the value you measured for the vertically hung system. Realizing an opportunity for extra credit, you wonder if the rotational inertia of the pulley is buried in your data. ••• CR

Figure P15.90

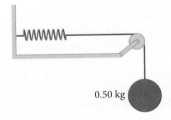

0.50 kg

91. You are aboard the International Space Station and are required to keep track of your mass because of your long stay in the free-fall conditions of deep space. To carry out this task, you hook a 215-kg cargo box to one end of a stiff spring and hook yourself to the spring's other end. With the spring and box initially at rest relative to you, you push off from the cargo box. Floating inside the cabin, you and the cargo box oscillate with a period of 2.21 s. You then unhook yourself from the spring and attach in your place a cargo box identical to the one at the other end of the spring. The oscillation period for this two-box system is 3.13 s. Amazingly, even with squeeze-tube food, you seem to be gaining mass. ••• CR

Answers to Review Questions

1. Motion is periodic if the motion repeats itself at regular time intervals.
2. The amplitude is the magnitude of the moving object's maximum displacement from the equilibrium position. The frequency is the number of cycles of the repetitive motion completed during each second.
3. The period is the time interval needed to complete one cycle: $(5.00 \text{ s})/(4 \text{ cycles}) = 1.25 \text{ s/cycle}$; $T = 1.25$ s. The frequency is the reciprocal of the period, $f = 1/T = 0.800$ Hz.
4. The kinetic energy is greatest at the equilibrium position and zero at the extremes of the motion. The magnitude of the potential energy is greatest at the extremes of the motion and zero at the equilibrium position.
5. As long as the amplitude is not too great, the period does not change because simple harmonic motion is isochronous.
6. All simple harmonic motion is isochronous. The acceleration for isochronous motion is directly proportional to the object's displacement from its equilibrium position. The restoring force is proportional to the object's acceleration by Newton's second law and hence directly proportional to the object's displacement. Direct proportionality guarantees a linear relationship.
7. A turning point is a point at which the object turns around—that is, changes its direction of motion. These points are the extrema of the motion, the points farthest from the equilibrium position.
8. Circular motion is two dimensional, and hence the position of an object in circular motion can be described using two orthogonal components. Either component may be expressed as a sinusoidal function of time. Any motion that can be expressed as a sinusoidal function of time is simple harmonic motion. Thus simple harmonic motion has the same functional behavior as one component of constant-speed circular motion.
9. Any periodic function, no matter how complex, can be written as a sum of sinusoidal simple harmonic functions of frequency $f_n = n/T$, where $n \geq 1$ is an integer and T is the period of the motion. This means that any periodic function can be treated as a superposition of simple harmonic motions, and so we need to understand only simple harmonic motion in order to deal with any periodic motion.
10. The fundamental frequency is the same as the frequency of the original periodic function.
11. The spectrum of a periodic function is a plot of the square of the amplitudes of each harmonic frequency in the Fourier series versus frequency.
12. Fourier analysis is the breaking down of a function into a Fourier series (a sum of sinusoidal functions with frequencies that are integer multiples of the fundamental frequency). Fourier synthesis is the inverse of Fourier analysis: the construction of periodic functions by adding together sinusoidal functions with frequencies that are integer multiples of some fundamental frequency.
13. Stable equilibrium allows periodic motion.
14. They have identical periods. Both the restoring force and the rotational inertia are proportional to m, and so the dependence on m cancels.
15. For sufficiently small displacements away from the equilibrium position, restoring forces are always linearly proportional to the displacement.
16. The period of the swing does not depend on the mass of the pendulum bob—in this case, the rider. Thus the two stay in synchrony, as your common experience tells you.
17. No, even the angular frequency ω associated with simple harmonic motion is merely a convenience that exploits the mathematical similarity between simple harmonic motion and circular motion at constant speed. (That is why we don't call it *rotational velocity* here,

although it is mathematically the same.) The reference circle helps you relate simple harmonic motion to circular motion, providing a visual interpretation for angular frequency, phase, and amplitude.
18. The variable x describes simple harmonic motion if (and only if) the second derivative of x with respect to time is equal to minus a constant (the square of the angular frequency) times the variable x.
19. You need to know the amplitude A because $\sin \phi_i = x_i/A$ (from Eq. 15.6, with $t = 0$).
20. Angular frequency is given in units of s^{-1}, just as for circular motion, while frequency is given in units of Hz (Hertz).
21. You need to know the spring constant k because $T = 2\pi \sqrt{m/k}$ (from Eqs. 15.1 and Eq. 15.22).
22. It decreases. After another person gets into the car, the mass of the load compressing the springs increases, and angular frequency ω is inversely proportional to the square root of that mass (Eq. 15.22).
23. You can control all except period and frequency, which are determined by the design (mass and spring constant) of the oscillating system.
24. When the friction in the horizontal system is negligibly small, the frequencies are identical.
25. B. Equation 15.33 shows the ratio ℓ/I in the expression for the angular frequency: $\omega = \sqrt{m\ell g/I}$. Length ℓ is double for B, but I for this pendulum is greater by a factor of 4 because it is proportional to the length squared (Equation 11.30; $I = mr^2$). Hence ω is smaller for B. Because period is inversely proportional to angular frequency ($\omega = 2\pi/T$), T is greater for this pendulum.
26. Because the clock is losing time, you need to increase the angular frequency ω of the oscillation. You should therefore shorten the length of the pendulum by moving the disk higher up on the rod to decrease the rotational inertia.
27. The small-angle approximation is required. The equation of motion for a pendulum equates the gravitational torque, which is proportional to $\sin \vartheta$, to the inertial term (which is proportional to the second derivative with respect to time of the angular position ϑ). The simple harmonic oscillator equation requires that the variable involved in the second derivative also be present as a linear term (first power only) in the restoring term. This is possible only if we replace $\sin \vartheta$ with ϑ. This is called the "small-angle approximation" because it is valid for only angular positions much less than 1, or angles much less than 1 radian.
28. In Eq. 15.25, the rotational displacement is unitless, so the torsional constant has same units as torque, newton-meters.
29. No. The overdamped condition occurs when the damping is so great that the system returns to equilibrium without oscillating.
30. The amplitude of the oscillations dies out more and more rapidly as the damping coefficient increases. When b reaches its greatest value, the system makes only a few complete oscillations before the motion becomes too small to measure. Also the angular frequency ω decreases as the damping increases, but this effect amounts to no more than a 5% change.
31. The energy decreases by a factor of e, or 2.718, so that approximately 37% of the energy at the beginning of the time interval remains at the end.

Answers to Guided Problems

Guided Problem 15.2 (*a*) $x(t) = (152 \text{ mm}) \sin[(3.49 \text{ s}^{-1})t]$; (*b*) $v_{max} = 532$ mm/s; (*c*) $a_{max} = 1.86 \text{ m/s}^2$

Guided Problem 15.4 $\ell = h$

Guided Problem 15.6 The rock begins to oscillate and it falls all the way through Earth, then returns. The time interval needed for the rock to return to you is one period, $T = 84$ min.

Guided Problem 15.8 (*a*) $b = 2m\omega$; (*b*) $t = 0.73T = 4.4$ s

Appendix A

Notation

Notation used in this text, listed alphabetically, Greek letters first.
For information concerning superscripts and subscripts, see the explanation at the end of this table.

Symbol	Name of Quantity	Definition	Where Defined	SI units
α (alpha)	polarizability	scalar measure of amount of charge separation occurring in material due to external electric field	Eq. 23.24	$C^2 \cdot m/N$
α	Bragg angle	in x-ray diffraction, angle between incident x rays and sample surface	Section 34.3	degree, radian, or revolution
α_ϑ	(ϑ component of) rotational acceleration	rate at which rotational velocity ω_ϑ increases	Eq. 11.12	s^{-2}
β (beta)	sound intensity level	logarithmic scale for sound intensity, proportional to $\log(I/I_{th})$	Eq. 17.5	dB (not an SI unit)
γ (gamma)	Lorentz factor	factor indicating how much relativistic values deviate from nonrelativistic ones	Eq. 14.6	unitless
γ	surface tension	force per unit length exerted parallel to surface of liquid; energy per unit area required to increase surface area of liquid	Eq. 18.48	N/m
γ	heat capacity ratio	ratio of heat capacity at constant pressure to heat capacity at constant volume	Eq. 20.26	unitless
Δ	delta	change in	Eq. 2.4	
$\Delta \vec{r}$	displacement	vector from object's initial to final position	Eq. 2.8	m
$\Delta \vec{r}_F, \Delta x_F$	force displacement	displacement of point of application of a force	Eq. 9.7	m
Δt	interval of time	difference between final and initial instants	Table 2.2	s
Δt_{proper}	proper time interval	time interval between two events occurring at same position	Section 14.1	s
Δt_v	interval of time	time interval measured by observer moving at speed v with respect to events	Eq. 14.13	s
Δx	x component of displacement	difference between final and initial positions along x axis	Eq. 2.4	m
δ (delta)	delta	infinitesimally small amount of	Eq. 3.24	
ϵ_0 (epsilon)	permittivity constant	constant relating units of electrical charge to mechanical units	Eq. 24.7	$C^2/(N \cdot m^2)$
η (eta)	viscosity	measure of fluid's resistance to shear deformation	Eq. 18.38	$Pa \cdot s$
η	efficiency	ratio of work done by heat engine to thermal input of energy	Eq. 21.21	unitless
θ (theta)	angular coordinate	polar coordinate measuring angle between position vector and x axis	Eq. 10.2	degree, radian, or revolution
θ_c	contact angle	angle between solid surface and tangent to liquid surface at meeting point measured within liquid	Section 18.4	degree, radian, or revolution
θ_c	critical angle	angle of incidence greater than which total internal reflection occurs	Eq. 33.9	degree, radian, or revolution
θ_i	angle of incidence	angle between incident ray of light and normal to surface	Section 33.1	degree, radian, or revolution

Symbol	Name of Quantity	Definition	Where Defined	SI units
θ_i	angle subtended by image	angle subtended by image	Section 33.6	degree, radian, or revolution
θ_o	angle subtended by object	angle subtended by object	Section 33.6	degree, radian, or revolution
θ_r	angle of reflection	angle between reflected ray of light and normal to surface	Section 33.1	degree, radian, or revolution
θ_r	minimum resolving angle	smallest angular separation between objects that can be resolved by optical instrument with given aperture	Eq. 34.30	degree, radian, or revolution
ϑ (script theta)	rotational coordinate	for object traveling along circular path, arc length traveled divided by circle radius	Eq. 11.1	unitless
κ (kappa)	torsional constant	ratio of torque required to twist object to rotational displacement	Eq. 15.25	$N \cdot m$
κ	dielectric constant	factor by which potential difference across isolated capacitor is reduced by insertion of dielectric	Eq. 26.9	unitless
λ (lambda)	inertia per unit length	for uniform one-dimensional object, amount of inertia in a given length	Eq. 11.44	kg/m
λ	wavelength	minimum distance over which periodic wave repeats itself	Eq. 16.9	m
λ	linear charge density	amount of charge per unit length	Eq. 23.16	C/m
μ (mu)	reduced mass	product of two interacting objects' inertias divided by their sum	Eq. 6.39	kg
μ	linear mass density	mass per unit length	Eq. 16.25	kg/m
$\vec{\mu}$	magnetic dipole moment	vector pointing along direction of magnetic field of current loop, with magnitude equal to current times area of loop	Section 28.3	$A \cdot m^2$
μ_0	magnetic constant	constant relating units of electric current to mechanical units	Eq. 28.1	$T \cdot m/A$
μ_k	coefficient of kinetic friction	proportionality constant relating magnitudes of force of kinetic friction and normal force between two surfaces	Eq. 10.55	unitless
μ_s	coefficient of static friction	proportionality constant relating magnitudes of force of static friction and normal force between two surfaces	Eq. 10.46	unitless
ρ (rho)	mass density	amount of mass per unit volume	Eq. 1.4	kg/m^3
ρ	inertia per unit volume	for uniform three-dimensional object, amount of inertia in a given volume divided by that volume	Eq. 11.46	kg/m^3
ρ	(volume) charge density	amount of charge per unit volume	Eq. 23.18	C/m^3
σ (sigma)	inertia per unit area	for uniform two-dimensional object, inertia divided by area	Eq. 11.45	kg/m^2
σ	surface charge density	amount of charge per unit area	Eq. 23.17	C/m^2
σ	conductivity	ratio of current density to applied electric field	Eq. 31.8	$A/(V \cdot m)$
τ (tau)	torque	magnitude of axial vector describing ability of forces to change objects' rotational motion	Eq. 12.1	$N \cdot m$
τ	time constant	for damped oscillation, time for energy of oscillator to decrease by factor e^{-1}	Eq. 15.39	s
τ_ϑ	(ϑ component of) torque	ϑ component of axial vector describing ability of forces to change objects' rotational motion	Eq. 12.3	$N \cdot m$
Φ_E (phi, upper case)	electric flux	scalar product of electric field and area through which it passes	Eq. 24.1	$N \cdot m^2/C$

Symbol	Name of Quantity	Definition	Where Defined	SI units
Φ_B	magnetic flux	scalar product of magnetic field and area through which it passes	Eq. 27.10	Wb
ϕ (phi)	phase constant	phase difference between source emf and current in circuit	Eq. 32.16	unitless
$\phi(t)$	phase	time-dependent argument of sine function describing simple harmonic motion	Eq. 15.5	unitless
Ω (omega, upper case)	number of basic states	number of basic states corresponding to macrostate	Sections 19.4, 19.1	unitless
ω (omega)	rotational speed	magnitude of rotational velocity	Eq. 11.7	s^{-1}
ω	angular frequency	for oscillation with period T, $2\pi/T$	Eq. 15.4	s^{-1}
ω_0	resonant angular frequency	angular frequency at which current in circuit is maximal	Eq. 32.47	s^{-1}
ω_ϑ	(ϑ component of) rotational velocity	rate at which rotational coordinate ϑ changes	Eq. 11.6	s^{-1}
A	area	length $\times$ width	Eq. 11.45	m^2
A	amplitude	magnitude of maximum displacement of oscillating object from equilibrium position	Eq. 15.6	m (for linear mechanical oscillation; unitless for rotational oscillation; various units for nonmechanical oscillation)
$\vec{A}$	area vector	vector with magnitude equal to area and direction normal to plane of area	Section 24.6	m^2
$\vec{a}$	acceleration	time rate of change in velocity	Section 3.1	m/s^2
$\vec{a}_{Ao}$	relative acceleration	value observer in reference frame A records for acceleration of object o in reference frame A	Eq. 6.11	m/s^2
a_c	magnitude of centripetal acceleration	acceleration required to make object follow circular trajectory	Eq. 11.15	m/s^2
a_r	radial component of acceleration	component of acceleration in radial direction	Eq. 11.16	m/s^2
a_t	tangential component of acceleration	component of acceleration tangent to trajectory; for circular motion at constant speed $a_t = 0$	Eq. 11.17	m/s^2
a_x	x component of acceleration	component of acceleration directed along x axis	Eq. 3.21	m/s^2
$\vec{B}$	magnetic field	vector field providing measure of magnetic interactions	Eq. 27.5	T
$\vec{B}_{ind}$	induced magnetic field	magnetic field produced by induced current	Section 29.4	T
b	damping coefficient	ratio of drag force on moving object to its speed	Eq. 15.34	kg/s
C	heat capacity per particle	ratio of energy transferred thermally per particle to change in temperature	Section 20.3	J/K
C	capacitance	ratio of magnitude of charge on one of a pair of oppositely charged conductors to magnitude of potential difference between them	Eq. 26.1	F
C_P	heat capacity per particle at constant pressure	ratio of energy transferred thermally per particle to change in temperature, while holding pressure constant	Eq. 20.20	J/K
C_V	heat capacity per particle at constant volume	ratio of energy transferred thermally per particle to change in temperature, while holding volume constant	Eq. 20.13	J/K
$COP_{cooling}$	coefficient of performance of cooling	ratio of thermal input of energy to work done on a heat pump	Eq. 21.27	unitless

Symbol	Name of Quantity	Definition	Where Defined	SI units
$COP_{heating}$	coefficient of performance of heating	ratio of thermal output of energy to work done on a heat pump	Eq. 21.25	unitless
c	shape factor	ratio of object's rotational inertia to mR^2; function of distribution of inertia within object	Tables 11.3, 12.25	unitless
c	wave speed	speed at which mechanical wave travels through medium	Eq. 16.3	m/s
c	specific heat capacity	ratio of energy transferred thermally per unit mass to change in temperature	Section 20.3	$J/(K \cdot kg)$
c_0	speed of light in vacuum	speed of light in vacuum	Section 14.2	m/s
c_V	specific heat capacity at constant volume	ratio of energy transferred thermally per unit mass to change in temperature, while holding volume constant	Eq. 20.48	$J/(K \cdot kg)$
$\vec{D}$	displacement (of particle in wave)	displacement of particle from its equilibrium position	Eq. 16.1	m
d	diameter	diameter	Section 1.9	m
d	distance	distance between two locations	Eq. 2.5	m
d	degrees of freedom	number of ways particle can store thermal energy	Eq. 20.4	unitless
d	lens strength	1 m divided by focal length	Eq. 33.22	diopters
E	energy of system	sum of kinetic and internal energies of system	Tables 1.1, 5.21	J
$\vec{E}$	electric field	vector field representing electric force per unit charge	Eq. 23.1	N/C
E_0	work function	minimum energy required to free electron from surface of metal	Eq. 34.35	J
E_{chem}	chemical energy	internal energy associated with object's chemical state	Eq. 5.27	J
E_{int}	internal energy of system	energy associated with an object's state	Eqs. 5.20, 14.54	J
E_{mech}	mechanical energy	sum of system's kinetic and potential energies	Eq. 7.9	J
E_s	source energy	incoherent energy used to produce other forms of energy	Eq. 7.7	J
E_{th}	thermal energy	internal energy associated with object's temperature	Eq. 5.27	J
$\mathscr{E}$	emf	in charge-separating device, nonelectrostatic work per unit charge done in separating positive and negative charge carriers	Eq. 26.7	V
$\mathscr{E}_{ind}$	induced emf	emf resulting from changing magnetic flux	Eqs. 29.3, 29.8	V
$\mathscr{E}_{max}$	amplitude of emf	amplitude of time-dependent emf produced by AC source	Sections 32.1, 32.1	V
$\mathscr{E}_{rms}$	rms emf	root-mean-square emf	Eq. 32.55	V
e	coefficient of restitution	measure of amount of initial relative speed recovered after collision	Eq. 5.18	unitless
e	eccentricity	measure of deviation of conic section from circular	Section 13.7	unitless
e	elementary charge	magnitude of charge on electron	Eq. 22.3	C
$\vec{F}$	force	time rate of change of object's momentum	Eq. 8.2	N
$\vec{F}^B$	magnetic force	force exerted on electric current or moving charged particle by magnetic field	Eqs. 27.8, 27.19	N
$\vec{F}^b$	buoyant force	upward force exerted by fluid on submerged object	Eq. 18.12	N
$\vec{F}^c$	contact force	force between objects in physical contact	Section 8.5	N

Symbol	Name of Quantity	Definition	Where Defined	SI units
$\vec{F}^d$	drag force	force exerted by medium on object moving through medium	Eq. 15.34	N
$\vec{F}^E$	electric force	force exerted between electrically charged objects or on electrically charged objects by electric field	Eq. 22.1	N
$\vec{F}^{EB}$	electromagnetic force	force exerted on electrically charged objects by electric and magnetic fields	Eq. 27.20	N
$\vec{F}^f$	frictional force	force exerted on object due to friction between it and a second object or surface	Eq. 9.26	N
$\vec{F}^G$	gravitational force	force exerted by Earth or any object having mass on any other object having mass	Eqs. 8.16, 13.1	N
$\vec{F}^k$	force of kinetic friction	frictional force between two objects in relative motion	Sections 10.4, 10.55	N
$\vec{F}^n$	normal force	force directed perpendicular to a surface	Sections 10.4, 10.46	N
$\vec{F}^s$	force of static friction	frictional force between two objects not in relative motion	Sections 10.4, 10.46	N
f	frequency	number of cycles per second of periodic motion	Eq. 15.2	Hz
f	focal length	distance from center of lens to focus	Sections 33.4, 33.16	m
f_{beat}	beat frequency	frequency at which beats occur when waves of different frequency interfere	Eq. 17.8	Hz
G	gravitational constant	proportionality constant relating gravitational force between two objects to their masses and separation	Eq. 13.1	$\text{N} \cdot \text{m}^2/\text{kg}^2$
g	magnitude of acceleration due to gravity	magnitude of acceleration of object in free fall near Earth's surface	Eq. 3.14	m/s^2
h	height	vertical distance	Eq. 10.26	m
h	Planck's constant	constant describing scale of quantum mechanics; relates photon energy to frequency and de Broglie wavelength to momentum of particle	Eq. 34.35	$\text{J} \cdot \text{s}$
I	rotational inertia	measure of object's resistance to change in its rotational velocity	Eq. 11.30	$\text{kg} \cdot \text{m}^2$
I	intensity	energy delivered by wave per unit time per unit area normal to direction of propagation	Eq. 17.1	W/m^2
I	(electric) current	rate at which charged particles cross a section of a conductor in a given direction	Eq. 27.2	A
I	amplitude of oscillating current	maximum value of oscillating current in circuit	Sections 32.1, 32.5	A
I_{cm}	rotational inertia about center of mass	object's rotational inertia about an axis through its center of mass	Eq. 11.48	$\text{kg} \cdot \text{m}^2$
I_{disp}	displacement current	current-like quantity in Ampère's law caused by changing electric flux	Eq. 30.7	A
I_{enc}	enclosed current	current enclosed by Ampèrian path	Eq. 28.1	A
I_{ind}	induced current	current in loop caused by changing magnetic flux through loop	Eq. 29.4	A
I_{int}	intercepted current	current intercepted by surface spanning Ampèrian path	Eq. 30.6	A
I_{rms}	rms current	root-mean-square current	Eq. 32.53	A
I_{th}	intensity at threshold of hearing	minimum intensity audible to human ear	Eq. 17.4	W/m^2
i	time-dependent current	time-dependent current through circuit; $I(t)$	Sections 32.1, 32.5	A
i	image distance	distance from lens to image	Sections 33.6, 33.16	m

Symbol	Name of Quantity	Definition	Where Defined	SI units
$\hat{\imath}$	unit vector ("i hat")	vector for defining direction of x axis	Eq. 2.1	unitless
$\vec{J}$	impulse	amount of momentum transferred from environment to system	Eq. 4.18	$kg \cdot m/s$
$\vec{J}$	current density	current per unit area	Eq. 31.6	A/m^2
J_ϑ	rotational impulse	amount of angular momentum transferred from environment to system	Eq. 12.15	$kg \cdot m^2/s$
$\hat{\jmath}$	unit vector	vector for defining direction of y axis	Eq. 10.4	unitless
K	kinetic energy	energy object has because of its translational motion	Eqs. 5.12, 14.51	J
K	surface current density	current per unit of sheet width	Section 28.5	A/m
K_{cm}	translational kinetic energy	kinetic energy associated with motion of center of mass of system	Eq. 6.32	J
K_{conv}	convertible kinetic energy	kinetic energy that can be converted to internal energy without changing system's momentum	Eq. 6.33	J
K_{rot}	rotational kinetic energy	energy object has due to its rotational motion	Eq. 11.31	J
k	spring constant	ratio of force exerted on spring to displacement of free end of spring	Eq. 8.18	N/m
k	wave number	number of wavelengths in 2π units of distance; for wave with wavelength λ, $2\pi/\lambda$	Eqs. 16.7, 16.11	m^{-1}
k	Coulomb's law constant	constant relating electrostatic force to charges and their separation distance	Eq. 22.5	$N \cdot m^2/C^2$
k_B	Boltzmann constant	constant relating thermal energy to absolute temperature	Eq. 19.39	J/K
L	inductance	negative of ratio of induced emf around loop to rate of change of current in loop	Eq. 29.19	H
L_ϑ	(ϑ component of) angular momentum	capacity of object to make other objects rotate	Eq. 11.34	$kg \cdot m^2/s$
L_m	specific transformation energy for melting	energy transferred thermally per unit mass required to melt substance	Eq. 20.55	J/kg
L_v	specific transformation energy for vaporization	energy transferred thermally per unit mass required to vaporize substance	Eq. 20.55	J/kg
ℓ	length	distance or extent in space	Table 1.1	m
ℓ_{proper}	proper length	length measured by observer at rest relative to object	Section 14.3	m
ℓ_v	length	measured length of object moving at speed v relative to observer	Eq. 14.28	m
M	magnification	ratio of signed image height to object height	Eq. 33.17	unitless
M_θ	angular magnification	ratio of angle subtended by image to angle subtended by object	Eq. 33.18	unitless
m	mass	amount of substance	Tables 1.1, 13.1	kg
m	inertia	measure of object's resistance to change in its velocity	Eq. 4.2	kg
m	fringe order	number indexing bright interference fringes, counting from central, zeroth-order bright fringe	Sections 34.2, 34.5	unitless
m_v	inertia	inertia of object moving at speed v relative to observer	Eq. 14.41	kg
N	number of objects	number of objects in sample	Eq. 1.3	unitless
N_A	Avogadro's number	number of particles in 1 mol of a substance	Eq. 1.2	unitless
n	number density	number of objects per unit volume	Eq. 1.3	m^{-3}

Symbol	Name of Quantity	Definition	Where Defined	SI units
n	windings per unit length	in a solenoid, number of windings per unit length	Eq. 28.4	unitless
n	index of refraction	ratio of speed of light in vacuum to speed of light in a medium	Eq. 33.1	unitless
n	fringe order	number indexing dark interference fringes, counting from central, zeroth-order bright fringe	Sections 34.2, 34.7	unitless
O	origin	origin of coordinate system	Section 10.2	
o	object distance	distance from lens to object	Sections 33.6, 33.16	m
P	power	time rate at which energy is transferred or converted	Eq. 9.30	W
P	pressure	force per unit area exerted by fluid	Eq. 18.1	Pa
P_{atm}	atmospheric pressure	average pressure in Earth's atmosphere at sea level	Eq. 18.3	Pa
P_{gauge}	gauge pressure	pressure measured as difference between absolute pressure and atmospheric pressure	Eq. 18.16	Pa
p	time-dependent power	time-dependent rate at which source delivers energy to load; $P(t)$	Eq. 32.49	W
$\vec{p}$	momentum	vector that is product of an object's inertia and velocity	Eq. 4.6	kg·m/s
$\vec{p}$	(electric) dipole moment	vector representing magnitude and direction of electric dipole, equal amounts of positive and negative charge separated by small distance	Eq. 23.9	C·m
$\vec{p}_{ind}$	induced dipole moment	dipole moment induced in material by external electric field	Eq. 23.24	C·m
p_x	x component of momentum	x component of momentum	Eq. 4.7	kg·m/s
Q	quality factor	for damped oscillation, number of cycles for energy of oscillator to decrease by factor $e^{-2\pi}$	Eq. 15.41	unitless
Q	volume flow rate	rate at which volume of fluid crosses section of tube	Eq. 18.25	m³/s
Q	energy transferred thermally	energy transferred into system by thermal interactions	Eq. 20.1	J
Q_{in}	thermal input of energy	positive amount of energy transferred into system by thermal interactions	Sections 21.1, 21.5	J
Q_{out}	thermal output of energy	positive amount of energy transferred out of system by thermal interactions	Sections 21.1, 21.5	J
q	electrical charge	attribute responsible for electromagnetic interactions	Eq. 22.1	C
q_{enc}	enclosed charge	sum of all charge within a closed surface	Eq. 24.8	C
q_P	dipole charge	charge of positively charged pole of dipole	Section 23.6	C
R	radius	radius of an object	Eq. 11.47	m
R	resistance	ratio of applied potential difference to resulting current	Eqs. 29.4, 31.10	Ω
R_{eq}	equivalent resistance	resistance that could be used to replace combination of circuit elements	Eqs. 31.26, 31.33	Ω
r	radial coordinate	polar coordinate measuring distance from origin of coordinate system	Eq. 10.1	m
$\vec{r}$	position	vector for determining position	Eqs. 2.9, 10.4	m
$\hat{r}_{12}$	unit vector ("r hat")	unit vector pointing from tip of $\vec{r}_1$ to tip of $\vec{r}_2$	Eq. 22.6	unitless
$\vec{r}_{AB}$	relative position	position of observer B in reference frame of observer A	Eq. 6.3	m
$\vec{r}_{Ae}$	relative position	value observer in reference frame A records for position at which event e occurs	Eq. 6.3	m

Symbol	Name of Quantity	Definition	Where Defined	SI units
$\vec{r}_{cm}$	position of a system's center of mass	a fixed position in a system that is independent of choice of reference frame	Eq. 6.24	m
$\vec{r}_{p}$	dipole separation	position of positively charged particle relative to negatively charged particle in dipole	Section 23.6	m
$r_{\perp}$	lever arm distance *or* lever arm	perpendicular distance between rotation axis and line of action of a vector	Eq. 11.36	m
$\Delta\vec{r}$	displacement	vector from object's initial to final position	Eq. 2.8	m
$\Delta\vec{r}_F$	force displacement	displacement of point of application of a force	Eq. 9.7	m
S	entropy	logarithm of number of basic states	Eq. 19.4	unitless
S	intensity	intensity of electromagnetic wave	Eq. 30.36	W/m^2
$\vec{S}$	Pointing vector	vector representing flow of energy in combined electric and magnetic fields	Eq. 30.37	W/m^2
s	arc length	distance along circular path	Eq. 11.1	m
s^2	space-time interval	invariant measure of separation of events in space-time	Eq. 14.18	m^2
T	period	time interval needed for object in circular motion to complete one revolution	Eq. 11.20	s
T	absolute temperature	quantity related to rate of change of entropy with respect to thermal energy	Eq. 19.38	K
$\mathcal{T}$	tension	stress in object subject to opposing forces stretching the object	Section 8.6	N
t	instant in time	physical quantity that allows us to determine the sequence of related events	Table 1.1	s
t_{Ae}	instant in time	value observer A measures for instant at which event e occurs	Eq. 6.1	s
Δt	interval of time	difference between final and initial instants	Table 2.2	s
Δt_{proper}	proper time interval	time interval between two events occurring at same position	Section 14.1	s
Δt_v	interval of time	time interval between two events measured by observer moving at speed v relative to an observer for whom the events occur at the same position	Eq. 14.13	s
U	potential energy	energy stored in reversible changes to system's configuration state	Eq. 7.7	J
U^B	magnetic potential energy	potential energy stored in magnetic field	Eqs. 29.25, 29.30	J
U^E	electric potential energy	potential energy due to relative position of charged objects	Eq. 25.8	J
U^G	gravitational potential energy	potential energy due to relative position of gravitationally interacting objects	Eqs. 7.13, Eq. 13.14	J
u_B	energy density of magnetic field	energy per unit volume stored in magnetic field	Eq. 29.29	J/m^3
u_E	energy density of electric field	energy per unit volume stored in electric field	Eq. 26.6	J/m^3
V	volume	amount of space occupied by an object	Table 1.1	m^3
V_{AB}	potential difference	negative of electrostatic work per unit charge done on charged particle as it is moved from point A to point B	Eq. 25.15	V
V_{batt}	battery potential difference	magnitude of potential difference between terminals of battery	Eq. 25.19	V
V_C	amplitude of oscillating potential	maximum magnitude of potential across circuit element C	Sections 32.1, 32.8	V
V_{disp}	displaced volume	volume of fluid displaced by submerged object	Eq. 18.12	m^3

Symbol	Name of Quantity	Definition	Where Defined	SI units
V_P	(electrostatic) potential	potential difference between conveniently chosen reference point of potential zero and point P	Eq. 25.30	V
V_{rms}	rms potential	root-mean-square potential difference	Eq. 32.55	V
V_{stop}	stopping potential	minimum potential difference required to stop flow of electrons from photoelectric effect	Eq. 34.34	V
$\mathcal{V}$	"volume" in velocity space	measure of range of velocities in three dimensions	Eq. 19.20	$(m/s)^3$
v	speed	magnitude of velocity	Table 1.1	m/s
$\vec{v}$	velocity	time rate of change in position	Eq. 2.23	m/s
$\vec{v}_{12}$	relative velocity	velocity of object 2 relative to object 1	Eq. 5.1	m/s
$\vec{v}_{AB}$	relative velocity	velocity of observer B in reference frame of observer A	Eq. 6.3	m/s
v_C	time-dependent potential	time-dependent potential across circuit element C; $V_C(t)$	Sections 32.1, 32.8	V
$\vec{v}_{cm}$	velocity, center of mass	velocity of the center of mass of a system, equal to the velocity of the zero-momentum reference frame of the system	Eq. 6.26	m/s
$\vec{v}_d$	drift velocity	average velocity of electrons in conductor in presence of electric field	Eq. 31.3	m/s
v_{esc}	escape speed	minimum launch speed required for object to reach infinity	Eq. 13.23	m/s
v_r	radial component of velocity	for object moving along circular path, always zero	Eq. 11.18	m/s
v_{rms}	root-mean-square speed	square root of average of square of speed	Eq. 19.21	m/s
v_t	tangential component of velocity	for object in circular motion, rate at which arc length is swept out	Eq. 11.9	m/s
v_x	x component of velocity	component of velocity directed along x axis	Eq. 2.21	m/s
W	work	change in system's energy due to external forces exerted on system	Eqs. 9.1, 10.35	J
$W_{P \to Q}$	work	work done along path from P to Q	Eq. 13.12	J
W_{in}	mechanical input of energy	positive amount of mechanical work done on system	Section 21.1	J
W_{out}	mechanical output of energy	positive amount of mechanical work done by system	Section 21.1	J
W_q	electrostatic work	work done by electrostatic field on charged particle moving through field	Sections 25.2, 25.17	J
X_C	capacitive reactance	ratio of potential difference amplitude to current amplitude for capacitor	Eq. 32.14	Ω
X_L	inductive reactance	ratio of potential difference amplitude to current amplitude for inductor	Eq. 32.26	Ω
x	position	position along x axis	Eq. 2.4	m
$x(t)$	position as function of time	position x at instant t	Section 2.3	m
Δx	x component of displacement	difference between final and initial positions along x axis	Eq. 2.4	m
Δx_F	force displacement	displacement of point of application of a force	Eq. 9.7	m
Z	impedance	(frequency-dependent) ratio of potential difference to current through circuit	Eq. 32.33	Ω
z	zero-momentum reference frame	reference frame in which system of interest has zero momentum	Eq. 6.23	

Math notation

Math notation	Name	Where introduced
$\equiv$	defined as	Eq. 1.3
$\approx$	approximately equal to	Section 1.9
Σ (sigma, upper case)	sum of	Eq. 3.25
$\int$	integral of	Eq. 3.27
$\parallel$	parallel	Section 10.2
$\perp$	perpendicular	Section 10.2
$\propto$	proportional to	Section 13.1
$\cdot$	scalar product of two vectors	Eq. 10.33
$\times$	vector product of two vectors	Eq. 12.35
$\dfrac{\partial f}{\partial x}$	partial derivative of f with respect to x	Eq. 16.47
$\vec{b}$	vector b	Eq. 2.2
$|\vec{b}|$ or b	magnitude of $\vec{b}$	Eq. 2.3
b_x	x component of $\vec{b}$	Eq. 2.2
$\vec{b}_x$	x component vector of $\vec{b}$	Eq. 10.5
$\hat{\imath}$	unit vector ("i hat")	Eq. 2.1
$\hat{r}_{12}$	unit vector ("r hat")	Eq. 22.6

Note concerning superscripts and subscripts

Superscripts are appended to forces and potential energies to indicate the type of force or energy. They may be found in the main list under F, for forces, and U, for potential energies. Uppercase superscripts are used for fundamental interactions.

Subscripts are used on many symbols to identify objects, reference frames, types (for example, of energy), and processes. Object identifiers may be numbers, letters, or groups of letters. Reference frames are indicated by capital letters. Object identifiers and reference frames can occur in pairs, indicating relative quantities. In this case, the main symbol describes a property of whatever is identified by the second subscript relative to that of the first. In the case of forces, the first subscript identifies the object that causes the force and the second identifies the object on which the force is exerted. Types and processes are identified in various ways; many are given in the main list. Here are some examples:

m_1	inertia of object 1
m_{ball}	inertia of ball
$\vec{v}_{\text{cm}}$	velocity of center of mass of system

$\vec{r}_{12}$	position of object 2 relative to object 1; $\vec{r}_{12} = \vec{r}_2 - \vec{r}_1$
$\vec{p}_1$	momentum of object 1
$\vec{p}_{Z2}$	momentum of object 2 as measured in zero-momentum reference frame
$\vec{v}_{AB}$	velocity of observer B as measured in reference frame of observer A
$\vec{v}_{Ao}$	velocity of object o as measured in reference frame A
$\vec{r}_{Ee}$	position of event e as measured in Earth reference frame
$\vec{F}^c_{pw}$	contact force exerted by person on wall
$\vec{F}^G_{Eb}$	gravitational force exerted by Earth on ball
E_{th}	thermal energy
K_{conv}	convertible kinetic energy
P_{av}	average power
a_c	centripetal acceleration
$W_{P \to Q}$	work done along path from P to Q

Initial and final conditions are identified by subscripts i and f, following other identifiers. For example:

$\vec{p}_{1i}$	initial momentum of object 1
$\vec{p}_{Zball,f}$	final momentum of ball as measured in zero-momentum reference frame

Italic subscripts are used to identify components of vectors. These include x, y, z, r (radial), t (tangential), and ϑ (angular, with respect to given axis). They are also used to enumerate collections, for example, as indices of summation, and to indicate that a subscript refers to another variable. In the context of optics (Chapter 33), the italic subscripts o and i represent object and image, respectively. Here are some examples:

r_x	x component of position
a_t	tangential component of acceleration
L_ϑ	ϑ component of angular momentum
$p_{Zball\,y,f}$	final y component of momentum of ball as measured in zero-momentum reference frame
$\delta m_n r_n^2$	contribution to rotational inertia of extended object of small segment n, with inertia δm_n at position r_n
c_P	specific heat capacity at constant pressure
W_q	electrostatic work
h_o	height of object
θ_i	angle subtended by image

Appendix B

Mathematics Review

1 Algebra

Factors

$$ax + bx + cx = (a + b + c)x$$
$$(a + b)^2 = a^2 + 2ab + b^2$$
$$(a - b)^2 = a^2 - 2ab + b^2$$
$$(a + b)(a - b) = a^2 - b^2$$

Fractions

$$\left(\frac{a}{b}\right)\left(\frac{c}{d}\right) = \frac{ac}{bd}$$
$$\left(\frac{a/b}{c/d}\right) = \frac{a}{b} \div \frac{c}{d} = \frac{a}{b} \cdot \frac{d}{c} = \frac{ad}{bc}$$
$$\left(\frac{1}{1/a}\right) = a$$

Exponents

$$a^n = \underbrace{a \times a \times a \times \cdots \times a}_{n \text{ factors}}$$

Any real number can be used as an exponent:

$$a^{-x} = \frac{1}{a^x}$$
$$a^0 = 1$$
$$a^1 = a$$
$$a^{1/2} = \sqrt{a}$$
$$a^{1/n} = \sqrt[n]{a}$$
$$a^x a^y = a^{x+y}$$
$$\frac{a^x}{a^y} = a^{x-y}$$
$$(a^x)^y = a^{x \cdot y}$$
$$a^x b^x = (ab)^x$$
$$\frac{a^x}{b^x} = \left(\frac{a}{b}\right)^x$$

Logarithms

Logarithm is the inverse function of the exponential function:

$$y = a^x \Leftrightarrow \log_a y = \log_a a^x = x \quad \text{and} \quad x = \log_a (a^x) = a^{\log_a x}$$

The two most common values for the base a are 10 (the common logarithm base) and e (the natural logarithm base).

$$y = e^x \Leftrightarrow \log_e y = \ln y = \ln e^x = x \quad \text{and} \quad x = \ln e^x = e^{\ln x}$$

Logarithm rules (valid for any base):

$$\ln(ab) = \ln(a) + \ln(b)$$
$$\ln\left(\frac{a}{b}\right) = \ln(a) - \ln(b)$$
$$\ln(a^n) = n \ln(a)$$
$$\ln 1 = 0$$

The expression $\ln(a + b)$ cannot be simplified.

Linear equations

A linear equation has the form $y = ax + b$, where a and b are constants. A graph of y versus x is a straight line. The value of a equals the slope of the line, and the value of b equals the value of y when x equals zero.

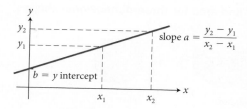

If $a = 0$, the line is horizontal. If $a > 0$, the line rises as x increases. If $a < 0$, the line falls as x increases. For any two values of x, say x_1 and x_2, the slope a can be calculated as

$$a = \frac{y_2 - y_1}{x_2 - x_1}$$

where y_1 and y_2 correspond to x_1 and x_2 (that is to say, $y_1 = ax_1 + b$ and $y_2 = ax_2 + b$).

Proportionality

If y is proportional to x (written $y \propto x$), then $y = ax$, where a is a constant. Proportionality is a subset of linearity. Because $y/x = a = $ constant for any corresponding x and y,

$$\frac{y_1}{x_1} = \frac{y_2}{x_2} \Leftrightarrow \frac{y_1}{y_2} = \frac{x_1}{x_2}.$$

Quadratic equation

The equation $ax^2 + bx + c = 0$ (the quadratic equation) has two solutions (called *roots*) for x:

$$x = \frac{-b \pm \sqrt{b^2 - 4ac}}{2a}$$

If $b^2 \geq 4ac$, the solutions are real numbers.

2 Geometry

Area and circumference for two-dimensional shapes

rectangle:
area $= ab$
circumference $= 2(a + b)$

paralellogram:
area $= bh$
circumference $= 2(a + b)$

triangle:
area $= \frac{1}{2}bh$
circumference $= a + b + c$

circle:
area $= \pi r^2$
circumference $= 2\pi r$

Volume and area for three-dimensional shapes

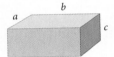

rectangular box:
volume $= abc$
area $= 2(a^2 + b^2 + c^2)$

sphere:
volume $= \frac{4}{3}\pi r^3$
area $= 4\pi r^2$

right circular cylinder:
volume $= \pi r^2 \ell$
area $= 2\pi r \ell + 2\pi r^2$

right circular cone:
volume $= \frac{1}{3}\pi r^2 h$
area $= \pi r^2 + \pi r \sqrt{r^2 + h^2}$

3 Trigonometry

Angle and arc length

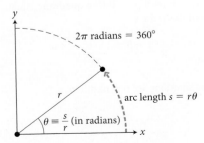

2π radians $= 360°$

arc length $s = r\theta$

$\theta \equiv \frac{s}{r}$ (in radians)

Right triangles

A right triangle is a triangle in which one of the angles is a right angle:

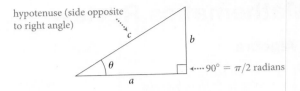

hypotenuse (side opposite to right angle)

$\cdots\cdots$ 90° $= \pi/2$ radians

Pythagorean theorem: $a^2 + b^2 = c^2 \Leftrightarrow c = \sqrt{a^2 + b^2}$

Trigonometric functions:

$$\sin\theta = \frac{b}{c} = \frac{\text{opposite side}}{\text{hypotenuse}}, \quad \theta = \sin^{-1}\left(\frac{b}{c}\right) = \arcsin\left(\frac{b}{c}\right)$$

$$\cos\theta = \frac{a}{c} = \frac{\text{adjacent side}}{\text{hypotenuse}}, \quad \theta = \cos^{-1}\left(\frac{a}{c}\right) = \arccos\left(\frac{a}{c}\right)$$

$$\tan\theta = \frac{b}{a} = \frac{\text{opposite side}}{\text{adjacent side}}, \quad \theta = \tan^{-1}\left(\frac{b}{a}\right) = \arctan\left(\frac{b}{a}\right)$$

General triangles

For any triangle, the following relationships hold:

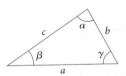

$$\alpha + \beta + \gamma = 180° = \pi \text{ rad}$$

Sine law: $\dfrac{\sin\alpha}{a} = \dfrac{\sin\beta}{b} = \dfrac{\sin\gamma}{c}$

Cosine law: $c^2 = a^2 + b^2 - 2ab\cos\gamma$

Identities

$$\tan\theta = \frac{\sin\theta}{\cos\theta}$$

$$\cot\theta = \frac{1}{\tan\theta} = \frac{\cos\theta}{\sin\theta}$$

$$\csc\theta = \frac{1}{\sin\theta}$$

$$\sec\theta = \frac{1}{\cos\theta}$$

Periodicity

$$\cos(\alpha + 2\pi) = \cos\alpha$$

$$\tan(\alpha + \pi) = \sin\alpha$$

Angle addition

$$\sin(\alpha \pm \beta) = \sin\alpha\cos\beta \pm \cos\alpha\sin\beta$$

$$\cos(\alpha \pm \beta) = \cos\alpha\cos\beta \mp \sin\alpha\sin\beta$$

Double angles

$$\sin(2\alpha) = 2\sin\alpha\cos\alpha$$

$$\cos(2\alpha) = \cos^2\alpha - \sin^2\alpha = 1 - 2\sin^2\alpha = 2\cos^2\alpha - 1$$

Other relations

$$\sin^2\alpha + \cos^2\alpha = 1$$

$$\sin(-\alpha) = -\sin\alpha$$

$$\cos(-\alpha) = \cos\alpha$$

$$\sin(\alpha \pm \pi) = -\sin\alpha$$

$$\cos(\alpha \pm \pi) = -\cos\alpha$$

$$\sin(\alpha \pm \pi/2) = \pm\cos\alpha$$

$$\cos(\alpha \pm \pi/2) = \mp\sin\alpha$$

The following graphs show $\sin\theta$, $\cos\theta$, and $\tan\theta$ as functions of θ:

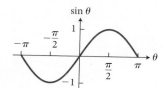
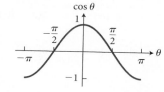

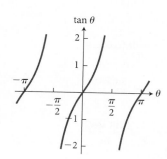

4 Calculus

In this section, x is a variable, and a and n are constants.

Derivatives

Geometrically, the derivative of a function $f(x)$ at $x = x_1$ is the slope of $f(x)$ at x_1:

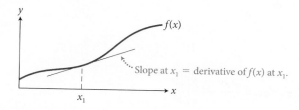

Derivatives of common functions

$$\frac{d}{dx}a = 0$$

$$\frac{d}{dx}x^n = nx^{n-1} \quad (n \text{ need not be an integer})$$

$$\frac{d}{dx}\sin x = \cos x$$

$$\frac{d}{dx}\cos x = -\sin x$$

$$\frac{d}{dx}\tan x = \frac{1}{\cos^2 x}$$

$$\frac{d}{dx}e^{ax} = ae^{ax}$$

$$\frac{d}{dx}\ln(ax) = \frac{1}{x}$$

$$\frac{d}{dx}a^x = a^x\ln a$$

Derivatives of sums, products, and functions of functions

Constant times a function: $\dfrac{d}{dx}[a \cdot f(x)] = a \cdot \dfrac{d}{dx}f(x)$

Sum of functions: $\dfrac{d}{dx}[f(x) + g(x)] = \dfrac{d}{dx}f(x) + \dfrac{d}{dx}g(x)$

Product of functions:

$$\frac{d}{dx}[f(x)\cdot g(x)] = g(x)\frac{d}{dx}f(x) + f(x)\frac{d}{dx}g(x)$$

Quotient of functions: $\dfrac{d}{dx}\left[\dfrac{f(x)}{g(x)}\right] = \dfrac{g(x)\dfrac{d}{dx}f(x) - f(x)\dfrac{d}{dx}g(x)}{[g(x)]^2}$

Functions of functions (the chain rule): If f is a function of u, and u is a function of x, then

$$\frac{d[f(u)]}{du} \cdot \frac{d[u(x)]}{dx} = \frac{d[f(x)]}{dx}$$

Second and higher derivatives The second derivative of a function f with respect to x is the derivative of the derivative:

$$\frac{d^2f(x)}{dx^2} = \frac{d}{dx}\left(\frac{d}{dx}f(x)\right)$$

Higher derivatives are defined similarly:

$$\frac{d^nf(x)}{dx^n} = \underbrace{\cdots \frac{d}{dx}\left(\frac{d}{dx}\left(\frac{d}{dx}f(x)\right)\right)}_{n \text{ uses of } \frac{d}{dx}} \quad \text{(where } n \text{ is a positive integer).}$$

Partial derivatives For functions of more than one variable, the partial derivative, written $\dfrac{\partial}{\partial x}$, is the derivative with respect to one variable; all other variables are treated as constants.

Integrals

Indefinite integrals Integration is the reverse of differentiation. An indefinite integral $\int f(x)dx$ is a function whose derivative is $f(x)$.

That is to say, $\dfrac{d}{dx}\left[\int f(x)dx\right] = f(x)$.

If $A(x)$ is an indefinite integral of $f(x)$, then so is $A(x) + C$, where C is any constant. Thus, it is customary when evaluating indefinite integrals to add a "constant of integration" C.

Definite integrals The definite integral of $f(x)$, written as $\int_{x_1}^{x_2} f(x)dx$, represents the sum of the area of contiguous rectangles that each intersect $f(x)$ at some point along one base and that each have another base coincident with the x axis over some part of the range between x_1 and x_2; the indefinite integral evaluates the sum in the limit of arbitrarily small rectangle bases. In other words, the indefinite integral gives the net area that lies under $f(x)$ but above the x axis between the boundaries x_1 and x_2.

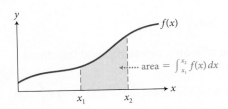

area $= \int_{x_1}^{x_2} f(x)\,dx$

If $A(x)$ is any indefinite integral of $f(x)$, then the definite integral is given by $\int_{x_1}^{x_2} f(x)dx = A(x_2) - A(x_1) \equiv A(x)|_{x_1}^{x_2}$. The constant of integration C does not affect the value of definite integrals and thus can be ignored (i.e., set to zero) during evaluation.

Integration by parts $\int_a^b u\,dv$ is the area under the curve of $u(v)$. If $\int_a^b u\,dv$ is difficult to evaluate directly, it is sometimes easier to express the area under the curve as the area within part of a rectangle minus the area under the curve of $v(u)$. In other words:

$$\int_a^b u\,dv = uv\Big|_a^b - \int_a^b v\,du.$$

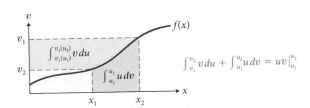

$\int_{v_1}^{v_2} v\,du + \int_{u_1}^{u_2} u\,dv = uv\Big|_{u_1}^{u_2}$

By choosing u and dv appropriately (both can be functions of x), this approach, called "integration by parts", can transform difficult integrals into easier ones.

Table of integrals In the following expressions, a and b are constants. An arbitrary constant of integration C can be added to the right-hand side.

$$\int x^n dx = \frac{1}{n+1}x^{n+1} \ \text{(for } n \neq -1)$$

$$\int x^{-1}dx = \ln|x|$$

$$\int \frac{1}{a^2 + x^2}dx = \frac{1}{a}\tan^{-1}\frac{x}{a}$$

$$\int \frac{1}{(a^2 + x^2)^2}dx = \frac{1}{2a^3}\tan^{-1}\frac{x}{a} + \frac{x}{2a^2(x^2 + a^2)}$$

$$\int \frac{1}{\sqrt{\pm a^2 + x^2}}dx = \ln\left|x + \sqrt{\pm a^2 + x^2}\right|$$

$$\int \frac{1}{\sqrt{a^2 - x^2}}dx = \sin^{-1}\frac{x}{|a|} = \tan^{-1}\frac{x}{\sqrt{a^2 - x^2}}$$

$$\int \frac{x}{\sqrt{\pm a^2 - x^2}}dx = -\sqrt{\pm a^2 - x^2}$$

$$\int \frac{x}{\sqrt{\pm a^2 + x^2}}dx = \sqrt{\pm a^2 + x^2}$$

$$\int \frac{1}{(\pm a^2 + x^2)^{3/2}}dx = \frac{\pm x}{a^2\sqrt{\pm a^2 + x^2}}$$

$$\int \frac{x}{(a^2 + x^2)^{3/2}}dx = -\frac{1}{\sqrt{a^2 + x^2}}$$

$$\int \frac{1}{a + bx}dx = \frac{1}{b}\ln(a + bx)$$

$$\int \frac{1}{(a + bx)^2}dx = -\frac{1}{b(a + bx)}$$

$$\int \sin(ax)dx = -\frac{1}{a}\cos(ax)$$

$$\int \cos(ax)dx = \frac{1}{a}\sin(ax)$$

$$\int \tan(ax)dx = -\frac{1}{a}\ln(\cos ax)$$

$$\int \sin^2(ax)dx = \frac{x}{2} - \frac{\sin 2ax}{4a}$$

$$\int \cos^2(ax)dx = \frac{x}{2} + \frac{\sin 2ax}{4a}$$

$$\int x\sin(ax)dx = \frac{1}{a^2}\sin ax - \frac{1}{a}x\cos ax$$

$$\int x\cos(ax)dx = \frac{1}{a^2}\cos ax + \frac{1}{a}x\sin ax$$

$$\int e^{ax}dx = \frac{1}{a}e^{ax}$$

$$\int xe^{ax}dx = \frac{e^{ax}}{a^2}(ax - 1)$$

$$\int x^2 e^{ax}dx = \frac{x^2 e^{ax}}{a} - \frac{2}{a}\left[\frac{e^{ax}}{a^2}(ax - 1)\right]$$

$$\int \ln ax\,dx = x\ln(ax) - x$$

$$\int_0^\infty x^n e^{-ax}dx = \frac{n!}{a^{n+1}}$$

$$\int_0^\infty e^{-ax^2}dx = \frac{1}{2}\sqrt{\frac{\pi}{a}}$$

5 Vector algebra

A vector $\vec{A}$ in three-dimensional space can be written in terms of magnitudes A_x, A_y, and A_z of unit vectors $\hat{\imath}$, $\hat{\jmath}$, and $\hat{k}$, which have length 1 and lie along the x, y, and z axes:

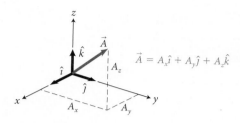

Dot products between vectors produce scalars:

$$\vec{A} \cdot \vec{B} = A_x B_x + A_y B_y + A_z B_z = |A||B|\cos\theta$$
(θ is the angle between vectors $\vec{A}$ and $\vec{B}$)

Cross products between vectors produce vectors:

$$\vec{A} \times \vec{B} = (A_y B_z - A_z B_y)\hat{\imath} + (A_z B_x - A_x B_z)\hat{\jmath} + (A_x B_y - A_y B_x)\hat{k}$$

$$|\vec{A} \times \vec{B}| = |\vec{A}||\vec{B}|\sin\theta \ (\theta \text{ is the angle between vectors } \vec{A} \text{ and } \vec{B})$$

The direction of $\vec{A} \times \vec{B}$ is given by the right-hand rule (see Figure 12.44).

6 Complex numbers

A complex number $z = x + iy$ is defined in terms of its real part x and its imaginary part y. Both x and y are real numbers. i is Euler's constant, defined by the property $i^2 = -1$.

Each complex number z has a "complex conjugate" z^* which has the same real part but an imaginary part with opposite sign: $z = x + iy \Leftrightarrow z^* = x - iy$.

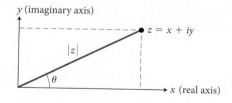

The real and imaginary parts can be expressed in terms of the complex number and its conjugate:

$$x = \tfrac{1}{2}(z + z^*)$$

$$y = \tfrac{1}{2}i(z - z^*)$$

A complex number is like a two-dimensional vector in a plane with a real axis and an imaginary axis. Thus, z can be described by a magnitude or length $|z|$ and an angle θ formed with the real axis (called the "phase angle"):

$$z = |z|(\cos\theta + i\sin\theta), \text{ where } |z| = \sqrt{zz^*} \text{ and}$$

$$\theta = \tan^{-1}\frac{y}{x} = \tan^{-1}\frac{i(z - z^*)}{(z + z^*)}.$$

Euler's formula says that $e^{i\theta} = \cos\theta + i\sin\theta$, allowing complex numbers to be written in the form $z = |z|e^{i\theta}$. This is a convenient form for expressing complex numbers. For example, it is easy to raise a complex number z to a power n: $z^n = |z|^n e^{in\theta}$.

7 Useful approximations

Binomial expansion

$$(1 + x)^n = 1 + nx + \frac{n(n-1)}{2}x^2 + \cdots$$

If $x \ll 1$, then $(1 + x)^n \approx 1 + nx$

Trigonometric expansions

$$\sin\alpha = \alpha - \frac{\alpha^3}{3!} + \frac{\alpha^5}{5!} - \frac{\alpha^7}{7!} + \cdots \ (\alpha \text{ in rad})$$

$$\cos\alpha = 1 - \frac{\alpha^2}{2!} + \frac{\alpha^4}{4!} - \frac{\alpha^6}{6!} + \cdots \ (\alpha \text{ in rad})$$

$$\tan\alpha = \alpha + \frac{1}{3}\alpha^3 + \frac{2}{15}\alpha^5 + \frac{17}{315}\alpha^7 + \cdots \ (\alpha \text{ in rad})$$

If $\alpha \ll 1$ rad, then $\sin\alpha \approx \alpha$, $\cos\alpha \approx 1$, and $\tan\alpha \approx \alpha$.

Other useful expansions

$$\frac{1}{1 - x} = 1 + x + x^2 + x^3 + \cdots \text{ for } -1 < x < 1$$

$$e^x = 1 + x + \frac{1}{2}x^2 + \frac{1}{6}x^3 + \frac{1}{24}x^4 + \cdots$$

$$\ln(1 + x) = x - \frac{1}{2}x^2 + \frac{1}{3}x^3 - \frac{1}{4}x^4 + \cdots \quad \text{for } -1 < x < 1$$

$$\ln\left(\frac{1 + x}{1 - x}\right) = 2x + \frac{2}{3}x^3 + \frac{2}{5}x^5 - \frac{2}{7}x^7 + \cdots \text{ for } -1 < x < 1$$

Appendix C

SI Units, Useful Data, and Unit Conversion Factors

The seven base SI units

Unit	Abbreviation	Physical quantity
meter	m	length
kilogram	kg	mass
second	s	time
ampere	A	electric current
kelvin	K	thermodynamic temperature
mole	mol	amount of substance
candela	cd	luminous intensity

Some derived SI units

Unit	Abbreviation	Physical quantity	In terms of base units
newton	N	force	$kg \cdot m/s^2$
joule	J	energy	$kg \cdot m^2/s^2$
watt	W	power	$kg \cdot m^2/s^3$
pascal	Pa	pressure	$kg/m \cdot s^2$
hertz	Hz	frequency	s^{-1}
coulomb	C	electric charge	$A \cdot s$
volt	V	electric potential	$kg \cdot m^2/(A \cdot s^3)$
ohm	Ω	electric resistance	$kg \cdot m^2/(A^2 \cdot s^3)$
farad	F	capacitance	$A^2 \cdot s^4/(kg \cdot m^2)$
tesla	T	magnetic field	$kg/(A \cdot s^2)$
weber	Wb	magnetic flux	$kg \cdot m^2/(A \cdot s^2)$
henry	H	inductance	$kg \cdot m^2/(A^2 \cdot s^2)$

SI Prefixes

10^n	Prefix	Abbreviation	10^n	Prefix	Abbreviation
10^0	—	—			
10^3	kilo-	k	10^{-3}	milli-	m
10^6	mega-	M	10^{-6}	micro-	μ
10^9	giga-	G	10^{-9}	nano-	n
10^{12}	tera-	T	10^{-12}	pico-	p
10^{15}	peta-	P	10^{-15}	femto-	f
10^{18}	exa-	E	10^{-18}	atto-	a
10^{21}	zetta-	Z	10^{-21}	zepto-	z
10^{24}	yotta-	Y	10^{-24}	yocto-	y

Values of fundamental constants

Quantity	Symbol	Value
Speed of light in vacuum	c_0	3.00×10^8 m/s
Gravitational constant	G	6.6738×10^{-11} N·m^2/kg^2
Avogadro's number	N_A	6.0221413×10^{23} mol^{-1}
Boltzmann's constant	k_B	1.380×10^{-23} J/K
Charge on electron	e	1.60×10^{-19} C
Permittivity constant	ϵ_0	$8.85418782 \times 10^{-12}$ C^2/(N·m^2)
Permeability constant	μ_0	$4\pi \times 10^{-7}$ T·m/A
Planck's constant	h	6.626×10^{-34} J·s
Electron mass	m_e	9.11×10^{-31} kg
Proton mass	m_p	1.6726×10^{-27} kg
Neutron mass	m_n	1.6749×10^{-27} kg
Atomic mass unit	amu	1.6605×10^{-27} kg

Other useful numbers

Number or quantity	Value
π	3.1415927
e	2.7182818
1 radian	57.2957795°
Absolute zero ($T = 0$)	-273.15 °C
Average acceleration g due to gravity near Earth's surface	9.8 m/s^2
Speed of sound in air at 20 °C	343 m/s
Density of dry air at atmospheric pressure and 20 °C	1.29 kg/m^3
Earth's mass	5.97×10^{24} kg
Earth's radius (mean)	6.38×10^6 m
Earth–Moon distance (mean)	3.84×10^8 m

Unit conversion factors

Length

1 in. = 2.54 cm (defined)

1 cm = 0.3937 in.

1 ft = 30.48 cm

1 m = 39.37 in. = 3.281 ft

1 mi = 5280 ft = 1.609 km

1 km = 0.6214 mi

1 nautical mile (U.S.) = 1.151 mi = 6076 ft = 1.852 km

1 fermi = 1 femtometer (fm) = 10^{-15} m

1 angstrom (Å) = 10^{-10} m = 0.1 nm

1 light − year (ly) = 9.461×10^{15} m

1 parsec = 3.26 ly = 3.09×10^{16} m

Volume

1 liter (L) = 1000 mL = 1000 cm^3 = 1.0×10^{-3} m^3
       = 1.057 qt (U.S.) = 61.02 $in.^3$

1 gal (U.S.) = 4 qt (U.S.) = 231 $in.^3$ = 3.785 L = 0.8327 gal (British)

1 quart (U.S.) = 2 pints (U.S.) = 946 mL

1 pint (British) = 1.20 pints (U.S.) = 568 mL

1 m^3 = 35.31 ft^3

Speed

1 mi/h = 1.4667 ft/s = 1.6093 km/h = 0.4470 m/s

1 km/h = 0.2778 m/s = 0.6214 mi/h

1 ft/s = 0.3048 m/s = 0.6818 mi/h = 1.0973 km/h

1 m/s = 3.281 ft/s = 3.600 km/h = 2.237 mi/h

1 knot = 1.151 mi/h = 0.5144 m/s

Angle

1 radian (rad) = 57.30° = 57°18′

1° = 0.01745 rad

1 rev/min (rpm) = 0.1047 rad/s

Time

1 day = 8.640×10^4 s

1 year = 365.242 days = 3.156×10^7 s

Mass

1 atomic mass unit (u) = 1.6605×10^{-27} kg

1 kg = 0.06852 slug

1 metric ton = 1000 kg

1 long ton = 2240 lbs = 1016 kg

1 short ton = 2000 lbs = 909.1 kg

1 kg has a weight of 2.20 lb where g = 9.80 m/s^2

Force

1 lb = 4.44822 N

1 N = 10^5 dyne = 0.2248 lb

Energy and work

1 J = 10^7 ergs = 0.7376 ft · lb

1 ft · lb = 1.356 J = 1.29×10^{-3} Btu = 3.24×10^{-4} kcal

1 kcal = 4.19×10^3 J = 3.97 Btu

1 eV = 1.6022×10^{-19} J

1 kWh = 3.600×10^6 J = 860 kcal

1 Btu = 1.056×10^3 J

Power

1 W = 1 J/s = 0.7376 ft · lb/s = 3.41 Btu/h

1 hp = 550 ft · lb/s = 746 W

1 kWh/day = 41.667 W

Pressure

1 atm = 1.01325 bar = 1.01325×10^5 N/m^2 = 14.7 lb/$in.^2$ = 760 torr

1 lb/$in.^2$ = 6.895×10^3 N/m^2

1 Pa = 1 N/m^2 = 1.450×10^{-4} lb/$in.^2$

Periodic Table of the Elements

Key/Legend:

Number of protons → 29
Symbol for element → **Cu**
63.546

Average atomic mass in g/mol. For elements having no stable isotope, value in parentheses is approximate atomic mass of longest-lived isotope.

Period \ Group	1	2	3	4	5	6	7	8	9	10	11	12	13	14	15	16	17	18
1	1 **H** 1.008																	2 **He** 4.003
2	3 **Li** 6.941	4 **Be** 9.012											5 **B** 10.811	6 **C** 12.011	7 **N** 14.007	8 **O** 15.999	9 **F** 18.998	10 **Ne** 20.180
3	11 **Na** 22.990	12 **Mg** 24.305											13 **Al** 26.982	14 **Si** 28.086	15 **P** 30.974	16 **S** 32.065	17 **Cl** 35.453	18 **Ar** 39.948
4	19 **K** 39.098	20 **Ca** 40.078	21 **Sc** 44.956	22 **Ti** 47.867	23 **V** 50.942	24 **Cr** 51.996	25 **Mn** 54.938	26 **Fe** 55.845	27 **Co** 58.933	28 **Ni** 58.693	29 **Cu** 63.546	30 **Zn** 65.409	31 **Ga** 69.723	32 **Ge** 72.64	33 **As** 74.922	34 **Se** 78.96	35 **Br** 79.904	36 **Kr** 83.798
5	37 **Rb** 85.468	38 **Sr** 87.62	39 **Y** 88.906	40 **Zr** 91.224	41 **Nb** 92.906	42 **Mo** 95.94	43 **Tc** (98)	44 **Ru** 101.07	45 **Rh** 102.906	46 **Pd** 106.42	47 **Ag** 107.868	48 **Cd** 112.411	49 **In** 114.818	50 **Sn** 118.710	51 **Sb** 121.760	52 **Te** 127.60	53 **I** 126.904	54 **Xe** 131.293
6	55 **Cs** 132.905	56 **Ba** 137.327	71 **Lu** 174.967	72 **Hf** 178.49	73 **Ta** 180.948	74 **W** 183.84	75 **Re** 186.207	76 **Os** 190.23	77 **Ir** 192.217	78 **Pt** 195.078	79 **Au** 196.967	80 **Hg** 200.59	81 **Tl** 204.383	82 **Pb** 207.2	83 **Bi** 208.980	84 **Po** (209)	85 **At** (210)	86 **Rn** (222)
7	87 **Fr** (223)	88 **Ra** (226)	103 **Lr** (262)	104 **Rf** (261)	105 **Db** (262)	106 **Sg** (266)	107 **Bh** (264)	108 **Hs** (269)	109 **Mt** (268)	110 **Ds** (271)	111 **Rg** (272)	112 **Uub** (285)	113 **Uut** (284)	114 **Uuq** (289)	115 **Uup** (288)	116 **Uuh** (292)	117 **Uus** (294)	118 **Uuo**

Lanthanoids

57 **La** 138.905	58 **Ce** 140.116	59 **Pr** 140.908	60 **Nd** 144.24	61 **Pm** (145)	62 **Sm** 150.36	63 **Eu** 151.964	64 **Gd** 157.25	65 **Tb** 158.925	66 **Dy** 162.500	67 **Ho** 164.930	68 **Er** 167.259	69 **Tm** 168.934	70 **Yb** 173.04

Actinoids

89 **Ac** (227)	90 **Th** (232)	91 **Pa** (231)	92 **U** (238)	93 **Np** (237)	94 **Pu** (244)	95 **Am** (243)	96 **Cm** (247)	97 **Bk** (247)	98 **Cf** (251)	99 **Es** (252)	100 **Fm** (257)	101 **Md** (258)	102 **No** (259)

Answers to Selected Odd-Numbered Questions and Problems

Chapter 1

1. Undetectable

3. That the sequence is linear, meaning the difference between any two adjacent digits is 1.

5. 12 ways

7. One

9. T, A: reflection symmetry across vertical line passing through letter center. E, B: reflection symmetry across horizontal line passing through letter center. L, S: no reflection symmetry.

11. 9 axes of reflection symmetry, 13 axes of rotational symmetry

13. Two axes of reflection symmetry

15. (a) 1.5×10^{14} mm (b) 12,000 Earths

17. 10^4 gastrotrich lifetimes/tortoise lifetime

19. 10^9 to 10^{10} books

21. (a) Either one order of magnitude or none. Performing $V = \ell^3$ first yields $V_1 = \ell_1^3$, $V_2 = (2\ell_1)^3 = 8\ell_1^3$, which rounds to $10\ell_1^3$ so that $V_2 = 10V_1$, one order of magnitude. (b) Yes, because of the rules of rounding numerical values. For example, if $V_1 = 3.5$ m³, that value would round to an order of magnitude of 10 m³. Then $V_2 = 8V_1 = 28$ m³, which also rounds to an order of magnitude of 10 m³.

23. 10^5 leaves

25. Not reasonable. Because light travels much faster than sound, any thunder peal is delayed compared to the light signal caused by the lightning bolt event. From the principle of causality, the lightning you see after you hear the peal cannot have caused the peal.

27. That the barrier lowers time after time 30 s before a train passes is consistent with a causal relationship between the two events. The single negative result, however, tells you that the lowering of the barrier cannot be the *direct* cause of the passing of the train. More likely, the lowering is triggered when the train passes a sensor quite a distance up the tracks from the barrier and the sensor sends an electrical signal to the lowering mechanism. A malfunction in either the sensor, the electrical connections, or the lowering mechanism would account for the one negative result you observed.

29. $E = mc^2$; E is type of energy described, m is object mass, c is speed of light.

31. If the 30° angles must be interior to adjacent sides, the resulting zigzag pattern gives a distance of 1.0ℓ (to 2 significant digits). If the 30° angles must be exterior to adjacent sides, the result ranges from 3.4ℓ to 3.7ℓ. If a mixture of interior and exterior angles are allowed, the distance ranges from zero (parallelogram) to 2.4ℓ (one interior angle at end of chain).

33. 1.32×10^3 s

35. (a) The position decreases linearly as a function of time, from an initial position $x = 4.0$ m to a final position $x = 0$, reaching this final position at $t = 8.0$ s. (b) $x(t) = mt + b$ with $m = -0.5$ m/s, and $b = 4.0$ m

37. 352 in

39. (a) and (b). In both cases, the density of each piece is the same as the density of the original block.

41. No, because there is a significant difference (14 percent) in their mass densities.

43. Meters

45. (a) 10^{21} kg (b) 10^{25} kg (c) 10^{39} kg

47. (a) 3.00×10^8 m/s (b) 8.99×10^{16} m²/s² (c) No. Because you should wait until the final answer to round off, the value in part b is $(2.99792 \times 10^8 \text{ m/s})^2 = 8.98752 \times 10^{16}$ m²/s² rounded to three significant digits. The square of the value in part a is $(3.00 \times 10^8 \text{ m/s})^2 = 9.00 \times 10^{16}$ m²/s².

49. Four significant digits

51. 35,987.1 km

53. 0.17 L. However, in the absence of information about volume measuring devices, mixing a standard one liter is probably best.

55. 7.4×10^{-3} g/s

57. 1.6 m

59. Place two coins on the balance, and hold the third in your hand. If the two coins balance, the one in your hand is the counterfeit. If the two coins do not balance, one must be the counterfeit. Swap the lighter of the two for the coin in your hand. If the two coins on the balance now balance, the counterfeit coin is the one you just removed, and it is lighter than real coins. If the two coins are still unbalanced, the counterfeit is the one that stayed on the balance during the whole experiment, and it is heavier than real coins.

61. 10^{56} mol

63. 2×10^3 boards

65. 10^4 m

67. 10^8

69. 0.349 mm

71. No. Atoms are typically 10^5 times larger than nuclei. If you make the nucleus diameter 500 mm, the atom diameter must be 50 km.

73. No. The swing will never get closer to the ground than about 0.80 m and will never rise above the ground more than about 2 m.

75. (a) 10^{17} kg/m³ (b) 13 orders of magnitude larger than Earth mass density, 14 orders of magnitude larger than water mass density (c) 10^{14} kg

Chapter 2

1. Time interval between adjacent frames and object size

3.

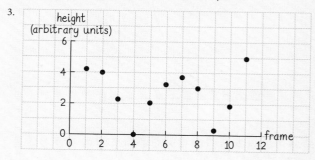

5. (*a*), (*c*), (*d*), (*e*), and (*f*)

7. Distance traveled 6.4 km, displacement zero

9. (*a*) Infinitely many (*b*) Infinitely many (*c*) Infinitely many (*d*) Two (one for each choice of positive direction)

11. If numerical values of time and distance are converted, but the scale of each axis still uses the same numerical labels (that is, "0.40 m" becomes "0.40 in"), then the curve would be much narrower and much taller. This is just a matter of perspective, though. If the scale of each axis is also converted, so that "0.40 m" becomes "16 in," then the shape of the graph is not changed by the conversion of units.

13. +1 block

15. The swimmer swims in the positive *x* direction at a constant speed (left sloping leg of curve, increasing *x* values). She stops briefly (horizontal leg, most probably at end of her lane) and then returns to the starting point (right sloping leg, decreasing *x* values) at a speed slightly lower than her initial speed (this leg not as steep as left leg).

17. Interpolation always gives a continuous path, but there is no reason to expect that the path is accurate everywhere. Suppose you are photographing a clock's pendulum at 1.0-s intervals, collecting data to use in a graph showing the pendulum's position as a function of time. Suppose further that the pendulum takes 1.0 s to swing from left to right and back left again. At this swing speed, the pendulum has just enough time between photographs to swing and return to its initial position, so that the photographs make it appear that the pendulum does not move at all. An interpolation of data points collected from the photographs would show a continuous horizontal line on a position-versus-time graph, which is certainly not correct.

19.

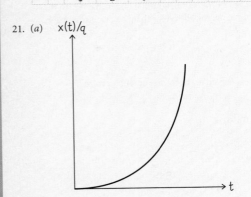

21. (*a*)

(*b*) $26qT^3\,\hat{\imath}$

25. (*a*) 10.2 m/s (*b*) 10.4 m/s (*c*) 9.24 m/s (*d*) 7.233 m/s (*e*) 6.3 m/s (*f*) 5.545 m/s

27. (*a*) Because the images were created at equal time intervals, the spacing between adjacent images is a function of the ball's speed. That the spacing between adjacent images has one value in the first five frames and a different value in the final five frames tells you that the ball moved at one speed at the beginning of the motion and at a different speed at the end of the motion. (*b*) During the first five frames.

29. (*a*) Just before instant t_2 and just after instant t_6. (*b*) From just before instant t_4 until instant t_6, which is where the bottom-curve slope is essentially the same as the top-curve slope.

31. B is closer to C than it is to A.

33. (*a*) 12 m/s (*b*) He likely treated the problem as though the cyclist rode at the two speeds for equal time intervals rather than for equal distances, and so just averaged the speeds: $(10\text{ m/s} + 16\text{ m/s})/2 = 13$ m/s.

35. 110 km/h

37. (1) No, because you could walk in either direction and so there are two possibilities for your final location. (2) Yes

39. (*a*) 3 m (*b*) 3 m/s (*c*) 3 m/s

41. (*a*) $A_x\hat{\imath}$ (*b*) $A_x\hat{\imath}$ (*c*) $A_x\hat{\imath}$

43. (*a*) 4.0 m (*b*) (+4.0 m) $\hat{\imath}$

45. (*a*) +0.52 m (*b*) +0.80 m (*c*) 0 (*d*) (+0.28 m)$\hat{\imath}$ (*e*) (−0.80 m)$\hat{\imath}$ (*f*) (−0.52 m)$\hat{\imath}$ (*g*) 0.66 m (*h*) 0.82 m (*i*) 1.5 m

47. $\vec{B} = -\vec{A}/2$

49. (*a*) $\vec{C} - \vec{A}$ (*b*) $\vec{A} - \vec{C}$

(*b*)

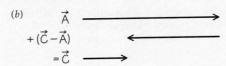

51. (*a*) (−0.42 m/s) $\hat{\imath}$ (*b*) 1.3 m/s (*c*) Because average velocity considers only actual distance between initial and final positions, but average speed considers distance traveled between these two positions; average speed is path-dependent, average velocity is not.

53. 87 km/h

55. (*a*)

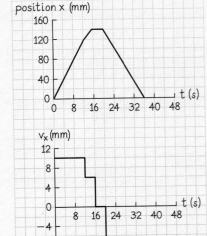

(*b*)

57. Object A

59.

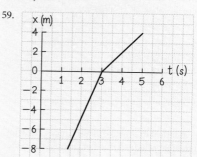

61. (a) 1.4 m/s (b) 10 m/s

63. 6.0×10^2 m

65. At 1.0 s: 45 m/s; at 4.0 s: 91 m/s

67. The fact that A passed B at 2:00 p.m. tells you that initially A was moving faster than 30 m/s. If A continued moving faster than 30 m/s eastward during the entire 1.0-h time interval, there is no way B could have caught up with A at 3:00 p.m. Therefore A must have slowed from a velocity above 30 m/s eastward to a velocity below 30 m/s eastward. Because a car cannot make discontinuous jumps in speed/velocity, B must have had a velocity of 30 m/s eastward at some instant. (Note that the speed could have always been greater than 30 m/s, but the velocity could not always have been greater than 30 m/s eastward.)

69. (a) Average speed of shadow leading edge (b) Yes, at the instant when the car is directly across from the light

71. (a) 21 m/s (b) 48 mi/h

73. (a) $x = -6.0$ m (b) $\vec{x} = (-6.0$ m$)\,\hat{\imath}$ (c) $x = 6.0$ m

75. By a time interval of $0.25(\Delta t)$, where Δt is A's time interval for the race

77. (a) 60 s (b) 3.6×10^2 m

79. (a) Greater than (b) Greater than

81. (a) 1:2 (b) 1:2

83. (a) $+0.39$ m/s (b) $+0.3603$ m/s (c) $\dfrac{x(t = 1.005) - x(t = 0.995)}{0.01 \text{ s}} =$

$(+0.360003$ m/s), whereas $v_x = \dfrac{dx}{dt} = 3ct^2$ such that

$v_x(t = 1.0 \text{ s}) = 0.36$ m/s

85. (a) $d/2\Delta t$ (b) $2\Delta t$ (c) The distance traveled by the runner is an infinite series that requires infinitely many terms to approach d, but the time intervals that correspond to these distance intervals also get smaller and smaller in the series. At higher and higher terms in the series, the runner travels almost no distance in each term but does so in a time interval that is almost zero. Because the two effects cancel each other, the runner travels the distance from starting line to finish line in a finite time interval.

87. Tortoise wins by 0.2 mi, even with Hare running at his top speed, 6.0 mi/h. In order to cross the finish line first, Hare must run at 10 mi/h once he gets up from his nap, far above his top speed.

89. (a) $x(t) = -p - qt - rt^2$ (b) $x(t) = (p - 2) + qt + rt^2$

(c)

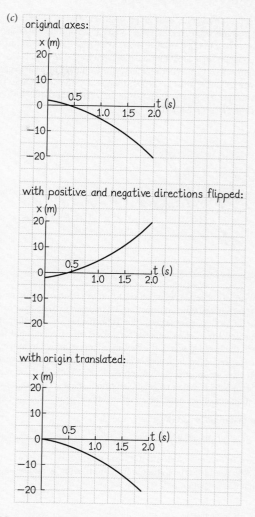

original axes:

with positive and negative directions flipped:

with origin translated:

(d) -74 m, $+74$ m, -76 m

(e) -35 m/s, $+35$ m/s, -35 m/s

(f) There is no physical difference because the only things changed are position chosen as origin and direction chosen as positive.

91. You pass 12 empty trucks along the way, with a thirteenth pulling out of the mill just as you enter. The next day, you pass 12 full trucks, with a thirteenth pulling out of the mine just as you enter.

Chapter 3

1. (a) From first dot to fifth dot, evidenced by fact that space between adjacent dots increases for these five dots. (b) From fifth dot to ninth dot, evidenced by fact that space between adjacent dots decreases for these five dots. (c) The answers would be the same because the dots would be designated "first" through "ninth" starting at the right end of the sequence instead of the left end.

3. Not accelerating, indicated by the horse appearing in same position in each frame. With cameras equally spaced and triggered at equal time intervals, the unchanging position in all frames means the horse's speed did not change as the photographs were taken.

5. No. In the position-versus-time graph in the Figure below your car is the solid line and your friend's is the dotted line. Point A, where the lines intersect for the first time, is where your friend passes. To catch up, you accelerate at a constant rate, indicated by the upward bend of the solid line after A. In order to intersect the dotted line again, the slope of the solid line must be steeper than that of the dotted line. This means your speed is greater than that of your friend at B, the position at which you catch up with him.

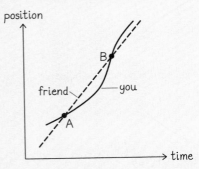

7. Initially you are at rest. When you start to move upward, your velocity has increased in the upward direction. If you take up to be the positive direction of motion, your acceleration is positive as you speed up from rest. This upward acceleration lasts for a short time interval, and soon, you reach the speed at which the elevator is programmed to move. Once that speed is reached, your acceleration is zero as you move at a constant velocity upward. As you approach the 19th floor, the elevator slows down, which makes your acceleration negative. This negative acceleration continues until the elevator stops at your floor.

9. The car that initially had the greater speed.

11. (a) 4.9 m/s (b) 15 m/s (c) 9.8 m/s

13. The curve in Figure 3.6a would still bend downward but not as much. Figure 3.6b would still be a straight line, but its slope would be about half what it is in that graph.

(a)

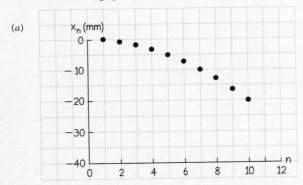

(b)

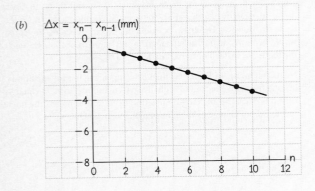

15. (a) Velocity 49 m/s upward, speed 49 m/s (b) Velocity 0, speed 0 (c) Velocity 49 m/s downward, speed 49 m/s (d) Velocity 98 m/s downward, speed 98 m/s

17. (a) Yes. A ball thrown upward stops for an instant at the peak of its path. At this instant, the velocity is zero but the acceleration is $g = -9.8$ m/s^2. (b) Yes. Any object moving at constant velocity has nonzero velocity and zero acceleration.

19. The two techniques cause the snowballs to hit the sidewalk at the same speed.

21. The curve would start out identical to the no-air-resistance case, with a slope of -9.8 m/s^2, but the magnitude of this slope would decrease with time. After some time interval, the curve would become horizontal, indicating a constant speed.

23.

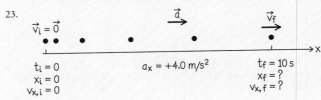

25. (b)

27.

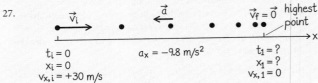

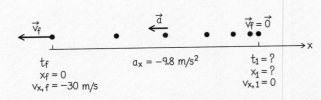

29. (a) 0.33 m/s^2 in direction of motion (b) 6.0×10^2 m

31. (a) 75 mm (b) 6.0×10^{13} m/s^2 in direction of motion

33. (a) You integrate. Determining the area under the curve is equivalent to adding up differential amounts of $v\delta t = \delta x$, and the sum of all such tiny steps is the displacement. (b) 1.0×10^2 m (c) No. In the figure below, the initial and final velocities are both zero, a case in which the equation from Worked Problem 3.3 yields zero for $v_{x,av}$. Yet the graph shows that the object has a positive velocity in the x direction for a significant time interval. Therefore the object must have a displacement in the x direction and its average velocity cannot be zero.

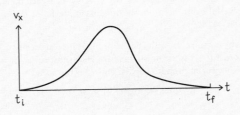

35. (a) 0.20 m downward (b) 0.22 m downward

37. (a) 4.2×10^{15} m/s^2 in direction of motion (b) 2.4 ns

39. (a) True. The fact that the curve is a straight line means the rate at which the velocity changes is constant. (b) Not necessarily true. The object could have started far enough from the origin that it never passes through position $x = 0$ on a position-versus-time plot. (c) True. The velocity is zero at the instant the curve crosses the time axis. (d) Not true. The fact that the curve crosses from positive v values to negative v values means the direction of the motion changed.

41. (a) 3.3×10^2 m (b) 38 s (c) 8.6 m/s

43. 13 m

45. 1.0 m

47. Factor of $\sqrt{2}$

49. The ball reaches its maximum height at $t = 3.0$ s, making the trajectory symmetric around that point. Between $t = 2.0$ s and $t = 3.0$ s, the ball moves upward and has a downward acceleration g; it moves some distance in this time interval. Between $t = 3.0$ s and $t = 4.0$ s, the ball, after having zero velocity for an instant, moves downward and has a downward acceleration g. Because the time intervals are equal, the distance the ball moves downward in 1.0 s equals the distance it moves upward in 1.0 s.

51. (a) 22 m/s (b) 2.3 s

53. (a) 0.65 s (b) 0.90 s

55. Let $\Delta y_{\Delta t} = \frac{1}{2} g (\Delta t)^2$ be the distance traveled after a time interval Δt and $\Delta y_{\Delta t+1} = \frac{1}{2} g (\Delta t + 1)^2$ be the distance traveled after time interval $\Delta t + 1$. The difference between these two distances is the distance traveled in the 1-s interval between these two intervals. Call this distance $h_{\Delta t} = g(\Delta t + \frac{1}{2})$, and because we are restricting ourselves to time intervals of 1 s, we could also write $h_N = g(N + \frac{1}{2})$ where N is an integer number of seconds. The ratio of the distance traveled in the first second to the distance traveled in the Nth second is

$$\frac{h_0}{h_N} = \frac{\frac{1}{2}}{N + \frac{1}{2}} = \frac{1}{2N + 1}.$$

Inserting the first few values for N yields $1/3$, $1/5$, $1/7$, $1/9$, . . .

57. 14 m

59. (a) 29 m/s upward (b) 36 m (c) 42 m

61. 13.6°

63. (a) Child on more-inclined slide (40°) (b) Child on more inclined slide (40°)

65. (a) 2.0 m/s² (b) 2.0 m/s² (c) 12 s (d) 20 m/s (e) 24 m/s

67. (a) 1.5 s (b) 3.1 m/s

69. (a) $\sqrt{2g \sin(\theta)\ell}$ (b) $\sqrt{g \sin(\theta)\ell}$

71. 35°

73. 3.1 m

75. $a_x(t) = 6bt$

77. (a) Yes; acceleration constant (b) 6.12 m/s in $+x$ direction (c) 6.08 m/s in $+x$ direction, 6.16 m/s in $+x$ direction (d) 0.400 m/s² in $+x$ direction (e) 0.400 m/s² in $+x$ direction at both instants

79. (a) 10.0 m/s² in $+x$ direction (b) 50.0 m/s in $+x$ direction (c) 5.00 m/s² in $+x$ direction (d) 167 m

81. (a) 5.1 m (b) 1.8 s

83. (a) $a_x(t) = -v_{max}\omega \sin(\omega t)$ (b) $x(t) = (v_{max}/\omega) \sin(\omega t)$

85. 63 m

87. (a) 6.4×10^5 m/s² in direction opposite direction of motion (b) 0.25 ms (c) 0.18 m

89. (a) 2.24 m/s² (b) 21 m/s (c) 1.0×10^2 m

91. (a) 44.3 m/s (b) 4.52 s

93. (a) $(20\text{ m})\hat{\imath}$, $(29\text{ m})\hat{\imath}$, $(29\text{ m})\hat{\imath}$, $(20\text{ m})\hat{\imath}$ (b) $(15\text{ m/s})\hat{\imath}$, $(4.9\text{ m/s})\hat{\imath}$, $(-4.9\text{ m/s})\hat{\imath}$, $(-15\text{ m/s})\hat{\imath}$ (c) 0 (d) 12 m/s

95. 9.6×10^2 m/s² upward

97. $4g$, or 4×10^1 m/s² upward

99. (a) 6.0 m/s² upward (b) 3.3 s

101. (a) 11 m/s (b) upward (c) 5.2 m

103. 1.2 s

105. "Cloud-scraper" goes highest (1.5×10^2 m).

107. $a_1 = \dfrac{2d}{(\Delta t)^2} = 20.2$ m/s², $a_2 = \dfrac{v_f^2}{2d} = 11.5$ m/s², $a_3 = \dfrac{v_f - v_i}{t} = 15.3$ m/s² (b) Student 3 calculated the average acceleration, and his value is approximately halfway between the values calculated by students 1 and 2. (c) Acceleration was not constant, but students 1 and 2 assumed it was.

109. Don't take the bet. You can throw a stone to a maximum height of 6.7 m.

Chapter 4

1. Puck 2 had twice the initial speed of puck 1.

3. No for object 1, yes for object 2. Friction between two objects opposes relative motion between them, but object 1 speeds up as it travels across the surface, meaning friction cannot account for the change in v_x. Object 2 slows down as it travels across the surface, meaning friction is one valid explanation for the change in v_x. (Though not the only explanation: A hockey stick slowing a puck sliding on ice, for instance, yields the same $v(t)$ curve as that shown for object 2.)

5. Object 2 has three times the inertia of object 1.

7.

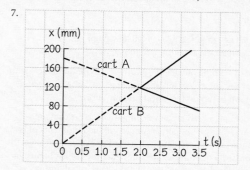

9. (a) Car A: upper solid line; car B: lower solid line (b) Car A

11. Inertia of A 2.5 times greater than inertia of B

13. Does not change

15. The inertia of just the bottle is the same in each case, but if the contents are included then the full bottle has the greater inertia because the water filling it has a greater inertia than the air filling the "empty" bottle.

17. Graph (a): smooth icy track; one wooden cart, one plastic cart; plastic cart initially at rest, wooden cart initially moving at 1.5 m/s. Graph (b): dusty unpolished track; carts made of same material but we cannot determine whether wood or plastic because we have no information about friction between each cart and surface; carts initially moving in same direction, one at 2.0 m/s and one at 5.0 m/s. Graph (c): rough damaged track; carts made of same material but we cannot determine whether wood or plastic because we have no information about friction between each cart and surface; one cart initially at rest, one initially moving.

19. (a) Extensive (c) No; number of passengers changes as people get on or off at each stop. In principle, people could also die or be born on the bus. (d) Yes, ignoring any deaths or births

21. 32 ways, assuming 5 objects named (person, truck, ball, friend, ground) cannot be subdivided.

23. 0.33 kg

25. 1.0 m/s to left

27. (a) $v_{1x,f} = -0.13$ m/s, $v_{2x,f} = +1.2$ m/s. (b) $\Delta v_{1x} = -1.6$ m/s, $\Delta v_{2x} = +1.2$ m/s. (c) 50 kg (d) $a_{1x} = -3.2$ m/s^2, $a_{2x} = +2.4$ m/s^2

29. Baseball

31. The two cars have the same change in momentum.

33. Yes. Momentum is a vector. Therefore the system has zero momentum if two conditions are met: The carts move in opposite directions and the absolute value of the product of inertia and velocity is the same for the two carts.

35. (a) $p_{1,i} = 0.196$ kg · m/s, $p_{1,f} = 0.131$ kg · m/s (b) $\vec{p}_{2i} = \vec{0}$, $\vec{p}_{2,f} = +0.327$ kg · m/s (c) 1.22×10^3 kg/m^3

37. 46 m

39. 1.8 kg · m/s

41. 2.0 m/s

43. (a) No (b) No (c) Yes

45. 2.4 m/s

47. (a) 6 kg · m/s to the right (b) Zero (c) No. The wall does not move because it is attached to the ground or some underlying structure.

49. (a) $+1.0$ kg · m/s (b) $+1.0$ kg · m/s (c) Yes

51. (a) $\Delta \vec{p}_A = -8.0$ kg · m/s $\hat{\imath}$, $\Delta \vec{p}_B = +8.0$ kg · m/s $\hat{\imath}$ (b) Zero (c) Yes, because change in momentum for system is zero

53. (a) No, provided nothing outside the two-object system interacts with the system. (b) Yes, the object initially moving can come to rest and impart all its momentum to the object initially at rest.

55. (a) It is not possible to predict the velocities of each car using momentum. (b) Yes, the system momentum (combined momenta of both cars) must be the same immediately afterward as before. More specifically, the momentum of both cars together after the collision must be 2000 kg · m/s in the direction of the 1200-kg car's initial motion.

57. Remove your clothing and throw it toward the shore, so that it has a nonzero momentum. You gain equal momentum directed toward the opposite shore and drift toward that shore.

59. 80 kg

61. (a) 0.40 m/s (b) 7.0×10^2 kg · m/s in direction of initial motion (c) higher

63. (a) Before collision, cue ball has momentum directed toward 8 ball and 8 ball has zero momentum. After collision, cue ball has zero momentum and 8 ball has momentum equal in magnitude and direction to momentum of cue ball before collision. (b) Call direc-

tion of motion the $+x$ direction. For the isolated system of two balls, $m_8(v_{8x,f} - v_{8x,i}) = -m_c(v_{cx,f} - v_{cx,i})$. With $v_{8x,i} = 0$, $v_{cx,f} = 0$, and $m_8 = m_c$, this yields $v_{8x,f} = v_{cx,i}$. Because $m_8 = m_c$, this corresponds to $p_{8x,f} = p_{cx,i}$, or zero change in system momentum, consistent with zero impulse and with the answer in (a). Considering each ball separately, the non-zero change in momentum is also equal to the non-zero impulse, because the impulse affecting each ball is equal in magnitude and in the opposite direction, just as the momentum changes of each ball are equal and opposite. Each of the four quantities has magnitude mv.

65. 2.0×10^4 kg

67. (a) $v_{1x,i} = +2.0$ m/s, $v_{1x,f} = 0$ (b) $v_{2x,i} = -0.33$ m/s, $v_{2x,f} = +0.33$ m/s (c) $\Delta p_{1,x} = -2.0$ kg · m/s, $\Delta p_{2,x} = +2.0$ kg · m/s (d) Yes

(e)

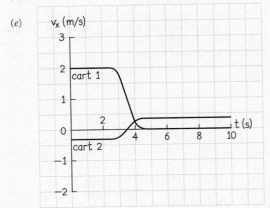

69. $m_{black} = 2m_{red}$

71. 2.0×10^2 kg

73. $p_{golf\,ball}/p_{baseball} = 1/2$

75. 5.6 kg

77.

	initial		final	
	A	B	A	B
golf balls	○	○	○	○
velocity	$\vec{v}_{A,i}\rightarrow$	$\vec{v}_{B,i} = \vec{0}$	$\vec{v}_{A,f} = \vec{0}$	$\vec{v}_{B,f}\rightarrow$
momentum	$\vec{p}_{A,i}\rightarrow$	$\vec{p}_{B,i} = \vec{0}$	$\vec{p}_{A,f} = \vec{0}$	$\vec{p}_{B,f}\rightarrow$
golf balls & basketball	A ○	B ◯	A ○	B ◯
velocity	$\vec{v}_{A,i}\rightarrow$	$\vec{v}_{B,i} = \vec{0}$	$\leftarrow\vec{v}_{A,f}$	$\vec{v}_{B,f}\rightarrow$
momentum	$\vec{p}_{A,i}\rightarrow$	$\vec{p}_{B,i} = \vec{0}$	$\leftarrow\vec{p}_{A,f}$	$\vec{p}_{B,f}\longrightarrow$

79. $v_{block,f} = \dfrac{m_{bullet}}{m_{block}}(v_{bullet,i} - v_{bullet,f})$

81. When a rocket is launched or a cannonball fired, the projectile has significant forward momentum. Because the momentum of the system is initially zero, after firing there must be momentum in the direction opposite the projectile direction. In a bazooka, material (largely burnt fuel) is expelled out the back, and so the required backward momentum need not come entirely from recoil. In a cannon, there is no outlet for air or debris at the back; consequently, the cannon must recoil with the same magnitude of momentum imparted to the cannonball.

83. In space, the rocket expels a huge number of gas particles at very high speeds. When these gas particles are ejected from the rear of the rocket, the rocket must acquire a forward velocity in order for system momentum to remain constant.

85. The two cars experience the same change in momentum $\Delta \vec{p} = m\Delta \vec{v}$ during the collision. If you are in the car of greater inertia, this momentum change is accomplished with a smaller velocity change. Because change in velocity over time is acceleration, being in the car of greater inertia means your body experiences the smaller acceleration in the collision, which is preferable in avoiding injury.

87. Don't drop the stereo because at the instant it reaches your friend its momentum is 53 kg · m/s downward. This could be harmful to both your friend and the stereo.

89. You get rid of the sandbags to decrease inertia and therefore momentum. If you simply drop them from the basket, your momentum when you reach the ground is 2.66×10^3 kg · m/s downward, below the critical value 2850 kg · m/s at which the basket is damaged. However, if you *throw* the sandbags downward, you increase their downward momentum and therefore decrease your downward momentum, making your momentum when you land far below the critical value.

91. (a) $3v_{ex}/4$ (b) $2v_{ex}/3$ (c) Two-stage, because each time a stage is detached, the inertia that must be accelerated in the next stage is decreased.

Chapter 5

1. (a) 3.0 m/s in negative x direction (b) 4.0 m/s in positive x direction

3. (a) Inelastic (b) Totally inelastic

5. Doubles momentum, quadruples kinetic energy

7. 2.5×10^6 kg

9. $m_A/m_B = 1/4$

11. For same momentum, $K_Y > K_X$. For same kinetic energy, $p_X > p_Y$.

13. The book slows down due to friction, and the book and table heat up slightly as the book's kinetic energy is converted to thermal energy.

15. Cannon: temperature, velocity (due to recoil), momentum. Ball: temperature, velocity, momentum, internal energy (due to slight deformation). Gunpowder: transformational energy (as solid converted to gas when ignited).

17. (a) I choose the system consisting of the two blocks only. It is not isolated; the blocks make contact with the spring outside the system, and this changes the momentum of the system.

(b)

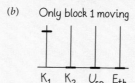

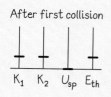

(c) The system starts out with kinetic energy. During the collision of the two blocks, this kinetic energy is changed into thermal energy. As the warmer blocks strike the spring, they momentarily come to a stop as their kinetic energy is changed to internal energy in the spring. Once the spring returns to its equilibrium length, the blocks regain kinetic energy. (d) $\frac{1}{2}v$

19. Kinetic energy of bike or car converted to thermal energy of brake pads and tires

21. The cars convert the same amount of kinetic energy to internal energy.

23. $|\Delta \vec{p}| = 2m|\vec{v}|$; $\Delta K = 0$; the answers are consistent because the change in momentum accounts for the change in velocity direction and the change in kinetic energy accounts for the change in velocity magnitude.

25. 6.3×10^{-3} J

27. (a) 5.0 m/s (b) −5.4 m/s (c) Yes (d) Zero

31. (a) No (b) Yes

33. $v_{1f} = -\dfrac{(m_1 - 2m_2)}{m_1 + m_2}v_{1i}$, $\quad v_{2f} = \dfrac{1}{2}\left(\dfrac{5m_1 - m_2}{m_1 + m_2}\right)v_{1i}$

35. Inelastic, $m_B = 0.16$ kg

39. $\dfrac{\text{Conversion at 34 m/s}}{\text{Conversion at 25 m/s}} = 1.8$

43. Collision in which initial momenta have same magnitude

45. $\sqrt{\frac{3}{4}}v_{end}$

47. (a) 5.0 m/s (b) No (c) 0.50 J

51. Answers between 1×10^8 and 3×10^8 megatons are reasonable.

53. The coefficients are each other's multiplicative inverse; reversing the time order reverses *initial* and *final* in definition of coefficient, thus inverting it.

55. (a) 0.20 m/s, to left (b) $e = \infty$ (c) 0.20 m/s, to right (d) $e = \infty$ (e) $\Delta K = +200$ J; from chemical energy stored in her body (f) 0.047 Cal

57. (a) 7000 m/s in direction of initial velocity (b) 1.013×10^{10} J

59. $v_{shuttle} = 890$ m/s in direction of initial motion; $v_{rocket} = 790$ m/s in direction of initial motion

61. −0.60 m/s

63. $v_i + \sqrt{\dfrac{3m_B E_{spring}}{2m_A(m_B + m_A)}}$

65. (a)

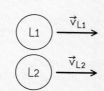

(b)

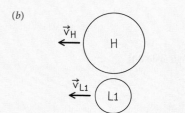

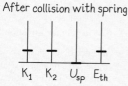

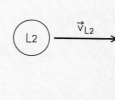

69. 1 m/s

71. (a) 2.0 m/s to right in Figure P5.70 (b) 0.098 J (c) $\vec{v}_{1f} = 0.66$ m/s to right, $\vec{v}_{2f} = 0.66$ m/s to right (d) 0.33 J (e) $e = 0$, as expected for a totally inelastic collision

77. Magnitude of momentum delivered by water each second is
$\frac{\Delta p}{t} = \frac{Q^2 \rho}{\pi r^2}$. You can determine whether the collision is elastic or
inelastic by comparing the rate at which the ball's momentum changes
with time to this momentum magnitude. You expect the collision to be
inelastic because water droplets stick to the ball.

Chapter 6

1. Stepping on, move quickly enough to have your speed match the sidewalk speed; stepping off, stop walking as you reach the sidewalk end to
minimize your speed relative to the ground.

3. 5.0 m/s in direction of motion of both cars

5. (a) Toward truck (b) Away from truck

7. The truck and the car traveling in the same direction: the truck driver
sees the car initially moving backwards at 30 m/s, then accelerating in
the direction of the truck travel, finally moving at zero velocity relative
to the truck. The car and truck traveling in opposite directions: the truck
driver sees the car initially approaching him at 30 m/s, then accelerating in the direction of the truck travel, finally approaching the truck at a
speed of 60 m/s. The direction of acceleration is the same in both cases.

9. Before the truck slows, the truck and the cans have the same initial
velocity. Because the cans are not attached to the truck, their velocity
cannot change when the truck slows; they continue moving in the
initial direction of motion until they hit up against the back of the cab.

11. (a) $\vec{p}_{A,i} = 2.5 \times 10^6$ kg·m/s west, $\vec{p}_{B,i} = 2.4 \times 10^6$ kg·m/s west,
$\vec{p}_{A,f} = 3.0 \times 10^6$ kg·m/s west, $\vec{p}_{B,f} = 2.0 \times 10^6$ kg·m/s west
(b) $\vec{p}_{A,i} = 1.7 \times 10^6$ kg·m/s east, $\vec{p}_{B,i} = 3.3 \times 10^5$ kg·m/s east,
$\vec{p}_{A,f} = 1.2 \times 10^6$ kg·m/s east, $\vec{p}_{B,f} = 8.0 \times 10^5$ kg·m/s east

13. 4.00 m/s, or 12.0 m/s, depending on direction of officer's motion

15. 0.75 m/s

17. (a) $+2.3$ kg·m/s $\hat{\imath}$ (b) $+3.8$ m/s $\hat{\imath}$

19. Yes, in reference frame moving at speed v toward cart A

23.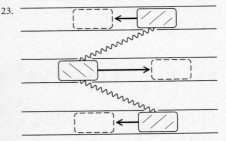

25. (a) 8.0 m/s (b) 5.0×10^4 kg·m/s in direction in which unit moves
(c) 8.0 m/s in direction in which unit moves

27. Lightweight object; no

29. Shorter

31. Yes

33. You do not walk fast enough to do this.

35. (a) $\Delta t_{calm} = 2d/v$

37. 4.66×10^3 km

39. Boy's raft; 2.5 m.

41. (a) 0.38 m to right of origin (b) 0.63 m to left of origin (c) 1.38 m to
right of origin. The center of mass location must be calculated for only
one origin choice; that location can then be expressed relative to the
other two origin choices with almost no calculation.

43. If the system's center of mass is not moving at constant velocity, the
reference frame from which motion is measured is not inertial.

45.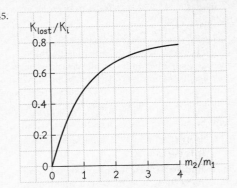

As m_2/m_1 approaches infinity, K_{lost}/K_i approaches 1, meaning all
kinetic energy is converted to other forms.

47. μ cannot be larger than m_1: $\mu_{max} = m_1$. Smallest possible value is
$\mu_{min} = m_1/2$.

49. Reference frame in which ball is at rest before collision

51. (a) 11 m/s to the right (b) 0.030 kg (c) 60% (d) 1.8 J

53. (a) 9.3 m/s (b) 19 m/s toward each other (c) 0.27 kg (d) 99.7%
(e) $\vec{v}_{1f} = 10$ m/s away from you; $\vec{v}_{2f} = 7.4$ m/s toward you

55. Collision between two objects of equal inertia m

57. (a) 0.25 m/s in direction in which ball is thrown (b) 4.6 m/s in direction in which ball is thrown (c) 16 J (d) 16 J

61. (a) 4.4 m/s (b) Before collision: $(1.0$ m/s$)m_{mother}$ toward swimming
penguin; after collision: $(2.3$ m/s$)m_{mother}$ toward swimming penguin

63. (a) $\vec{v}_{FG} = \dfrac{\vec{v}_{orange}}{2}\left(\dfrac{2m_{orange} + m_{apple}}{m_{orange} + m_{apple}}\right)$

65. (a) $\vec{v}_{rubber,f} = 18$ m/s west, $\vec{v}_{soft,f} = 8.9$ m/s east. (b) 12-J increase.
(c) Observer measures same 12-J conversion from kinetic energy to
internal energy. (d) Observer measures same 12-J conversion.

67. (a)   car's frame   (b)   guard's frame

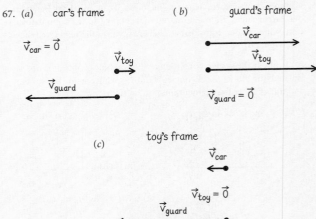

69. 0.86 m/s in the direction man and dog travel

71. (a) Each object has an equal apparent acceleration in your reference
frame. (b) The rate of change of momentum of each object is in the
same direction, and of magnitude proportional to the object's
inertia.

73. Ball: 23 m/s in direction opposite its direction before collision; pot:
1.8 m/s in direction of ball's velocity before collision.

75. (a) 1.52×10^9 kg·m/s in direction in which A1 moves before collision (b) 317 m/s in direction in which A1 moves before collision (e) No dipping down, then rising again in either kinetic energy curve because all kinetic energy is converted to internal energy; none is converted back to kinetic energy after collision; kinetic energy drops to zero and remains there.

77. $\vec{v}_{0.30\text{ kg,f}} = 0.17$ m/s in direction of initial velocity, $\vec{v}_{0.50\text{ kg,f}} = 2.1$ m/s in direction of initial velocity

79. Call disk radius R. Draw horizontal x axis in plane of page, with axis origin at disk center and axis passing through hole center; center of mass is at $x_{cm} = -R/6$.

81. (a) 1.0 m/s (b) 6.9 m (c) 3.7 m from front end of car (d) 3.9 m from front end of car (e) Because speed changes as grain is added, grain is not evenly distributed over length of car.

Chapter 7

1. More than one (one between each piece of bread and salami)

3.

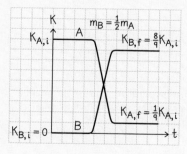

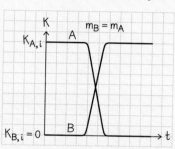

5. 1.80 m/s² to the left

7. (a)   (b)   (c)

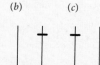

9. Case (b)

11. $K_m / K_{3m} = 3$

13. Trip down

15. (a) Into deforming metal (by breaking and re-forming chemical bonds) and heating it up (b) No, in fact you have to put in more energy to do a similar rearrangement of molecules to fix the slinky.

17. (a)

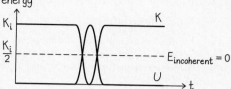

(b)

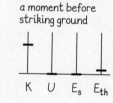

19. (a) System: ball, spring, air, Earth; air resistance not ignored

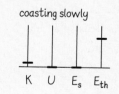

(b) System: ball, air, Earth; air resistance not ignored

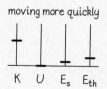

(c) System: bicycle, cyclist, air, road

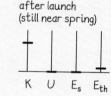

(d) System: car, air, road, fuel

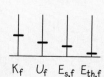

21. (a)

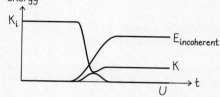

23. Yes. The copper atoms are likely joined to one another by chemical bonds, whereas contact between macroscopic objects typically involves repulsion at a distance. So the copper atoms may be pictured as being more "in contact" than your shoe is with the ground, but only because the atoms in copper are closer together. However, even the chemical bonds between atoms are interactions over a distance.

25. (a) Beginning of flight (b) End of flight

27. (a) All charged particles in the universe would attract one another and collapse into a compact charged center. (b) All charged particles in the universe would repel one another and spread out, making the universe ever more and more diffuse.

29. The strong interaction is responsible for holding atomic nuclei together against the electrical repulsion between protons. These interactions currently balance at a radius of about $10^{-14} - 10^{-15}$ m. If the strong interaction increased in strength by 20 orders of magnitude, but the electrostatic repulsion did not increase, atomic nuclei would be crushed to much smaller radii, perhaps becoming black holes! Atomic size scales are determined by electrical interactions among electrons and nuclei, so atomic sizes would remain more or less unaffected. However, with such a large strength, the attraction between neutrons and protons would be large enough so that nuclei in adjacent atoms could be drawn together. All matter might therefore be crushed. This new strong interaction would still not be large enough to affect interplanetary distances, where gravity would still dominate.

31. (a) 0.39 kg (b) $\vec{a}_{0.66\,kg} = 3.2 \times 10^2$ m/s^2 to the left, $\vec{a}_{0.39\,kg} = 5.4 \times 10^2$ m/s^2 to the right (c) Acceleration ratio and inverse of inertia ratio equal (Chapter 4): $|\vec{a}_{0.66\,kg}/\vec{a}_{0.39\,kg}| = 0.59$, $|m_{0.39\,kg}/m_{0.66\,kg}| = 0.59$

33. (a) −0.0020 (b) 250 m/s (c) 0.50 m/s

35. (a) 0.14 kg (b) $\vec{a}_{glob,av} = 25$ m/s^2 to the left, $\vec{a}_{cart,av} = 6.7$ m/s^2 to the right

37. 0.297 s

39. (a) $\vec{a}_{goalie,av} = 1.7 \times 10^{-2}$ m/s^2 in puck's initial direction of motion, $\vec{a}_{puck,av} = 5.3 \times 10^3$ m/s^2 in direction opposite puck's initial direction of motion (b) 90 kg. (c) 1.3×10^2 J

41. Position x_1:

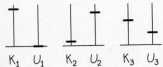

43. (a) Positive (b) Positive

45. 0.20 J

47. (a) 2.9 m/s (b) 3.2 m/s

49. (a) 0.59 J (b) $\vec{v}_{0.36\,kg,f} = 1.1$ m/s to the left, $\vec{v}_{0.12\,kg,f}$ 3.0 m/s to the right

51. If air resistance is ignored, the cap reaches your hands at speed v; if air resistance is not ignored, the cap reaches your hands at a speed slightly less than v because some initial kinetic energy is converted to incoherent energy of molecules in the air. Both answers come from the law of conservation of energy.

53. 4.59 km

55. 2.2 m/s

57. (a) 24 m/s (b) 49 J (c) 49 J

59. (a) 0.95 (b) 10 m/s, either upward or downward

61. $v_f = \frac{1}{4}\sqrt{15g\ell}$

63. Configuration (b)

65. 9.4%

67. (c) and (d)

69. (a) 57 J (b) 55 J

71. (a) Before the fall, the system's energy is all potential. During the fall, the system's potential energy is converted to kinetic energy. Just before impact, the system's energy is all kinetic. After impact, the system's kinetic energy has converted to thermal energy as Humpty Dumpty breaks.

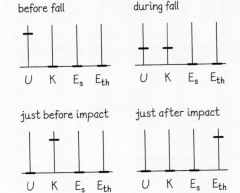

(b) Because conversion of coherent energy (kinetic, potential) into incoherent energy (thermal) is irreversible. If the men try to put Humpty Dumpty together, they constitute source energy. Although this can raise his pieces up to gain (coherent) potential energy, it cannot reverse the irreversible conversion of coherent kinetic energy to incoherent thermal energy. No matter how much source energy the men put in, Humpty Dumpty cannot be put back together.

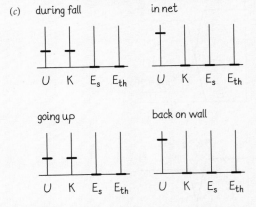

73. No, because some system coherent energy (kinetic) is converted to incoherent energy (mainly sound, small amount thermal energy).

75. 0.092 candy bars

77. Not impossible. The steam engine converts thermal energy (incoherent) to kinetic energy (coherent): Thermal energy causes liquid water to vaporize and expand, and the expanded volume moves pistons.

79. 0.0756 J

81. $a = 0.38$ m/s^2 in the direction of the bicycle's initial motion

83. $\vec{v}_{1-kg,f} = \left(-\dfrac{2v}{3} - \sqrt{\dfrac{7}{9}v^2 + \dfrac{E}{2m}}\right)\hat{\imath}$,

$\vec{v}_{2-kg,f} = \left(+\dfrac{v}{3} - \sqrt{\dfrac{7}{9}v^2 + \dfrac{E}{2m}}\right)\hat{\imath}$, where $m = 1.00$ kg, $\hat{\imath}$ points to the right

Chapter 8

1. Neither; vector sum of forces exerted on truck = vector sum of forces exerted on cycle = 0

3.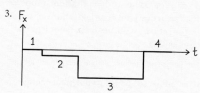

5. Object inertia is constant, which means change in momentum came from change in velocity, but knowing change in velocity does not tell you values of initial and final velocities; both are needed in order to determine change in kinetic energy.

7. (a) Magnitudes equal (b) Magnitudes equal (c) Magnitudes equal

9. (a)–(f) can all exert force.

11. (a) No. The force exerted by the pitcher on the ball results from contact between the pitcher's hand and the ball; once the ball leaves his hand, the pitcher no longer exerts force on it. (b) Drag force is exerted by the air on the ball, gravitational force is exerted by Earth on the ball.

13. No; at the top of the bounce, the cord has stopped exerting upward force but Earth's downward gravitational force is still exerted. The baby was moving upward, and gravitational force is changing the velocity's direction to downward. (The instant of zero velocity is the instant at which the velocity's direction changes from upward to downward.)

15. Because the vector sum of the forces exerted on the refrigerator is not zero. A force opposes your pushing force, probably frictional force exerted by the floor or, if the refrigerator back is against the wall, contact force exerted by the wall.

17. Downward contact force exerted by you on scale and upward contact force exerted by scale on you

19. $\vec{F}_{\text{by floor on you}}$; magnitude decreases

21.

The crate moves because the magnitude of frictional force exerted by the floor on the crate is smaller than the magnitude of the contact force exerted by you on the crate. You do not move because the magnitude of the frictional force exerted by the floor on you is equal to the magnitude of the contact force exerted by the crate on you. (If you try to exert a force exceeding the available friction between your feet and the floor, your feet will slip.)

23. (a)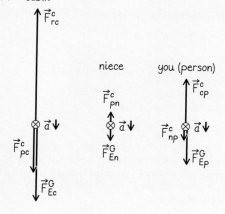

(b) $\vec{F}_{\text{En}}^{G} = 2.0 \times 10^2$ N downward, $\vec{F}_{\text{pn}}^{c} = 1.8 \times 10^2$ N upward, $\vec{F}_{\text{Ep}}^{G} = 4.9 \times 10^2$ N downward, $\vec{F}_{\text{np}}^{c} = 1.8 \times 10^2$ N downward, $\vec{F}_{\text{cp}}^{c} = 6.2 \times 10^2$ N upward, $\vec{F}_{\text{Ec}}^{G} = 9.8 \times 10^2$ N downward, $\vec{F}_{\text{pc}}^{c} = 6.2 \times 10^2$ N downward, $\vec{F}_{\text{rc}}^{c} = 1.5 \times 10^3$ N upward. Interaction pairs: $\vec{F}_{\text{pn}}^{c} - \vec{F}_{\text{np}}^{c}$, $\vec{F}_{\text{cp}}^{c} - \vec{F}_{\text{pc}}^{c}$ Gravitational force magnitudes given by mg. Interaction pairs are equal in magnitude, opposite in direction. Vector sum of forces exerted on each object must equal object's inertia times system acceleration (1.0 m/s² downward).

25. No; the tension in the upper portion is greater than the tension in the lower portion because the upward tensile force exerted by the upper portion on the lower portion must counteract the downward gravitational force exerted on the lower portion.

27. Forces equal in magnitude

29. 5.30×10^{17} m/s² in electron's initial direction of motion

31. Yes. Knowing the history of all the forces exerted on an object allows you to obtain the object's acceleration history and therefore the history of its velocity changes. Coupling this history with the known present position and velocity, you can work backwards to obtain the position and velocity at all previous instants.

33. (a) $\sum F_x(t) = \alpha - \beta t$, where $\alpha = 1.5 \times 10^3$ N and $\beta = 6.0 \times 10^3$ N/s (b) $\sum F_x > 0$ at $t < 0.25$ s, $\sum F_x < 0$ at $t > 0.25$ s, $\sum F_x = 0$ at $t = 0.25$ s

35. 59 m

37. (a) 9.8 N (b) 9.8 N

39. (a) 1500 N (b) 750 N

41. (a) 5.18×10^3 N (b) 3.2×10^2 N

43. (a) Tension equal to magnitude of gravitational force exerted by Earth on hanging block (b) Tension less than magnitude of gravitational force exerted on hanging block

45. (a) 4.9 m/s² to the right (b) 8.2 m/s² to the right

47. $a = b < d < c$

49. $m_{\text{W}} : m_{\text{M}} : m_{\text{P}} = \frac{1}{2} : 3 : \frac{3}{2}$

51. (a) 1.8×10^3 N/m (b) 0.41 m

53. $k_{\text{combination}} = k_1 k_2 (k_1 + k_2)$, therefore smaller than k_1

55. 60 mm

57. $\vec{a} = \dfrac{dk - gm}{3m}$ downward

59. (a) 5.4×10^2 N/m (b) 2.7 kg (c) 2.1 m/s² downward

63. (a) In segment attached to ceiling (b) In segment attached to tassel

65. (a) 4.2×10^4 N · s forward (b) 6.9×10^2 N forward

67. (a) 2.33 m/s (b) 381 N

69. (a) 0.22 N · s upward (b) 9.0 ms (c) 24 N (d) 6.1 N

71. (a) 0.45 N · s (b) 1.5×10^2 N (c) 2.3×10^2 N (d) 1.7 m/s in the direction of the serve

73. (a) 0.70 m/s² in direction of push (b) 1.4 m/s² in direction of push (c) 0.20 m/s² in direction of push

75. (a) $F/3m$ (b) $F/3$ to the right (c) $F/3$ to the left (d) (a) remains $F/3m$, (b) changes to $2F/3$ to the right, (c) changes to $2F/3$ to the left.

77. (a) Zero center-of-mass acceleration (b) $\sum \vec{F}_{\text{car}} = 1.2 \times 10^3$ N in direction car faces, $\sum \vec{F}_{\text{truck}} = 1.2 \times 10^3$ N in direction truck faces (c) 0.80 m/s² in direction truck faces

79. (a) 5.0 m/s² in direction of 50 N force (b) 18 m/s in direction away from 4.0-kg block

81. As you accelerate the child upward, the magnitude of the upward force you exert on her is greater than the magnitude F_{Ec}^G of the gravitational force exerted by Earth on her. This means that the magnitude of the downward force she exerts on you is also greater than F_{Ec}^Gr. Hence, as you accelerate her upward, the scale reading increases for an instant or two. Once she is on your shoulders, however, the scale reading is the same as when the two of you stood side by side.

83. (a) $\vec{F}$/3000 kg toward the winch (b) 4.0×10^3 N toward the winch (c) 2.4×10^3 N toward the winch (which is now in front of trailer)

85. (a) 2.7×10^3 N forward (b) 3.0×10^3 N toward the tractor

87. (a) 0.25 m/s² to the right (b) $\vec{a}_{10\text{ kg red}} = 1.0$ m/s² in direction of your push, $\vec{a}_{20\text{ kg}} = \vec{a}_{10\text{ kg blue}} = \vec{0}$ (c) $\Sigma\vec{F}_{10\text{ kg red}} = 4.0$ N in direction of your push, $\Sigma\vec{F}_{20\text{ kg}} = 4.0$ N in direction of your push, $\Sigma\vec{F}_{10\text{ kg blue}} = 2.0$ N in direction of your push

89. For a car to stop from a given speed requires a fixed momentum change and therefore a fixed impulse. If we approximate the force magnitude as being constant in time, we can write $F = \Delta p/\Delta t$. If we allow a collision to occur over a greater distance (and therefore a longer time interval) the force magnitude at any given instant is smaller than the force magnitude when the car must stop in a shorter distance. The crumple zone allows more of the car to be destroyed, but it slows the passengers down over a greater distance, thereby exerting smaller-magnitude forces on them.

91. (a) 3.3 kg (b) The greatest tensile force is exerted when the elevator accelerates upward, which means when the elevator is either moving upward and increasing its speed or moving downward and decreasing its speed.

93. (a) 0.33 m/s² in direction of your push (b) $\vec{a}_{10\text{ kg}} = 1.0$ m/s² in direction of your push, $\vec{a}_{20\text{ kg}} = \vec{0}$ (c) 0.33 m/s² in direction of your push

95. The heavy pulley is attached to the beam, the light pulley is attached to the load. Run the rope twice around each pulley. Pull on the rope with force $\frac{9}{32}(m_\ell + m_p)g$.

97. There are at least two possibilities: (1) Use your arm strength to accelerate yourself up the free end of the rope at greater than 1.1 m/s². The tension in the rope will lift your friend off the ground. (2) Tie yourself to the same side of the rope as your friend, and pull on the other (free) end of the rope, lifting both of you with your amazing arm strength and the mechanical advantage of two rope segments to exert tension forces on your combined inertia.

99. The acceleration increases even though the force exerted on the rocket remains constant. In this case $\vec{a} = \vec{F}_{\text{thrust}}/m_{\text{rocket}}$, but the inertia is the inertia of everything attached to the rocket, including payload and fuel. Because the fuel is being expelled, the inertia is decreasing. This makes the effect of the constant force (the acceleration) increase.

Chapter 9

1. No

3. When you drop the brick from the greater height, the force of gravity is exerted on the brick over a greater distance, thus doing more work on the brick. When more work is done on the brick, it has more kinetic energy when it hits your foot and thus hurts more.

5. Initially, your hand exerts an upward force on the ball. The force displacement is nonzero because the point of application of the force is at the ball and therefore moves. Hence this force does work on the ball. Once you release the ball, the only force exerted on it is the gravitational force. The point of application is again at the ball, which is moving; thus the force displacement is nonzero, and the gravitational force does work on the ball as it rises. At the top of its path the ball reverses its motion and the gravitational force again does work on it because again the force displacement is nonzero. Finally, the laundry exerts an upward force on the ball as the ball moves downward; this force does work on the ball because again the force displacement is nonzero.

7. No. You push down on the floor, but the floor doesn't move, and so the force displacement is zero; you do no work on the floor. (You rise because your torso and hips push down on your legs, and your leg muscles push up on your torso and hips, lifting them. Your legs do positive work on your upper body.)

9. Both observers are right in their own reference frames.

11. No, work done on a system could also change its potential energy.

13. (c) The block alone should not be used as the system because friction occurs at the block-incline interface and it is not possible to know how much thermal energy produced by friction goes into the block (i.e., into the system) and how much goes into the incline (i.e., out of the system).

15. You push a block up an incline, with the block and incline part of a system that also includes Earth (but not you). The block increases in speed as you push (positive ΔK and positive ΔU), and some energy is lost to friction (positive ΔE_{th}). All these changes constitute work you do on the system (positive W).

17. The downward trip takes longer because the ball's initial kinetic energy is converted to other forms, making the downward speed lower than the upward speed. System 1—ball: Gravity does negative work as the ball rises but does an equal amount of positive work as the ball falls. The air does negative work on the ball both as it rises and as it falls, decreasing the system's energy. System 2—ball, Earth: As the ball rises, kinetic energy is converted to gravitational potential energy, and as the ball falls that potential energy is converted to kinetic energy. The air does negative work on the ball, making the system's final energy lower than its initial energy. System 3—ball, Earth, air: The same kinetic-potential-kinetic interconversion as in system 2, plus collisions between the ball and molecules in the air cause the molecules and ball to heat up, converting some mechanical energy to thermal energy.

19. No, the work magnitudes are the same, but the work done is positive for one ball and negative for the other.

21. The time interval doubles.

23. (a) 8.6×10^2 N (b) 2.4 m/s

27. (a) Yes (b) Gravitational force, tensile force exerted by rope, resistive force exerted by water , normal force exerted by bay floor

29. 13 J

31. (a) 3.1×10^2 N (b) 6.2×10^2 N (c) $(+9.2 \times 10^2$ N)$\hat{\imath}$ where $\hat{\imath}$ points up (d) 1.5×10^2 J (e) 31 J (f) 1.5×10^2 J (g) 0 (h) -1.2×10^2 J

33. (a) 2.5×10^4 N (b) -1.3×10^4 J (c) -1.3×10^4 J

35. (a) 0.30 J (b) 75 mm (c) 0.15 J

37. They require the same amount of work.

39. (a) 2.0 J (b) 2.0 J (c) 1.0 J (d) 1.0 J (e) 1.0 J

41. 3 snowballs

43. $\dfrac{1}{2d}\left(\dfrac{mv^2}{5} - kd^2\right)$

45. They require the same amount of work.

47. 54 J

49. (a) 5.5 m/s (b) Yes, if the dart is fired vertically, some of the spring potential energy is converted to gravitational potential energy as the dart moves upward. Hence, not all the initial energy is converted to kinetic energy, making the vertical launch speed slightly slower than the horizontal launch speed.

51.

	Compression distance				
	0	0.050 m	0.10 m	0.15 m	0.20 m
K (J)	0.59	2.9	4.0	3.8	2.3
U^G (J)	0	−2.9	−5.9	−8.8	−12
U (J)	0	0.63	2.5	5.6	10

53. 6.1 J

55. (a)

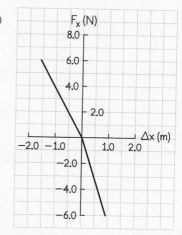

(b) Moves 0.77 m if pushed left, 0.63 m if pushed right (c) 1.2 J

57. 1.3×10^2 W

59. Because the zigzag path is spread out over a greater distance, the climber expends the same quantity of energy over a greater time interval, for a lower average power rating.

61. No. The work increases with time.

63. (a) 0.63 J (b) 1.3 W

65. (a) 9.0 W from the 3.0-N force, 6.0 W from the 2.0-N force (b) 3.0 W (c) Yes. The box is accelerating because the vector sum of the forces exerted on it is nonzero. Because power is equal to the product of force and speed, if the box moves at a higher speed, then the power from the same forces will increase in magnitude.

67. mgh

69. (a) If the initial velocity of the cart is in positive x direction, $\vec{F}_{avg} = -1.0 \times 10^3$ N $\hat{\imath}$ (b) Zero, because the point of application does not move (c) $\Delta K_{cm} = -1.0 \times 10^2$ J

71. (a) 1.9×10^3 N (b) 1.2×10^{-3} candy bars

73. The block is moving at constant speed.

75. (a) With no dissipation, it doesn't matter which way you roll the ball. (b) With energy dissipation, you want to roll the ball down hill A, so that it travels a shorter distance. The shorter the path, the less energy dissipated, so the ball has a better chance of having sufficient mechanical energy to get over hill B.

77. Your center of mass must be 0.23 m above the floor. This is not practical.

79. 20 ± 10 N

Chapter 10

1. Parabolic, starting tangent to the vertical axis and curving toward the horizontal axis

3. 0.214 m

5. Ignoring air resistance, the acceleration is downward. Not ignoring air resistance, the acceleration is mostly down and a little bit in the direction opposite the velocity.

7. (a) No (b) Yes

9. (a) Acceleration to right as car speed increases (b) Very small downward acceleration but almost constant speed (c) Downward acceleration with small component to the left as car slows slightly (d) Very small acceleration downward and left, perpendicular to velocity (e) Acceleration upward and left, opposite velocity as car slows (f) Acceleration upward and left, opposite velocity as car stops

11. No

13. Five forces are exerted:

	Parallel to roof ridge	Normal to roof surface	Tangential to roof surface
Gravitational force by Earth, $\vec{F}^G_{Ep}$		X	X
Normal force by roof, $\vec{F}^n_{roof,p}$		X	
Tensile force by left rope, $\vec{F}^c_{left,p}$	X		X
Tensile force by right rope, $\vec{F}^c_{right,p}$	X	X	X
Force of static friction by roof, $\vec{F}^s_{rp}$	X		X

15. Static

17. (a) Force of static friction (b) The magnitude is equal to the magnitude of the gravitational force Earth exerts on the eraser. (c) No, unless you push with insufficient force to hold the eraser up at all. (d) It decreases to zero.

19. (a)

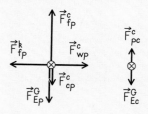

(b)

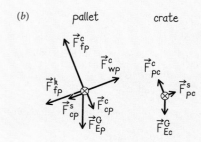

21. The normal force does positive work on the saw as the board and saw are lifted, no work as they are transported horizontally, and negative work as they are lowered onto the sawhorses; the algebraic sum of these work values is zero. The force of static friction does no work on the saw as the board and saw are lifted, positive work as their speed increases from zero to carrying speed, negative work as their speed decreases from carrying speed to zero at the sawhorses, and no work as they are lowered onto the horses; the algebraic sum of these work values is zero.

23. (a) 28° (b) 0.54

25. (a) They have the same speed at the bottom, but the child on the steeper slide arrives first. (b) The child on the steeper slide moves faster.

27. (a) $1.0\hat{\imath} + 4.0\hat{\jmath}$ (b) 4.1 units

29. (a) (30, −53°) (b) (18, −24)

31. 1.7 h

33. $(A, \omega t)$

35. (a) 0.93 km (b) 1.0 m/s at 31° north of east (c) 1.4 m/s

37. (a) One (b) Two (clockwise and counterclockwise equilateral triangles) (c) Three (d) N-1 distinct patterns for N arrows

39. Path (a) is the best representation. Path (b) is a poor representation because it ends up pointing directly downward, indicating no horizontal component to the velocity; as long as it moves, the cantaloupe always has a horizontal velocity component. Path (c) has two big issues: no vertical acceleration at first, then abrupt acceleration at the upper right corner.

41. 0.116 m

43. (a) 0.40 s (b) 1.2 m/s

45. (a) Never (b) Never (c) $t = 0.70$ s

47. A position that is more than half the range

49.
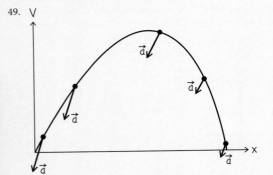

51. 6.0 m/s

53. (a) 1.5 s (b) 1.4 m (c) $v_{x,f} = 13$ m/s, $v_{y,f} = -7.6$ m/s (d) Yes

55. 15 m/s

57. (a) $\vec{a} = (-g\sin\phi)\hat{\jmath}$ (b) $\vec{v}(t) = (v_i\cos\theta)\hat{\imath} + (v_i\sin\theta - gt\sin\phi)\hat{\jmath}$

(c) $\Delta y = \dfrac{v_i^2\sin^2\theta}{2g\sin\phi}$ (d) $\Delta x = \dfrac{2v_i^2\sin\theta\cos\theta}{g\sin\phi}$

59. Assuming elastic collisions, (a) $v/16$ (b) $mv/16$ in the original direction of motion

61. (a) $v_1 = 0.28$ m/s, $v_2 = 0.44$ m/s (b) No

63. (a) $\vec{v} = 4.0$ m/s at $\theta = 44°$ (b) 69 mm

65. -7.2 m²

67. (a) 23° (b) 35 units²

69. (a) 2.0 J (b) 79°

71. (a) 2.5 m/s (b) 49 mm

73. (a) 13 m/s (b) 34

77. (a) 1.0 m/s (b) 4.9 kJ (c) 2.0 m/s (d) 4.9 kJ

79. Pulling

81. $\mu_s = 0.40$, $\mu_k = 0.20$

83. (a) 4.3×10^2 N (b) 43 kJ

85. (a) 11 N (b) 13 N

87. (a) 3.4 s (b) 11 s (c) 1.5×10^2 m

89. 1.0 m

91. (a) 0.12 m (b) 0.15 m

93. (a) 14 m/s (b) 13 m/s (c) 13 m

95. 22 kW

97. 53 m/s

99. range

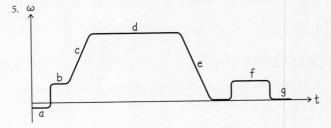

101. (a) 8.0×10^2 m (b) $(-20$ m/s$)\hat{\jmath}$ (c) -4.0×10^2 m

103. (a) $(1.49$ N$)\,\hat{\imath} + (5.33 \times 10^{-3}$ N$)\,\hat{\jmath}$ (b) 55.5 J

105. 75°

107. In order for a 100-kg pair to reach the top at a maximum speed of 5.0 m/s, the counterweight must have an inertia of 6.6×10^1 kg. For a 200-kg pair, the counterweight inertia must be 1.3×10^2 kg. The plan won't work. Either the 200-kg pair will not be accelerated or the 100-kg pair will be accelerated to a speed higher than 5.0 m/s.

109. (a) $y(x) = (\tan\theta)x - \dfrac{g}{2v^2\cos^2\theta}x^2$ (b) 23 m/s (c) 70 N

111. 51 mm

Chapter 11

1. Innermost track

3. The distance around the curved track is shorter in the inner lane, meaning the distance the passing car has to travel in order to pass is shorter in the inner lane than it would be in an outer lane.

5.

7. $p = 2, q = -1$

9.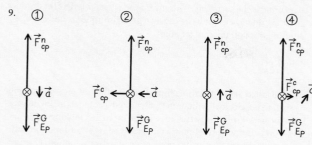

11. As the spinner rotates, the water and lettuce, because of their inertia, tend to move in a straight line tangent to the spinner circumference. Because it cannot fit through the holes in the spinner wall, the lettuce can't move in a straight line. The contact force exerted on it by the wall keeps it inside. Water drops can pass through the holes and do so, separating from the lettuce and leaving it dry. Initially, the drops might be held to the lettuce by cohesive forces, but as the spinner speeds up, these forces are insufficient to keep the drops moving in a circle, and they spin out of the spinner.

13. Ignoring friction between the car and track, the car climbs up the right wall as it enters the left turn. Because the student's body and the metal chunk climb this wall by the same amount as the car, the chunk stays between his knees, just as it was before the ride began.

15. (a)

(b) Tensile force exerted by string. Acceleration has a radial and a tangential component. (c) Ignoring air resistance: hand stays at circle center; tensile force is directed toward circle center and acceleration has no tangential component.

17. Because the body's rotational inertia is greater in the layout position than in the tucked position. The radius of the circular path the head and feet travel during the flip is larger in the layout position than in the tucked position. Because the body remains airborne for roughly the same time interval in either position, the gymnast must have much greater kinetic energy in the layout position to complete the backflip.

19. (a) No, the inertia of the bowling ball is so much greater than that of the baseball that even placing the rotation axis on the outer surface of the baseball would not make it as difficult to spin as the bowling ball. (b) Yes. You could choose to revolve the baseball around an axis that passes no closer than, say, 5 m to the baseball, while choosing an axis passing just centimeters from the edge of the bowling ball. The rotational inertias are $I_{bowl} = m_{bowl} r_{bowl}^2$ and $I_{base} = m_{base} r_{base}^2$, where r is distance from either ball to the axis of revolution. Choose an arbitrary location for this axis and place the two balls at distances r_{bowl} and $r_{base} \gg r_{bowl}$ from the axis. Once the inequality $r_{base} \gg r_{bowl}$ is sufficiently large to make the difference between m_{bowl} and m_{base} insignificant, $I_{base} > I_{bowl}$.

21. (a) Rotational inertia decreases. (b) Carousel rotational speed increases.

23. $0.10 \text{ s}^{-1}, 1.5 \times 10^{-4} \text{ s}^{-1}$

25. $8.7 \times 10^{-3} \text{ s}^{-2}$

27. Figure P11.27(b)

29. (a) $3mg \sin \theta$ (b) $3mg$

31. (a) $5.95 \times 10^{-3} \text{ m/s}^2$ (b) 3.55×10^{22} N toward the center of mass of the Earth-sun system, which is essentially the center of the sun.

33. 19.6 m/s^2

35. (a) 58 s^{-1} (b) 2.9×10^2 (c) 19 m/s (d) 96 m

37. (a) $\sqrt{2g(h - d)}$ (b) $mg\left(1 + \dfrac{2(h - d)}{R}\right)$ (c) $\sqrt{2g(h - d - R)}$

(d) $mg\left(\dfrac{2(h - d - R)}{R}\right)$ (e) $g\left(\dfrac{2(h - d - R)}{R}\right)$

39. (a) 18 m/s^2 (b) 0 (c) The vertical component of the normal force exerted by the cone on the ball (d) 0.92 m

41. 31 J

43. (a) $12 \text{ kg} \cdot \text{m}^2/\text{s}$ (b) $4.0 \text{ kg} \cdot \text{m}^2/\text{s}$

45. In order to continue moving in a circle at constant speed along with the pail, the water must have centripetal acceleration. This acceleration must be due to some force directed toward the center of the circle. When you whirl the pail rapidly, the forces exerted on the water are the force of gravity and a large contact force exerted by the bottom of the pail. When you whirl the pail at just the right speed, the gravitational force alone is sufficient to keep the water from spilling out. If you whirl too slowly, the acceleration due to gravity is larger than the necessary centripetal acceleration and the water spills out.

47. (a) $\dfrac{4v_i}{5\ell}$ (b) $\left(-\dfrac{3v}{5}\right)\hat{\imath}$ where $\hat{\imath}$ points in the direction of $\vec{v}_i$

49. 3.0 s^{-1}

51. $\omega_{\vartheta,i}/2$

53. $2mr^2 \sin^2 \theta$

55. $\frac{5}{3} mR^2$

57. 0.16 s^{-1}

59. (a) $\frac{1}{2} mR^2$ (b) $\frac{1}{12} ma^2$

61. The day would lengthen.

63. 20 to 35 $\text{kg} \cdot \text{m}^2$, depending on your inertia

65. $2.0 \times 10^2 \text{ s}^{-1}$

67. $\frac{13}{8} mR_{outer}^2$

69. $0.28 \text{ kg} \cdot \text{m}^2$

71. 8.2×10^{-2} J

73. (a) $\dfrac{2m\ell^2}{3}$ (b) $\dfrac{m\ell^2}{6}$

75. $0.25 \text{ kg} \cdot \text{m}^2$

77. (a) $0.535°$ (b) $0.517°$

79. The liquid inside the egg continues to rotate once you stop the egg. When stopped and released, viscous forces between the liquid and shell soon cause the shell to move along with the liquid, causing the entire egg to spin. In a hard-boiled egg, there is no liquid to continue rotating once you stop the egg. When you stop the outside of the hard-boiled egg, you stop everything.

81. (a) No work is done. (b) At the lowest point

83. (a) Force independent of r (b) Force proportional to $1/r^2$ (c) Force proportional to r

85. $\dfrac{6m_b v_b}{(3m_d + 4m_b)\ell_d}$

87. The acceleration due to gravity at her head is about 6.5 m/s². To avoid lightheadedness, you figure that the acceleration due to gravity at her head must be within 5% of the value at her feet. To accomplish this, the radius of the cylinder must be no less than 40 m.

89. You should maintain a constant tension of at least 74 N in the string while you unwind the entire string.

91. $\omega(x) = \dfrac{\sqrt{\left(\dfrac{3}{4}v\right)^2 - \dfrac{3}{2}g\left(\sqrt{d^2 - 4dx - 4x^2} - d\right)}}{d}$

Chapter 12

1. You can exert a greater torque by exerting a greater force tangential to the lid. The coefficient of static friction between the rubber and the lid is likely to be greater than the coefficient of static friction between your bare hand and the lid.

3. Torque is a better tightening specification because giving the specification in terms of the force exerted on the wrench you use does not provide a unique specification for how tight you fasten the bolts. You could use a wrench of any length, and the force necessary to get the correct tightness would be different for each different wrench length.

5. c and e

7. Option a. The baton's center of mass is closer to the larger sphere, making the lever arm distance between your hand and the center of mass longer in option a than in option b. The longer lever arm distance gives you finer control.

9. $\tau_B < \tau_A < \tau_C < \tau_D$

11. $C_A/C_B = 1/4$

13. The stick does not rotate, which means the torques caused by the contact forces exerted on it by your two fingers cancel each other. The farther a finger is from the stick's center of mass, the greater its lever arm distance and the smaller the contact force it exerts on the stick. So as you begin sliding your fingers, any slight difference in lever arm distance causes the finger farther from the center of mass—the left finger, say—to exert a smaller contact force and so slide more easily than the right finger. Once your left finger moves far enough toward the center of the stick to make the right lever arm distance greater than the left lever arm distance, the right finger exerts the smaller contact force and so speeds up, but only until its lever arm distance is less than that of the left finger. This tradeoff continues until the fingers meet, always at the 0.5-m mark.

15. c

17. (a)

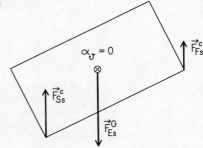

(b) The son carries the greater share of the load. (c) They would carry equal loads in the case of a sheet of plywood.

19. As the painter climbs, he moves farther from the point where the ladder contacts the carpeting. This gives him a greater lever arm distance, meaning that as he climbs he causes a greater torque around the axis running through the contact point.

21. $\tan^{-1}\left(\dfrac{1}{2\mu_s}\right)$

23. In the United States, the threading of nearly all bolts, screws, and nuts is such that a clockwise rotation tightens and a counterclockwise rotation loosens. From the perspective of the mechanic looking at his or her work, clockwise rotation is comparable to rightward motion and counterclockwise rotation is comparable to leftward motion. (Picture your hand placed on the 12 on a clock face: moving to the 1 is to the right, moving to the 11 is to the left.) Note that this clockwise-counterclockwise convention is consistent with the right-hand rule. With a nut on a bolt, for instance, the direction of the torque is the direction that the nut should move (down the body of the bolt to tighten, for example).

25. Positive x direction

27. 90° around the y axis followed by −90° around x axis

29. (a) 0.75 m from either end of the rod (b) 0.30 m from the heavier bucket

31. $8.2 \times 10^{-1}\,\text{s}^{-1}$

33. You fall backwards. The axis around which you are free to rotate (your toes) is right up against the wall. Your center of mass is not right up against the wall; it is near the (back-to-front) center of your body. Hence the force of gravity exerted on you causes a torque that tends to rotate you backwards about your toes.

35. $10^{-10}\,T_E$, where T_E is Earth's period of rotation, 24 h.

37. (a) 45 N (b) 30 N

39. If there are no external torques on the helicopter, its angular momentum must remain constant. With only a main rotor, the body of the helicopter would therefore rotate in the direction opposite the direction in which the rotor rotates. The purpose of the tail rotor is to exert a force on the air surrounding the craft, causing an external torque that keeps the body from rotating.

41. (a) $17\,\text{s}^{-1}$ (b) $1.2 \times 10^2\,\text{J}$ (c) The additional energy comes from work done by the skater's muscles as she pulls her arms and weighted hands in toward her body. This force is not perpendicular to the motion because the hands and weights spiral in toward the center of her body.

43. (a) Stays the same (b) increases (c) increases

45. (a) The center of mass is a distance $\ell/3$ from the ball. (b) $v_f = v_i/3$ in the direction of the ball's initial motion. (c) $\omega_f = v/\ell$

47. (a) $7/17\,\omega$ clockwise (b) −0.42

49. The object continues to roll without slipping. Nothing changes when the coefficient of friction increases. Only the maximum possible force of static friction increases, not the actual force of static friction exerted.

51. (a) $3.5\,\text{m/s}^2$ (b) 4.2 N

53. $\sqrt{3}/6$

55. The two cans reach the bottom at the same instant.

57. $t_h/t_s = \sqrt{25/21}$

59. (a) $\dfrac{2}{7}g$ (b) $\dfrac{2}{5}g$

61. 1.7×10^{-1} J

63. 1.4 m

65. 0.15 MJ

67. 5.9 s^{-2}

69. 43 s^{-1}

71. (a) 5.0 J (b) 5.0 J (c) $7.6 \times 10^2 \text{ s}^{-1}$

73. (a) 3.4 m/s (b) $\mathcal{T} = \dfrac{mv_i^2 r_i^2}{r^3} = 2.8$ N (c) 2.3×10^{-1} J

75. 54°

77. (a) Because $L = mrv_i$, treat the cube as a particle located at the cube's center, and insert $r = d/2$. This yields $L = mdv_i/2$. (b) At the instant of collision, the point of application of the contact force exerted by the lip on the cube is at the axis of rotation. The lever arm distance for this force is zero and thus this force cannot cause a torque about the axis. Hence there is no external torque on the cube and its angular momentum must be constant. (c) $3g/4d$ (d) $\sqrt{\dfrac{16gd}{3\sqrt{2}}}$

79. To change your angular momentum, there must be a torque on you. Otherwise your body continues moving in a straight line and you fall off the bike.

81. (a) 30 N·m (b) 26 N·m (c) 21 N·m (d) 15 N·m (e) 0 (f) 19 N·m

83. Writing $\vec{A} = A_x \hat{\imath} + A_y \hat{\jmath}$ and $\vec{B} = B_x \hat{\imath} + B_y \hat{\jmath}$, you can compute the vector product one component at a time, noting that for any two components along the same axis the vector product is always zero. Thus you have $\vec{A} \times \vec{B} = A_x B_y \hat{k} + A_y B_x (-\hat{k}) = (A_x B_y - A_y B_x)\hat{k}$.

85. (a) 20 N·m in the direction opposite the direction of motion (b) 68 N·m in the direction of motion

87. (a) 3.0 s (b) 4.9 rev

89. 0.37 N·m

91. $\hat{\jmath}$

93. (a) The engine is connected to the rear wheels, and so the force that accelerates the car forward is exerted at the bottom of the rear wheels. Because the line of action of this force goes below the center of mass of the car, the force causes a torque on the car that tends to lift the front wheels off the road. (b) With front-wheel drive, the force that accelerates the car forward is exerted at the bottom of the front wheels. The line of action of this force also goes below the center of mass of the car causing a torque on the car, but now the rear wheels are pushed into the road.

95. (a) 7.8×10^2 N (b) 6.0×10^2 N (c) 8.6×10^2 N and 6.6×10^2 N

97. $1 \times 10^4 \text{ s}^{-1}$

99. The torque is the gravitational force exerted on the cyclist times the lever arm distance, which is the length of the metal rod attaching each pedal to the bicycle. For a typical human ($m = 75$ kg) and bicycle (lever arm distance 0.3 m), this yields a torque of 10^2 N·m.

101. (a) 5.3 m/s^2 (b) 0.27

103. 4.0 m

105. (a) 3.1×10^2 N (b) 5.5×10^2 N

107. (a) 2.5 m/s^2 (b) 9.1 s^{-1} (c) 9.1 s^{-1} (d) 22 N (e) 60 W

109. The yo-yo accelerates downward, and the magnitude of the acceleration is $\dfrac{g}{1 + a/2b}$.

Chapter 13

1. The acceleration due to gravity increases linearly with radius.

3. 18 km

5. 6 : 1

7. $\dfrac{m_1}{m_2} = 1$

9. Torque can be interpreted as the second time derivative of the same area whose first derivative represents angular momentum.

11. (a) Yes, when you jump the scale shows a higher number at first. (b) No, the magnitude of the gravitational force exerted by Earth on you has not changed.

13. (a) 9.8 m/s^2 downward (b) 1.6 m/s^2 downward (c) Conclusion would not change.

15. The airplane moves downward at $2g$. During the time interval in which this motion takes place, the person is in free fall, accelerating downward at g. This means that, relative to the plane, he is accelerating *upward* at g! This relative motion of plane and person is exactly like the person falling onto his head from an upside-down position on solid ground.

17. At least two possibilities to obtain a second scale reading: (1) Hold the tube horizontally and mark the unstretched (0 g) position, or (2) Add a second identical bob to find the 2 g position.

19. (a) The milk climbs the sides of the bowl, and the milk surface becomes concave. (b) Either a massive hemisphere placed near the milk or the combination of a toroid around the bowl and a massive object beneath the bowl.

21. 6.7×10^{-15} N

23. $\dfrac{4}{9}$

25. $3 \times 10^{11} g$

27. 3.2 km, 32 km, 3.5×10^2 km

29. 1×10^1 h

30. (a) Force exerted by Sun (b) $F_{SE}^G / F_{ME}^G = 1.8 \times 10^2$

31. 1.05×10^{18} kg

33. $Gm_{\text{test}}^2 \sqrt{\left(\dfrac{1}{d^2} - \dfrac{2d}{(d^2 + \ell^2)^{3/2}}\right)^2 + \left(\dfrac{2\ell}{(d^2 + \ell^2)^{3/2}}\right)^2}$

35. $\dfrac{Gm_E r}{R_E^3}$

37. No. A 70-kg person can jump high enough to bring his center of mass to a height of 1.0 m to 1.5 m. That means the legs can exert forces capable of increasing the gravitational potential energy by about 10^3 J On the surface of Toro, this person's gravitational potential energy is -1.9×10^3 J, meaning he is bound to Toro by more energy than he can deliver in a jump.

39. Closer to object 1; specifically, a distance $(\sqrt{2} - 1)d$ from object 1 and a distance $(2 - \sqrt{2})d$ from object 2

41. (a) $\dfrac{Cm_E m_m}{2h^2}$ (b) $\sqrt{\dfrac{Cm_E}{h^2}}$

43. 9.4×10^2 km

45. 2.3×10^{30} N·m²

47. (a) $\vec{0}$ (b) -1.00×10^{-12} J/g (c) -5.0×10^{-13} J

49. Over a larger range

51. Elastic

53. 19 km/s

55. Being in very low orbit means that the radial distance from the satellite to the center of the moon is approximately the moon radius R_{moon}. The projectile reaches a distance $2R_{moon}$ from the moon's center, which is a height R_{moon} above the moon surface.

57. $v_{comet}/v_{Mercury} = \sqrt{2}$

59. $\sqrt{2C/a}$

61. 7.5 km/s

63. (a) $2.2 \times 10^{-7}\,\text{s}^{-1}$ (b) 89 km/s

67. To the right in the figure

69. $\dfrac{Gm_{ring}m_{obj}s}{(R_{ring}^2 + s^2)^{3/2}}$

70. (a) $0.54\dfrac{Gm_{inner}m_{obj}}{R^2}$ (b) $0.83\dfrac{Gm_{inner}m_{obj}}{R^2}$ (c) 0

71. (a) $-\dfrac{2Gm_{part}m_{disk}}{R^2}\left(1 - \dfrac{y}{\sqrt{R^2 + y^2}}\right)\hat{\jmath}$

73. From a theoretical standpoint, it does not matter. However, there may be logistical considerations, such as how much atmosphere you must traverse before reaching deep space, where air resistance is negligible. With these secondary considerations, it is best to launch vertically.

75. The acceleration vector is directed towards one focus of the ellipse. The tangential component could be greater than the perpendicular component of acceleration when the satellite is crossing a semiminor axis, provided $e^2 > \frac{1}{2}$. If the eccentricity is much greater, this condition could be satisfied elsewhere in the orbit. However, it can never be satisfied when the satellite is crossing a semimajor axis.

77. For $R_{planetoid} \approx 5$ m, mass density must be 6×10^9 kg/m^3.

79. 4.24×10^7 m above the surface of Jupiter

81. Satellite is 2.7×10^2 km above Earth's surface.

83. (a) $\rho_{max} = \dfrac{3H^2}{8\pi G} = 9.6 \times 10^{-27}$ kg/m^3 (b) The estimate is about the same as what we found for part (a). As it is, the universe seems balanced between being open and being closed. If there were no dark matter or dark energy in the universe, it would be much less dense than estimated, and we would expect it to be open.

Chapter 14

1. (b), (e), (f)

3. A, 33 μs after noon; B, 67 μs after noon; C, 75 μs after noon; D, 94 μs after noon.

5. (a) Later (b) Same time (c) Earlier

7. c_0

9. 1.4×10^2 m

11. Straight calculation shows the speed of each end larger than the speed of light. This cannot happen because the concept of a rigid object is an approximation that breaks down under such extreme conditions: The rod would bend into a spiral shape so that no portion of it moves faster than the speed of light.

13. Longer than.

15. 1.4×10^3 m

17. 2.88×10^3 s

19. 0.5 km^2

21. Zero

23. Energy, mass, and inertia

25. 3.59×10^6 m

27. (a) Lightlike (b) Lightlike (c) Timelike (d) Timelike

29. $v_A = 0.48c_0$, $v_B = 0.96c_0$

31. $0.99c_0$

33. $(1 - 2.47 \times 10^{-7})c_0$

35. (a) 8.00 km (b) 6.14×10^{-5} s

37. 2.13

39. (a) 7.62×10^3 km (b) 1.27×10^4 km

41. (a) 3.625×10^{14} m (b) 8.108×10^{15} m

43. (a) 59.5° (b) Length 5.00 m, wingspan 8.00 m, opening angle 77.3°.

47. (a) $x = 0$ (b) $t = 0$ (c) $x = 1.77 \times 10^8$ m (d) $t = 0.591$ s

49. (a) 5.93×10^3 s (b) 1.43×10^{12} m (c) 4.70×10^3 s

51. (a) 11.3° (b) 24.6°

53. (a) 1.00 (no increase) (b) 1.64

57. 9.46×10^{-20} kg·m/s

59. Defining the direction of motion of the 150-kg probe to be the $+x$ direction, $\vec{v}_{150\,kg} = (+0.764c_0)\,\hat{\imath}$, $\vec{v}_{250\,kg} = (-0.578c_0)\,\hat{\imath}$

61. (a) $\gamma_C > \gamma_B > \gamma_A$ (b) $K_C > K_B > K_A$ (c) $|\vec{v}_C| > |\vec{v}_B| > |\vec{v}_A|$ (d) $p_C > p_B > p_A$

63. 1.73×10^{12} J; $E_{1\,kg\,U}/E_{1\,kg\,coal} = 5.8 \times 10^4$

65. $E_1 = \dfrac{m_1 c_0^2}{\sqrt{1 - \dfrac{(m_1 - m_2 - m_{orig})(m_1 + m_2 - m_{orig})(m_1 - m_2 + m_{orig})(m_1 + m_2 + m_{orig})}{(m_1^2 - m_2^2 + m_{orig}^2)^2}}}$

$p_1 = \dfrac{m_1 c_0 \sqrt{(m_1 - m_2 - m_{orig})(m_1 + m_2 - m_{orig})(m_1 - m_2 + m_{orig})(m_1 + m_2 + m_{orig})}}{\sqrt{(m_1^2 - m_2^2 + m_{orig}^2)^2 - (m_1 - m_2 - m_{orig})(m_1 + m_2 - m_{orig}) \times \sqrt{(m_1 - m_2 + m_{orig})(m_1 + m_2 + m_{orig})}}}$

67. (a) $v = (1 - 8.98 \times 10^{-9})c_0$ (b) 1.40×10^4 GeV/c_0^2 or 2.49×10^{-23} kg

69. (a) 3.01×10^{-10} J (b) 9.0×10^8 m

71. 30.1 m, 24.2° away from vertical.

73. Yes, an observer in any reference frame moving along the perpendicular bisector of a line drawn between the two events sees the events as simultaneous.

75. (a) $0.986c_0$ (b) 91.1 s

77. If Orion's initial position is corrected for light travel time, the station will have just under 45 minutes to evacuate, so no re-write is required, just some hurry. If Orion's position is not corrected for travel time, so that the ship is seen at one location but has continued to travel toward the station while the light signal is in transit, there is no hope and a rewrite is definitely in order.

79. You can complete the race in the shortest time interval (measured in the Earth reference frame) by flying at speed $c_0/\sqrt{2}$.

Chapter 15

1. Least blurry at highest and lowest wing positions because at these positions vertical motion of wings is instantaneously zero.

3.

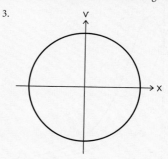

5. $0.5 < x < 3.5$

7. 5 s

9. Motion is periodic motion because it repeats itself over time but is not simple harmonic motion because its position and velocity cannot be described by a simple sine or cosine function. As yo-yo rises and falls, not all of its potential energy is converted to translational kinetic energy; some of it becomes rotational kinetic energy.

11. 127 Hz

13. Three harmonics

15. $\dfrac{4}{\pi} \sum\limits_{n=1}^{\infty} \dfrac{1}{2n-1} \sin\left[(2n-1)\dfrac{2\pi t}{T}\right]$, where T is the period of one square "oscillation"

21. $x(t) = A \sin\left(\dfrac{2\pi t}{T} + \phi_i\right)$

$v_x(t) = \dfrac{2\pi A}{T} \cos\left(\dfrac{2\pi t}{T} + \phi_i\right)$

$a_x(t) = -\left(\dfrac{2\pi}{T}\right)^2 A \sin\left(\dfrac{2\pi t}{T} + \phi_i\right)$

23. (a) Yes (b) $-\pi/2$

29. 0.30 J

33. 0.69

35. 1/3

37. (a) Increase (b) It decreases.

39. (a) Yes (b) No (c) Yes (d) Yes (e) No (f) No (g) If the cup is glued to the table, the answers do not change. If the cup is free to move, then the coffee would not fly out of the cup even if the cup flew into the air in free fall (at least until it hit the table again).

41. (a) 0.059 m (b) 2.7 s^{-1} (c) 4.6 mJ (d) 2.6 N/m (e) 5.6 rad
(f) $x(t) = A \sin(\omega t + \phi_i)$ where $A = 0.059$ m, $\omega = 2.7$ s^{-1}, $\phi_i = 5.6$ rad

43. 0.062 m

45. (a) 300 N/m (b) $x(t) = A \sin(\omega t + \phi_i)$, where $A = 0.020$ m, $\omega = 71$ s^{-1}, $\phi_i = 1.6$

47. 0.060 m

49. 0.25 m

51. Pendulum on left

53. $v = \sqrt{2g\ell(\cos\vartheta_{max} - \cos\vartheta)}$

55. (a) 9.11×10^{-4} kg·m² (b) 0.766 s

57. (a) 0.15 rad (b) 0.10 m/s

59. 0.90 s

61. $\Delta T = -\dfrac{1}{12}\dfrac{m_{rod}}{m_{bob}} + \dfrac{11}{288}\left(\dfrac{m_{rod}}{m_{bob}}\right)^2$ plus higher order terms

63. ω_d increases.

65. (b) 4.24 Hz (c) 0.0800 s (d) 2.13 rad

67. (a) 25 s (b) Instant $t = 18$ s

69. (a) 5.00 s^{-1} (b) 2.00 kg/s (c) $y(t) = Ae^{-tb/2m}\sin(\omega_d t + \phi_i)$ where $A = 0.100$ m, $b = 2.00$ kg/s, $m = 0.500$ kg, $\omega_d = 4.58$ s^{-1}

71. (a) 0.58% (b) 0.69

73. (a) $2\sqrt{mk}$ (b) Damping occurs so quickly that the spring returns to equilibrium position before being fully compressed. In the car, this means the system returns quickly to normal level rather than oscillating for a long time interval; the negative aspect is that springs do not absorb road-bump shocks as well as a spring system having $b \le b_{crit}$.

75. $x(t) = Ae^{-t/2\tau}\sin(\omega t + \phi)$ where $A = 67$ mm, $\tau = 7.2$ s, $\omega = 2\pi$ s^{-1}, and $\phi = 3\pi/2$

77. (a) Energy increases by factor of 4. (b) Maximum speed increases by factor of 2. (c) Period does not change.

79. $f_{half}/f_{whole} = \sqrt{2}$

83. 0.14 kg·m²

85. (a) 0.522 kg·m² (b) 0.129 kg·m² (c) 1.74 s

87. $mv^2/6$

89. You cut the wire to a length equal to the distance the ball stretches vertical spring.

91. Your mass is 71.4 kg.

Chapter 16

1. Yes, water can carry both transverse and longitudinal waves.

3. (a)

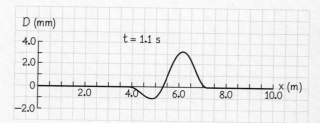

(b)

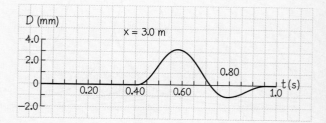

5.

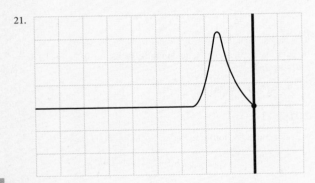

7. The definition is the same as for harmonic periodic wave: number of cycles per second executed by each medium particle.

9. The first crate must have struck when the cable was under minimal tension, resulting in low wave speed; the second crate must have struck when the cable was under significant tension, resulting in high wave speed. The tension change was probably the result of the ship's drifting away from the dock in the time interval between the two incidents.

13. (a) 0.10 m (b) 0.20 m (c) 0.30 m

15. Zero

17. Reflected wave not inverted

19. (a) Thinner string

(b)

21.

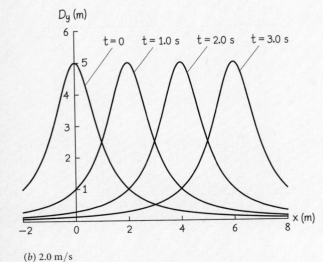

23. 30 m

25. (a)

Dy (m)

t = 0   t = 1.0 s   t = 2.0 s   t = 3.0 s

x (m)

(b) 2.0 m/s

27. (a) $f(x, t) = \dfrac{a}{b^2 + (x - ct)^2}$, where $c = 1.75$ m/s

(b)

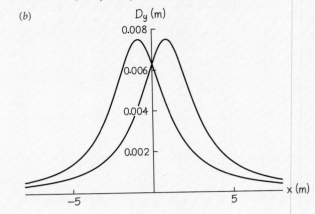

(c)

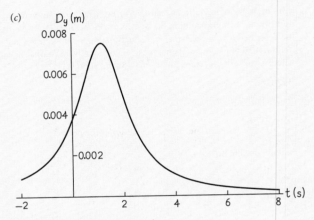

29. (a) 2.79 m (b) 6.09 Hz (c) 38.3 s^{-1}

31. 382 m/s in negative x direction

33. (a) 0.50 m (b) 5.7×10^2 s^{-1} (c) 90 Hz (d) 0.011 s

35. 0, 0.16 m, 0.32 m from tank edge

37. $f/5$

39. (a) 110 Hz (b) At frequencies $n(110$ Hz), where n is any integer

41. $A = 3.00 \times 10^{-2}$ m for both components; components travel in opposite directions at 1.20 m/s.

43. (a) 0.60 m (b) 10 mm in the y direction

(c) $f(x,t) = (0.010$ m$) \sin\left(\dfrac{2\pi}{0.60 \text{ m}}x + \dfrac{\pi}{2}\right) \cos\left(\dfrac{2\pi}{0.60 \text{ s}}t\right)$

45. 0.0578 m

47. 4.4 kg

49. (1) Shorten string by pinching off part of it; (2) replace string with one having lower linear mass density; (3) increase string tension.

51. 22 m/s

53. Fundamental frequency decreases by factor of 4

55. 4.62×10^6 N

57. 1.22 kW

59. (a) No change in power required. (b) Power required increased by factor of 16. (c) Power required decreased by factor of 0.25.

61. 2.3×10^5 J

63. 14.3 m/s

65. (a) $f(x,t) = a \sin(bx - qt)$, where $a = 0.0725$ m, $b = 2.09$ m^{-1}, $q = 377$ s^{-1} (b) 89.6 W

67. The fact that the speed at which the surfer moves forward toward the wave base equals the speed at which the wave moves forward.

71. (a) No, because the pulse needs a nonzero time interval to travel the length of the rope. (b) No, again because the pulse needs a nonzero time interval to travel the length of the bar, though the pulse speed is much greater in the bar than in the rope.

73. $\Delta t_B = 0.87 \Delta t_A$

75. (a) 0.51 m/s (b) 12 s

77. (a) 0.500 m (b) 260 Hz (c) $f(x,t) = A \sin\left[\dfrac{2\pi}{\lambda}x\right]\cos[\omega t]$, where $A = 0.0200$ m, $\lambda = 0.500$ m, $\omega = 1.63 \times 10^3$ s^{-1}

79. 1.9 m/s

81. For the pulse to travel end to end through a rope of length ℓ, $\Delta t = 2\sqrt{\ell/g}$.

Chapter 17

1. The approaching train emits waves that travel through the air and the track; those traveling through the air spread out spherically, those traveling in the track remain (largely) confined to the track, making the vibrations they cause detectable at greater distance. Also, the speed of vibrations through steel is greater than the speed of sound through air.

3. Yes, as long as the sound waves are free to spread out spherically from the source (professor's mouth). However, large lecture halls are often designed to have sound waves reflect off the walls and ceiling; in such cases, the amplitude may not drop off as $1/r$.

7. Because the bat feels air compressions created by your shout.

9. (a)

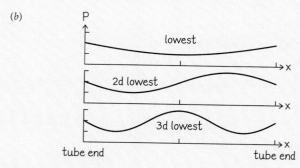

(b)

11. Nodal lines become antinodal lines, and vice-versa.

13. Because waves reflecting off the wall behind the speakers can travel to parts of the room that might otherwise be at the nodes.

15. The ripples are two-dimensional surface waves. The vibration causes the cup interior to act like a chain of sources. Because waves propagate away from sources, the wavefronts are parallel to the cup interior and therefore circular.

17. (a) $\theta = \sin^{-1}\left[\dfrac{(m + \frac{1}{2})c}{df}\right]$, $m = 0, 1, 2, \ldots$

(b) $\theta = \sin^{-1}\left[\left(m + \dfrac{1}{2} - \dfrac{\phi}{2\pi}\right)\dfrac{c}{df}\right]$, $m = 0, 1, 2, \ldots$

19. The sound waves she emitted diffracted around the corner and reached your ears.

21. (c) Nodal line locations in (a) are antinodal line locations in (b); antinodal line locations in (a) are nodal line locations in (b).

23. 5.0×10^{-6} W/m^2

25. 35 dB

27. 4.8 dB

29. 0.60 mW

31. (a) 3.2×10^{-3} W/m^2 (b) 7.9 W

33. 67 dB

35. (a) 70 dB (b) 6.0×10^2 W (c) 1.1×10^3 whales

37. The possible intensities at point R are $\dfrac{P}{4\pi}\left(\dfrac{1}{9\lambda} - \dfrac{1}{d}\right)^2$ where $d = 7.5\lambda, 8.5\lambda, 9.5\lambda$, or 10.5λ. The possible intensities at point Q are $\dfrac{P}{4\pi}\left(\dfrac{1}{6\lambda} + \dfrac{1}{d}\right)^2$ where $d = 4\lambda, 5\lambda, 6\lambda, 7\lambda$, or 8λ.

39. You hear both frequencies because the ear cannot resolve beats above $\Delta f = 20$ Hz.

41. 194 Hz, 198 Hz

43. $C < B < D < A$

45. 2 Hz, 4 Hz, 6 Hz

47. 6.6 Hz

49. Shift magnitude decreases.

51. 405 Hz

53. (a) 3 m/s (b) 34 m/s (c) 172 m/s

55. (a) 299 Hz (b) 298 Hz

57. (a) 450 Hz (b) 350 Hz

59. (a) 50 pulses (b) 6.2 s

61. 16 m/s away from whistle

63. 800 Hz

65. 4.7×10^2 m/s

67. 18.0 km/h

69. 50.3°

71. The reasoning is invalid because it is based on the boom being created at only one instant, the instant the plane attains Mach 1 speed. The boom is continuous at all speeds above Mach 1.

73. (a) 28° (b) 0.82 m

75. 19 m

77. No

79. (a) $f_1 = 399.8$ Hz, $f_2 = 409.8$ Hz (b) 10.0 Hz

81. Hyperbola

83. $f_b = 12.4$ Hz

85. Your current setup would only allow you to pick up radio stations 3.4 km down the road. You might try using a larger dish.

Chapter 18

1. 3.33 N

3. (a) $P_{C(top)} < P_{A,av} < P_{B,av} < P_{C(bottom)}$ (b) $F_{C(top)} < F_A < F_B < F_{C(bottom)}$

5. (a) $\vec{F}^c_{\ell w}$ at A directed to the right (b) No direction; pressure is scalar (c) $\vec{F}^c_{w\ell}$ at B directed upward (d) $\vec{F}^c_{\ell p}$ directed downward when piston pushed (e) $\vec{F}^c_{\ell p}$ directed upward when piston pulled (f) Answers (d), (e)

7. $\frac{1}{20}$

9. $45Gm^2_{planet}/64\pi R^4$

11. 25 N

13. (a) 125 kg/m³ (b) Yes. (c) 1.75×10^7 kg

15. Water levels are identical.

17. Four friends

19. $\rho_1/\rho_2 = 2$

21. 2.99×10^{-3} m³

23. (a) 0.133 kg/m³ (b) $He/H_2 = 0.479/0.521 = 0.919$

25. 5.21 m/s

27. $v_B > v_A$, $P_B < P_A$, diameter at $B <$ diameter at A

29. The speed of the air outside the window is greater than the speed of the air inside, making the outside pressure lower than the inside pressure, so that the smoke is pulled out of the car.

31. (a) 3.56 m/s (b) 16.4 s

33. 2.53 N

35. (a) $T_B < T_C < T_A$ (b) $P_A = P_B = P_C$

37. 2.92 mm

39. Wetting: $R_{tube} = R_{men} \cos \theta_c$; nonwetting: $R_{tube} = -R_{men} \cos \theta_c$

43. To top of tube

45. 6.3×10^4 N

47. (a) 1.5×10^4 N (b) 2.2×10^4 N (c) Because capsule mass increases as water enters through the leak

51. (a) 25 km (b) No, because at that height air resistance would not be negligible and because for water passing through such a small opening, viscosity would affect flow rate.

53. $F = P_{atm}wh/2 + gwh^2\rho/6$

55. $P = P_{atm} + mg/\pi R^2$

57. $\dfrac{P_A/P_B}{P_{A'}/P_{B'}} = 1.0$

59. 9.9×10^2 N

61. (a) 4.7×10^5 Pa (b) 80 m

63. 759 kg/m³

69. 21.6 m/s

71. $v_2 = \sqrt{\dfrac{2gh}{1 - \left(\frac{d_2}{d_1}\right)^4}}$

73. 93%

75. (a) 4.84 m/s (b) 0.0974 m³/s (c) 1.38 m/s

77. 2.3×10^{-5} kg

79. 2.05×10^{-7} m³/s

81. 4.66 μJ

83. 5.43 m/s

85. $R_2/R_1 = \sqrt[4]{2}$

87. (a) 7.52 days (b) 51.5 years

89. $Q_2 = Q_1/256$

91. 6.09 m

93. (a) 1.68×10^3 N (b) 1.09×10^3 N

97. 4.88×10^{24} kg

99. $T_A = 269$ N, $T_B = 419$ N

101. (a) 0.0608 Hz (b) 0.0542 Hz

Chapter 19

1. (a) 1/13 (b) 4/13

3. 6.31×10^{-7} s

5. (a) 24 configurations (b) 624 configurations

7. 1/4

9. (a) HHHHH and HTHTH equally likely (b) Heads-up three times

11. 11 s

13. 3/10

15. $v_{av,N_2}/v_{av,O_2} = 1.069$

17. 1/14

19. 158 m/s

21. (a) 3.32×10^{-22} J (b) 102 m/s

23. (a) 1 basic state (b) 7 basic states

25. (a) 20 basic states (b) 64 basic states

27. 5 particles

29. 88 times

31. 9.79 times more likely

33. (a) C (b) A and E (c) 16 (d) 1/4

35. (a) 6 basic states (b) 30 basic states

37. 1.67×10^6 basic states

39. (a) No (b) No (c) Yes

41. The system containing 15.0-mm beads has 1.09 times more entropy.

43. 3.00×10^{18} particles

45. $S_B > S_C > S_D > S_E > S_A$

47. (a) 1.13×10^{15} (b) 34.7

49. 8 particles

51. 90 K

53. (a) 11.2 m/s (b) 12.5 m/s

55. 9

57. $\frac{7}{2}N\frac{1}{E_{th}} + \frac{2}{15}NE_{th}^{-13/15}$

59. 453 m/s

61. 1.1×10^4 m/s

63. (a) 5.05×10^{-20} J (b) $K_{particle}/K_{bacterium} = 1.0 \times 10^9$, $K_{particle}/K_{slug} = 1.0 \times 10^{-12}$

65. $m_B = 36m_A$

67. 181 kg·m/s

69. (a) Pressure due to helium 95.44%, pressure due to krypton 4.560%

71. (a) 4.13 m/s (b) 2.56×10^{-3} m/s

73. +0.432

75. $+8.73 \times 10^{24}$

77. $+3.50 \times 10^{23}$

79. +1.30

81. Increased by factor of 3.16

83. 8.59 K

85. (a) 2.50×10^{24}, mixed state (b) Zero

87. 1.07×10^9 basic states

89. $+2.08 \times 10^3$

91. (a) 3.69×10^{-25} kg (b) 6.38×10^{-21} J

93. 0.64%. Uranium-235

95. (a) 95.8 K (b) 1.19×10^{-21} J

97. (a) 18.6 (b) 4.31

99. 94.7 km

101. (a) 0.447 (b) 0.200

103. This is an unsafe container; its estimated life is less than 5 hours.

Chapter 20

1. Positive

3. (a) Neither (b) Quasistatic (c) Both

5. Ignoring collisions between molecules, the upper limit is about 10^3 cycles/s, because the piston must move more slowly in the x direction than the gas molecules. You can increase this frequency by increasing the gas temperature, by reducing the stroke length, or by using less massive gas particles.

7. (a) 100 K (b) 100 °C (c) 180 °F

11. (a) 22.1 °C (b) Yes

13. 2.14×10^3 °C

15. 1.5×10^7 J

17. 4.4×10^{14} J

19. (a) 1.46×10^{10} J (b) 1.5×10^2 km

21. 0.19 °C

23. 5.3×10^{-22} J (b) 6.11×10^{-21} J

25. 403 J/K³·kg

27. B

29. 0

31. $T_A = 314$ K, $T_B = 176$ K, $T_C = 102$ K

33. $W_{on\ gas} = -P_i V_i$

35. -5.0×10^2 J

37. 0.0459 m

39. 14.1 kJ

41. 0

43. (a) 0.062 K, assuming $\rho_{air} = 1.2$ kg/m³ (b) 5.2×10^5 J

45. 1.09×10^{23} atoms

47. 2.68×10^3 J

49. Reason 1: Because $C_P = C_V + k_B$, $C_P/C_V > 1$ always, regardless of number of degrees of freedom. Reason 2: 6/9 corresponds to $d = 6/11$, which is not valid because d must be integer.

51. -2.70×10^3 J (b) $+2.70 \times 10^3$ J

53. $\frac{P_1 V_1 T_2}{T_1 V_3}$

55. (a) 928 J (b) 928 J

57. $W = -Nk_B T \ln\left(\frac{V_f - nb}{V_i - nb}\right) - an^2\left(\frac{1}{V_f} - \frac{1}{V_i}\right)$.

$W(a = b = 0) = -Nk_B T \ln\left(\frac{V_f}{V_i}\right)$

59. 5.07×10^4 Pa

61. $4N(\ln 3)$

63. $C_P = 2.5k_B$

65. 1.9×10^2 K

67. (a) $d = 3$ (b) Particles are monatomic.

69. $W = (21/20)P_i V_i$

71. (a) 5.08×10^4 J (b) 1.36 km

73. 2.4×10^{23}

75. (a) °F (b) Either °F or °C

77. -5.28×10^3 Pa

79. In the process undergone by sample A, because $Q_A = 6Nk_B T_{tp}$ and $Q_B = 0$, $Q_A > Q_B$ by $6Nk_B T_{tp}$.

81. Half a dozen tubes, but better plan on 10 because of assumptions made.

Chapter 21

1. (a)

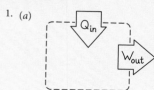

(b) System energy does not change; system entropy increases. (c) No

3. $\Delta S_3 < \Delta S_1 < \Delta S_2$

5. 425 J

7. (a) 9.6 J (b) 5.5×10^6 J (c) 2.8×10^4 kg

9. Material 2

11. 1.34×10^3 K

15. 4.09×10^3 J

17. (a) Reservoir 1 (b) Yes; device A

19. Cycle in (a)

21. 1.8 J

23. $\Delta S_{dev} = 0$, $\Delta S_{env} = 1.25 \times 10^{26}$

25. 1.0×10^{22}

27. 0.268

29. 18.8 MJ

31. $Q_{in} = 195$ J, $\eta = 0.44$

33. 0.396

35. System is reversible.

37. (a) 16 (b) 6.1

39. (a) 6.9 (b) 12 (c) 1.6×10^7 J

41. 3.1 kJ

43. 163 °C

45. $P_1 = 58.2$ kPa, $P_2 = 19.4$ kPa, $P_3 = 2.33$ kPa, $P_4 = 6.99$ kPa, $V_3 = 1.07$ m^3, $V_4 = 0.356$ m^3

47. 187 K

49. (a) 215 W (b) 6.28

51. 4.15×10^3 J

53. $\left(\dfrac{\Delta S}{\Delta t}\right)_{ocean} = -3.4 \times 10^{23}$ s^{-1}, $\left(\dfrac{\Delta S}{\Delta t}\right)_{cabin} = +3.4 \times 10^{23}$ s^{-1}

55. (a) 228 s (b) 6.38 (c) 1.93×10^5 J (d) -4.43×10^{25} (e) $+3.1 \times 10^{24}$ (not zero, because not all parts of the process are reversible).

57. 0.15

59. 0.0149

61. 431 K

63. 1.09×10^3 W

65. 7

67. 1.14×10^{24}

69. 4

71. (a) 478 K (b) 0.562 MJ

73. 2.2%

75. (a) 10.8 W (b) 1.17 W

77. (a) 0.71 (b) 12 km

79. (a) $COP_{cooling} = \dfrac{1}{\eta} - 1$ (b) $COP_{heating} = \dfrac{1}{\eta}$ (c) $COP_{cooling} > 0$, $COP_{heating} > 1$

81. (a) 20.5 kJ (b) 13.7 kJ

83. Five degrees of freedom

85.

	Car 1, Brayton cycle	Car 2, Carnot cycle
Efficiency	0.415	0.500
Work per cycle per unit mass (J/kg)	4.2	5.1
Power per unit mass (W/kg)	34	5.1
Speed after 11.5 s (m/s)	28	11
Drag force per unit mass due to air resistance at top speed[1] (N/kg)	1.2	0.33
Top speed[1] (m/s)	29	16

[1] Assuming drag force proportional to v^2.

Credits

Unit conversion factors

Length

1 in. = 2.54 cm (defined)

1 cm = 0.3937 in.

1 ft = 30.48 cm

1 m = 39.37 in. = 3.281 ft

1 mi = 5280 ft = 1.609 km

1 km = 0.6214 mi

1 nautical mile (U.S.) = 1.151 mi = 6076 ft = 1.852 km

1 fermi = 1 femtometer (fm) = 10^{-15} m

1 angstrom (Å) = 10^{-10} m = 0.1 nm

1 light − year (ly) = 9.461×10^{15} m

1 parsec = 3.26 ly = 3.09×10^{16} m

Volume

1 liter (L) = 1000 mL = 1000 cm^3 = 1.0×10^{-3} m^3
  = 1.057 qt (U.S.) = 61.02 $in.^3$

1 gal (U.S.) = 4 qt (U.S.) = 231 $in.^3$ = 3.785 L = 0.8327 gal (British)

1 quart (U.S.) = 2 pints (U.S.) = 946 mL

1 pint (British) = 1.20 pints (U.S.) = 568 mL

1 m^3 = 35.31 ft^3

Speed

1 mi/h = 1.4667 ft/s = 1.6093 km/h = 0.4470 m/s

1 km/h = 0.2778 m/s = 0.6214 mi/h

1 ft/s = 0.3048 m/s = 0.6818 mi/h = 1.0973 km/h

1 m/s = 3.281 ft/s = 3.600 km/h = 2.237 mi/h

1 knot = 1.151 mi/h = 0.5144 m/s

Angle

1 radian (rad) = 57.30° = 57°18′

1° = 0.01745 rad

1 rev/min (rpm) = 0.1047 rad/s

Time

1 day = 8.640×10^4 s

1 year = 365.242 days = 3.156×10^7 s

Mass

1 atomic mass unit (u) = 1.6605×10^{-27} kg

1 kg = 0.06852 slug

1 metric ton = 1000 kg

1 long ton = 2240 lbs = 1016 kg

1 short ton = 2000 lbs = 909.1 kg

1 kg has a weight of 2.20 lb where g = 9.80 m/s^2

Force

1 lb = 4.44822 N

1 N = 10^5 dyne = 0.2248 lb

Energy and work

1 J = 10^7 ergs = 0.7376 ft · lb

1 ft · lb = 1.356 J = 1.29×10^{-3} Btu = 3.24×10^{-4} kcal

1 kcal = 4.19×10^3 J = 3.97 Btu

1 eV = 1.6022×10^{-19} J

1 kWh = 3.600×10^6 J = 860 kcal

1 Btu = 1.056×10^3 J

Power

1 W = 1 J/s = 0.7376 ft · lb/s = 3.41 Btu/h

1 hp = 550 ft · lb/s = 746 W

1 kWh/day = 41.667 W

Pressure

1 atm = 1.01325 bar = 1.01325×10^5 N/m^2 = 14.7 $lb/in.^2$ = 760 torr

1 $lb/in.^2$ = 6.895×10^3 N/m^2

1 Pa = 1 N/m^2 = 1.450×10^{-4} $lb/in.^2$